ADVANCES IN WATER DESALINATION TECHNOLOGIES

Materials and Energy

Print ISSN: 2335-6596
Online ISSN: 2335-660X

Energy and sustainability are keywords driving current science and technology. Concerns about the environment and the supply of fossil fuel have driven researchers to explore technological solutions seeking alternative means of energy supply and storage. New materials and material structures are at the very core of this research endeavor. The search for cleaner, cheaper, smaller and more efficient energy technologies is intimately connected to the discovery and the development of new materials.

This collection focuses onmaterials-based solutions to the energy problem through a series of case studies illustrating advances in energy-related materials research. The research studies employ creativity, discovery, rationale design and improvement of the physical and chemical properties of materials leading to new paradigms for competitive energy-production. The challenge tests both our fundamental understanding of material and our ability to manipulate and reconfigure materials into practical and useful configurations. Invariably these materials issues arise at the nano-scale!

For electricity generation, dramatic breakthroughs are taking place in the fields of solar cells and fuel cells, the former giving rise to entirely new classes of semiconductors; the latter testing our knowledge of the behavior of ionic transport through a solid medium. Inenergy-storage exciting developments are emerging from the fields of rechargeable batteries and hydrogen storage. On the horizon are breakthroughs in thermoelectrics, high temperature superconductivity, and power generation. Still to emerge are the harnessing of systems that mimic nature, ranging from fusion, as in the sun, to photosynthesis, nature's photovoltaic. All of these approaches represent a body of materials–based research employing the most sophisticated experimental and theoretical techniques dedicated to a commongoal. The aim of this series is to capture these advances, through a collection of volumes authored by leading physicists, chemists, biologists and engineers that represent the forefront of energy-related materials research.

Published

Vol. 17 *Advances in Water Desalination Technologies*
edited by Yoram Cohen (University of California, Los Angeles, USA)

For further details, please visit: http://www.worldscientific.com/series/mae

(Continued at the end of the book)

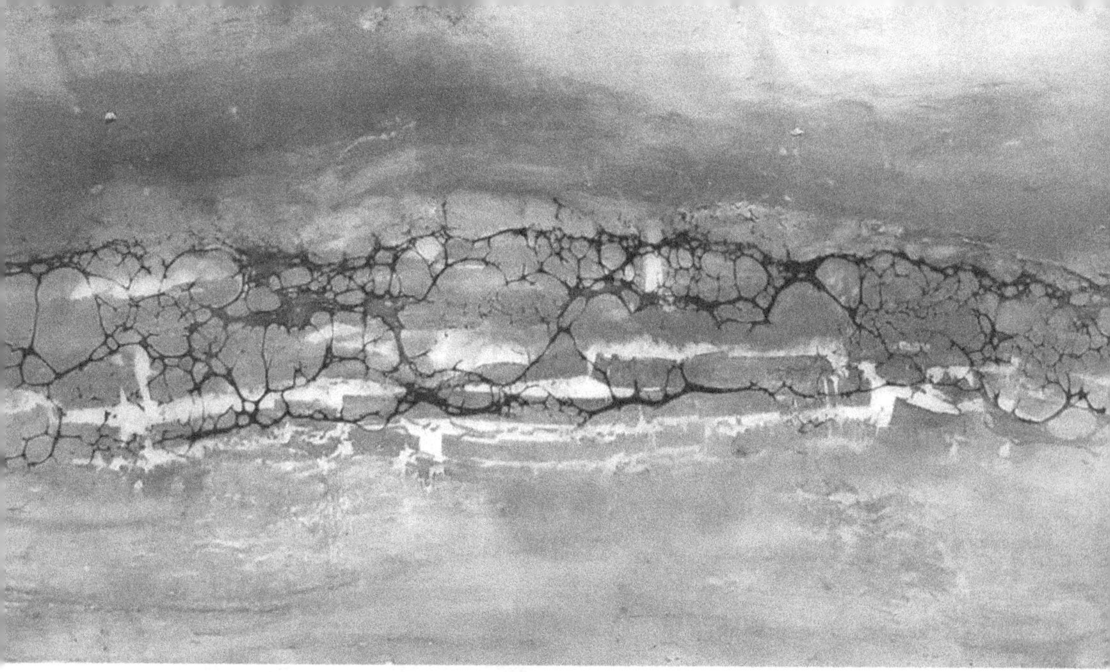

MATERIALS AND ENERGY – Vol.17

ADVANCES IN WATER DESALINATION TECHNOLOGIES

Editor

Yoram Cohen

University of California, Los Angeles, USA

World Scientific

NEW JERSEY • LONDON • SINGAPORE • BEIJING • SHANGHAI • HONG KONG • TAIPEI • CHENNAI • TOKYO

Published by

World Scientific Publishing Co. Pte. Ltd.
5 Toh Tuck Link, Singapore 596224
USA office: 27 Warren Street, Suite 401-402, Hackensack, NJ 07601
UK office: 57 Shelton Street, Covent Garden, London WC2H 9HE

British Library Cataloguing-in-Publication Data
A catalogue record for this book is available from the British Library.

Materials and Energy — Vol. 17
ADVANCES IN WATER DESALINATION TECHNOLOGIES

ISBN 978-981-122-697-7 (hardcover)
ISBN 978-981-122-698-4 (ebook for institutions)
ISBN 978-981-122-699-1 (ebook for individuals)

For any available supplementary material, please visit
https://www.worldscientific.com/worldscibooks/10.1142/12009#t=suppl

Typeset by Stallion Press
Email: enquiries@stallionpress.com

Preface

The development of new water resources is of increasing interest around the globe given population growth, shrinking freshwater resources, and impeding impacts of global climate change. Given that about 97% of the world water is saline, it is not surprising that major efforts have been devoted to developing both thermal and athermal technologies for desalination of seawater, as well as inland brackish water. Desalination technologies have advanced significantly during the last six decades being deployed at both large and small scales. Desalination is now utilized for developing new potable water sources, and in various water reuse applications (e.g., municipal, industrial, irrigation and aquifer recharge).

Given the expected increase on reliance on desalination technologies, efforts have been mounting to improve process reliability while also reducing the cost of water desalination. Accordingly, the book on "Advances in Water Desalination Technologies" addresses various aspects of existing and emerging desalination technologies. Book chapters from world leaders present a balanced discussion of water desalination advances with respect to processes, separations fundamentals, materials, as well as energy and environmental considerations. Detailed accounts are provided on the fundamental and practical aspects of modeling design and deployment of reverse osmosis (RO), nanofiltration (NF), electrodialysis, membrane distillation, capacitive deionization, reactive membranes, new materials (including carbon nanotubes and metal organic frameworks) such as carbon nanotubes and nanomaterials for desalination applications, in addition to considerations of energy generation from salinity gradients.

This book has come to fruition owing to the persistence of Dr. Zvi Ruder (Senior Executive Publisher) of World scientific Publishing Company (WSPC), the wonderful attention to details by editor Rhaimie Wahap and his team at WSPC, and the encouragement of Professor Mathew Tirrell (University of Chicago). Writing a book chapter is a daunting task and the efforts of my colleagues who contributed to this book is much appreciated. Their time investment and continued refinement of the chapters are reflected in this volume. We thank the families of all the contributors for their support and patience over the past two years. Lastly, the contribution of the original art for the book cover by Lila Terbescuis greatly appreciated.

Yoram Cohen
Distinguished Professor

Chemical and Biomolecular Engineering Department
Water Technology Research Center
Henry Samueli school of Engineering and Applied Science
University of California, Los Angeles, USA

4 May 2021

Contents

CHAPTER 1

Review of Thermal- and Membrane-based Water Desalination Technologies and Integration with Alternative Energy Sources

Yan Zhao and Bart Van der Bruggen*

Department of Chemical Engineering, KU Leuven, Celestijnenlaan 200F, B-3001 Leuven, Belgium

1.1. Introduction

Water is the most precious and important resource for human life. The current water scarcity is one of the most serious global challenges. Many countries or regions are augmenting their water supply by desalination to meet the increased water requirement from the increased population, industrial expansion, tourism and agriculture. Water desalination, which is considered as one of the main sources of producing clean water, refers to the process of removing the salts and minerals from either seawater or brackish water to obtain pure water.[1,2] Especially, some large-scale seawater desalination plants have been built in water stressed countries to augment available water resources. Therefore, there are substantial research interests in finding effective technologies for pure water production. Nowadays, desalination plants operate in more than 120 countries in the world. Figure 1.1 shows the global desalination capacities (m^3/d) and the source water types.[3]

*Corresponding author: bart.vanderbruggen@kuleuven.be

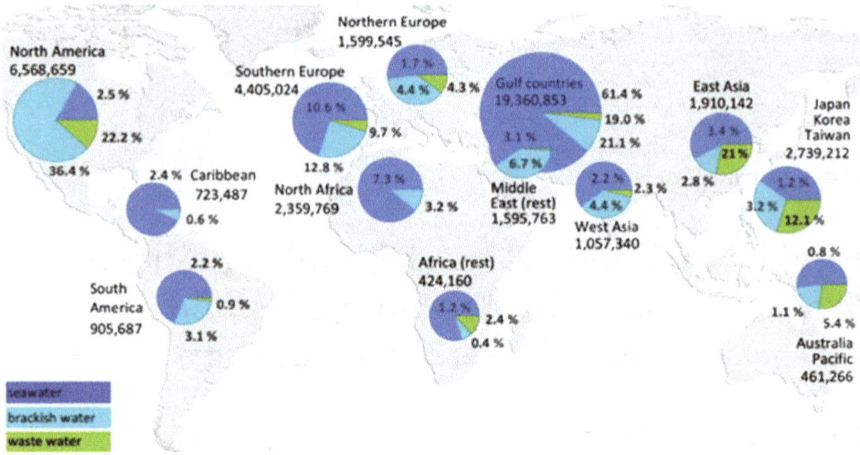

Fig. 1.1. The global desalination capacities (m^3/d) and source water types.[3]

With the development of the desalination technologies, they can be classified into non-membrane based technologies (thermal treatment for water desalination) mostly based on thermal evaporation, and membrane based technologies in which the driving force can be pressure, an electrical potential, a concentration difference or a temperature difference, leading to pressure-driven membranes, electro-driven membranes, concentration-driven membranes and temperature-driven membranes, as shown in Fig. 1.2.[4]

Thermal treatment for water desalination utilizing phase change processes, such as evaporation, precipitation, crystallization, salting out, freezing, solvent extraction and ion-exchange, have been used to produce fresh water. These processes, however, are more energy intensive, and have high capital and operational costs due to their inherent reliance on energy.[5]

Membrane technology for desalination has rapidly developed in recent years due to its separation performance, eco-friendliness due to the lack of production of a secondary pollutant, and its high energy efficiency.[6,7] With different driving forces, membranes can be classified into pressure-driven membranes (microfiltration, ultrafiltration, nanofiltration and reverse osmosis), electro-driven membranes (electrodialysis and fuel cells), concentration-driven

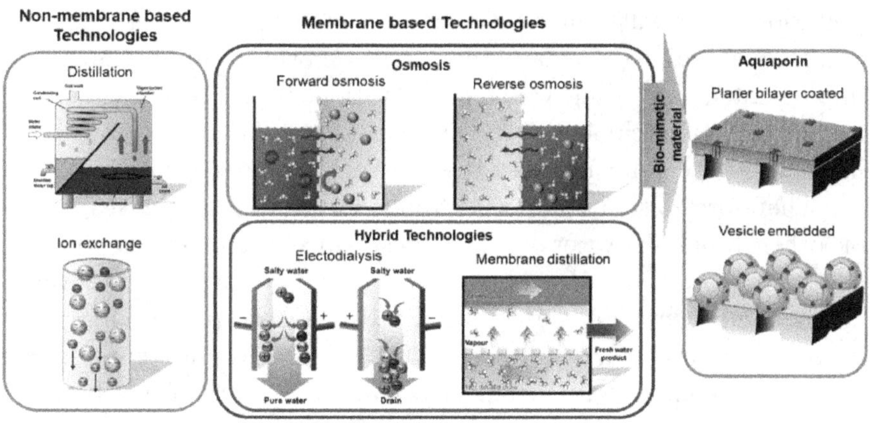

Fig. 1.2. Different desalination technologies.[4]

membranes (dialysis, pervaporation and forward) and temperature-driven membranes (membrane distillation).

In this chapter, thermal methods for water desalination, including evaporation, precipitation, crystallization, salting out, freezing, solvent extraction and ion exchange are summarized. Besides, advanced membrane technologies in water desalination are discussed, with emphasis on nanofiltration and reverse osmosis, electrodialysis and membrane capacitive deionization, forward osmosis and membrane distillation.

1.2. Water Resources

Water is a critical resource of all life. Fresh water is necessary for the economic development of every country of the world, and it is consumed by all economic sectors such as agriculture, industrial, commercial and dwellings.[8] However, clean water is a scarce and stressed resource since only 2.5% of all the water present on Earth is suitable for direct human use and consumption.[9] Presently, more than one third of the global population lives in water-stressed countries and it will reach to two thirds in 2025.[10,11] In addition, in 2030, the latest statistical forecasts have revealed that about half of the Earth's population may face a water stressed condition.[12] These challenges come from the exponentially growing population, continuing industrialization, expanding agricultural activities, water pollution, poor water management, and climate change. Up to now, many measures to alleviate the stress on water supply should be implemented, including water conservation, repair of infrastructure and improved catchment and distribution systems.[13] However, these methods still cannot increase the pressure on water availability.

Therefore, desalination of seawater, water from salty lakes, brines or other brackish water is of interest since these sources provide a seemingly unlimited and steady supply of high-quality water. The advances in water desalination technologies are developing fast in recent years. Currently, the water source for desalination are split with about 58.9% from seawater and 21.2% from brackish groundwater sources, and the remaining percentage from surface water and saline wastewater.[14] Numerous large-scale seawater desalination plants have been built in water-stressed countries to augment available water resources. The notable examples are the large-scale membrane separation technologies, such as seawater reverse osmosis (SWRO) desalination plants constructed in Spain and Israel, Qatar and Kuwait rely 100% on desalinated water for domestic and industrial supplies.[9] The conceptual drawing of seawater reverse osmosis desalination plants is shown in Fig. 1.3.[15] According to

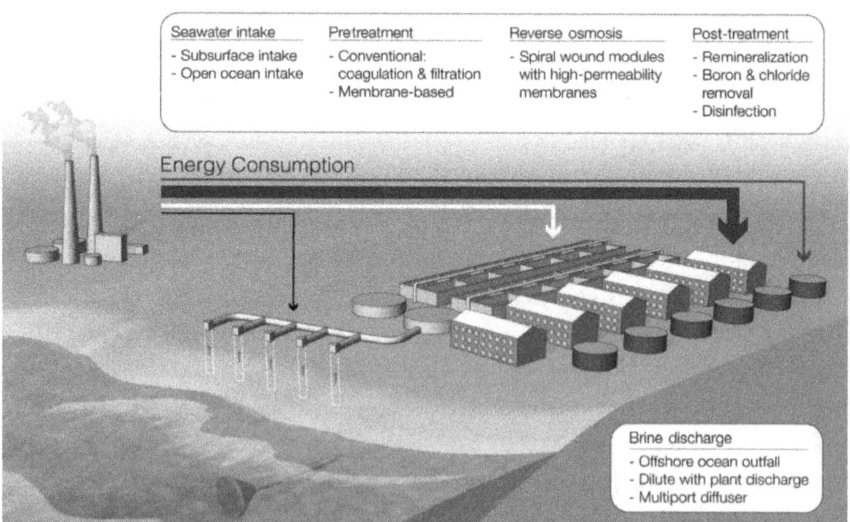

Fig. 1.3. Conceptual drawing of a seawater reverse osmosis desalination plants.[15]

the data, the global water production by desalination is over 38 billion m^3/year in 2016, and construction of new desalination plants is expected to increase in the near future.[15] According to the literature, a variety of desalination technologies on the basis of thermal distillation, membrane separation, freezing, electrodialysis, etc. has been developed over the years.

1.3. Thermal Treatment for Water Desalination

The thermal treatment for water desalination mainly includes evaporation and freezing.

1.3.1. *Evaporation*

The evaporation process for water desalination includes multi-stage flash (MSF) distillation, multiple-effect distillation (MED) and vapor compression distillation (VCD).

1.3.1.1. *Multi-stage flash distillation*

The MSF concept allied with parallel development in technology coincided with an increasing demand for water in arid regions.[9] The common evaporation method in desalination usually in an MSF distillation process, which is based on the principle of flash evaporation. Different from the approach of raising the temperature, the MSF desalination process is evaporated by reducing the pressure. As shown in Fig. 1.4, the economies of this process are achieved by regenerative heating where the seawater flashing in each flash chamber or stage gives up some of its heat to the seawater going through the flashing process. The heat of condensation released by the condensing water vapor at each stage gradually raises the temperature of the incoming seawater. The MSF plant consists of heat input, heat recovery, and heat rejection sections. These MSF plants usually operate at top brine temperatures of 90−120°C. When the operating temperature is higher than 120°C, the efficiency may be increased, but so is the potential for scale formation and accelerated corrosion of metal surfaces.

In this process (Fig. 1.4), the pressure should remain below the corresponding saturation temperature of the heated inflowing

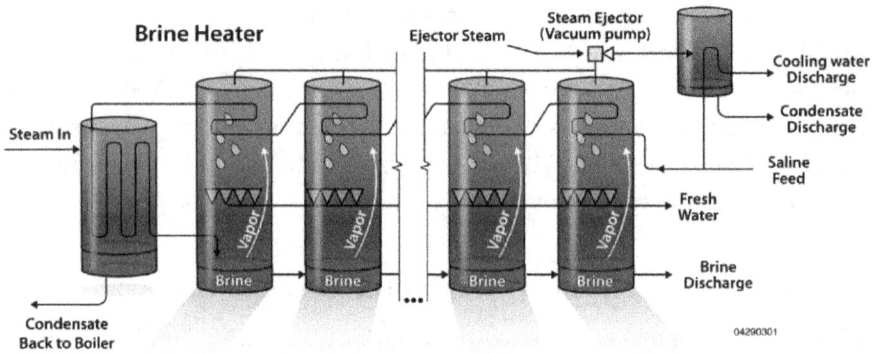

Fig. 1.4. Diagram of a multi-stage flash distillation unit through evaporation.[16]

seawater in each stage.[16] The introduction of the salt solutions into the flash chamber causes them to boil rapidly and vigorously. Baffles and orifices installed between stages make the brine's pressure reduce to that of the equilibrium vapor pressure required for boiling at the brine's temperature. Boiling continues until the seawater temperature reaches the boiling point at the stage. The unflushed brine passes from one stage to the next lower pressure stage for further flashing. Thus, the salt solution can be evaporated repeatedly without adding more heat. In this way, each stage of the evaporator is provided with demisters to minimize carryover of brine droplets into the distillate. The flashed water vapor is finally cooled and condensed by colder seawater flowing in tubes of the condenser to produce distillate. Therefore, the fresh water is obtained. Currently, the desalinated water produced by the MSF process in industry contains typically 2–10 ppm dissolved solids.

1.3.1.2. *Multiple-effect distillation*

Multiple-effect distillation (MED) is an important large-scale thermal process offering significant potential for water cost reduction.[9] The MED process takes place in a series of effects, which are a kind of evaporators, and uses the principle of reducing the ambient pressure in the various effects. This process permits the salt solution to undergo multiple boiling without supplying additional heat after the first effect.

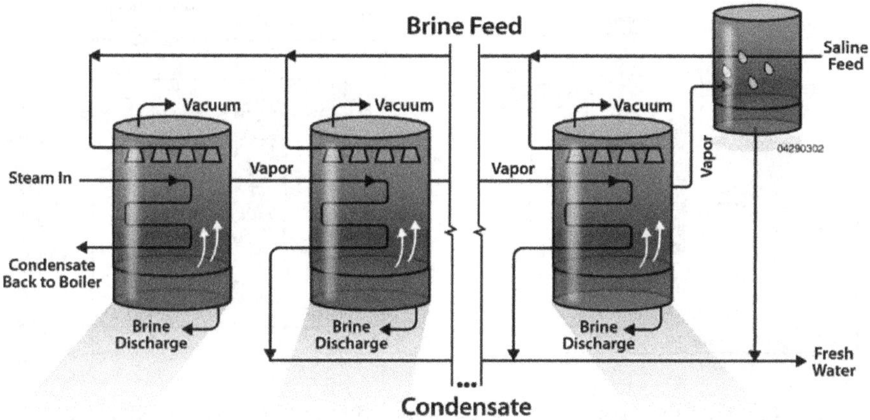

Fig. 1.5. Diagram of a multi-effect desalination unit.[16]

In this process (Fig. 1.5), the salt solution enters the first effect and is raised to the boiling point after being preheated in tubes.[16] These tubes are heated by externally supplied steam. The steam is condensed on the opposite side of the tubes. Then, the steam condensate is recycled to the power plant for its boiler feedwater. The MED plant's steam economy is proportional to the number of effects. The total number of effects is limited by the total temperature range available and the minimum allowable temperature difference between each of the effects. In this process, the evaporation and condensation are repeated from effect to effect each at a successively lower pressure and temperature. The energy consumption of the MED plant is significantly lower than that of an MSF plant. The performance ratio of the MED plant is higher than for an MSF plant.

1.3.1.3. *Vapor compression distillation*

Vapor compression distillation (VCD) includes mechanical vapor compression desalination (Fig. 1.6(a)) and thermal vapor compression desalination as two similar methods (Fig. 1.6(b)), in which the heat for evaporating the seawater comes from the compression of vapor.[16] Through reducing the pressure, the VCD plants take advantage of the principle of reducing the boiling point temperature.

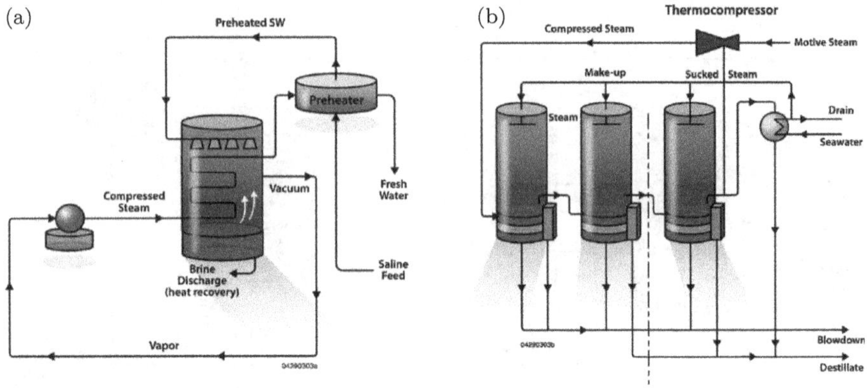

Fig. 1.6. Diagram of a mechanical vapor compression (a) and diagram of a thermal vapor compression (b) for desalination.[16]

The mechanical vapor compression desalination process is usually electrically driven. The VCD units are built in a variety of configurations to promote the exchange of heat to evaporate the salt solution. The compressor creates a vacuum in the evaporator and then compresses the vapor taken from the evaporator and condenses it inside a tube bundle. The salt solution is sprayed on the outside of the heated tube bundle where it boils and partially evaporates, producing more vapor.

The thermal vapor compression desalination process with the steam-jet type of VCD unit, a venturi orifice at the steam jet creates and extracts water vapor from the evaporator, and then creates a lower ambient pressure. The extracted water vapor is compressed by the steam jet. VCD units are often used for resorts, industries and drilling sites where fresh water is not readily available.

1.3.2. *Freezing*

In the freezing process, the salts are removed from salty water by the formation of ice crystals, which do not contain any salt.[17] Before water has been frozen, the mixture is usually washed and rinsed to remove the salts in the remaining water or adhering to the ice and then the ice is melted to produce fresh water. The freezing process is made up of cooling of the salt solution feed, partial

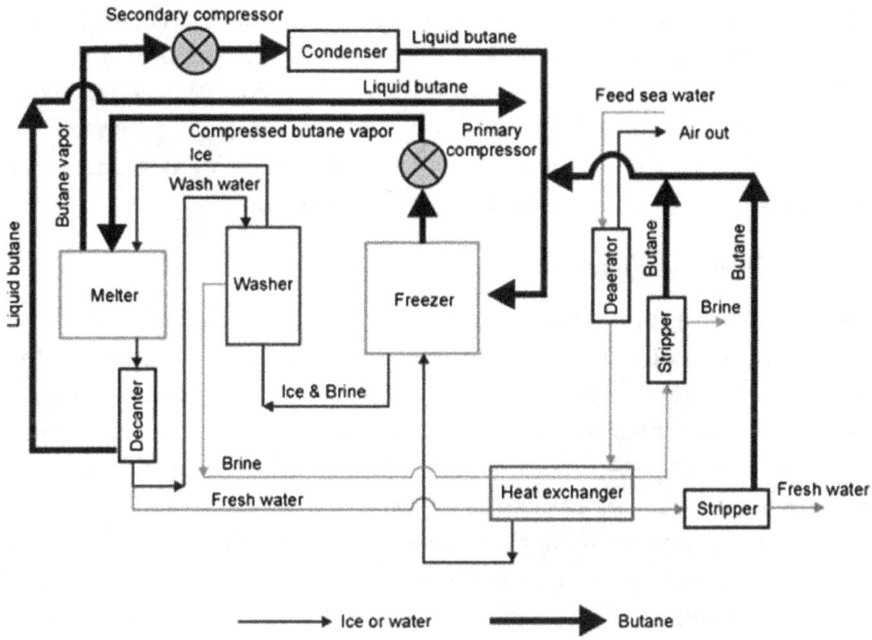

Fig. 1.7. Schematic diagram of a direct contact freezing process for desalination.[19]

crystallization of ice, separation of ice from seawater, melting of ice, refrigeration and heat rejection.[18] Usually, it contains the triple point freezing process, secondary refrigerant freezing process, eutectic freezing process, hydrate freezing process and direct contact freezing process (Fig. 1.7).[19] The advantages of freezing desalination are in the lower theoretical energy requirement, minimal potential corrosion, and little scaling or precipitation. The disadvantage of freezing involves handling ice and water mixtures, which are mechanically complicated to move and process.

1.4. Pressure-Driven Membranes Process in Water Desalination

Reverse osmosis (RO) and nanofiltration (NF) are the two pressure-driven membrane processes whereby a semi-permeable membrane rejects dissolved constituents present in the feed water.[20] They are gaining worldwide acceptance in both water treatment and desalination applications. One of the most important indices is rejection, which is due to membrane size exclusion, charge exclusion and physical/chemical interactions between solute, solvent and membranes. The process efficiency depends on operational parameters, membrane performance and feed water properties.[21]

1.4.1. *Reverse osmosis*

Osmosis, in simple terms, can be defined as a natural process in which water molecules spontaneously move from a solution of low osmotic pressure to a solution of high osmotic pressure across a semipermeable membrane.[22,23] The membrane, being semipermeable, rejects the solutes and only allows water molecules to pass through and until a state of osmotic equilibrium is reached where the chemical potentials across the membrane become balance.[24] In case the applied pressure difference is greater in magnitude than the osmotic pressure difference across the membrane, water molecules are forced to flow in a direction opposite to that of the natural osmosis phenomenon, as shown in Figs. 1.8(a) and 1.8(b).[13]

Presently, seawater or salt lake water RO dominate the global desalination market based on installed capacity. The major components of RO are four major units, the pretreatment, pumping, membrane assembly and post treatment units.

In the RO process, pretreatment is needed to eliminate undesirable constituents in the seawater, which would otherwise cause membrane fouling. Therefore, raw feed flows into the intake structure through trash racks and traveling screens to remove debris and finally obtain the cleaned salt solution after a filter process. As shown in Fig. 1.9, the filtered feed provides a protection to the high-pressure pumps and the RO section of the plant. The high-pressure pump raises the pressure of the pretreated feed water to the pressure appropriate for the RO membrane.[15] The osmotic pressure is overcome by applying an external pressure higher than the osmotic pressure on the seawater and then the water flows in the reverse direction to the natural flow across the RO membrane, leaving the dissolved salts behind with an increase in salt concentration. In this

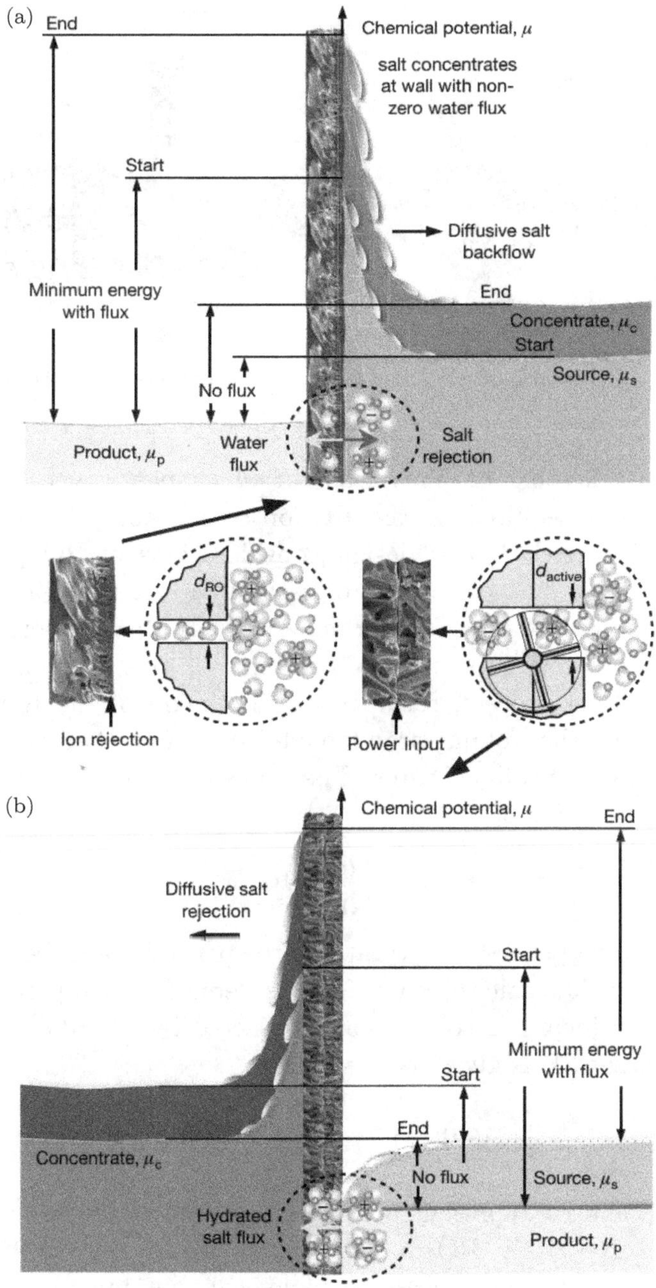

Fig. 1.8. Concentration gradients in RO (a) and active (b) desalination membranes.[13]

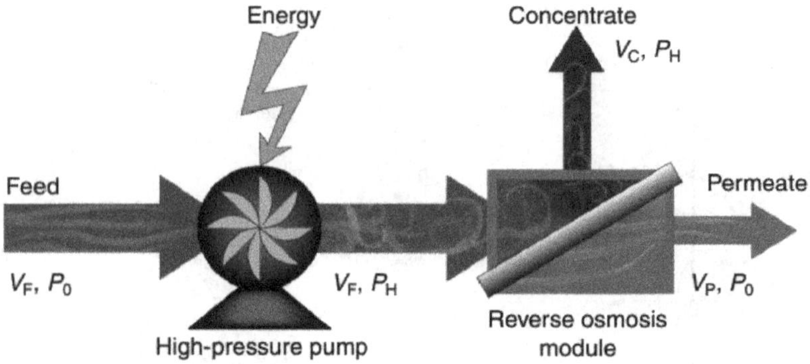

Fig. 1.9. A schematic of flows and pressures in a reverse osmosis desalination system.[15]

process, no heating or phase separation change is necessary. The major energy required in the RO process is for pressurizing the salt solution feed. Major design considerations of RO plants are the quantity of flux, conversion or recovery ratio, permeate salinity, membrane life, power consumption, and feedwater temperature.

The RO membrane is the core part of this technology. In a continuous RO process, recovery or conversion of the RO process is defined as the volume fraction of feed water that is recovered as permeate or product water. The percentage recovery (r) can be calculated as follows

$$r = \frac{Q_P}{Q_F} \times 100\%$$

where Q_p and Q_F are the flow rates of the permeate and feed streams, respectively. The salt rejection (SR) is defined as the percentage of a particular incoming solute that is rejected by the RO membrane. The apparent SR is given as

$$SR = \left(1 - \frac{c_P}{c_F}\right) \times 100\%$$

where, c_p and c_F represent permeate and feed solute concentration (mg/L), respectively. RO membranes can easily remove 95 to 99% of dissolved salts, solid particles, biological contaminants and other organic contaminants from the feed. The rejection efficiency relies on two factors, the charge of the membrane surface and the pore size of the membrane structure. Typically, the water flux is proportional to the net pressure driving force across the membrane. The permeate or water flux (J_w) is the volumetric flow rate of permeate per unit surface area of the RO membrane.

As RO works against osmosis, a variety of polymer-based membranes that can withstand this high pressure are employed, and they can be broadly categorized. According to the data, non-cellulose polyamide is the most commonly used material for RO membranes because of its high-water permeability and salt rejection. These RO membranes consist of a porous support with an active thin polyamide layer, which is responsible for salt rejection and water permeability. Additionally, these membranes offer a wide range of operational pH and temperatures but are highly sensitive to oxidizing agents. The major drawback of these polyamide membranes is fouling, which is mainly related to the hydrophobicity of the membrane surface.

There are four different types of RO membranes modules: plate and frame, tubular, spiral wound and hollow fiber. Among these, the plate and frame modules are the earliest RO membrane modules.[25] As shown in Fig. 1.10(a), a flat sheet RO membrane is attached to the two sides of a rigid plate. A number of plates are used that are stacked within a pressurized support framework. These plates contain grooved channels that provide a path for the permeate flow. As the feed solution enters the module from one end, water molecules permeate the membrane and are collected as permeate solution in a central permeate collection manifold and the concentrate salt solution leaves the module from the other end. Tubular modules, as shown in Fig. 1.10(b), are relatively simple in construction, and consist of a porous tube with an inserted or surface coated RO membrane. The tubes are made of ceramic, carbon, paper, plastic, or fiberglass. As high-pressurized feed water enters the tube from one end, water molecules permeate radially through the membrane to produce the product water. The concentrate salt solution leaves from the other end of the tube. Multiple tubes can be arranged in series or parallel to increase the system capacity. The hollow fiber module is composed of many small-diameter fibers contained within

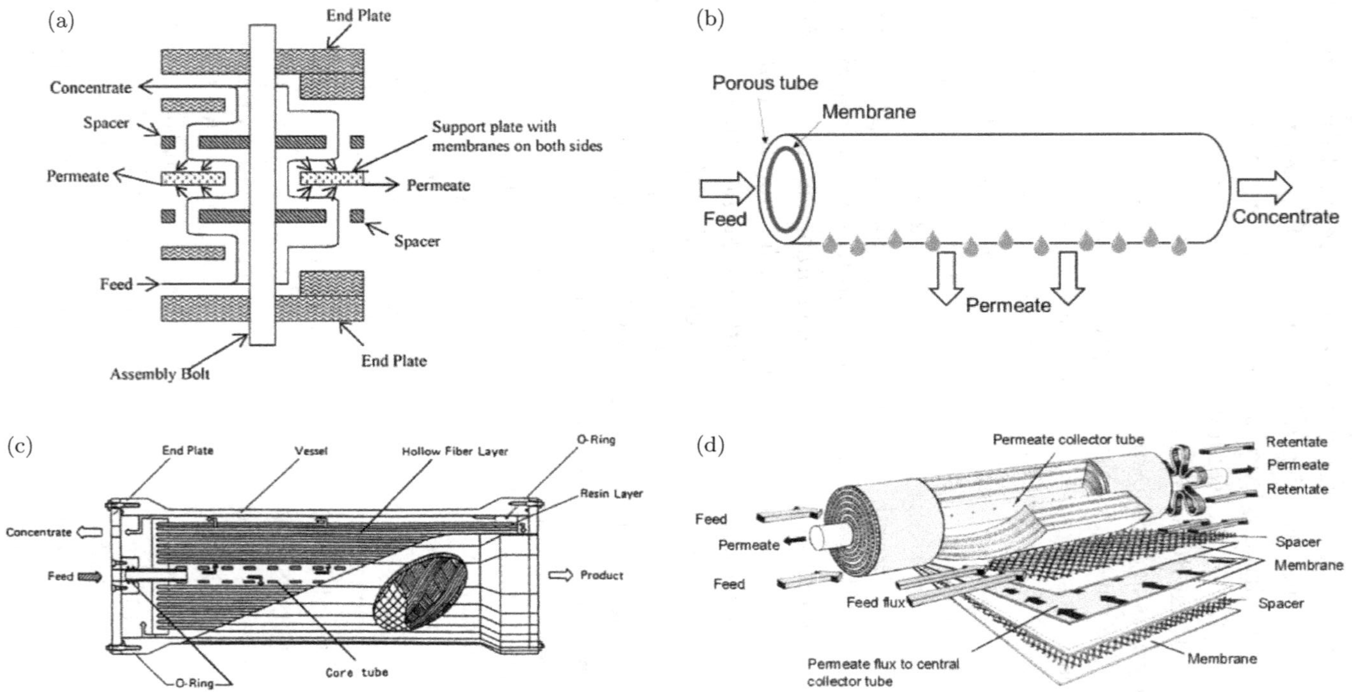

Fig. 1.10. Plate and frame RO membrane module (a), tubular RO membrane module (b), hollow fiber RO membrane module (c) and spiral wound RO membrane module (d).[25]

a pressure vessel. As shown in Fig. 1.10(c), the module consists of an epoxy tube sheet where the fibers ends are potted in epoxy while keeping them open for permeate flow, and the fiber ends are sealed in epoxy to form an epoxy nub, which prevents bypassing of the feed to the concentrate outlet. In this structure of a hollow, as pressurized feed water enters the module through the core tube, water molecules permeate radially into the fibers and exit pass the open fiber ends in the epoxy tube sheet while the concentrate leaves the module at the same end as the feed inlet. Figure 1.10(d) shows the spiral wound modules, which currently are the most common type of module applied for RO desalination. In this module, two membrane sheets are placed together with a permeate spacer in between to form a leaf. The membrane sheets are glued from three sides with the fourth side left open and connected to a central perforated permeate collector tube. In this module, the feed solution is introduced from one end of the module and travels axially along the length of the module. Water molecules are forced cross the membrane and are collected as permeate through the perforated permeate collector tube. The concentrate leaves the module at the end opposite to the feed.

1.4.2. *Nanofiltration*

Nanofiltration (NF) as an effective pressure-driven membrane process has a pore size and cut off ability between RO and ultrafiltration (UF).[26,27] Compare to RO, the NF process operates not only at lower operation pressures, higher water fluxes, and lower investment costs, but also with comparably high rejection rates for scale formation bivalent ions.[28] NF is dependent on the micro-hydrodynamic and interfacial events occurring at the membrane surface and within the membrane nanopores.[29] As the key compartment of NF technology, typically, the formation of NF membranes involves phase inversion followed by interfacial polymerization to produce thin-film composite (TFC) membranes, as shown in Fig. 1.11.[27,30,31] Current progress in preparation methods includes the incorporation of additives such as nanoparticles

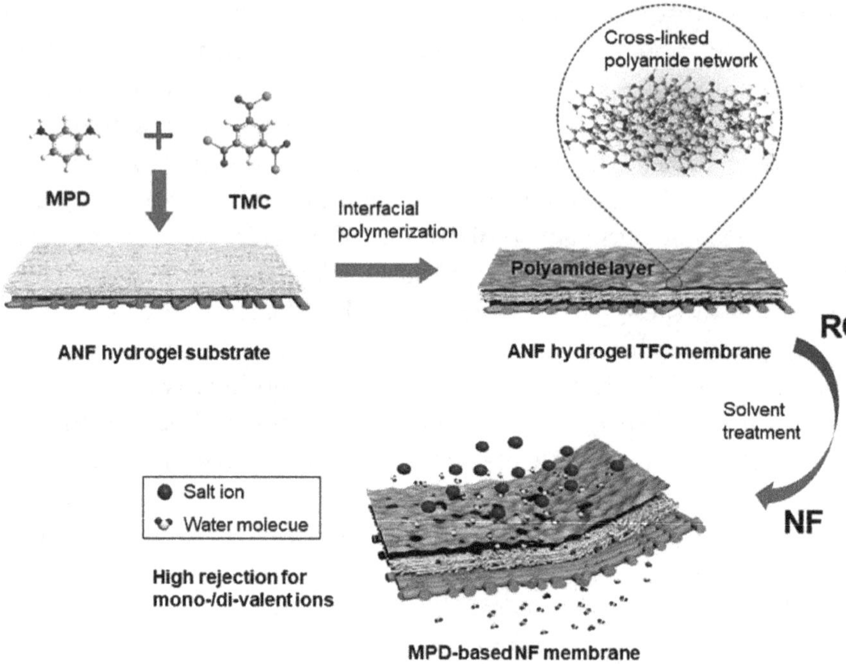

Fig. 1.11. Polyamide/Kevlar aramid nanofiber nanofiltration membranes for water desalination.[27]

in the thin film layer to enhance the separation properties. As shown in Fig. 1.12, the recent development of NF membrane technology has been in terms of creating better filtration technologies in wastewater treatment, food industry, solvent nanofiltration, desalination etc.[32]

The rejection obtained with NF membranes may be attributed to a combination of steric (membrane pore size sieving), Donnan (describe the equilibria and membrane potential interactions between a charged species and the interface of the charged membrane), dielectric and transport effects.[33] In application, NF is often combined with other membrane-based desalination processes, including dual stage NF desalination, integrated NF and RO desalination processes, integrated NF and forward osmosis (FO) desalination processes, and integrated NF and electrodialysis (ED) desalination processes.

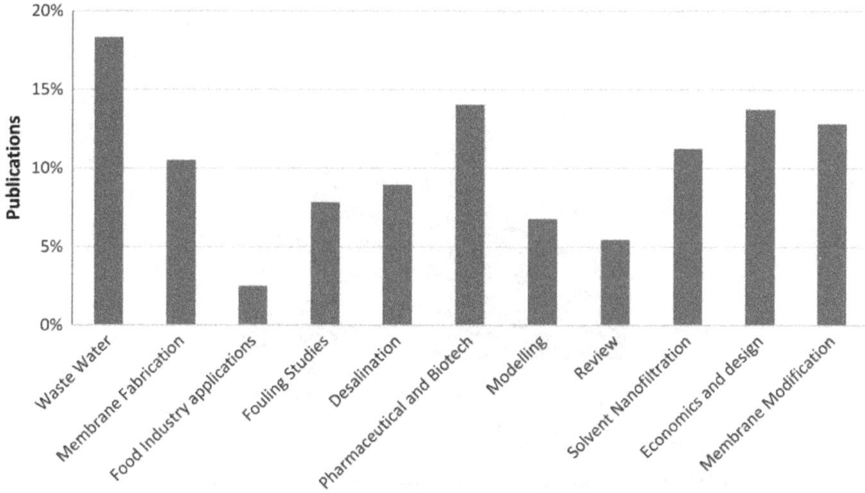

Fig. 1.12. The research topics of nanofiltration membrane process covered in the articles reviewed (2007–2017).[32]

1.5. Concentration-Driven Membranes Process in Water Desalination

Forward osmosis (FO) is an osmotically driven membrane separation process, which depends on the concentration gradient between the two sides of the membrane; water moves from a feed solution across a semipermeable membrane to a draw solution.[34] The driving force for mass transport is coming from the difference in osmotic pressure between the feed solution and a draw solution, which is caused by the concentration of solutions. Water diffuses from the feed solution of higher chemical potential (lower concentration) to a draw solution of lower chemical potential (higher concentration). With different concentration, an osmotic pressure difference builds up. This is used for water recovery enhancement in desalination of salt solutions, as shown in Fig. 1.13. As water diffuses through the membrane, the feed solution becomes concentrated and the draw solution is diluted; and thus, the draw solution must be re-concentrated in order to maintain the osmotic pressure driving force.[35] This process was considered by some scientists for the industrial purification of water due to the low energy composition. Without the energy driven advantage, FO draws

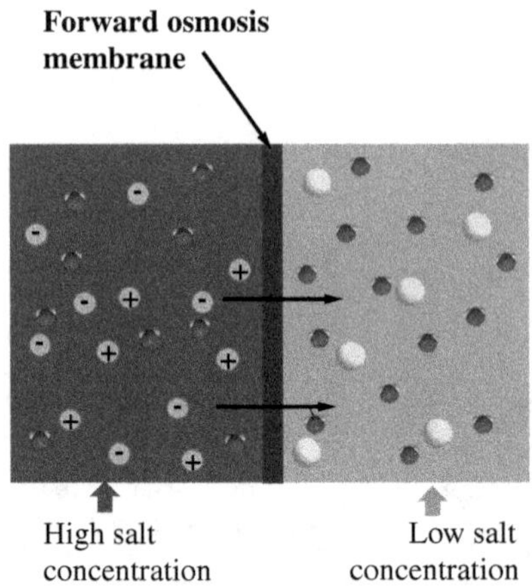

Fig. 1.13. Scheme of the osmosis desalination system.

a large portion of water from a feed, making it suitable for water contaminated with particulate and biological agents, especially for desalination.

However, the implementation of FO in water desalination has been limited by a few technical challenges, not the least of which is the draw solute (Fig. 1.13). In an FO system, when using conventional draw solutes, either the draw solute is incorporated into the final product or remove it through other separate processes. A standing challenge to the success of industrially relevant FO is the development of a next generation draw solute that can provide a significant osmotic pressure but be separated from water by means other than RO.[36]

The performance of the FO process was evaluated using a bench-scale membrane test unit coupled with a pilot-scale RO system. In this system, the pilot-scale RO system supplied the FO system with a constant-concentration draw solution, as shown in Fig. 1.14. The draw solution was recirculated on the support side of the FO membrane and its conductivity was measured by using a conductivity meter.[37] Therefore, the salt rejection (SR) is defined as the percentage of a particular incoming solute that is rejected by the

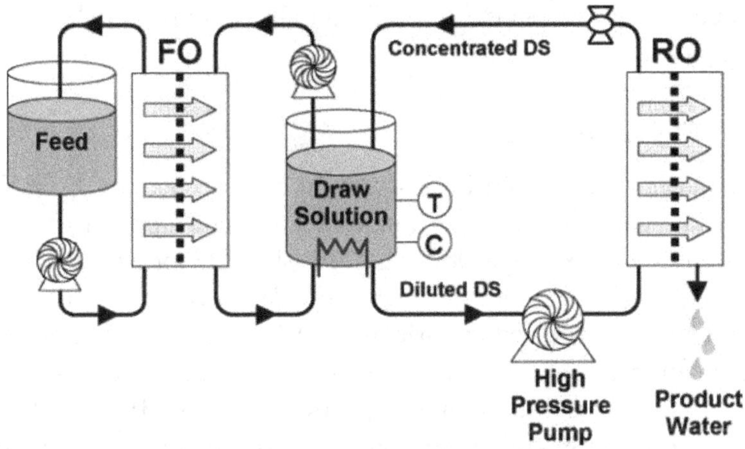

Fig. 1.14. Schematic drawing of the combined bench-scale FO and pilot-scale RO systems.[37]

FO membrane, which is given as

$$SR = \left(1 - \frac{c_P}{c_F}\right) \times 100\%$$

where, c_P and c_F represent permeate and feed solute concentration, respectively.

1.6. Electro-Driven Membranes Process in Water Desalination

Electrodialysis (ED) and membrane capacitive deionization (MCDI) are the two electro-driven membrane separation technologies, which are widely used in water desalination. Over the past decade, the development of electro-driven membranes processes has attracted increasing research attention in terms of materials, preparation and application, because of their potential in water treatment.[38–40] Different from pressure-driven membranes and concentration-driven membranes, the electro-driven membranes are composed of polymeric materials bearing covalently bound ionic fixed charges.[41,42] Only the ions with opposite charges to the surface charge of the membranes are transported through the membrane; while ions with the same charge are electrostatically repelled and retained in the feed compartment.

1.6.1. *Electrodialysis*

ED has been widely used for separation of cation/anion by different ion exchange membranes (IEMs) and has been applied on a large industrial scale for over 60 years. In the ED process, ions from the diluted compartment flow through IEMs to the concentrated compartment under the electric field. ED is used to produce pure water from brackish water, industrial waste and seawater.[43] The ED technology has four main parts, which is IEMs, electrode chamber, diluted chamber and concentrated chamber, as shown in Fig. 1.15. The key element of the ED process is the IEM, including anion exchange membranes (AEMs), cation exchange membranes (CEMs) and bipolar membranes (BMs). Cations/anions are selectively transported from one compartment (the diluted compartment) to another compartment (the concentrated compartment) through electro-driven membranes. Water molecules split into H^+ and OH^- by BMs in this process.

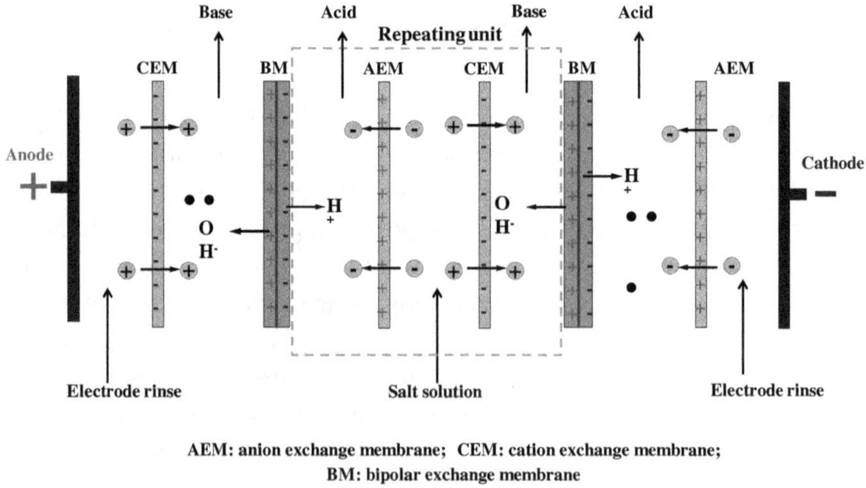

Fig. 1.15. The scheme of electrodialysis.

Based on the theories of Donnan equilibrium, chemical potential equilibrium and the electric double layer, anions and cations selective separation under the electric field force. The Donnan potential φ_{Don} and is given as

$$\varphi_{\text{Don}} = \varphi^m - \varphi^s = \frac{1}{z_i F}\left(RT \ln \frac{a_i^s}{a_i^m} + V_{m_i}(p^s - p^m)\right)$$
$$= \frac{1}{z_i F}\left(RT \ln \frac{a_i^s}{a_i^m} + V_{m_i}(\Delta\pi)\right)$$

where F is the Faraday constant; z_i is the charge number of the ions i in the solution; R is the gas constant and T is the temperature; V_{m_i} is the partial molar volume of component I; $\Delta\pi$ is the osmotic pressure difference between the membrane and the adjacent solution, which is also referred to as the swelling pressure of the membrane. The Donnan potential φ_{Don} cannot be measured directly. It can, however, be calculated from the ion activities in the solution and the membrane and by the swelling pressure $\Delta\pi$.

In the ED process, the desalination property is defined by the desalination rate (D),[44] which is defined as

$$D = \left(1 - \frac{c_t}{c_0}\right) \times 100\%$$

where c_t represents the concentration of the ion in dilute cell at time, t in the time and c_0 is its initial dilute in concentrated cell.

It is also can be illustrated by concentrated rate (C), which is defined as

$$C = \left(\frac{c_t'}{c_0'} - 1\right) \times 100\%$$

where c_t' represents the concentration of the ion in concentrated cell at time, t in the time and C_0 is its initial concentration in concentrated cell.

The energy consumption (EC) required in an ED process is the sum of the electrical energy to transfer the ionic species through ion selective membranes and the energy required to pump various solutions through the ED stack. It is a measure of the power consumed in Watthour per liter (Wh/L) for transport of ions and is calculated by

$$EC = \frac{E \int_0^t I dt}{V_D}$$

where E is the potential (V), I is current (A), V_D is the volume of the diluate (m^3), and t is time (s).

Usually, ED uses an assembly of alternatively stacked anion and cation exchange membranes to separate ions or produce pure water, as shown in Fig. 1.16, a series of AEMs and CEMs arranged in an alternating pattern between two electrodes.[45] In this process, ions from the diluted compartment pass through AEMs/CEMs to the concentrated compartment under the electric field. In an industrial size ED stack, hundreds of cell pairs are arranged between the electrodes. Different constructions of spacers and stacks such as sheet flow or tortuous path flow stack designs are applied.

In this process, the degree of desalination that can be achieved by passing the feed solution through a stack is a function of the solution

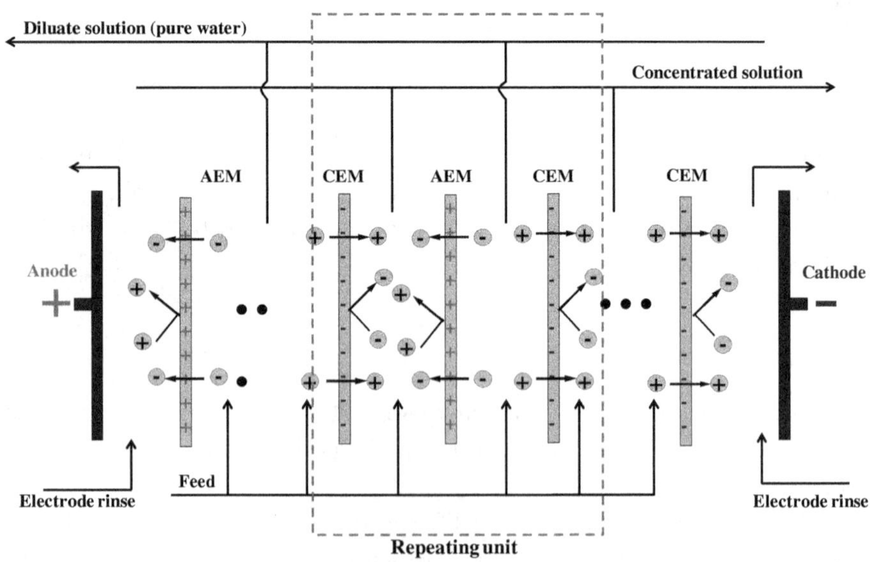

Fig. 1.16. Desalination by electrodialysis.

concentration, the applied current density, and the residence time of the solution in the stack. In the feed and bleed operating mode, both the brine and the product concentration can be determined independently, and very high recovery rates or brine concentrations can be obtained. In conventional ED, several indices should be measured:

(1) Concentration polarization and limiting current density, which is the electrochemical behavior of electro-driven membranes in ED separation process;
(2) The total energy required in electrodialysis for the actual desalination process is given by the current passing through the electrodialysis stack multiplied by the total voltage drop encountered between the electrodes.[46,47]

1.6.2. *Membrane capacitive deionization*

Membrane capacitive deionization (MCDI) consists of a porous electrode of an inexpensive commercial activated carbon fiber (ACF) and AMEs and CEMs into the cell compartment, as shown in Fig. 1.17. MCDI operation requires electricity, which makes it an electro-driven membranes process. It is an energy-efficient and environment-friendly desalination technology. In the MCDI cell, under a direct current electric field between two porous electrodes arranged in parallel, ions are removed from salt solutions by electrical migration. It is especially advantageous for treatment of low concentration salt solutions (lower than 5 g/L). Recent MCDI research is focused on exploiting high

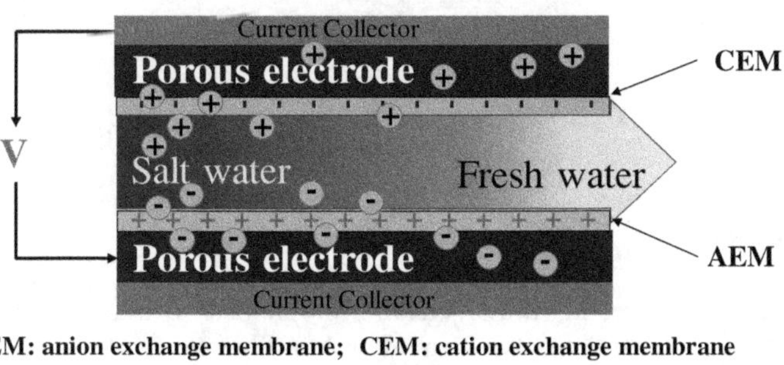

Fig. 1.17. Desalination by capacitive deionization.

performance porous electrode materials, such as carbon nanofibers, mesoporous carbon, graphene and carbide-derived carbon.

Currently, three main strategies are proposed for further development of MCDI electrode materials:

(1) developing novel electrode materials with high electric chemical performances and lower cost;
(2) enhancing the commercial electrode materials performance through surface modifications or optimization of the MCDI cell design;
(3) adding the suitable AEMs and CEMs into the cell compartment to increase the desalination rate, salt removal efficiency and charge efficiency.

Figure 1.18 shows the scheme of MCDI and the desalination process in the lab. The cell is assembled by a retaining plate, titanium

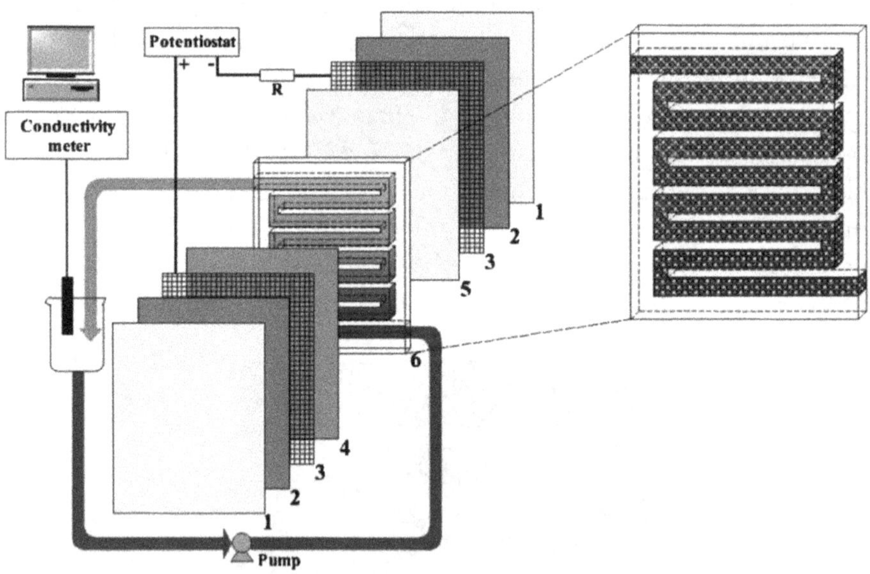

Fig. 1.18. Schematic diagram of a membrane capacitive desalination.[48] 1. retaining plate; 2. titanium plate; 3. activated carbon fiber electrode; 4. anion exchange membrane; 5. cation-exchange membrane; 6. rubber gasket with an S-shape channel.

plate, activated carbon fiber (ACF) electrode, anion exchange membrane, rubber gasket with an S-shaped channel, cation exchange membrane, ACF electrode, titanium plate and retaining plate. The 2-mm thick rubber gasket serves as the spacer between two ACF electrodes to avoid short circuit and allows the salt solution to flow through its S-shaped channel. In this process, the salt solution is pumped into the tested MCDI cell at a constant flow by using a peristaltic pump and the effluent is continuously recycled.[48] The electrical conductivity of the salt solution is monitored at the outlet of the MCDI cell using an online conductivity meter.

The desalination performance of MCDI is evaluated by the salt removal efficiency and the desalination rate. The salt removal efficiency is defined as,

$$\eta = \left(1 - \frac{c_e}{c_0}\right) \times 100\%$$

where c_0 and c_e refer to the initial and final concentration of the salt solution, respectively.

The desalination rate is calculated as,

$$v_1 = \frac{c_0 - c_e}{T_1}$$

where T_1 is the duration of the adsorption stage in an adsorption/desorption cycle.

The desorption rate is defined as,

$$v_2 = \frac{c_f - c_e}{T_2}$$

where T_2 is the duration of the desorption stage in an adsorption/desorption cycle, c_f is the salt solution concentration at the end of the desorption stage.

The charge efficiency is calculated as

$$\eta = \frac{(c_0 - c) \times V \times F}{M \times \sum it} \times 100\%$$

where V is the volume of the salt solution (m^3), F is the Faraday constant, which is 96,485 C/mol; M is the molecular weight of the salt; and i is the current (A) at time t (s).

1.7. Temperature-Driven Membranes Process in Water Desalination

Membrane Distillation (MD) is a temperature-driven membrane separation process in which only vapors can pass through a porous hydrophobic membrane, as shown in Fig. 1.19. The separation is achieved due to the vapor pressure differences between the membrane surfaces. As the core part of MD process, the membranes used for membrane distillation must be highly hydrophobic, highly porous and must have low thermal conductivity. Besides, the permeate flux increases with the increase in pore size and reduction in membrane thickness. The operating parameters including feed water temperature, flow rate, thickness of air gap, thickness of membrane, thermal conductivity of membrane, porosity, tortuosity and long-term operations have an effect on the distillate yield.

For these membranes, membrane porosity is given by

$$\varepsilon = 1 - \frac{\rho_m}{\rho_{\text{pol}}}$$

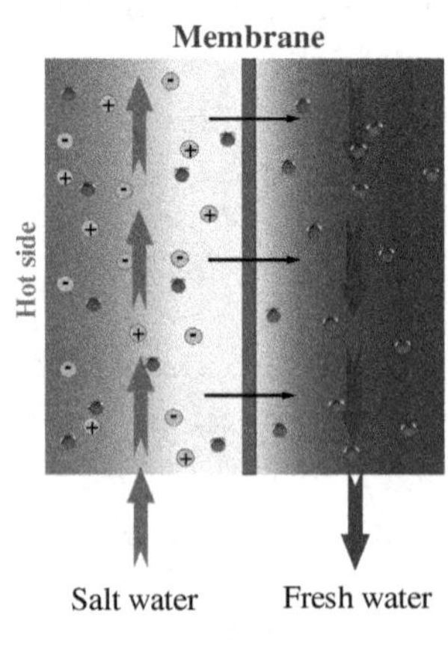

Fig. 1.19. Membrane distillation process for desalination.

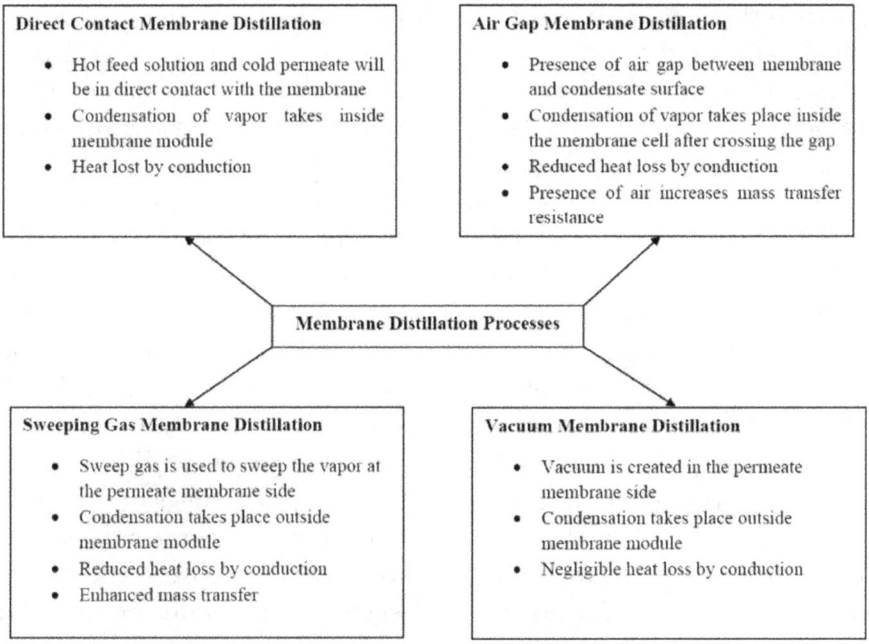

Fig. 1.20. Types of membrane distillation processes.

membrane tortuosity is given by

$$\tau = \frac{(2-\varepsilon)^2}{\varepsilon}$$

There are four main types of membrane distillation processes, namely air gap membrane distillation, sweeping gas distillation, direct contact membrane distillation and vacuum membrane distillation. The salient features of each membrane distillation process are shown in Fig. 1.20. All these processes, the hot feed solution is in direct contact with the membrane surface.

1.8. New Energy-Drive Process in Water Desalination

1.8.1. *Solar energy process in water desalination*

Solar energy is a common and most promising source for boosting the evolution of renewable energy technology. In past decades, solar-to-thermal technologies, such as domestic heating, salt-water desalination and power generation, have been the subject of both academic research and industrialization efforts.[49] Despite the unmatched resource potential of solar energy, the restricted utilization efficiency presents an enormous challenge.[50]

Solar vapor generation, which can directly transfer heat to facilitate evaporation, is an efficient way of harvesting solar energy for water desalination and obtain fresh water.[51] The use of direct solar energy for desalinating salt-water has been investigated extensively. It is similar to a part of the natural hydrologic cycle in which the salt-water is heated by solar energy to produce water vapor. Then, the water vapor is condensed on a cool surface, and the condensate collected as product water.[52] A common example of this type of process is the green house solar still, in which the salt water is heated in a basin on the floor and the water vapor condenses to be a fresh water on the sloping glass roof that covers the basin, as shown in Fig. 1.21. Solar-still designs can generally be grouped into four categories, which are the basin still, tilted-wick solar still, multiple-tray tilted still and concentrating mirror still. The basin

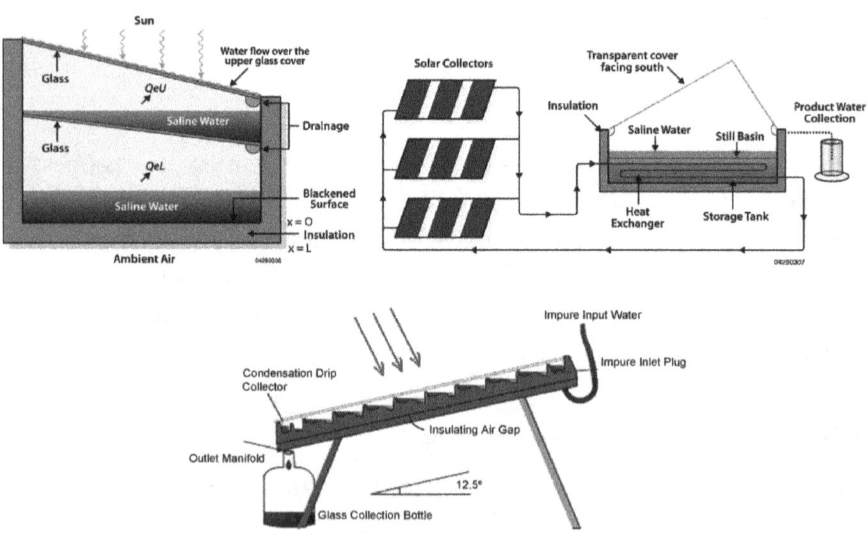

Fig. 1.21. Schematic diagrams of a simple solar still.

still consists of a basin, support structure, transparent glazing, and distillate trough.

Because solar radiation is the only power input for vapor generation, a lot of materials, including ultra-black absorbers, plasmonic nanoparticles and thermal-concentrating ceramics, have been used to enhance the conversion efficiency. In addition, variations of this type of solar still have been made in an effort to increase efficiency.[53] However, the requirement of a large solar collection area, high capital cost, vulnerability to weather-related damage and the inefficient utilization of converted heat imposes challenges. Of the solar energy process in water desalination, the three major heat consumption processes, which are water heating, parasitic thermal loss and water vaporization, only water vaporization is effective. Therefore, adequate thermal management in the solar vapor generation system is essential. The heat localization strategy, which restricts heat into a small amount of water in the evaporating surface, may substantially improve the efficiency of solar energy utilization.

1.8.2. *Nuclear energy in water desalination*

Compare to fossil fuels, nuclear power in particular received a considerable attention with its potential of less expensive nuclear fuel costs and highly efficient energy. Currently, energy production from all of the operating nuclear power plants utilizes the process of nuclear fission. During the nuclear fission, a huge amount of energy is released due to the split of heavy atomic nuclei in order to form lighter atomic nuclei. The nuclear energy offers a higher energy density compared to other conventional and renewable energy sources. With the continuous depletion of fossil fuels, continuous population growth, and the increase demand for fresh water, developing countries are currently interested in the development of nuclear reactors. According the International Atomic Energy Agency (IAEA) data, the nuclear power generation is expected to increase in the coming decades along with the applied policies towards the reduction in carbon dioxide emissions.

Nuclear energy in water desalination appears to be a feasible and a promising technical option to power desalination plants at reasonable costs. It is the process of producing fresh water by using an on-site nuclear reactor from salt-water. Plant capacity and water quality are among the several factors that can significantly affect the energy demand in any desalination process. In 1970s, in Kazakhstan and Japan, the nuclear desalination plants were first established. According to the type of coolant used, generally, there are two types of nuclear reactors that are used in desalination:

(1) light water reactors (LWR);
(2) the heavy water reactors (HWR).

For LWR category, it is also including boiling water reactors and pressurized water reactors; For HWR category, it is including pressurized heavy water reactors. There are other types such as the Liquid Metal Fast Breeder Reactor and high temperature gas cooled reactor. In general, the water-cooled reactors are preferred because of the well-established technology. In the literature, Pressurized Water Reactor, Pressurized Heavy Water Reactor and Liquid Metal Fast Breeder Reactor are the most common nuclear reactors coupled with desalination processes.

Figure 1.22 shows the nuclear energy process in water desalination, including a multi-effect distillation (MED) and an RO. The purpose is to generate electricity as well as utilizing the waste heat to produce steam that will be fed into the MED unit. To design a nuclear energy process in water desalination, the following steps should be performed:

(1) proper modeling for the reactor-desalination systems;
(2) careful evaluation for the nuclear plant safety;
(3) technical outcomes from the desalination process itself.[54]

Up to now, ten main projects around the world have launched to perform study and optimization for nuclear energy utilization in water desalination: INVAP in Argentina, CANDESAL in Canada, INET in China, NPPA in Egypt, BARC in India, KAERI in the republic of Korea, CNESTEN in Morocco, OPPE, OKBM, JSC in Malaya, Energetica in Russia and CNSTN in Tunisia. In addition, more and more countries are currently considering the nuclear energy process in water desalination. Whether this is a good option, taking

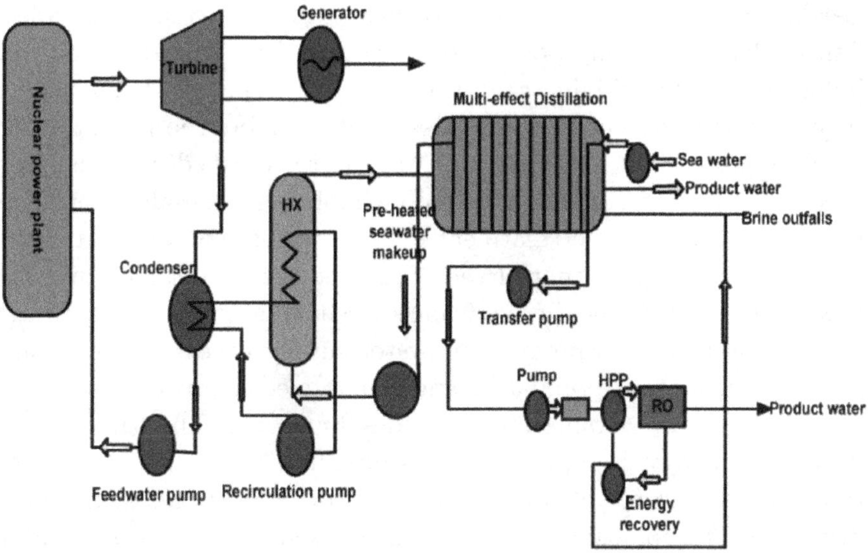

Fig. 1.22. A schematic diagram for a nuclear desalination process. HX: Heat Exchanger, and HPP: High Pressure Pump.[54]

the full environmental picture and the limited availability of uranium into account, remains however questionable.

1.8.3. *Wind energy in water desalination*

Wind turbines can be used to supply electricity or mechanical power in water desalination. Few applications have been implemented using wind energy to drive a mechanical vapor compression unit. Usually, the possible combination with wind energy process in water desalination are the mechanical compressor (MVC) and RO by the shaft, and ED, MVC and RO by the electricity. An example of the first world pilot plant was installed in 1991 at Borkum, an island in Germany, where a wind turbine with a nominal power of 45 kW was coupled to a 48 m^3/day MVC evaporator.[16]

1.9. Conclusions and Outlook

Recent years, numerous large-scale salt lakes or seawater desalination plants have been built in water-stressed countries and regions to augment available water resources, and construction of new desalination plants is expected to increase in the coming decades. Especially, the surging population growth, urban development, and industrialization will increase worldwide demand for fresh water, requiring new sources of water. Desalination offers the potential for an abundant and steady source of fresh water purified from the vast oceans or salt lakes. Many advanced technologies in desalination have been developed for the treatment of fresh water. For water-stressed countries and regions that already implement all other measures for freshwater generation, these advanced technologies may serve as the only viable means to provide the water supply necessary to sustain agriculture, support population and promote economic development.

Among these advances in water desalination technologies, membrane technology, especially the RO desalination process, has more than half a century of industrial operation and is one of the best performing technologies for desalting brackish water and sea water so far. Coupling this with the ability to handle a wide range of water sources makes it a strong candidate to tackle current and future water shortage problems. Alongside the advancements in other aspects of RO technology, the development of membrane materials has undeniably made RO desalination more economic by increasing performance and efficiency. Nevertheless, the search for multifunctional membrane materials that offer higher permeability, high ion and organic contaminant rejection, and operational robustness is still ongoing.

At the same time, research is expected to benefit the desalination industry by lowering the energy cost and membrane area required; simplifying pre-treatment processes; providing lower membrane maintenance costs; potentially achieving single pass RO desalination; and increasing plant capacity.[55,56] Therefore, combining the advantages of multiple desalination technologies in desalination will become the future research and developments efforts.[57,58] Such as,

(1) Combining the RO and ED process. As shown in Fig. 1.23, two extreme configurations of the generic design, representing designs with Seawater Diluate Ratio (SDR), which SDR $= 0$ and SDR $= 1$.[59]

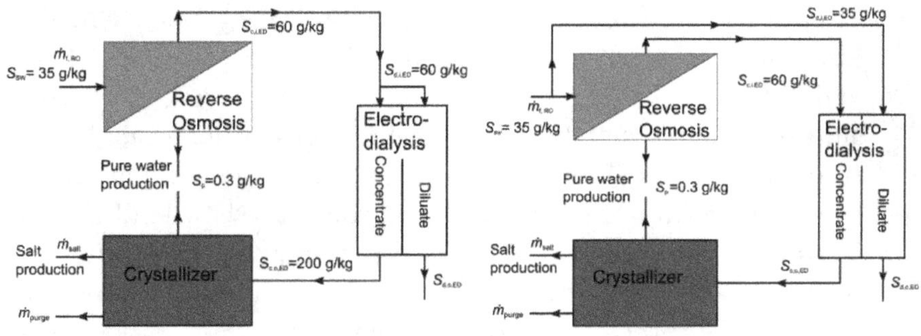

Fig. 1.23. Flow diagram of a RO-ED-crystallizer with (a) SDR = 0 (no seawater feed in ED dilute) and, (b) SDR = 1 (all of the ED dilute is from seawater feed).[59]

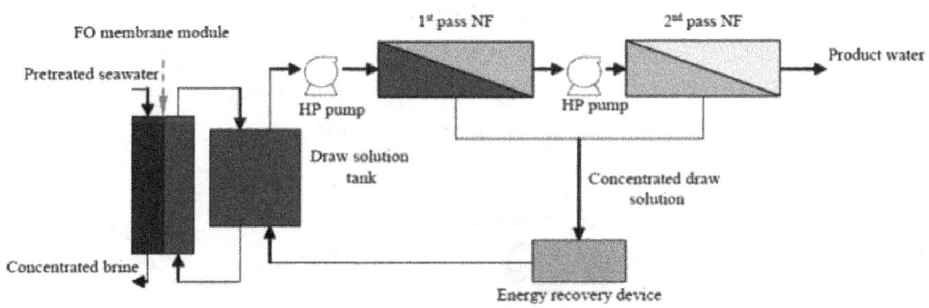

Fig. 1.24. Schematic diagram of hybrid FO-NF-NF seawater desalination process.[14]

(2) Integrated RO and NF desalination processes. Integrating RO with NF desalination process may increase the complexity and cost of desalination plant. However, NF pretreatment has shown effectiveness in removing divalent ions and reducing osmotic pressure from the RO feed water, and integrated NF with RO could join the advantages from both kinds of membranes.

(3) Integrated NF and FO desalination processes. In an FO-RO desalination system, the role of FO pass is drawing water from seawater, and the RO produced freshwater by concentrating the diluted FO draw solution. A FO-NF-NF system is shown in Fig. 1.24, water was first driven from seawater to draw solution through the FO membrane and then the two-stage NF process removed salt from diluted draw solution to form fresh water.[14]

(4) Integrated NF and ED desalination processes. ED technology also has potential to concentrate the brine of RO, NF-RO and NF-RO-multi-effect distillation (MED) processes to become suitable feed for membrane electrolysis for desalination.[14]
(5) Integrated RO, NF and multistage flash (MSF) desalination processes. A NF-RO-MSF desalination process as shown in Fig. 1.25, in which NF membranes were used to remove scale forming ions from seawater, allowing higher top brine temperature operation of the MSF processes. Not only was the water productivity improved but also the service life of cascade staging of MSF distillers was extended.[14]

Using desalination technologies driven by renewable energy sources is a viable way to produce fresh water in many locations today. Combining these renewable energy advances in water desalination technologies has unique synergies in regions where desalination

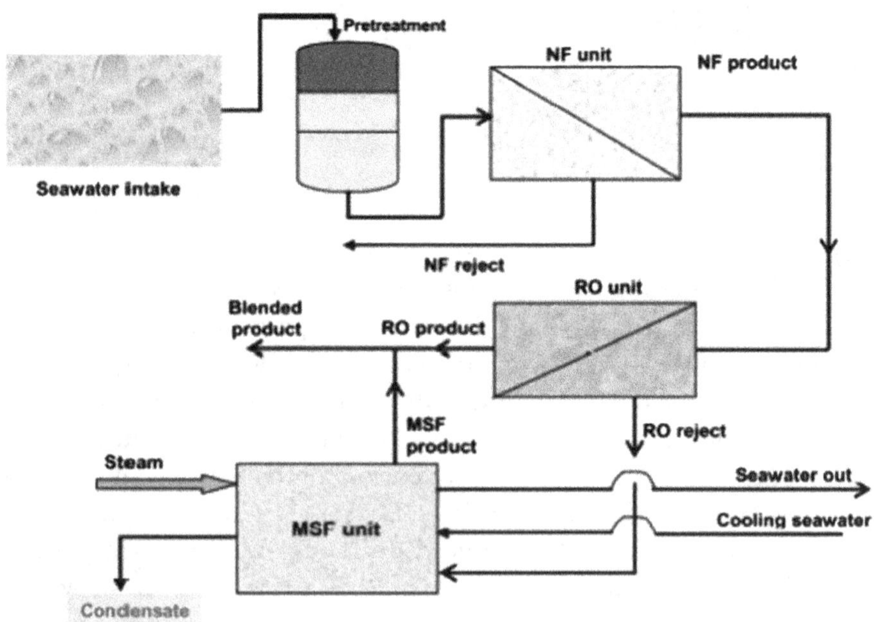

Fig. 1.25. Schematic diagram of hybrid NF-RO-MSF seawater desalination process.[14]

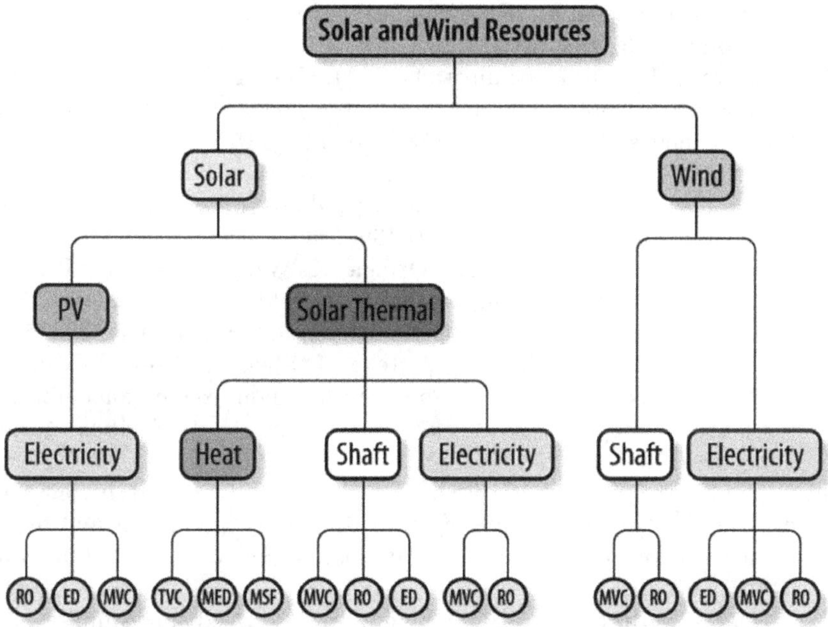

Fig. 1.26. Possible combination of solar and wind systems with desalination processes.[16]

is most needed.[60] Figure 1.26 shows the possible combination of solar and wind systems with desalination process.

Wither better water resource management, improved efficiencies, and conservation are vital for moderating demand and improving availability, it is our belief that improving the science and technologies of water desalination can help provide cost effective and robust solutions.

References

1. Y. Zhao, K. Tang, H. Liu, B. Van der Bruggen, A. Sotto Díaz, J. Shen, C. Gao, An anion exchange membrane modified by alternate electrodeposition layers with enhanced monovalent selectivity, *J. Membr. Sci.*, **520** (2016) 262–271.
2. A. F. Ismail, M. Padaki, N. Hilal, T. Matsuura, W. J. Lau, Thin film composite membrane — Recent development and future potential, *Desalination*, **356** (2015) 140–148.

3. M. G. Buonomenna, Nano-enhanced reverse osmosis membranes, *Desalination*, **314** (2013) 73–88.
4. A. Fuwad, H. Ryu, N. Malmstadt, S. M. Kim, T.-J. Jeon, Biomimetic membranes as potential tools for water purification: Preceding and future avenues, *Desalination*, **458** (2019) 97–115.
5. C. He, C. Liu, H. Gao, H. Xie, Y. Li, S. Wu, J. Xu, The optimal evaporation temperature and working fluids for subcritical organic Rankine cycle, *Energy*, **38** (2012) 136–143.
6. Y. Zhao, Y. Liu, C. Wang, E. Ortega, X. Wang, Y. F. Xie, J. Shen, C. Gao, B. Van der Bruggen, Electric field-based ionic control of selective separation layers, *J. Mater. Chem. A*, **8** (2020) 4244–4251.
7. Y. Zhao, Z. Mai, P. Shen, E. Ortega, J. Shen, C. Gao, B. Van der Bruggen, Nanofiber based organic solvent anion exchange membranes for selective separation of monovalent anions, *ACS Appl. Mater. Interfaces*, **12** (2020) 7539–7547.
8. N. Ghaffour, T. M. Missimer, G. L. Amy, Technical review and evaluation of the economics of water desalination: Current and future challenges for better water supply sustainability, *Desalination*, **309** (2013) 197–207.
9. A. D. Khawaji, I. K. Kutubkhanah, J.-M. Wie, Advances in seawater desalination technologies, *Desalination*, **221** (2008) 47–69.
10. A. Antony, J. H. Low, S. Gray, A. E. Childress, P. Le-Clech, G. Leslie, Scale formation and control in high pressure membrane water treatment systems: A review, *J. Membr. Sci.*, **383** (2011) 1–16.
11. Y. H. Teow, A. W. Mohammad, New generation nanomaterials for water desalination: A review, *Desalination*, **451** (2019) 2–17.
12. K. C. Ng, K. Thu, S. J. Oh, L. Ang, M. W. Shahzad, A. B. Ismail, Recent developments in thermally-driven seawater desalination: Energy efficiency improvement by hybridization of the MED and AD cycles, *Desalination*, **356** (2015) 255–270.
13. M. A. Shannon, P. W. Bohn, M. Elimelech, J. G. Georgiadis, B. J. Marinas, A. M. Mayes, Science and technology for water purification in the coming decades, *Nature*, **452** (2008) 301–310.
14. D. Zhou, L. Zhu, Y. Fu, M. Zhu, L. Xue, Development of lower cost seawater desalination processes using nanofiltration technologies — A review, *Desalination*, **376** (2015) 109–116.
15. M. Elimelech, W. A. Phillip, The future of seawater desalination: Energy, technology, and the environment, *Science*, **333** (2011) 712–717.
16. A. Al-Karaghouli, D. Renne, L. L. Kazmerski, Solar and wind opportunities for water desalination in the Arab regions, *Renew. Sustain. Energy Rev.*, **13** (2009) 2397–2407.

17. C.-S. Luo, W.-W. Chen, W.-F. Han, Experimental study on factors affecting the quality of ice crystal during the freezing concentration for the brackish water, *Desalination*, **260** (2010) 231–238.
18. M. S. Rahman, M. Ahmed, X. D. Chen, Freezing-melting process and desalination: I. Review of the state-of-the-art, *Sep. Purif. Rev.*, **35** (2006) 59–96.
19. P. M. Williams, M. Ahmad, B. S. Connolly, D. L. Oatley-Radcliffe, Technology for freeze concentration in the desalination industry, *Desalination*, **356** (2015) 314–327.
20. Z. Yang, X.-H. Ma, C. Y. Tang, Recent development of novel membranes for desalination, *Desalination*, **434** (2018) 37–59.
21. C. D. Peters, N. P. Hankins, Osmotically assisted reverse osmosis (OARO): Five approaches to dewatering saline brines using pressure-driven membrane processes, *Desalination*, **458** (2019) 1–13.
22. A. Perez-Gonzalez, A. M. Urtiaga, R. Ibanez, I. Ortiz, State-of-the-art and review on the treatment technologies of water reverse osmosis concentrates, *Water Res.*, **46** (2012) 267–283.
23. L. Malaeb, G. M. Ayoub, Reverse osmosis technology for water treatment: State-of-the-art review, *Desalination*, **267** (2011) 1–8.
24. K. P. Lee, T. C. Arnot, D. Mattia, A review of reverse osmosis membrane materials for desalination — Development to date and future potential, *J. Membr. Sci.*, **370** (2011) 1–22.
25. M. Qasim, M. Badrelzaman, N. N. Darwish, N. A. Darwish, N. Hilal, Reverse osmosis desalination: A state-of-the-art review, *Desalination*, **459** (2019) 59–104.
26. A. W. Mohammad, Y. H. Teow, W. L. Ang, Y. T. Chung, D. L. Oatley-Radcliffe, N. Hilal, Nanofiltration membranes review: Recent advances and future prospects, *Desalination*, **356** (2015) 226–254.
27. Y. Li, E. Wong, Z. Mai, B. Van der Bruggen, Fabrication of composite polyamide/Kevlar aramid nanofiber nanofiltration membranes with high permselectivity in water desalination, *J. Membr. Sci.*, **592** (2019) 117396.
28. A. Al-Amoudi, R. W. Lovitt, Fouling strategies and the cleaning system of NF membranes and factors affecting cleaning efficiency, *J. Membr. Sci.*, **303** (2007) 4–28.
29. F. Y. Zhao, Y. L. Ji, X. D. Weng, Y. F. Mi, C. C. Ye, Q. F. An, C. J. Gao, High-flux positively charged nanocomposite nanofiltration membranes filled with poly(dopamine) modified multiwall carbon nanotubes, *ACS Appl. Mater. Interfaces*, **8** (2016) 6693–6700.
30. J. Yin, B. Deng, Polymer-matrix nanocomposite membranes for water treatment, *J. Membr. Sci.*, **479** (2015) 256–275.

31. X. Li, A. Sotto, J. Li, B. Van der Bruggen, Progress and perspectives for synthesis of sustainable antifouling composite membranes containing in situ generated nanoparticles, *J. Membr. Sci.*, **524** (2017) 502–528.
32. D. L. Oatley-Radcliffe, M. Walters, T. J. Ainscough, P. M. Williams, A. W. Mohammad, N. Hilal, Nanofiltration membranes and processes: A review of research trends over the past decade, *J. Water Process Eng.*, **19** (2017) 164–171.
33. G. Gong, P. Wang, Z. Zhou, Y. Hu, New insights into the role of an interlayer for the fabrication of highly selective and permeable thin-film composite nanofiltration membrane, *ACS Appl. Mater. Interfaces*, **11** (2019) 7349–7356.
34. M. L. Stone, C. Rae, F. F. Stewart, A. D. Wilson, Switchable polarity solvents as draw solutes for forward osmosis, *Desalination*, **312** (2013) 124–129.
35. G. Amy, N. Ghaffour, Z. Li, L. Francis, R. V. Linares, T. Missimer, S. Lattemann, Membrane-based seawater desalination: Present and future prospects, *Desalination*, **401** (2017) 16–21.
36. M. Xie, L. D. Nghiem, W. E. Price, M. Elimelech, Toward resource recovery from wastewater: Extraction of phosphorus from digested sludge using a hybrid forward osmosis–membrane distillation process, *Environ. Sci. Tech. Lett.*, **1** (2014) 191–195.
37. C. R. Martinetti, A. E. Childress, T. Y. Cath, High recovery of concentrated RO brines using forward osmosis and membrane distillation, *J. Membr. Sci.*, **331** (2009) 31–39.
38. Y. Zhao, Y. Li, S. Yuan, J. Zhu, S. Houtmeyers, J. Li, R. Dewil, C. Gao, B. Van der Bruggen, A chemically assembled anion exchange membrane surface for monovalent anion selectivity and fouling reduction, *J. Mater. Chem. A*, **7** (2019) 6348–6356.
39. Y. Zhao, W. Shi, B. Van der Bruggen, C. Gao, J. Shen, Tunable nanoscale interlayer of graphene with symmetrical polyelectrolyte multilayer architecture for lithium extraction, *Adv. Mater. Interf.*, **5** (2018) 1701449.
40. Y. Zhao, J. Zhu, J. Li, Z. Zhao, S. I. Charchalac Ochoa, J. Shen, C. Gao, B. Van der Bruggen, Robust multilayer graphene-organic frameworks for selective separation of monovalent anions, *ACS Appl. Mater. Interfaces*, **10** (2018) 18426–18433.
41. Y. Zhao, K. Tang, Q. Liu, B. Van der Bruggen, A. S. Díaz, J. Pan, C. Gao, J. Shen, Recovery of chemically degraded polyethyleneimine by a re-modification method: Prolonging the lifetime of cation exchange membranes, *RSC Advances*, **6** (2016) 16548–16554.

42. Y. Zhao, C. Zhou, J. Wang, H. Liu, Y. Xu, J. W. Seo, J. Shen, C. Gao, B. Van der Bruggen, Formation of morphologically confined nanospaces via self-assembly of graphene and nanospheres for selective separation of lithium, *J. Mater. Chem. A*, **6** (2018) 18859–18864.
43. Y. Zhao, Y. Li, J. Zhu, A. Lejarazu-Larranaga, S. Yuan, E. Ortega, J. Shen, C. Gao, B. Van der Bruggen, Thin and robust organic solvent cation exchange membranes for ion separation, *J. Mater. Chem. A*, **7** (2019) 13903–13909.
44. Y. Zhao, Y. Qiu, Z. Mai, E. Ortega, J. Shen, C. Gao, B. Van der Bruggen, Symmetrically recombined nanofibers in a high-selectivity membrane for cation separation in high temperature and organic solvent, *J. Mater. Chem. A*, **7** (2019) 20006–20012.
45. Y. Zhao, K. Tang, H. Ruan, L. Xue, B. Van der Bruggen, C. Gao, J. Shen, Sulfonated reduced graphene oxide modification layers to improve monovalent anions selectivity and controllable resistance of anion exchange membrane, *J. Membr. Sci.*, **536** (2017) 167–175.
46. Y. Zhao, C. Gao, B. Van der Bruggen, Technology-driven layer-by-layer assembly of a membrane for selective separation of monovalent anions and antifouling, *Nanoscale*, **11** (2019) 2264–2274.
47. Y. Zhao, J. Zhu, J. Ding, B. Van der Bruggen, J. Shen, C. Gao, Electric-pulse layer-by-layer assembled of anion exchange membrane with enhanced monovalent selectivity, *J. Membr. Sci.*, **548** (2018) 81–90.
48. P. Liang, L. Yuan, X. Yang, S. Zhou, X. Huang, Coupling ion-exchangers with inexpensive activated carbon fiber electrodes to enhance the performance of capacitive deionization cells for domestic wastewater desalination, *Water Res.*, **47** (2013) 2523–2530.
49. K. Sampathkumar, T. V. Arjunan, P. Pitchandi, P. Senthilkumar, Active solar distillation — A detailed review, *Renew. Sust. Energ. Rev.*, **14** (2010) 1503–1526.
50. F. Zhao, X. Zhou, Y. Shi, X. Qian, M. Alexander, X. Zhao, S. Mendez, R. Yang, L. Qu, G. Yu, Highly efficient solar vapour generation via hierarchically nanostructured gels, *Nature Nanotech.*, **13** (2018) 489–495.
51. Y. Liu, S. Yu, R. Feng, A. Bernard, Y. Liu, Y. Zhang, H. Duan, W. Shang, P. Tao, C. Song, T. Deng, A bioinspired, reusable, paper-based system for high-performance large-scale evaporation, *Adv. Mater.*, **27** (2015) 2768–2774.
52. O. Mahian, A. Kianifar, S. Z. Heris, D. Wen, A. Z. Sahin, S. Wongwises, Nanofluids effects on the evaporation rate in a solar still equipped with a heat exchanger, *Nano Energy*, **36** (2017) 134–155.

53. V. Velmurugan, M. Gopalakrishnan, R. Raghu, K. Srithar, Single basin solar still with fin for enhancing productivity, *Energy Convers. Manag.*, **49** (2008) 2602–2608.
54. A. Al-Othman, N. N. Darwish, M. Qasim, M. Tawalbeh, N. A. Darwish, N. Hilal, Nuclear desalination: A state-of-the-art review, *Desalination*, **457** (2019) 39–61.
55. E. Drioli, E. Curcio, G. Di Profio, F. Macedonio, A. Criscuoli, Integrating membrane contactors technology and pressure-driven membrane operations for seawater desalination, *Chem. Eng. Res. Des.*, **84** (2006) 209–220.
56. Y. Ibrahim, H. A. Arafat, T. Mezher, F. AlMarzooqi, An integrated framework for sustainability assessment of seawater desalination, *Desalination*, **447** (2018) 1–17.
57. K.-N. Park, S. Y. Hong, J. W. Lee, K. C. Kang, Y. C. Lee, M.-G. Ha, J. D. Lee, A new apparatus for seawater desalination by gas hydrate process and removal characteristics of dissolved minerals (Na^+, Mg^{2+}, Ca^{2+}, K^+, B^{3+}), *Desalination*, **274** (2011) 91–96.
58. M. Pearce, F. Brennan, Novel findings in desalination, *Desalination*, **360** (2015) 13–18.
59. K. G. Nayar, J. Fernandes, R. K. McGovern, B. S. Al-Anzi, J. H. Lienhard, Cost and energy needs of RO-ED-crystallizer systems for zero brine discharge seawater desalination, *Desalination*, **457** (2019) 115–132.
60. H. Sharon, K. S. Reddy, A review of solar energy driven desalination technologies, *Renew. Sust. Energ. Rev.*, **41** (2015) 1080–1118.

CHAPTER 2

Desalination, Energy and Environment Nexus

Hilla Shemer and Raphael Semiat*

Rabin Desalination Laboratory, Grand Water Research Institute
Wolfson Faculty of Chemical Engineering, Technion — Israel Institute of Technology Technion City, Haifa 3200002, Israel

2.1. Introduction

The need for sustainable sources of water is ever growing, especially with increasing global population. Desalination of seawater (SW) and brackish (BW) water is already viable solution to save life in a thirsty world. Desalination techniques consist of membrane separation processes such as reverse osmosis (RO) and electrodialysis or thermal processes such as multi-stage flash (MSF) and multi-effect distillation (MED). At present, RO and MSF are the prevailing techniques. Freezing desalination, solar evaporation, forward osmosis, membrane distillation and humidification-dehumidification have been developed for desalination. However, these techniques are not commercially implemented due to their much higher energy demand, and equipment costs compared with the commercial techniques.

General public uncertainties and concerns persist with respect to the energy demand of desalination as well as the potential impacts of desalination on coastal and marine environments. Salt laden waste i.e., brine is one of the biggest concerns of desalination, as it potentially raises the level of the receiving water body salinity

*Corresponding author: cesemiat@technion.ac.il

and hence, poses a risk to marine ecosystems. It is established that salinity tolerances of marine organisms are highly variable as it depends on the salinity level and exposure time. Relative abundances and growth rates are not significantly impacted by elevated salinities of 10% above ambient, while community structure may change.

The degree of the environmental impacts, by brine discharge, is highly dependent on the volume and characteristics of the brine, the dilution ratio, the properties of the receiving water body, and the discharge method. Seawater desalination brine is discharged into coastal waters directly at the shoreline, via dedicated outfall, or through sub-sea multiport diffuser system farther from the coastline. Diffuser systems offer increased dilution and mixing as compared to open channels outfall. In a well-mixed environment, adverse impacts of the brine discharge are confined to an area of few hundred meters from the discharge point.

Sound environment and sustainability principles from planning through operation are essential. Therefore, desalination projects, worldwide, undergo environmental impact assessments from planning to design, construction and operation. On-going monitoring programs are conducted to assess the impact of brine discharge on the receiving marine environment by comparing sampling pre- and post-operation of the desalination plant. Both chemical and biological aspects are being evaluated. To date, these monitoring programs revealed that brine discharge have localized minimal impacts on the marine environment.

The energy demand of a desalination plant is site specific. It is affected by the desalting technology, engineering, energy recovery devices, water recovery level, source water characteristics such as temperature and salinity, and required product water quality. It should be realized that the quality of the desalinated water necessarily affects the quality of treated wastewater designated for different applications. Field experience along with scientific research showed that the energy consumption of reverse osmosis desalination constitutes only a small fraction of a national cumulative energy demand. Properly sited, designed and operated desalination plants necessarily reduce the energy demand and environmental footprint.

2.2. Global Brine Production

Global desalinated water production, as of 2019, stands at 95.37 million m^3/day resulting in 141.5 million m^3/day brine production. As half of the desalination capacity is located in the Middle East and North Africa with 70.3% of global brine production (100 million m^3/day) is concentrated in these areas.[1] Figure 2.1 displays the share in desalinated water and brine productions as a function of the feed water and desalination technology. As seen, reverse osmosis is the leading technology for desalinating seawater as well as other water types including brackish water, wastewater (WW) and river water (RW). The WW and RW water types are displayed in Fig. 2.1 under "Other RO" column. The Thermal desalination technologies, multi-stage flash distillation and multi-effect distillation, only desalinate SW. These technologies accounts for 24% of the global desalinated water capacity but to 55% of the global brine production.[1]

Almost 50% of the desalination plants worldwide are located within 1 Km of the coastline and about 80% are located within 10 Km range. Despite environmental concerns, these plants discharge

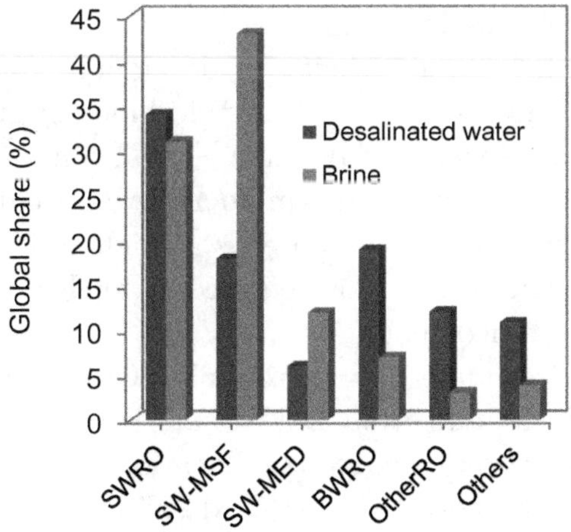

Fig. 2.1. Global share of desalinated water and brine productions (data obtained from Ref. 1) Total desalinated water and brine productions are 95.4 and 141.5 million m^3/day, respectively.

the brine back to the sea/ocean, as this is the most economical disposal method. Desalination plants located inland, produced about 22 million m^3/day of brine.[1] For these operations, several disposal methods have been practiced including sewer discharge, deep-well injections, evaporation ponds, and land applications. However, these techniques are unsustainable and hindered by high capital costs.[2] Development of sustainable brine management systems are based on minimization techniques.[3]

2.3. The True Cost of Water

Lack of safe and unreliable drinking water is a globally recognized problem, which may result in desertification, forced migration, hunger, diseases, and domestic or regional conflicts. An example for the latter is the Atatürk dam, on the Euphrates River, which enabled extensive irrigation within Turkey, while in Iraq and North-Eastern Syria there is a decline in water quantity and quality.[4] Desalination is increasingly crucial in a world where freshwater resources are progressively strained by population growth, development, droughts, climate change and more. Yet, the misconception of high cost and high energy demand often merits strong opposition to global implementation of desalination.

The impact of water scarcity in developing countries is more challenging than in the western world. Millions of girls and women carry heavy loads of water daily, often several times a day. This task results in lost opportunities of employment and education. It was estimated that the cost of this water is as high as 70 US $/m^3$. This estimation is based on the time required to fetch the water and the subsequent loss of income.

The cost of desalinated water from SWRO process range between 0.5 to 1.2 US $/m^3$ and from BWRO 0.2–0.4 US $/m^3$ depending on the technology and geographic location.[5] Energy contributes much of the total cost of SWRO desalination (20–35%) other constitutes are CAPEX (30–40%) and O&M (15–30%). Specifically, production cost of desalinated water is optimized based on the cost of land, electric energy for pumping, cost of pre- and post-treatments, equipment, brine disposal, labor, maintenance (including membranes replacement) and financial charges.[6]

The water price in Europe range between 1.53–1.99 euro/m^3.[7] These prices are higher than the prices of desalinated water. However, household customers pay more for desalinated water than the values listed above. In Israel, where close to 80% of the urban water supply is desalinated seawater, the total cost for a household customer is slightly above 2.2 US \$/m^3. This include cost of water transportation, wastewater treatment and taxes. Nonetheless, farmers pay about 0.6 US \$/m^3 for drinking water and around 0.3 US \$/m^3 for reclaimed/tertiary treated wastewater.[8] This example indicates that the cost of water is not directly related to the source of water and production technique. It is mainly site-specific and rest on governments' ability to maintain/control sustainable high-quality water at affordable prices.

2.4. Energy Consumption of Desalination

According to the rules of Thermodynamics, a minimal energy is needed to derive any separation process including desalination. Below this minimal energy, the driving force is negative, hence, separation is not possible. This concept was demonstrated for desalination.[9–13] In brief, at a temperature T a minimum reversible work W, for any separation process is given by[9]:

$$-W = \Delta H - T\Delta S = \Delta F \tag{1}$$

where, H is the enthalpy changes, S is the entropy changes, and F is the free energy change. The free energy relations in the form of molar concentration of the solute (salt) in water solution is as follows:

$$-W = \int \Delta F dn = \int RT \ln a_w dn = \int_{n_1}^{n_2} RT \ln \frac{p}{P^0} dn \tag{2}$$

here, n is the number of water moles in the solution, a_w is the water activity, R is the gas constant, 1 and 2 are the initial and final stages of the water separation, respectively, and P, P^0 are the vapor pressure of the solution and of pure water, respectively.

The final expression for the minimal energy of separation is:

$$-W = \frac{0.296T}{100 - n_2} \int_{100}^{n_2} \log_{10} \frac{p}{P^0} dn \tag{3}$$

The energy needed to separate fresh water at 25°C from an infinite source of seawater (3.5% salinity) is 0.79 KWh/m^3. For SW desalination, at a water recovery of 50%, the minimum separation energy is 1.1 KWh/m^3. Further discussion can be found elsewhere.[14] Practically, SWRO desalination consumes 1.7–2.5 KWh/m^3 for the process itself including energy recovery.[15] These values are in fact among the closest to ideal energy processes known in the chemical industry. This is due to high production and low cost of the product (i.e., desalinated water). Pumping, pre-treatment, brine discharge, and electric power used within the plant total to additional 0.3–1.5 kWh/m^3. Hence, overall, the energy demand of a SWRO desalination is 2–4 kWh/m^3.

The overall energy consumption of thermal desalination processes i.e., MED and MSF range between 5.5 to 16.0 KWh/m^3. These values include thermal energy (4–12 KWh/m^3) and electric energy (1.5–4.0 KWh/m^3) consumption.[5]

The energy consumption for pumping and transporting water over 200 Km is in same order as SW desalination. As mentioned before, close to 80% of the urban water usage in Israel is supplied by seawater desalination. The yearly energy cost of desalination, divided by the yearly national energy consumption for electricity and transportation, is about 1.3%. There is no doubt that it is easier to save 1.3% of the national energy consumption than to cut 80% of the urban water consumption. It should also be mentioned that above 60% of the desalinated water production in Israel is attained during the night using energy that would have been wasted otherwise. Before the desalination era in Israel, water was pumped from north to south, using 3% of the national energy consumption. Since the desalination plants are much closer to the large cities, the distribution systems are much shorter reducing significantly the national overall energy consumption. Therefore, on national basis, the energy consumption of a large-scale desalination is negative, namely has no influence on the carbon footprint.

At present, fossil fuel is the main energy source for desalination processes. Alternatives to fossil fuel power stations include nuclear power, solar power, geothermal power, wind power, tidal power, hydro-electric power and industrial waste.[16,17] For example, the Sydney desalination plant's (New South Wales, Australia) energy requirements are offset by a 67-turbine wind farm. It should be noted that renewable energies provide inconsistent power and therefore, the wind energy is fed to the grid while the desalination plant, that needs a stable energy source, utilizes the grid.

Some of the alternative energy sources are commercially available technologies (i.e., nuclear, wind and hydro-electric power) while others are still in prototype form, due to high cost and inability to produce energy on a continuous basis.[16] The existing renewable energy desalination systems are implemented mostly in small capacities plants, from a few cubic meters per day up to $100\,m^3$/day. The oil market itself is developing towards increased usage of natural gas and shale oil. Natural gas-fired electrical generation offers a variety of environmental benefits including fewer emissions, reduced sludge, combined cycle generation and higher efficiencies.

In addition to renewable energy, there are several approaches aimed to reduce the energy consumption of RO desalination processes among them:

(i) Improvements of membrane technology by developing highly permeable membranes and/or low fouling composites and by increasing the lifespan of the membrane;[18–21]
(ii) Energy recovery devices (reducing the total power consumed by high pressure pumps);[22,23]
(iii) Efficient high-pressure pumps (reducing the electrical power consumption);[24–26]
(iv) Optimization of the RO process (operating conditions and configuration);[18] and
(v) Intermediate chemical de-mineralization to obtain higher water recovery.[27]

2.5. Potential Environmental Impacts of Desalination

Throughout the years concerns were raised regarding the potential environmental impacts of desalination plants. These can be divided into three categories. The first is related to the construction stage. Much like any other development projects, construction of a desalination plant generates terrestrial effects such as alteration of the coastal environment and obstruction of free passage along the seashore, noise, vibration, native fauna, sediment discharge to waterways resulting from soil erosion or spoil from earthworks, accidental spillage or leakage of hazardous chemicals, etc.[28]

The second category is related to high carbon footprint. As explained previously, desalination plants are largely powered by fossil fuel sources, stemming air pollution and greenhouse gases emission concerns. It is estimated that in 2015 the installed and operational desalination plants worldwide emitted around 76 million tons of CO_2 per year.[29] In fact, desalination plants are indirectly linked to the generation of greenhouse gases as most of the emission occurs during electricity production from fossil fuels. As discussed earlier, a reduction of energy consumption, on a national basis, will necessarily reduce the greenhouse gases emission. Alternative energy sources will further minimize the carbon footprint.

The third category is related to the impact of desalination on the marine environment. This includes brine discharge, chemicals disposal, and entrainment and impingement of marine organisms from the SW intake. Brine is discharged into coastal waters either directly at the shoreline, via dedicated outfall, or by co-disposal with cooling water from power stations and/or with domestic wastewater. Alternatively, brine is discharged through sub-sea multi-port diffuser farther from the coastline.[30] In open discharge outfalls the mixing between the brine and ambient seawater is mainly driven by the velocity of the discharge and the fact that the discharge is located in an area with naturally occurring underwater currents of high intensity.[31] The key advantages of co-discharge with wastewater effluent are accelerated dissipation of both the wastewater and the brine and diminished concentration of metals, organics and pathogens which are originated solely from the effluent, in the combined stream. It is important to note that direct disposal of wastewater effluent to the sea, may cause adverse effects on the marine environment due to contamination by organics and nutrient such as phosphates and nitrate as well as a reduction in the osmotic pressure.

In a sub-sea diffusers outfall (multi-port or rosette pattern) the mixing depends on the discharge velocity, diffuser system design (i.e., angle of discharge, depth below sea surface, number and spacing between nozzles) brine salinity, and prevailing marine currents.[32]

The brine discharge of the Tuaspring desalination plant, Singapore is displayed in Fig. 2.2. The intake is located 50 m from the plant (Fig. 2.2(a)). The brine is discharged through an outfall diffuser 120 m from shore to a water depth of approximately 8 m. Sea surface at the upward jet of the brine discharge is shown in Fig. 2.2(b). The short distance between the intake and discharge points, of about 150 m, demonstrates the lack of impact of the brine on the feed concentration.

Seawater intake may be harmful for marine species by impingement and entrainment. It is well established that the intake is not threatening for the large marine species. Nonetheless smaller species such as phytoplankton, zooplankton may be affected.[33] Therefore, the water intake is placed well above the sea floor. The system is composed of large openings, covered by mesh to lower the flow

Fig. 2.2. (a) Feed intake point and (b) brine discharge location, Tuaspring desalination plant, Singapore. (Images were taken from the roof of the power station by R. Semiat).

velocities in order to avoid impinging of even small organisms such as larvae and eggs.

It was postulated that brine disposal affects the physical, chemical and biological characteristics of the receiving water body. Major concerns included increased salinity and the discharge of toxic chemicals such as heavy metals, scale inhibition additives (polymeric substances such as polyphosphates, phosphonates and polycarbonic acids), bio-fouling agents, antifoaming agents, coagulants/flocculants (ferric chloride, ferric sulfate, polyelectrolytes, and cationic polymer), cleaning chemicals, chlorine and bisulfite.[2,34–37] Typical characteristics of brine originated from reverse osmosis and thermal desalination is listed in Table 2.1.

Brine generated by desalination of seawater is typically twice as salty as the SW. Therefore, its density is higher than the density of the seawater. This causes the brine to fall on the sea floor near the outlet, a phenomenon defined as plume effect. The plume effect can be reduced/eliminated by

(i) Discharging the brine within a strong sea current to achieve good mixing;
(ii) Discharging the brine in an upward angle (jet) hence, minimizing the spatial extent and intensity of brine plumes and improving the mixing; High salinity, which can reduce dissolved oxygen levels, may impact the benthic organisms, resulting in ecological effects throughout the food chain.[39,40] It is important to note that fish appear to avoid waters above background salinities.[41]
(iii) Diluting the brine prior to discharge such as with cooling water of power stations.

The degree of the environmental impacts is highly dependent on the brine's characteristics and volume, dilution ratio, properties of the receiving water body, and discharge technique. It is recognized that in a well-mixed environment, adverse impacts are limited to a few hundred meters from the discharge point. While, in shallow and/or semi-closed bays the impacts are more pronounced.[42]

All desalination projects worldwide undergo thorough Environmental Impact Assessments (EIA) from planning to design, construction and operation. On-going monitoring is carried out to evaluate the impact of brine discharge on the receiving environment. The monitoring programs evaluate the effects of discharge through comparison to sampling before the plant began to discharge and/or to control areas.

Table 2.1. Typical characteristics of brine from RO and thermal seawater desalination.[38]

	Reverse osmosis	MED and MSF
Salinity (g/L)	Up to 65–80	About 50
Temperature	Ambient	5–15°C above ambient
Dissolved oxygen	Ambient	Below ambient
Chlorine (Biofouling control)	Typically neutralized	Approximately 10–25% of source water feed is dosage
Halogenated organics	Typically, below harmful levels	Typically, below harmful levels
Coagulants	May be present	Not present
Scale inhibitors	Typically, below toxic levels	Typically, below toxic levels
Antifoaming agents	Not present	Typically, below toxic levels
Heavy metals	Traces of copper, iron, chromium, nickel, molybdenum and zinc	Traces of copper and nickel
Cleaning chemicals	Alkaline or acidic solutions with additives such as detergents, complexing agents, oxidants, and biocides	Acidic solutions containing corrosion inhibitors

2.6. Worldwide Experience

2.6.1. *Middle East-Ashkelon and Sorek/Palmachim (Israel)*

The Rabin Desalination Laboratory researchers, at the Technion, along with Israeli water experts founded the Israel Desalination Society (IDS). With the advocacy of the IDS, in 1999, the Israeli government initiated a long-term desalination program. The program is designed to provide for the growing demands on Israel's scarce water resources, and to mitigate the drought conditions that are characteristics of the area since the mid-1990's.

2.6.1.1. *Ashkelon*

Since 2004 (a year before the plant was commenced), the SW quality in the area of Ashkelon desalination plant is surveyed by the Israeli Electric Company (IEC) in collaboration with Israel Oceanographic and Limnological Research (IOLR) center. Monitoring of the water quality parameters (salinity, temperature, turbidity, heavy metals concentration, nutrients etc.), planktonic microalgae, benthic infauna and sediment quality is carried out. The monitoring data, of the period between 2004 to 2018, was obtained from the yearly reports of the Israeli Electric Company.[43]

The Ashkelon desalination plant (127 million m^3/year) discharge its brine by mixing it with the cooling water of the adjacent power stations (Rutenberg and Dorad) and brine of three inland BWRO desalinated plants (Gat, Granot and Lahat). The brine of the three BWRO desalination plants is disposed through a 30 km underground pipeline (15–50 cm diameter). The open outfall has three discharge points located within 400 m of each other (Fig. 2.3). The first discharge point contains the cooling water of Rutenberg (unit 2) and Dorad power stations. The second, the cooling water of Rutenberg (unit 1 and 4) and Dorad power stations and the brine of the SW and BW desalination plants. The third discharge point is of cooling water of unit 3 of the Rutenberg power station.

The overall mixing is affected by the motion of the seawater as dictated by the wind and the sea currents. Israel is located on the eastern boundary of the Mediterranean Sea (a mid-latitude semi-enclosed sea with an area of $2.5 \times 106\,km^2$). In this region, the background current circulates counterclockwise and parallel to the coastline (south to north along Israel's coast) with a mean velocity of about 15 to 25 cm/s. Its activity is observed mainly way offshore, in the region where the water depth is 20 m or more. Most of the time (90%), the speed of the near-surface current does not exceed 15–35 cm/s. The largest near-surface and near-bottom speeds recorded occur during winter (December-March).[44]

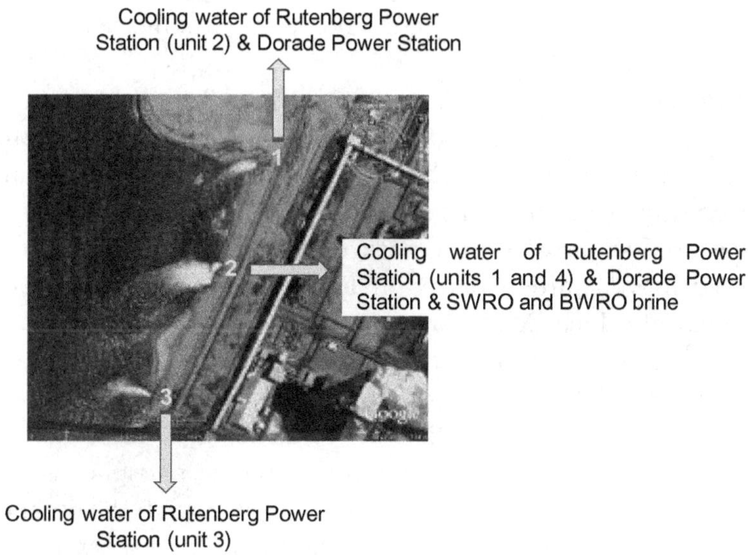

Fig. 2.3. Outfall of the Ashkelon desalination plant.

During 2005–2018 the main trend of bathymetric changes i.e., sand accumulation or erosion has changed almost yearly. The variability of the beach width over the years indicates natural changes without a uniform trend. These results indicate that the desalination plant does not affect the state of the coast and seabed. Additionally, particulate matter derived from the desalination plant did not accumulate in the sediments.

The maximal temperature increase was detected up to 500 m from the cooling water outlets. The temperature increase is defined as the difference of the measured temperature from the natural background temperatures (ΔT). The elevated temperatures in the receiving marine areas are due to the temperature of the power station's cooling water while the SWRO brine temperature is ambient. The maximal differences at the sea surface and near the seabed ranged from 5.1 to 9.9°C and 2.0 to 5.3°C in the spring and fall, respectively. The reasons for the range in the ΔT values are attributed to variation in the operating conditions of Ashkelon Desalination Plant and the Rutenberg power station (number of operating units, level of electricity production of each unit, and rate of cooling water discharge), and to different environmental conditions mainly wind velocity.

Since the commence of the Ashkelon desalination plant in 2005, excess salinity of more than 10% relative to the natural background (ΔS) was recorded only once (fall 2017, 12.8% excess), at a distance of 80 m from the brine outlet, near the sea bed. The differences between the measured salinity and the natural background near the seabed, during 2013–2018, are displayed in Table 2.2. As seen, the area near the seabed at which $\Delta S > 5\%$ was small and limited to an area of 0–0.4 Km^2 from the brine outlet. While, $\Delta S > 2.5\%$ was measured at an area of 0–2.3 km^2.

It was shown that turbidity, dissolved oxygen, total suspended solids (TSS), nutrients (except for total organic phosphorus (TOP) and nitrate), total organic carbon (TOC), particulate heavy metals, and particulate iron were not affected by the brine discharge i.e., the

Table 2.2. Summary of the salinity measurements at the seabed during 2013–2018. In brackets are the disabled power station unit number.

Year	Season	Number of power station units disabled	Maximum ΔS (%)	Area $\Delta S > 5\%$ (Km^2)	Area $\Delta S > 5\%$ (Km^2)
2013	Spring	1 [unit 4]	7.0	0.06	1.40
	Autumn	2[units 1,4]	8.3	0.20	1.30
2014	Spring	0	3.1	0.00	0.01
	Autumn	1 [unit 3]	<0.9	0.00	0.00
2015	Spring	0*	4.0	0.00	0.03
	Autumn	3 [units 1,2,4]	9.0	0.04	1.20
2016	Spring	1 [unit 2]	4.7	0.00	0.09
2017	Spring	1 [unit 1]	8.2	0.35	>1.20
	Autumn	1 [unit 1]	12.8	0.40	1.60
2018	Spring	1 [unit 1]**	7.1	0.24	2.30
	Autumn	1 [unit 2]	8.3	0.24	1.50

*Only 50% of the flowrate of the cooling water of unit 2

**No electricity production in unit 2 with full capacity of the cooling water of this unit values of the above listed parameters were found to be within the range of the natural concentrations. The TOP is originated from the use phosphonate-based scale inhibitors applied in the desalination process. During 2008–2018 there was no significant differences in the TOP load (<20%). The added TOP did not significantly contribute to the quantities of phosphate to the monitored area. Its concentration did not exceed the Israeli Ministry of Environmental Protection Standard (MOEP) of 0.1 mg/L.

In 2008–2018, an increase in nitrate and silicic acid loads was observed within 500 m (~0.4 km^2) from the brine outlet. This is due to a stepwise rise in silicic acid/nitrate concentrations in the inland BWRO desalination plants. During the 10 years period the nitrogen and silicic acid loads increased by approximately 10 and 9-folds, respectively. Since 2011 more than 90% of the total nitrogen was originated from the BWRO desalination plants with 98% in 2018, mostly as nitrate. The range of nitrate concentrations along with the average and background values are displayed in Fig. 2.4. It is evident that the nitrate concentrations are higher than the background values. Yet, these values are lower than the concentrations which cause eutrophication in the open sea.

The multiannual (2012–2018) chlorophyll (Chl-a) data did not reveal any trend of temporal changes, exhibiting median chlorophyll concentrations lower or close to the background values. Occasion high chlorophyll concentrations, over the years, were attributed to sewage discharge from the Gaza Strip which caused enhanced local growth of algae. Chlorophyll patches (algae) are also transported northwards from the Nile Delta (Egypt) by the sea currents. The general conclusion drawn from the multiannual monitoring is that the discharge of the cooling water of the Rutenberg power station and the brine of the Ashkelon desalination plant has no significant impact on the biomass of the algae.

Throughout the monitoring years, the area affected by the discharges on the infaunal community (reduction in the number of specimens and species) was small. This area was limited to a strip of shallow water (up to 5 m depth) extending up to a distance of 600 m from the outlet and 350 m from shore. Sampling of five taxonomic

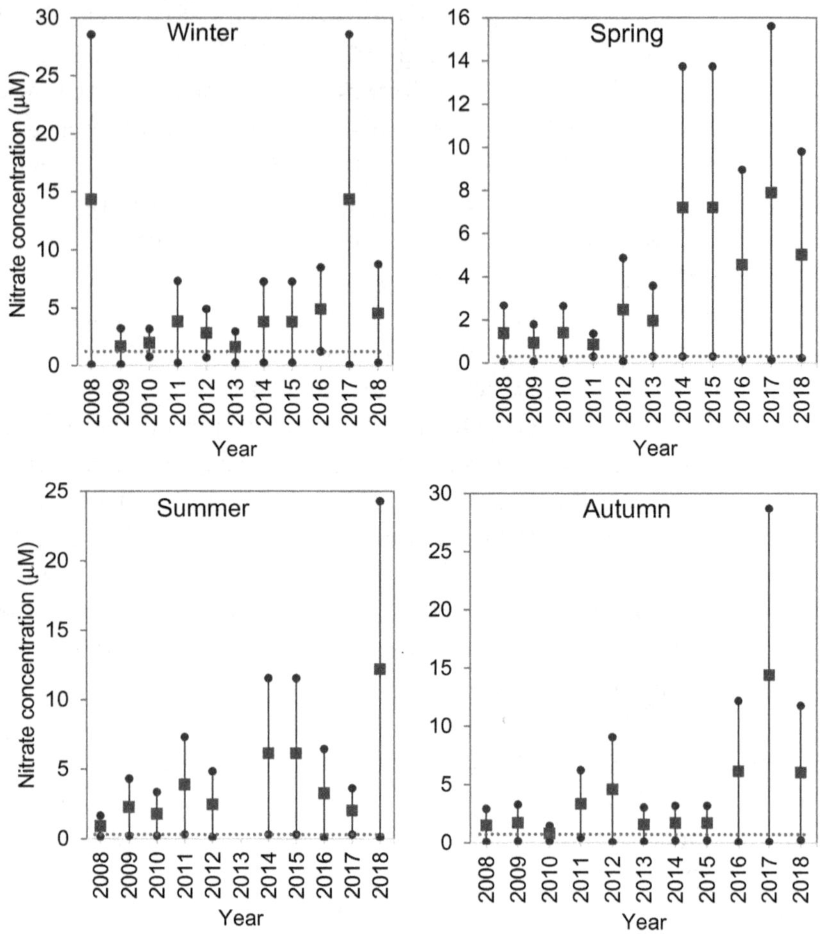

Fig. 2.4. Nitrate concentration range in 2008–2018. Dashed line represents the background values (data obatined from Ref. 43).

groups of the microalgae was conducted in the spring and the autumn of 2018, at three points located perpendicular to the coastline at distances of 250 (RUT3), 800 (RUT9), and 1,160 (RUT17) m from discharged point 2. The five groups, summarized in Table 2.3 are:

(i) Cells smaller than 2 mm (this group includes mostly Cyanobacteria dominated by *Synechococcus sp.* Genera);
(ii) Microalgae smaller than 5 μm, mostly Eukaryotic algae;

Table 2.3. Concentrations (Cells/L) of microalgae sampled in 2018.

	Spring			Autumn		
Microalgae	RUT 3	RUT 9	RUT 17	RUT 3	RUT 9	RUT 17
Dinoflagellates	1.36×10^4	1.90×10^4	1.28×10^4	1.80×10^4	1.80×10^4	1.50×10^4
Diatoms	3.86×10^5	3.28×10^5	3.54×10^5	6.40×10^5	5.70×10^5	5.00×10^5
Cyanobacteria	1.04×10^8	1.66×10^8	1.64×10^8	8.30×10^7	7.80×10^7	7.30×10^7
Chlorophyceae	16	4	0	580	620	150
Microalgae $> 5\mu m$	2.74×10^6	1.88×10^6	1.79×10^6	8.50×10^6	6.63×10^6	6.52×10^6
Total	1.08×10^8	1.68×10^8	1.66×10^8	9.26×10^7	8.55×10^7	8.09×10^7

(iii) Algae larger than 5 μm, typically Dinoflagellates;
(iv) Diatoms larger than 5 μm;
(v) Chlorophyceae.

The species richness, in all the microalgae taxonomic groups except from the Dinoflagellates, was higher in the autumn than in the spring. While, the total concentration of the cells in the autumn was about half of the concentration in the spring. These differences were attributed to the abundance of the *Synechococcus sp*. In the spring the *Synechococcus sp* were $98.2 \pm 0.1\%$ of the total cells while in the autumn they were $90.9 \pm 1\%$. In the spring, three potentially toxic species *(Chaetoceros spp.*, *Pseudo-nitzschia spp*, and *Cylindrotheca closterium)* were detected at concentrations far below the toxicity threshold. It should be mentioned that these species are indigenous to the Israeli shore. In the autumn, the concentrations of the microalgae with toxicity potential were up to 10^4 cells/L. These low concentrations are not considered harmful.

2.6.1.2. *Sorek and Palmachim*

The Sorek and Palmachim SWRO desalination plants have a combined production capacity of 240 million m^3/year of freshwater water. The brine is discharged through two separate outfalls with capacities of 104 and 187 million m^3/year, for Palmachim and Sorek plants, respectively. The outfalls are located approximately 850 m apart, at a depth of 20 m. The Sorek two intake points are located at distances of 584 and 608 m from the discharge points. The discharge of the Palmachim plant is located 680 m from its intake point. The locations of the intake and discharged points are displayed in Fig. 2.5.[45]

The Sorek outfall is located 1.85 km from the shore. The brine is discharged, at a velocities of 4.0 m/s, using four diffusers, spaced at 2.5 m in alternating pattern (two discharge the brine to the north and two to the south) 4 m above the seabed in upward angle of 45°. Palmachim outfall is located 1.98 km from the shore. It discharges brine at a flow rate of 1.4 m/s using three risers at an angle of 45° upward. The risers, spaced 6 m apart, are located 6 m above the seabed.

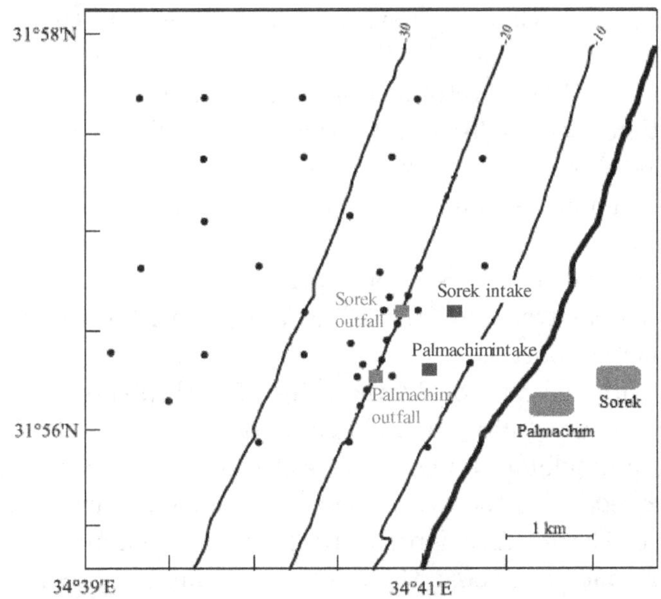

Fig. 2.5. Intake and outfall points of Palmachim and Sorek SWRO desalination plants (black points represents sampling points) adopted from Ref. 45.

As the plant's outfalls are so close, the overall effects of the brine discharge are reported. Six years monitoring by the IOLR, as of the start-up of the Sorek desalination plant in 2013, revealed no impact on seawater quality.[45] Specifically, analysis of the seawater quality, at the intake of the two desalination plants, showed no traces of the brine. Activated sludge discharge into the sea, from the Shafdan sewage treatment plant, stopped in January 2017 resulting in improved seawater quality at the former discharge site. To date, the Shafdan discharge treated effluent which are not expected to impact the seawater quality at the intakes.

The monitoring data presented herein was obtained from the yearly reports of the IOLR.[46,47] Increased salinity by 5% was small and localized near the outfalls ($0.1\,\text{km}^2$) close to the seabed. The area at which the salinity was higher by 1% above the background ($0.4\,\text{g/L}$) was 3–$5\,\text{km}^2$. It should be noted that a $0.4\,\text{g/L}$ salinity variation is within the range of the natural annual variability of salinity in coastal waters. Depth cross-sections of the excess salinity (i.e., above the background) are displayed in Fig. 2.6. Higher salinity is identified adjacent to the Sorek outlet, in a vertical pattern, up to a depth of 10 m. A total area of less than 2 km is affected by both outlets and is limited to the seabed (Fig. 2.6(a)). Similar results are shown on Fig. 2.6(b). In the direction of east to west the excess salinity is limited to an area of about 0.5 Km with slightly lower salinity found eastern to the Sorek outlet. The salinity drops to the background values at a depth of about 22 m. Figure 2.6(c) displays similar trends of Fig. 2.6(a) along with the contribution of the Palmachim outlet.

As mentioned previously, the brine of both plants is discharged, in upward angle of 45°, 16 and 14 m above the seabed in Sorek and Palmachim desalination plants, respectively. It is therefore clear that the upper section of the water column is not significantly affected by the brine discharge. At a depth of 10–30 m excess salinity is measured, due to the fact that near the seabed, less mixing occurs. Below a depth of 30 m changes in the salinity are negligible. Improved mixing, i.e., minimize the spatial extent and intensity of brine plumes, may be achieved by increasing the discharge angle or elongate the pipes to area at which the seabed depth is 30 m. Another option is to change the number/size/position/angle of the outfall diffusers.

The temperature of the seawater was increased by 0.3–0.7°C. The affected area is smaller than the area affected by excess salinity. Dissolved oxygen saturation ranged from 76 to 103%. These values are within the MOEP Standards. The values of pH, turbidity, suspended particulate matter (SPM), nitrate+nitrite, ammonium, total nitrogen, phosphate, TOC, and silicic acid were all within the natural range and did not exceed the regulatory guidelines. The concentration range of total phosphorus and total organic phosphorus was wide. Linear correlation was found between the concentration of total organic phosphorus and salinity in the near bottom water samples.

Chlorophyll concentrations were seasonal and temporal dependent, and uncorrelated to water salinity. The concentrations of metals and organic carbon concentrations in the sediments were low and natural for the area. There was no impact of the brine discharge

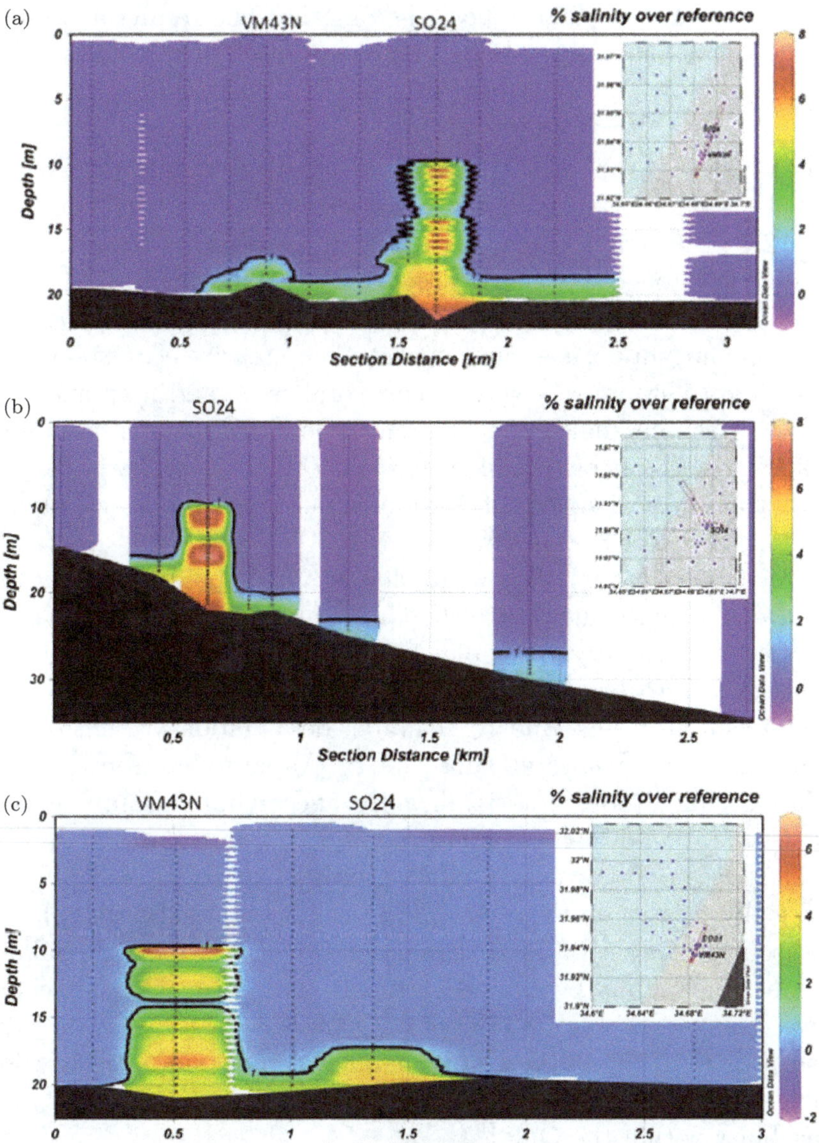

Fig. 2.6. Depth cross-sections of the salinity excess above the background (survey of 2017). The red line on the sampling points map (top right corner) is the area presented. VM43N SO24 are the outlets of the Palmachim and Sorek desalination plants, respectively. (a) May 2017, 3 km from south to north, through the outlets; (b) from east to west through Sorek outlet; (c) October 2017, 3 km from south to north, through the outlets.[46]

on the infauna at ≥200 m from the outfalls. The results analysis is hindered by different granulometric composition of the sediment.

2.6.2. *Persian Gulf*

The Persian (Arabian) Gulf countries produce about 60% of the global desalinated water. This Gulf is a marginal and semi-enclosed sea ($2.4 \times 105\,km^2$) characterized by extreme environmental conditions due to geographic position, relative shallowness and high evaporation rates. Despite these harsh environmental conditions, the Arabian Gulf supports a range of coastal and marine ecosystems such as mangrove swamps, seagrass beds, coral reefs, and mud and sand flats.[48] Weak circulation and extremely low freshwater input result in surface currents of typical speeds of 10–20 cm/s while the bottom flow attain typical speeds of 5–10 cm/s but magnifies to 20–30 cm/s past the Strait of Hormuz.[49]

Most desalination plants discharge the hypersaline brine via surface nearshore outfall into the Gulf. Salt buildup increased mean annual basin salinity by about 1%, confirming that basin salinity is insensitive to brine. This can be explained by the fact that the Gulf freshwater sink due to seawater desalination by far smaller than the Gulf freshwater sink due to evaporation. Consequently, brine discharge does not significantly affect Gulf flushing time by the evaporation-driven residual circulation between the Gulf and the Indian Ocean. However, regional sensitivity to brine is significant, especially in the southwestern Gulf region around Bahrain Island and southern coast of Saudi Arabia, at which maxima salt buildup increase the salinity by around 11%. This indicates that the regional-scale dynamics of the residual circulation is more sensitive to brine discharge than the basin-scale dynamics of the residual circulation. Thus, regional flushing patterns reflect the residual circulation spatial variability within the Gulf.[50]

Monitoring also revealed a drop of >0.2 pH units predominantly attributed to CO_2 sequestration in the Gulf's waters. Interestingly, the acidifying waters of the Gulf have utilized the elevated salinity to buffer the pH effect.[51] Heavy metals concentration in the sediments were reported to be within the natural background levels.[52] While localized elevated concentrations of Cu, As, Cr, Mn, and Zr were measured in proximity to desalination plants along the coastline of the Arabian Gulf.[53,54] These metals are most likely corrosion products originates from the Thermal desalination plants.

2.6.3. *Europe-Mediterranean coast of Spain*

Spain accounts for over half of the total desalination in Western Europe.[1] There are 50 SWRO desalination plants, located between the Mediterranean Arc and the Canary Islands. In addition, there are 100 small brackish water desalination plants. Around 30 SWRO facilities, of low (5.0–9.5 million m^3/year), medium (10–23 million m^3/year) and high capacity (25–80 million m^3/year) are located along the Spanish Mediterranean coast (Iberian Peninsula). Practically, none of the plants operate at full capacity.[55,56]

Requirements of the environmental monitoring plans have increased significantly over time in Spain and are higher in areas with the presence of protected species in comparison with areas of low ecological value.[56] As a result, the management of brine in Spanish desalination plants was modified with time. For example, in the San Pedro del Pinatar (I and II) desalination plants (Murcia, Spain), brine discharge started in January of 2006, 5 km offshore at a depth of 33 m. It resulted in a significant decrease in abundance, richness and diversity of polychaete families at the location closest to the discharge point, at which salinity reached high levels of 49 g/L. Consequently, in May 2010, a diffuser system was added at the end of the discharge pipeline. Discharge of the brine, at a 60^{0} to the horizontal, facilitated the mixing hence reduced the salinity (to less than 38.5 g/L) and dispersion area. The application of the diffuser system enabled recovery of the benthic communities. Within months, a significant recovery in polychaete richness and diversity was detected, reaching levels similar to those before the discharge. A less evident recovery in abundance was also observed, probably due to different recovery rates of polychaete families.[57]

The Jávea desalination plant (Alicante, Spain) discharges brine through 16 diffusers heads (3.5 cm in diameter) into an open channel (Fontana) to facilitate dilution of 4:1.[58] The Jávea facility has been continuously monitored since startup in 2002. In peak periods (summer, 2–4 trains in operation), the area affected by the brine discharge plume is limited to 300 m from the channel. In winter (1 train in operation) complete dilution is achieved a few meters away from the channel. No significant effects over marine sensible species (*echinoderms*), with high environmental value (*Posidonia Oceanica*), artisanal fisheries or soft-bottom fauna have been detected through 2002–2016.[59] While, some fish species were attracted to the area. Additionally, a recovery of the benthic fauna in the previously anoxic Fontana channel was observed.[58,59]

Águilas desalination plant (Murcia, Spain) was originally designed for 65 million m^3/year capacity after which it was extended to 77 m^3/year.[59] Brine is discharge, at a depth of 30–34 m through a 2.1 Km pipeline with 8 diffusers at the end. This part is made of 24 ports of (3.1 cm in diameter) at an angle of 60°. It should be realized that the marine surroundings of Águilas desalination plant are a protected area included in the European Union program-Red Natura 2000 i.e., a submerged coastal strip of the Murcia region. In this area the seagrass meadows and rock communities are predominantly *Posidonia oceanica* beds and *Cymodocea nodosa*.[60] Results, at a distance of 500 m form the outfall, indicated salinity of 38.5 g/L with the continental waters salinity values range between 36 and 38 g/L. This meets with environmental requirements which states a value of 38.5 g/L in more than 25% of the continuous analysis or 40 g/L in more than 5% for *P. oceanica*. As this seagrass have very low tolerance to increased salinity.[60]

Alicante (I and II) desalination plant (Valencia, Spain) was extended with time. Following startup in 2003 at a capacity of 18 million m^3/year the plant was expended in 2006 to a capacity of 24 million m^3/year. In 2008, Alicante II was commenced bringing the total capacity to 55 million m^3/year.[59] The management of brine was modified based on the monitoring of the marine ecosystem and changes of the environmental requirements. At the commence of the plant in 2003, brine was discharged directly to the shoreline, resulting in reduced growth of the *Posidonia Oceanica* community and the disappearance of *Echinoderm* (indicators) in the area affected by brine discharge.[61] This effect along with the increased capacity led to a dilution of the brine prior to the discharge at a ratio of 1.5-to-5.0.[62] The *Echinoderm* population was recovered after the brine dilution was started. Additionally, 14 years of monitoring revealed that the *Posidonia* meadows are not affected by the brine discharge.[59] Overall, monitoring of the water quality and marine ecosystem indicated that the size and time for dispersion of the salinity plume varied seasonally. These variations, however, did not affect the benthic organisms inhabiting the seafloor.[62]

2.6.4. *North America-Carlsbad and Tampa Bay (US)*

2.6.4.1. *Carlsbad*

The 70 million m^3/year Carlsbad desalination plant, located in Southern California, USA, started operation in 2015. The plant shares the intake and discharge outfall with the adjacent Encina power station. The distance between the intake and discharge points is approximately 1 km. The brine is mixed with the power station cooling water at a ratio of 1-to-10, resulting in final salt concentration about 20% higher than the ocean water. The brine is discharged to the ocean through a 10–15 m wide outfall about 50 m offshore.

Californian regulations require that the salinity of the discharged brine would not exceed two salinity units above background levels within 100 m offshore from the discharge point. Due to the plant's capacity, the salinity impact zone of the Carlsbad Desalination Plant was extended to 200 m. A study characterizing the chemical and biological spatial footprint of the discharge plume from Carlsbad Plant, revealed that the salinity level in the discharge zone exceeded the permitted level (by up to 2.7 g/L), and the salinity plume extended farther than permitted (600 m). Yet, no significant impact was found on the benthic epifaunal or infaunal composition and abundance in the same area.[63] It should be realized that better dilution of the brine prior to discharge would most likely resolve the high salinity level in the discharge zone.

The background salinity at Carlsbad Beach was measured at 33.2 ± 0.6 g/L pre-operation of the plant. The salinity was measured

Table 2.4. Mean, maximum and minimum salinity for surface and bottom water in 50–200 m around the outfall mouth and 250–1000 m the surrounding area measured at pre- (Dec. 2014, Sept. 2015) and post (May 2016, Nov. 2016) operation.[63]

Distance from outfall	Water level	Salinity	Pre-operation		Post-operation	
			Dec. 2014	Sept. 2015	May 2016	Nov. 2016
50–200 m	Surface	Min.	33.5	31.5	33.6	33.4
		Max.	33.6	33.5	35.5	33.9
		Mean	33.6	33.1	33.9	33.5
	Bottom	Min.	33.5	33.0	33.9	34.8
		Max.	33.6	33.9	35.7	35.9
		Mean	33.5	33.5	34.9	34.4
250–1000 m	Surface	Min.	33.3	30.4	32.5	33.3
		Max.	33.5	33.5	33.8	33.9
		Mean	33.4	33.4	33.4	33.5
	Bottom	Min.	33.4	33.3	33.5	33.6
		Max.	33.5	33.5	35.2	35.9
		Mean	33.7	32.8	34.4	34.4

pre- and post-operation of the plant (Table 2.4). The surface water salinity, 50–1000 m from the outfall mouth, did not differ significantly between pre- and post-operation, except is maximum salinity of 35.5 g/L about 50 m from the channel mouth measured in May 2016. Bottom water salinity in the 50–1000 m area post-operation was higher by 1.1 g/L than pre-operation salinity. Maximum values of 35.9 were measures 200 and 250 m offshore in Nov. 2016. The mean values did not exceed the regulatory limit of two units over the background. Nonetheless, the maximum measured values of 35.9 g/L were 2.7 g/L over ambient hence, exceeding the regulatory limit.

Temperature increase of 1–2°C was measured in both surface and bottom waters around the outfall. A visible halocline developed close to the discharge channel since the higher temperature of the plume water did not lower the density sufficiently to refute the higher salinity. The concentrations of the dissolved organic carbon, particulate organic matter, chlorophyll, and nutrients (nitrate, phosphate and silica) exhibited no significant temporal or spatial patterns. The concentrations of Li, Na, Mg, K, Ca, Mn, Fe, Sr and Ba in the plume also did not differ from background values. No correlation between the salinity and the above listed parameters was observed.

No significant changes in the organisms living on the seafloor and other biological indicators were observed. The researchers attributed this to the long history of industrial activities in the area along with the cooling water discharge from a power station adjacent to the desalination plant, which had already perturbed the natural setting. It was concluded that the already disturbed area and the generally low biodiversity of organisms in the sandy environment (Fig. 2.7), resulted in minimal direct impacts of Carlsbad Desalination Plant on the ecology of benthic organisms still inhabiting this area, despite the salinity plume extending >200 m offshore.[63]

2.6.4.2. *Tampa Bay*

Tampa Bay desalination plant (35 million m^3/year) is located in the southeastern part of Tampa Bay, Florida, USA. The plant begun operation in March of 2003 and was taken offline in 2005 for remediation. It resumed production in March of 2007. In the facility,

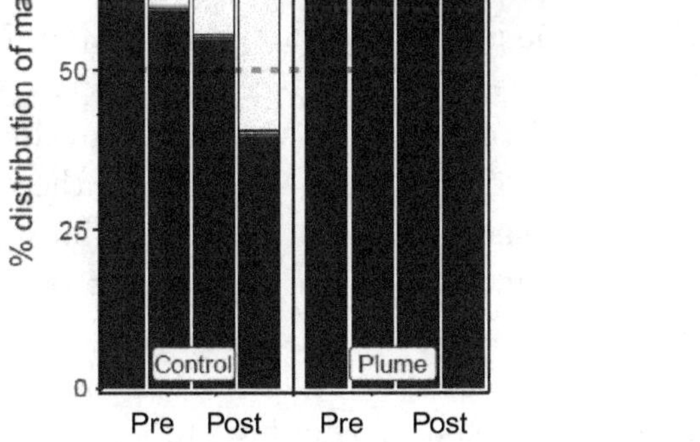

Fig. 2.7. Distribution of benthic epifauna at the northern (control) and southern (plume) end of the beach at pre- and post-operation.[63]

brine is mixed with the cooling water of the adjacent power station (Big Bend) achieving a blending ratio of up to 70-to-1. The brine is discharged into the cooling water outfall downstream from the desalination plant intake connection.

Monitoring commenced about a year before the initial operation of the plant. Data from 2002 to 2010[64] indicated that the Tampa Bay SWRO plant had no adverse impact on the water quality as well as on the biological abundance and diversity. Changes in SW salinity in the vicinity of the discharge point were within or below the maximum regulatory limit.

Diversity differences of benthic, fish, and seagrass communities were similar between operational and non-operational periods. An example for the diversity and abundance of Benthic Invertebrates during operational and non-operational periods of the plant is given in Fig. 2.8. Benthic assemblages varied spatially in terms of dominant taxa, diversity, and community structure throughout the monitoring period. Yet, the spatial heterogeneity of marine life distribution were directly related natural variation in bay salinity.

2.6.5. *Australia-Sydney and Adelaide*

2.6.5.1. *Sydney*

Sydney desalination plant is located on the coast of Kurnell, New South Wales, Australia. The 90 million m^3/year facility commenced operation in 2010. Four intake risers are located about 300 m offshore at a depth of 25 m. Brine is discharged, approximately 700 m south of the intake, using two rosette risers spaced 25 m apart with an angle of 60° to the horizontal plane. The two outlets are located above a rocky reef about 300 m offshore and 25 m below the ocean surface. The high-pressure diffusers system achieves a dilution factor of 30 at 75 m from the outfall. The intake and outlet of the plants are displayed in Fig. 2.9.

A seven-year study[66,67] was conducted to test the impacts and subsequent recovery of sessile marine invertebrate recruitment near the outfall of Sydney desalination plant. Monitoring of the outlet area (30 and 100 m) and multiple reference sites was carried out before,

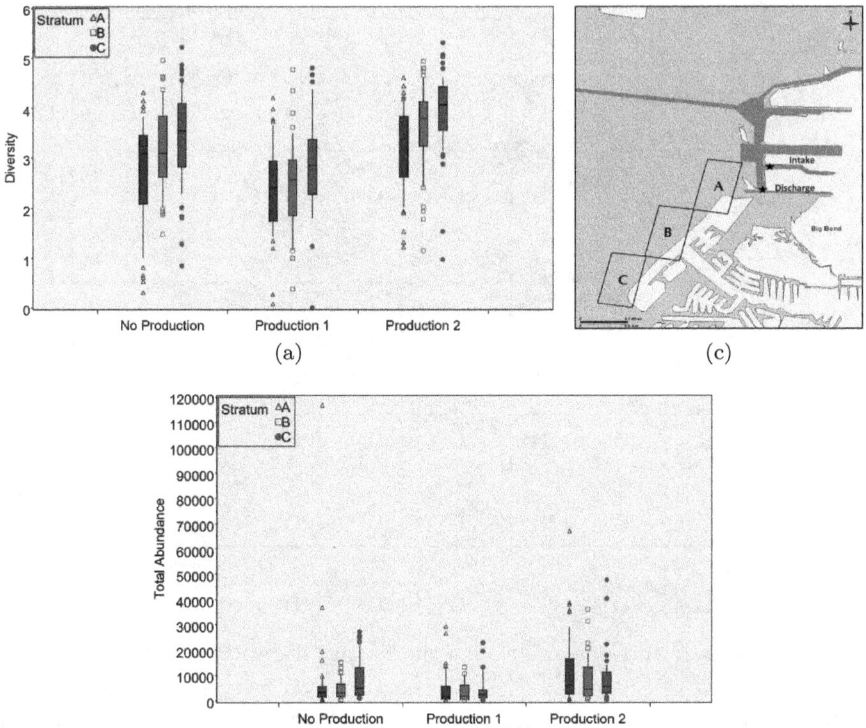

(b)

Fig. 2.8. Summary of (a) Shannon diversity index and (b) abundance for Benthic Invertebrates by stratum (c) and operational period (Production 1: March 2003–May 2005; Production 2: March 2007–June 2010).[64]

after and during the plant's shutdown periods. The results of this study debunked the prevailing understanding that high salt levels in the outfall brine are toxic to marine life. The findings showed that the main effect occurred over a small area within 100 m of the outfall, likely the result of changes to water flow. Interestingly, an increase in abundance reef fishes was observed. Overall, the study demonstrates that a well-designed SWRO brine discharge minimize the local effect of brine discharge and can also enhance local fish abundances and species richness.

The regulatory framework in the Sydney plant requires a salinity limit of $\leq$1 g/L at 75 m (as a maximum). The average increase in

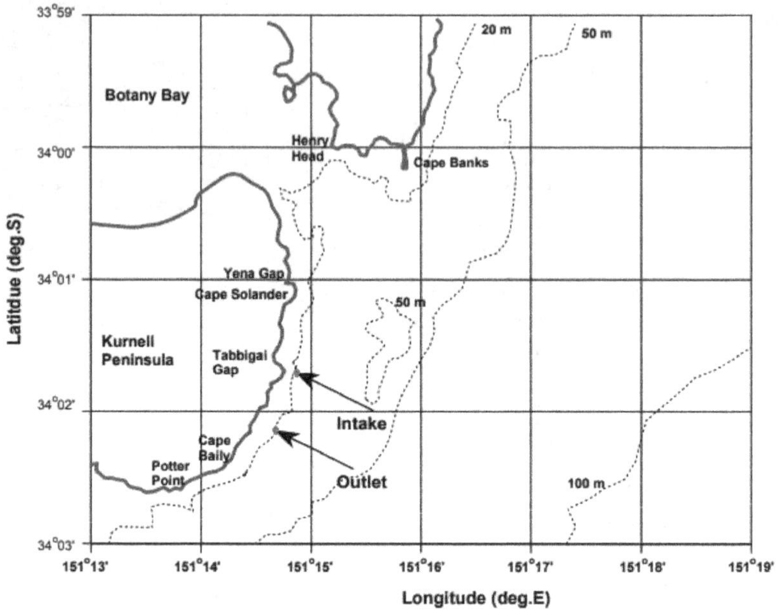

Fig. 2.9. Intake and outlet of the Sydney desalination plant.[65]

salinity at the seabed was approximately 1 g/L above the background at 30 m distance from the outfall, and 0.8 g/L above ambient at 100 m from the outfall. It was concluded that the diffusers system achieved good mixing as the target of less than 1 salinity unit difference to surrounding ambient waters within 100 m of the discharge outfall was obtained. No increased temperature at outfall locations was measured during the discharging period, relative to background locations.[66]

The sessile invertebrates were nonetheless impacted by the brine discharge. Polychaetes, bryozoans and sponges reduced in cover as far as 100 m from the outfall, while barnacles showed the opposite pattern and were more abundant near the discharging outfall. Since the ecological impacts were disproportionate to the minor change in salinity it was propose that impacts were primarily driven by changes to water flow (changed hydrodynamics) caused by the diffusers. In the small 100 m impact zone, some invertebrates increased in numbers, the ones that benefit from high-flow conditions such as barnacles. While others were reduced in quantity or did not recover after shutdown of the outflow. These were the species with slow swimming larvae, such as tubeworms, lace corals and sponges.

Despite the changes in communities near the outfall, there was not a large or persistent loss of biodiversity, but rather a localized redistribution of species in response to environmental conditions that are within the range of natural variability. The change in flow was analogous to the natural difference experienced by reefs that occur on exposed headlands versus those on sheltered reefs, and thus support different communities.[66]

Following the commence of the desalination plant the overall abundance of fishes was increased in the numbers by 279% while, fish abundances increased by only 30% in the reference sites over the same time period. It included substantially greater abundances of pelagic and demersal fish, as well as fishes targeted by recreational and commercial fishers such as silver sweep (*S. lineolata*) and grey morwong (*Nemadactylus douglasii*). In contrast, benthic fishes were not significantly affected by the hypersaline discharge, similarly to fish assemblages over soft sediment habitats. Following the cessation of plant, abundances of fishes mostly returned to levels such that there was no longer a significant effect compared to the period prior to the commencement of the desalination plant's operations. The authors[67] suggested that the increased number of fish around the desalination discharge outlet would locally influence the structure of overall reef communities within ~50 m of the outlet.

2.6.5.2. *Adelaide*

The Adelaide desalination plant is located in Lonsdale, South Australia. The 100 million m^3/year facility has been operating operational since 2012. The intake structure is located 1.2 km offshore 12 m below mean sea level and 8 m above the seabed. Brine is discharged through a 1.2 km tunnel terminating in six risers with the outermost riser being located approximately 850 m offshore. Each riser has a rosette of four duckbill valves. The duckbill valves are raised off the seabed, at a water depth of approximately 16.5 m.

The distance between the intake and the offshore outfall diffuser is approximately 300 m.

A study was targeted at validating the design of the diffuser showed that mixing of the hypersaline brine, on discharge, is generally sufficiently rapid to achieving a 50-to-1 dilution within 100 m of the outfall.[68] Under conditions of a dodge tide with low wind such dilution was not achieved. Nonetheless, such periods are of limited duration and ecotoxicology data suggested a minimum dilution of between 15-to-1 and 32-to-1 is sufficient to avoid environmental impact.[69]

The regulatory framework stipulates a salinity concentration $\leq$1.3 g/L above natural background concentrations at 100 m from the outfall. Comparing the average salinity with the ambient, the 1.3 g/L threshold is not reached, indicating that near-field mixing of the discharge is providing the necessary dilution across a range of ambient current flow and plant operating conditions.[68,69] Monitoring data revealed that at maximal production of the plant the salinity is typically elevated by an average of 0.7 g/L at 100 m from the outfall diffusers. Overall, it was concluded that there is no evidence of adverse impacts on water quality in the surrounding environment associated with brine discharge. This conclusion was based on measurements of salinity, temperature, dissolved oxygen, chlorophyll, pH, heavy metals, and turbidity.[69,70]

In terms of biota characterization, overall, no significant differences were observed in the benthic infauna communities between the discharge area and control areas. It is therefore concluded that the infaunal communities were affected by the spatial, and temporal variation of benthic habitats, rather than by the brine discharge from the desalination plant. Natural changes in sediment composition and seasons appear to be the main drivers for infaunal communities. Specifically, fish communities were consistent with those found more broadly along the coast. Some species such as Port Jackson sharks used the brine discharge area as a breeding/nursery area. Softbottom communities reflected the natural character of communities in this region of Gulf St Vincent. Similarly, plankton communities were largely unremarkable when compared to those elsewhere although some potentially harmful/toxic phytoplankton bloom species have been identified though in low numbers.[69] There was no significant difference in the spatial and temporal interactions for macrofauna diversity and evenness, with a high diversity of taxa observed at all locations and times.[71]

2.7. Summary

Global climatic changes together with the increase in the global population have significantly impacted the quantity and quality of the water sources. These trends have made desalination one of the world's most important non-conventional water resources. Public acceptance is typically slow to emerge, with key issues in the desalination debate of high energy consumption and environmental impacts. Therefore, sound environment and sustainability principles from planning to design, construction and operation are essential.

Energy consumption of desalination plants and the related greenhouse gas emissions are site specific, as it is affected by the technology, engineering, energy recovery devices, water recovery, source and product water characteristics. Measures are taken to reduce the energy consumption of RO processes including increase of the efficiency and lifespan of the membranes, improved equipment, optimization of the RO process; apply intermediate chemical de-mineralization techniques to gain higher water recoveries; and use of renewable energy resources.

Disposal and management to the hypersaline brine is a major environmental concern. Brine is characterized by high salinity and density. Its environmental impacts, dependent on the discharge location and technique, marine species tolerance, volume and characteristics of the brine, dilution ratio, and properties of the receiving water body. Environmental monitoring programs are applied to identify marine environmental impacts of brine discharges, and mitigate them when they are identified, allowing to sustainable management of RO effluents. Long term operation, of more than a decade, debunks the prevailing understanding that high salt levels in the brine outfall have adverse environmental impacts. It was shown that the brine discharge has localized minimal impacts on the marine environment.

The independent overview presented herein highlight the facts that desalinated water processes are becoming increasingly competitive, consume less energy, and that a well-designed brine discharge minimize the local impacts on the coastal and marine ecosystems and can also enhance local fish abundances and species richness. This will further establish desalination as a reliable and drought-proof alternative worldwide. It may be concluded that a significant increase in desalination production, on a global scale, is not expected to affect the environment.

References

1. E. Jones, M. Qadir, M. T. H. van Vliet, V. Smakhtin, S.-m. Kang, The state of desalination and brine production: A global outlook, *Sci. Total Environ.*, **657** (2019) 1343–1356.
2. A. Panagopoulos, K. J. Haralambous, M. Loizidou, Desalination brine disposal methods and treatment technologies — A review, *Sci. Total Environ.*, **693** (2019) 133545.
3. A. Giwa, V. Dufour, F. Al Marzooqi, M. Al Kaabi, S. W. Hasan, Brine management methods: Recent innovations and current status, *Desalination*, **407** (2017) 1–23.
4. E. Ayboga, Policy and impacts of dams in the Euphrates and Tigris basin. Mesopotamia Water Forum 6–8 April 2019, Sulaimani, Kurdistan Region of Iraq. https://www.savethetigris.org/wp-content/uploads/2019/01/Paper-Challenge-B-Dams-FINAL-to-be-published.pdf.
5. N. Ghaffour, T. M. Missimer, G. L. Amy, Technical review and evaluation of the economics of water desalination: Current and future challenges for better water supply sustainability, *Desalination*, **309** (2013) 197–207.
6. Y. Cohen, R. Semiat, A. Rahardianto, A perspective on reverse osmosis water desalination: Quest for sustainability, *AIChE J.*, **63** (2017) 1771–1784.
7. F. Lauruschkus, Comparison of European water prices, *Our Future Water Conference*, 8, Nov. 2018 Berlin, Germany.
8. R. Semiat, Desalination — present and future, *Water Int.*, **25** (2000) 54–65.
9. G. W. Murphy, R. C. Taber, H. S. Hauser, The minimum energy requirements for seawater conversion processes. Office of Saline Water Report No. 9, 1956.

10. B. F. Dodge, A. M. Eshaya, Thermodynamics of some desalting processes. Chapter 3 pp. 7–20, In *Advances in Chemistry Vol. 27 Saline Water Conversion*, American Chemical Society 1960.
11. L. Dresner, J. S. Johnson, J. S. 1980. Hyperfiltration (Reverse Osmosis). Chapter 8 pp. 401–56, In K. S. Spiegler, and A. D. K. Laird, *Principles of Desalination.* 2nd Edition, Academic Press, New York 1980.
12. Y. M. El-Sayed, Designing desalination systems for higher productivity, *Desalination*, **134**(1–3) (2001) 129–158.
13. K. S. Spiegler, Y. M. El-Sayed, The Energetics of desalination processes, *Desalination*, **134**(1–3) (2001) 109–128.
14. R. Semiat, Energy issues in desalination processes, *Env. Sci. Tech.*, **42**(22) (2008) 8193–8201.
15. H. Shemer, R. Semiat, Sustainable RO desalination- Energy demand and environmental impact, *Desalination* (2017) 10–16.
16. A. M. K. El-Ghonemy. Water desalination systems powered by renewable energy sources: review, *Renew. Sust. Energ. Rev.*, **16** (2012) 1537–1556.
17. N. Ghaffour, J. Bundschuh, H. Mahmoudi, M. F. A. Goosen, Renewable energy-driven desalination technologies: A comprehensive review on challenges and potential applications of integrated systems, *Desalination*, **356** (2015) 94–114.
18. M. Li, Reducing specific energy consumption in Reverse Osmosis (RO) water desalination: An analysis from first principles, *Desalination*, **276** (2011) 128–135.
19. A. Zhu, P. D. Christofides, Y. Cohen, On RO membrane and energy costs and associated incentives for future enhancements of membrane permeability, *J. Membr. Sci.*, **344** (2009) 1–5.
20. L. Song, J. Y. Hu, S. L. Ong, W. J. Ng, M. Elimelech, M. Wilf, Emergence of thermodynamic restriction and its implications for full-scale reverse osmosis processes, *Desalination*, **155** (2003) 213–228.
21. M. Wilf, S. Alt, Application of low fouling RO membrane elements for reclamation of municipal wastewater, *Desalination*, **132** (2000) 11–19.
22. T. H. Chong, S. L. Loo, A. G. Fane, W. B. Krantz, Energy-efficient reverse osmosis desalination: Effect of retentate recycle and pump and energy recovery device efficiencies, *Desalination*, **366** (2015) 15–31.
23. B. Penate, L. Garcia-Rodriguez, Energy optimization of existing SWRO (seawater reverse osmosis) plants with ERT (energy recovery turbines): technical and thermoeconomic Assessment, *Energy*, **36** (2011) 613–626.
24. V. G. Gude, Energy consumption and recovery in reverse osmosis, *Desalin. Water Treat.*, **36** (2011) 239–260.

25. C. Fritzmann, J. Lowenberg, T. Wintgens, T. Melin, State-of-the-art of reverse osmosis desalination, *Desalination*, **216** (2007) 1–76.
26. N. M. Eshoul, B. Agnew, A. Anderson, M. S. Atab, Exergetic and economic analysis of two-pass RO desalination proposed plant for domestic water and irrigation, *Energy*, **122** (2017) 319–328.
27. X. Li, D. Hasson, R. Semiat, H. Shemer, Intermediate concentrate demineralization techniques for enhanced brackish water reverse osmosis water recovery — A review, *Desalination*, **466** (2019) 24–35.
28. F. Lokiec, Sustainable desalination: Environmental approaches, The International Desalination Association World Congress on Desalination and Water Reuse, 20–25 Oct. Tianjin, China, 2013.
29. T. Michou, Water and greenhouse gases: The desalination challenge, *ABA J.*, **19**(2) (2016) 14–20.
30. K. L. Petersen, N. Heck, B. G. Reguero, D. Potts, A. Hovagimian, A. Paytan, Biological and physical effects of brine discharge from the Carlsbad desalination plant and implications for future desalination plant constructions, *Water*, **11** (2019) 208.
31. N. Voutchkov, Overview of seawater concentrate disposal alternatives, *Desalination*, **273**(1) (2011) 205–219.
32. T. M. Missimer, R. G. Maliva, Environmental issues in seawater reverse osmosis desalination: Intakes and outfalls, *Desalination*, **434** (2018) 198–215.
33. A. Musfique, A. Rifat, An assessment of the environmental impact of brine disposal in marine environment, *IJMER*, **24** (2012) 2756–2761.
34. R. Einav, K. Harussi, D. Perry, The footprint of the desalination processes on the environment, *Desalination*, **152**(1–3) (2002) 141–154.
35. A. Ketsetzi, A. Stathoulopoulou, K. D. Demadis, Being "green" in chemical water treatment technologies: Issues, challenges and developments, *Desalination*, **223**(1–3) (2008) 487–493.
36. D. A. Roberts, E. L. Johnston, N. A. Knott, Impacts of desalination plant discharges on the marine environment: A critical review of published studies, *Water Res.*, **44**(18) (2010) 5117–5128.
37. P. Palomar, I. J. Losada, Impacts of brine discharge on the marine environment. Modelling as a Predictive Tool. Chapter 13, In M. Schorr, *Desalination Trends and Technologies*. IntechOpen (2011).
38. M. A. Dawoud, M. M. Al Mulla. Environmental Impacts of Seawater Desalination: Arabian Gulf Case Study, *International Journal of Environment and Sustainability*, **1**(3) (2012) 22–37.
39. O. Rabinowitz, Nuclear energy and desalination in Israel, *Bull. At. Sci.*, **72**(1) (2016) 32–38.

40. H. Frank, E. Rahav, E. Bar-Zeev. Short-term effects of SWRO desalination brine on benthic heterotrophic microbial communities, *Desalination*, **417** (2017) 52–59.
41. S. Iso, S. Suizu, A. Maejima, The lethal effect of hypertonic solutions and avoidance of marine organisms in relation to discharged brine from a desalination plant, *Desalination*, **97** (1994) 389–399.
42. I. Alameddine, M. El-Fadel, Brine discharge from desalination plants: A modeling approach to an optimized outfall design, *Desalination*, **214**(1–3) (2007) 241–260.
43. Monitoring of the Marine and Coastal Environment of the Rutenberg Power Station, the Ashkelon desalination plant, the brackish water desalination plants, and the power station of Dorad Energy. Israel Electric Co. RELP-5-2019 (in Hebrew).
44. E. Kit, U. Kroszynski, Marine policy plan for Israel: Physical oceanography, deep sea and coastal zone overview. P. N. 800/14 http://msp-israel.net.technion.ac.il/files/2014/12/ATOS-report-final.pdf.
45. N. Kress, Y. Gertner, E. Shoham-Frider, Seawater quality at the brine discharge site from two mega size seawater reverse osmosis desalination plants in Israel (Eastern Mediterranean), *Water Res.*, **171** (2020) 115402.
46. N. Kress, E. Shoham-Frider, H. Lubinevsky, Monitoring the effect of brine discharge on the marine environment of the Palmachim and Sorek desalination plants. IOLR report H22/2018.
47. N. Kress, E. Shoham-Frider, H. Lubinevsky, Monitoring the effect of brine discharge on the marine environment of the Palmachim and Sorek desalination plants. IOLR report H22/2019.
48. H. A. Naser, Marine ecosystem diversity in the Arabian Gulf: Threats and conservation, Chapter 13, in: O. Grillo (Ed.), *Biodiversity — The Dynamic Balance of the Planet*, InTech Publisher, 2014.
49. J. Kampf, M. Sadrinasab, The circulation of the Persian Gulf: A numerical study, *Ocean Science*, **2** (2006) 27–41.
50. H. D. Ibrahim, E. A. B. Eltahir, Impact of brine discharge from seawater desalination plants on Persian/Arabian Gulf salinity, *J. Env. Eng.*, **145**(12) (2019) 04019084.
51. S. Uddin, Environmental impacts of desalination activities in the Arabian Gulf, *Int. J. Environ. Sci. Dev.*, **5**(2) (2014) 114–117.
52. H. A. Naser, Assessment and management of heavy metal pollution in the marine environment of the Arabian Gulf: A review, *Marine Pollution Bulletin*, **72** (2013) 6–13.
53. M. Sadiq, Metal contamination in sediments from a desalination plant effluent outfall area, *Sci. Total Environ.*, **287** (2002) 37–44.

54. F. Alshahri, Heavy metal contamination in sand and sediments near to disposal site of reject brine from desalination plant, Arabian Gulf: assessment of environmental pollution, *Environ. Sci. Pollut. Res.*, (2017) 1821–1831.
55. A. F. Morote, A. M. Rico, E. Moltó, Critical review of desalination in Spain: A resource for the future? *Geogr. Res.*, **55**(4) (2017) 412–423.
56. A. Pulido-Bosch, A. Vallejos, F. Sola, Methods to supply seawater to desalination plants along the Spanish Mediterranean coast and their associated issues, *Environ. Earth Sci.*, **78** (2019) 322.
57. Y. Del-Pilar-Ruso, E. Martinez-Garcia, F. Giménez-Casalduero, A. Loya-Fernández, L. M. Ferrero-Vicente, C. Marco-Méndez, J. A. de-la-Ossa-Carretero, J. L. Sánchez-Lizaso, Benthic community recovery from brine impact after the implementation of mitigation measures, *Water Res.*, **70** (2015) 325–336.
58. Y. Fernandez-Torquemada, J. L. Sanchez-Lizaso, Monitoring of brine discharges from seawater desalination plants in the Mediterranean, *Int. J. Environ. Health Res.*, **1**(3) (2007) 449–461.
59. J. L. Sanchez, D. Zarzo, Environmental challenges on design, construction and operation of desalination plant worldwide. TexasDesal. 2017 Developing a drought-proof water supply conference, Sep. 21&22, Austin TX, USA.
60. B. Arconada, P. Delgado, A. Garcia, Minimizing environmental risks on constructing marine pipelines: Aguilas desalination plant, *Desalin. Water Treat.*, **51** (2013) 246–261.
61. Y. Fernández-Torquemada, J. Sánchez-Lizaso, Effects of salinity on leaf growth and survival of the Mediterranean seagrass Posidonia oceanica (L.) Delile, *J. Exp. Mar. Biol. Ecol.*, **320**(1) (2005) 57–63.
62. N. Voutchkov, G. Kaiser, R. Stover, J. Lienhart, L. Awerbuch, Sustainable management of desalination plant concentrate-desalination industry position. The International Desalination Association World Congress on Desalination and Water Reuse 2019 Dubai, UAE. IDAWC19.
63. K. L. Petersen, N. Heck, B. G. Reguero, D. Potts, A. Hovagimian, A. Paytan. Biological and physical effects of brine discharge from the Carlsbad desalination plant and implications for future desalination plant constructions, *Water*, **11** (2019) 208.
64. PBS&J. Tampa Bay desalination facility hydrobiological monitoring data summary 2002–2010. https://www.tampabaywater.org/documents/supplies/monitoringprograms/surfacewatermonitoring/TampaBayDesalHBMPDataSummary2002-2010.pdf
65. Marine & Estuarine Monitoring Program- detailed design (Version 3) Preferred Project Report for Sydney's Desalination Project (Sydney

Water). https://www.sydneydesal.com.au/media/1092/marine-and-estuarine-monitoring-program.pdf

66. G. F. Clark, N. A. Knott, B. M. Miller, B. P. Kelaher, M. A. Coleman, S. Ushiama, E. L. Johnston, First large-scale ecological impact study of desalination outfall reveals trade-offs in effects of hypersalinity and hydrodynamics, *Water Res.*, **145** (2018) 757–768.
67. B. P. Kelaher, G. F. Clark, E. L. Johnston, M. A. Coleman, Effect of desalination discharge on the abundance and diversity of reef fishes, *Environ. Sci. Technol.*, **54**(2) (2020) 735–744.
68. V. Ayala, T. Kildea, J. Artal, Adelaide desalination plant outfall diffuser performance validation, *Desalin. Water Treat.*, **55** (2015) 2543–2554.
69. A. C. Cheshire, Assessment of the environmental performance of the Adelaide desalination plant based on a review of all documentation provided in compliance with EPA Licensing Requirements (June 2014). Report 22 in a series prepared for Adelaide Aqua Pty. Ltd. by Science to Manage Uncertainty, 37 pp.
70. T. N. Kildea, L. Andreacchio, Adelaide desalination project: water quality characterization study (April 2013). Australian Water Quality Centre, Adelaide, 19 pp.
71. S. Dittmann, R. Baring, H. Jessup-Case, O. Lam, Adelaide Aqua Desalination Plant; Infauna Survey 2017. Final Report (December 2017), Flinders University, Adelaide.

CHAPTER 3

Application of Reverse Osmosis (RO) and Nanofiltration (NF) Processes for Desalination and Reuse of Membrane Bioreactor (MBR) Effluent as Irrigation and Process Water

Nalan Kabay[a,*], Mert Can Hacıfazlıoğlu[a], İlker Parlar[a], Horacio R. Tomasini[a,d], Nasim Jalilnejad Falizi[a,b], Taylan Ö. Pek[c], L. Bertin[d], and Mithat Yüksel[a]

[a] *Ege University, Department of Chemical Engineering, Faculty of Engineering, 35100 Izmir, Turkey*
[b] *Ege University, Graduate School of Natural and Applied Sciences, Division of Biotechnology, 35100 Izmir, Turkey*
[c] *ITOB Organized Industrial Zone, Tekeli-Menderes, Izmir, Turkey*
[d] *Alma Mater Studiorum, University of Bologna, School of Engineering and Architecture, Bologna, Italy*

3.1. Introduction

According to Food and Agriculture Organization of the United Nations, 1,800 million people will be living in countries or regions with absolute water scarcity, and two-thirds of the world population could be under stress conditions by 2025. A 70% of water consumption belongs to agriculture as world average[1] while the respective value is around 80% for developing countries and 60% in developed countries. Increasing water scarcity problems tend to penalize the

*Corresponding author: nalan.kabay@ege.edu.tr

water usage for agricultural irrigation purposes as drinking water is even more essential.[2] This creates a need to reclaim water for irrigation purposes from alternative resources such as wastewaters. Another advantage is the relative easiness of reclamation of water for irrigation rather than for drinking purposes as much higher quality of water is required for drinking.[3]

Membrane bioreactor (MBR) effluents can be potential sources for irrigation, however their high salinity (over 3 mS/cm) makes them unsuitable for irrigation of most plants except salinity-tolerant ones. Secondary treatment effluents, e.g. MBR effluents, may also include heavy metals, some micropollutants and pathogens depending on its feed water properties and type of membranes employed for MBR operations.[4] Salt accumulation lowers water infiltration rate in the soil[5] and heavy metals may accumulate in soil which will slowly poison the plants.[6]

Tertiary treatment methods can be used to further treatment of secondary effluents in order to make them suitable for reuse in irrigation. Nanofiltration (NF) and reverse osmosis (RO) membranes can reduce salinity and organic content in water and can be used in reclamation for irrigation purposes if heavy metal concentrations of secondary effluents are not critical.[7]

There are active desalination plants in Israel, Spain and Jordan that desalinate seawater for agricultural irrigation purposes.[8–10] NF, RO, MBR-NF and MBR-RO hybrid processes are used to produce irrigation water from wastewaters by some researchers. Chon *et al.*[11,12] used integrated MBR-NF and MBR-RO processes for the treatment of urban wastewaters and production of irrigation water. Bunani *et al.*[13,14] worked on NF and RO processes to recover irrigation water from biologically treated municipal wastewater.

Shanmuganathan *et al.*[15] considered the parameter so-called sodium adsorption ratio (SAR) when producing suitable irrigation water by treatment of urban wastewater using NF and RO processes. It was indicated that NF permeates have very high SAR values due to the relatively low sodium rejections of NF membranes compared to their high calcium and magnesium rejections while RO permeate has insufficient amount of minerals to sustain plant life. Mixing NF and RO permeates were also found to be applicable in some cases.[11,15–17]

Although MBR process seems to be an alternative solution by providing a high quality effluent for discharge, it is still not possible to use this water directly for irrigation and as process water. The MBR effluent can still contain dissolved salts that have to be removed before the water can be reused. Therefore, additional treatment would be needed. This can be accomplished by pressure driven membrane processes such as NF or RO. Membrane processes such as RO, ultra-low pressure RO (ULPRO) and NF are becoming increasingly widespread in water treatment and wastewater reclamation applications where a high product quality is desired.

Sert *et al.*[18] studied on the applicability of some NF and RO membranes in advanced treatment of MBR effluent obtained from a wastewater treatment plant in an organized industrial zone. Typical characteristics of MBR effluent were summarized in Table 3.1. For this, NF and RO membranes were tested with a laboratory scale, cross-flow membrane test unit (Fig. 3.1) using RO membrane (GE-Osmonics AG membrane) and NF membranes (GE-Osmonics HL and DL membranes). The qualities of MBR effluent and permeates of RO and NF membranes were compared with irrigation water standards for their reuse in agricultural irrigation. Among the membranes tested, AG-BWRO membrane showed a good performance in terms of water quality (Table 3.2).

However, considering irrigation standards, product water of RO membrane may cause an infiltration problem due to low conductivity of permeate water. Thus, high rejection of sodium ions by AG-BWRO membrane destroyed the balance between SAR and conductivity. It was considered that the infiltration problem can be resolved by blending permeate of AG-BWRO membrane with MBR effluent to produce water for agricultural irrigation.[18]

Elsewhere, Sert *et al.*[19] investigated utilization of membrane separation methods such as NF and RO for reuse of wastewater treated with MBR process using a mini-pilot scale NF/RO test system installed at wastewater treatment plant of an organized

Table 3.1. Properties MBR effluent. (Reproduced with permission from Ref. 18. Copyright of Elsevier.)

Parameter	
Parameter	MBR effluent quality
TDS (g/L)	3.87 ± 0.19
EC (mS/cm)	7.73 ± 0.39
pH	7.96 ± 0.06
Salinity (psu)	3.86 ± 0.17
Turbidity (NTU)	0.98 ± 0.04
Na^+(mg/L)	1168 ± 32
Ca^{2+}(mg/L)	199 ± 7
K^+(mg/L)	139 ± 4
Mg^{2+}(mg/L)	37.0 ± 1.10
NH_4-N (mg/L)	0.14 ± 0.01
HCO_3^-(mg/L)	276 ± 10
SO_4^{2-}(mg/L)	616 ± 30
Cl^-(mg/L)	1863 ± 30
PO_4-P (mg/L)	0.24 ± 0.04
NO_3-N (mg/L)	44.4 ± 2.61
NO_2-N (mg/L)	0.01 ± 0.00
Si (mg/L)	2.96 ± 0.18
COD (mg/L)	17.7 ± 2.11
Color (Hazen)	20.4 ± 0.69
TOC (mg/L)	15.0 ± 7.0

industrial zone (Fig. 3.2). Product water qualities obtained by BWRO membrane (BW30) and NF membrane (NF90) were assessed by comparing the permeate water qualities with irrigation water standards. According to their results, it was reported that NF90 and BW30 membranes can be used for tertiary treatment of MBR effluent for agricultural irrigation purposes.[19]

Elsewhere, Gündoğdu *et al.*[20] also conducted some field tests at wastewater treatment plant of an organized industrial zone where MBR process has been used for wastewater treatment. A mini-pilot scale membrane test system having three different NF membranes (NF90, NF270 and TR60) and two different RO membranes (BW30 and AG) with spiral wound configurations was applied for demineralization of MBR effluent (Fig. 3.3). The system was operated at 10 bar

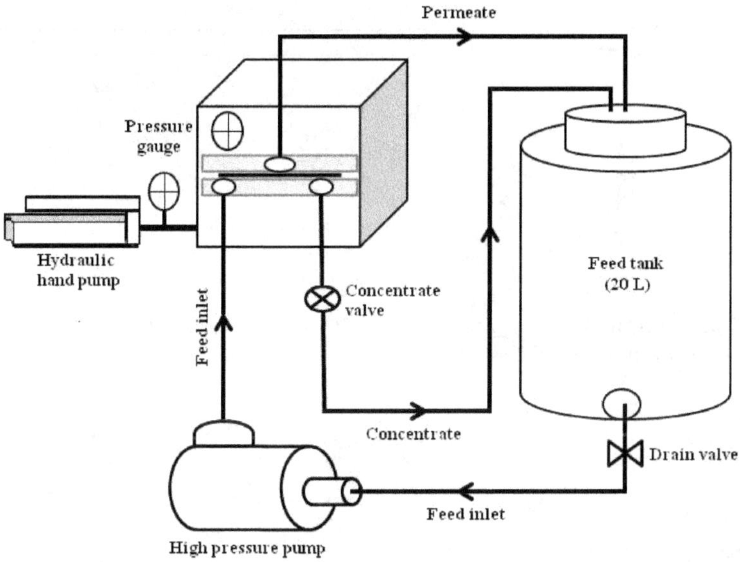

Fig. 3.1. Cross-flow membrane test unit. (Reproduced with permission from Ref. 18. Copyright of Elsevier.)

Table 3.2. Comparison of different membranes in terms of permeate quality and water quality standards. (Reproduced with permission from Ref. 18. Copyright of Elsevier.)

Parameters		HL	DL	AG
SAR		29.3	25.8	7.48
Sodyum (mg/L)		629	690	40
Chloride (mg/L)		1530	1603	73
TDS (g/L)		2.42	2.67	0.13
EC (mS/cm)		4.84	5.33	0.25
Salinity		3rd class	3rd class	1st class
Infiltration		2nd class	1st class	3rd class
Specific ion toxicity	Na	3rd class	3rd class D	1st class D
	Cl	3rd class Y	3rd class Y	1st class Y, D
	B	—	—	—

Y: Surface irrigation, D: Sprinkler irrigation

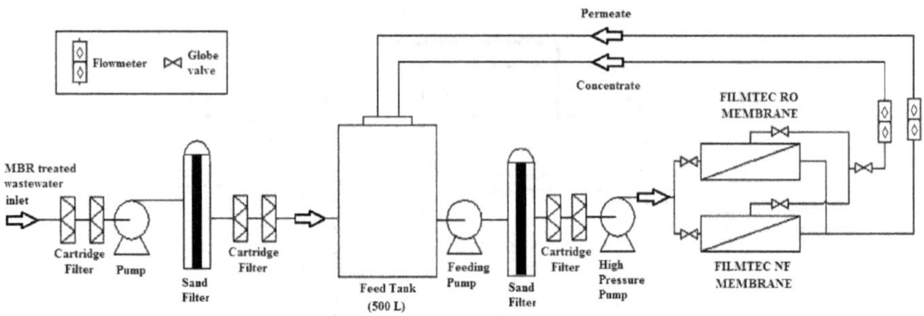

Fig. 3.2. Mini-pilot scale spiral-wound NF/RO membrane test system. (Reproduced with permission from Ref. 19. Copyright of Elsevier.)

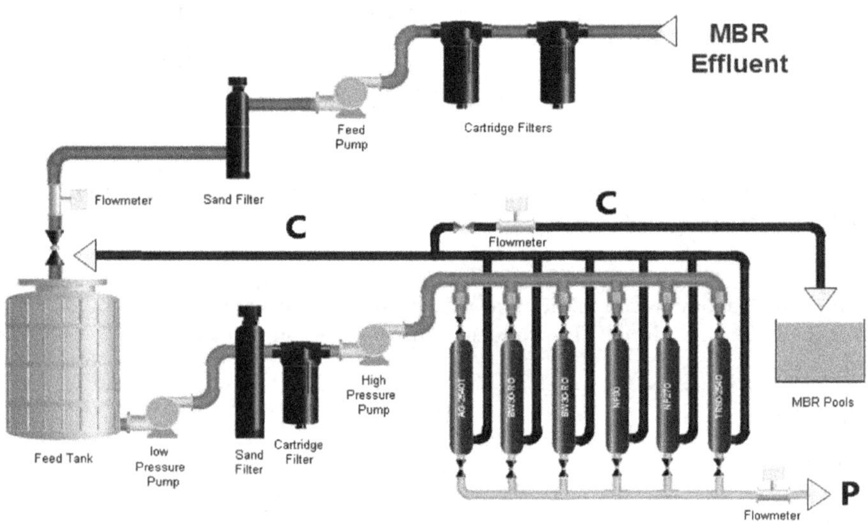

Fig. 3.3. Flow diagram of mini-pilot scale NF/RO test system (P: Permeate stream, C: Concentrate stream). (Reproduced with permission from Ref. 20. Copyright of Elsevier.)

of pressure for NF membranes and at 20 bar for RO membranes. The average water recoveries of NF membranes were found as 50%, 56% and 52% for NF90, NF270 and TR60 membranes, respectively while average water recoveries of RO membranes were 64% for BW30 membrane and 62% for AG membrane.

The qualities of product waters were analyzed and compared with water standards to reuse them for agricultural irrigation and as process water. The permeates of NF270 and TR60 membranes agreed well with the third class water quality for salinity and the second class quality for the rest of parameters. On the other hand, the permeates of NF90, BW30 and AG membranes provided with the first class water for all parameters except for infiltration. Mixing the permeates of NF and RO membranes with MBR effluent in appropriate ratios was considered to be helpful to get acceptable quality values for product water and a good permeability for agricultural irrigation. Therefore, an optimum theoretical mixture of MBR effluent to permeates was determined. When the qualities of product water obtained by NF and RO membranes were compared to the standards of cooling and boiling feed waters for industrial usage and with the process water characteristics for pulp and paper, textile industries, it was considered that the permeates of NF90, BW30 and AG membranes obeyed well the quality standards required.[20]

Although increasing pressure increases irreversible fouling due to compression of foulants, increased flux and better salt rejection are often desired. The effect of pressure on the produced permeate from NF and RO membranes and evaluation of the product water for reuse in agricultural irrigation were investigated by Hacıfazlıoğlu *et al.*[21] They used a mini-pilot scale membrane test system having NF270 and BW30 membranes in spiral wound configuration (Fig. 3.4).

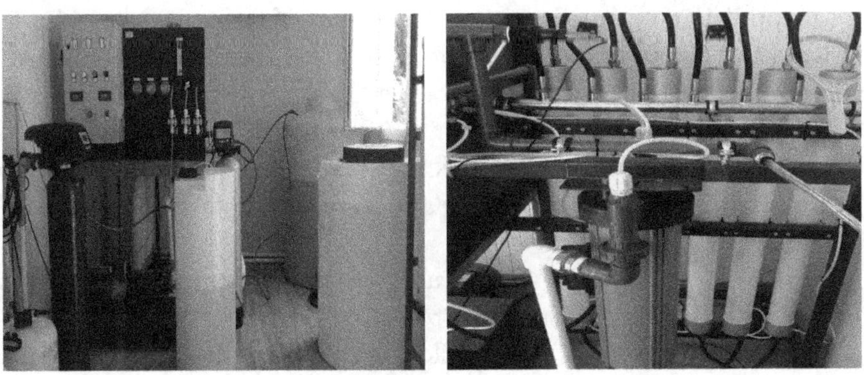

Fig. 3.4. Mini-pilot NF/RO test system. (Reproduced with permission from Ref. 21. Copyright of Elsevier.)

3.2. Desalination of MBR Effluent by NF and RO Processes for Agricultural Irrigation

The reuse of water reclaimed for irrigation is more important than other purposes due to the highest portion of water consumption in agricultural irrigation. It is much easier to regain water as irrigation water than drinking water because it is hard to fit all required very high water quality criteria for drinking water. Different plants require different quality criteria and it is easier to fit all parameters for certain plants.[3]

It is important to treat the secondary treatment effluents of wastewater because they cannot be used directly due to their generally considerable contents of dissolved matter, heavy metals, micropollutants and pathogens.[4] There are concerns over their reuse in irrigation due to some factors explained below:

- Heavy metal accumulation in soil which will harm plant growth in middle to long term irrigation.[6]
- Salinity related salt accumulation and lowered water infiltration rate in the soil.[5]
- Quality and safety of produced food from plants irrigated with secondary treated wastewaters.[22]
- Possible health problems for agricultural workers due to pesticides and pathogens.[23]

In addition to investigation of suitability of treated wastewater to irrigation standards, those effluents should be also examined for their reuse in irrigation of various plants according to their tolerance to salinity. It is also noted that salinity tolerance of a plant generally is similar to its tolerance to ion toxicity and boron content of irrigation water. Salinity tolerances of some agricultural plants are summarized in Table 3.3.

Different criteria for the evaluation of agricultural irrigation water stated by Environmental Protection Agency,[24] Food and Agriculture Organization[25] and World Health Organization.[26] Sodium hazard is identified by a parameter so-called SAR which indicates the effect of sodium ion on soil hydraulic properties, calculated by

Table 3.3. Salinity tolerances of agricultural plants.[27]

Tolerance level (TDS, mg/L)	Field crops	Vegetables	Fruit plants
Non-Tolerant (500–1000)	Bean	Carrot, onion	Almond, apricot, blackberry, orange, peach, plum, strawberry
Slightly Non-Tolerant (1000–1500)	Corn, pea, linen, rice, sugar cane	Cabbage, celery, cucumber, lettuce, potato, spinach, tomato, turnip	Grape
Slightly Tolerant (1500–2000)	Oat, rye, sorghum, soya bean	Artichoke, red beet, pumpkin	—
Tolerant (>2000)	Barley, cotton, sugar beet	Asparagus, safflower	Date palm

Eq. (1).

$$\mathrm{SAR} = \frac{[\mathrm{Na}^+]}{\sqrt{\frac{[\mathrm{Ca}^{2+}]+[\mathrm{Mg}^{2+}]}{2}}} \tag{1}$$

Similarly, PAR (potassium adsorption ratio) parameter indicates the effect of potassium ion on soil properties. There is no agreeement or guideline on suitable PAR values, however maximum PAR of 5 seems to be suitable for irrigation.[28] It is calculated by Eq. (2).

$$PAR = \frac{[K^+]}{\sqrt{\frac{[\mathrm{Ca}^{2+}]+[\mathrm{Mg}^{2+}]}{2}}} \tag{2}$$

in which all concentration units for SAR and PAR calculations are meq/L.

Table 3.4 shows the evaluation of chemical quality for irrigational water and includes standards stated by Ministry of Environment and Urbanization of Turkish Republic. Table 3.4 includes the data given in Table 3.2 and also additional criteria stated by US Environmental Protection Agency (EPA), UN Food and Agriculture Organization (FAO), and World Health Organization (WHO). Given SAR-EC data

Table 3.4. Criteria for evaluation of agricultural irrigation. (Reproduced with permission from Ref. 21. Copyright of Elsevier.)

			Degree of restriction on use		
Type	Parameter	Unit	None	Slight-moderate	Severe
Salinity	Electrical Conductivity (EC)	mS/cm	<0.7	0.7–3.0	>3.0
	Total Dissolved Solid (TDS)	g/L	<0.45	0.45–2.0	>2.0
Nutrients	Nitrate-Nitrogen	mg/L	<5	5–30	>30
	Total Nitrogen (TN)	mg/L	<5	5–30	>30
	Phosphate-Phosporus	mg/L		<2(<5)	
	Potassium (PAR)	—	<5	5–10	>10
Ion Toxicity	Bicarbonate	mg/L	<90	90–500	>500
	Boron	mg/L	<0.7	0.7–3.0	>3.0
	Chloride	mg/L	<106.5	>106.5	
	Sodium	mg/L	<69	>69	
	Free Chlorine	mg/L	<1	1–5	>5
	Permeability Criteria				
SAR	0–3		EC $\geq$ 0.7	0.7 − 0.2	$\leq$0.2
	3–6		$\geq$1.2	1.2 − 0.3	$\leq$0.3
	6–12		$\geq$1.9	1.9 − 0.5	$\leq$0.5
	12–20		$\geq$2.9	2.9 − 1.3	$\leq$1.3
	20–40		$\geq$5.0	5.0 − 2.9	$\leq$2.9
Other	Total Suspended Solid (TSS)	mg/L	<50	50–100	>100
	pH	—		6.0–9.0	
	Turbidity	NTU		<2	

in Table 3.4 for determination of the effect of irrigational water on soil permeability was visualized in Fig. 3.5. In Fig. 3.5, Class III area represents that this water may cause serious drop in infiltration and it is not useable for irrigation. In Class II area, this water may have some negative effects on infiltration but it can be used with caution. On the other hand, the water in Class I area will have no harmful effects on infiltration. It is noted that heavy metal concentrations

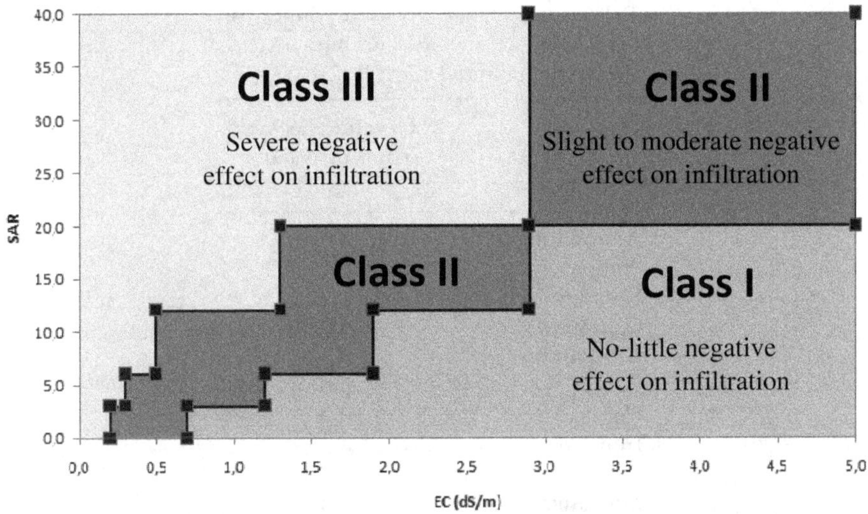

Fig. 3.5. Illustration of SAR-EC classification for irrigation water. (Reproduced with permission from Ref. 30. Copyright of Elsevier.)

should be also considered before decided about the suitability of water to be used for irrigation because they can accumulate and pose health risks for both plants and consumers of those plants.[29] Hence, they are limited by regulations. Table 3.5 summarized the data obtained from EPA guideline.[24]

NF and RO processes can be used to reduce salinity and organic content in water and they can solve the problem with secondary effluents if heavy metal concentrations of the feed is not critical.[7] Falizi *et al.*[30] reported on reclamation of MBR effluent of industrial wastewater by means of NF and RO processes to produce water for agricultural irrigation. The pilot-scale RO and NF systems installed at wastewater treatment plant were depicted in Fig. 3.6. The flow sheet of NF/RO treatment process was shown in Fig. 3.7. TM620 NF and TM720D RO membranes were employed in that pilot-scale system.

As shown in Table 3.6, MBR effluent was not suitable to be reused in irrigation for most of the parameters. Although permeability of MBR effluent seemed to be suitable for irrigation, its salinity, nutrient amounts and toxic ions contents make its reuse not possible unless

Table 3.5. Concentration limits of some elements in water for agricultural irrigation (adapted from Ref. 24).

Element	Maximum allowable concentration (mg/L)
Aluminum	5.0
Arsenic	0.1
Beryllium	0.1
Cadmium	0.01
Chromium	0.1
Cobalt	0.05
Copper	0.2
Iron	5.0
Lithium	2.5
Manganese	0.2
Molybdenum	0.01
Nickel	0.2
Lead	5.0
Selenium	0.02
Vanadium	0.1
Zinc	2.0

the plants to be irrigated with MBR effluent are highly resistant to salinity and toxicity. Sodium content of MBR permeate was relatively high but calcium and magnesium contents of it balanced the harmful effects of sodium on soil infiltration capability. It must be also noted that quality of MBR permeate at organized industrial zone varied highly depending on time since different industrial plants produce different qualities of wastewaters in various seasons.

Falizi *et al.*[30] checked the suitability of NF permeates, RO permeates of this pilot-scale treatment system along with tap water, MBR effluent, the mixture of RO permeate and MBR effluent (2:1) for irrigation. Table 3.6 showed the data obtained in that research.

MBR permeate obtained in certain seasons may be used for irrigation while MBR permeate obtained in another season may not be. It depends on the quality of wastewater collected from various industrial plants where the process run is stopped at some seasons.

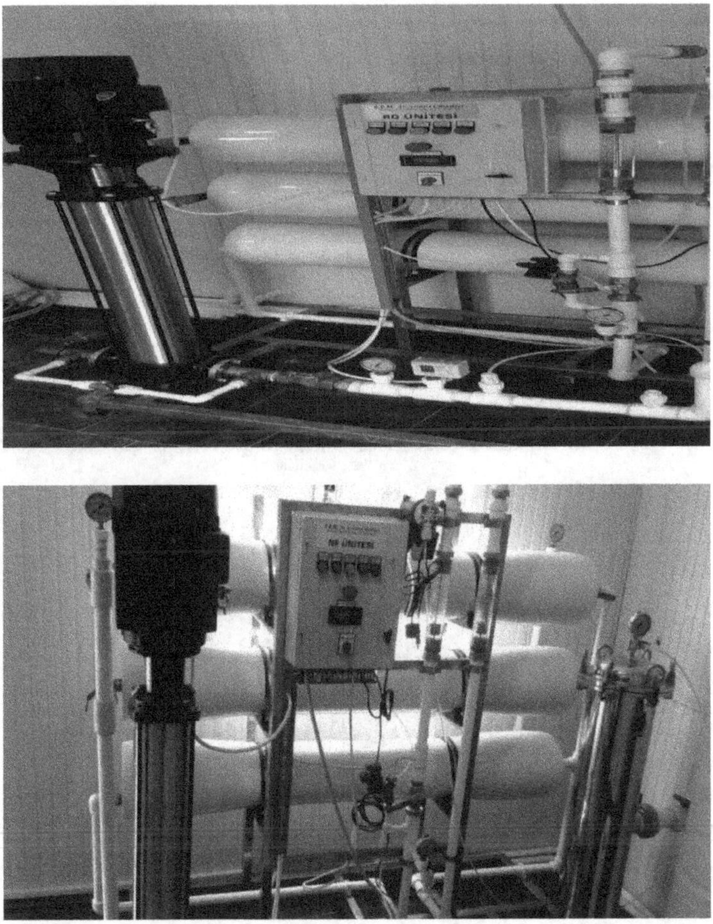

Fig. 3.6. Pilot scale RO and NF systems. (Reproduced with permission from Ref. 31. Copyright of Elsevier.)

Both NF and RO membranes produced permeates that fit most of the basic irrigation parameters. Average properties of permeate samples of NF and RO membranes were given in Table 3.6.

Although NF membranes (TM620 NF) reduced the toxicity significantly, salinity and most of toxicity parameters were still above the criteria stated. It may seem as an improvement to MBR permeate anyway; however its soil infiltration ability was worse than MBR permeate which makes NF permeate not an applicable option. NF

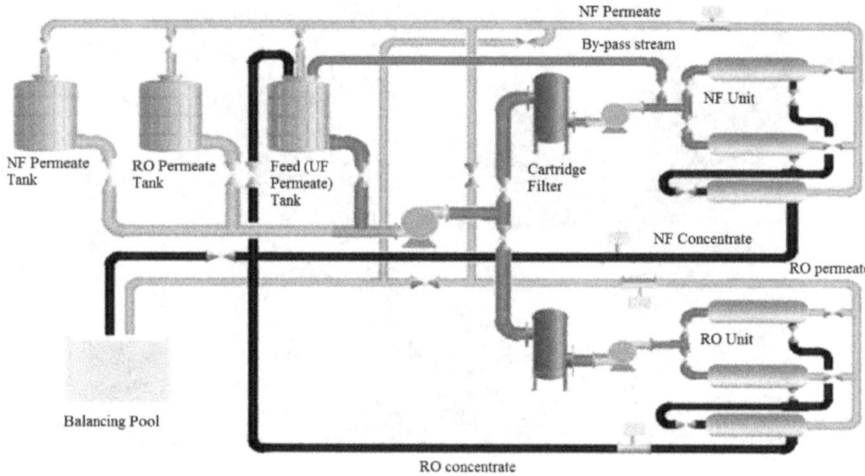

Fig. 3.7. Flow scheme of pilot NF/RO system. (Reproduced with permission from Ref. 30. Copyright of Elsevier.)

membranes removed divalent ions much better than monovalent ions and this situation resulted in a high SAR and hence, it is second class of water at SAR-EC classification graph.

RO permeate produced by RO membranes (TM720D RO) seemed to fit all agricultural irrigation standards given; yet as it is stated by other studies, too low ion content of this permeate would not sustain enough nutrients for plant life. There are two other problems related with the quality of RO permeate. First one of these is the very low hardness (smaller than 2 mg/L $CaCO_3$) of water that may cause corrosion in transfer pipes. Second one is its harsh effect on soil infiltration since its SAR-EC classification is third class. This is caused by its very low salinity which is not enough to balance its sodium hazard.

Due to the inapplicability of RO permeate and the fact that NF permeate is not much superior to MBR effluent, a theoretical mixture of MBR effluent with RO permeate was also considered. In this mixture, RO permeate would suppress the negative effects of toxic ions in MBR effluent and reduce salinity. Also MBR effluent would increase the nutrition properties and balances sodium hazard due to its SAR-EC values of RO permeate. Both SAR and EC

Table 3.6. Evaluation of different types of water for agricultural irrigation. (Reproduced with permission from Ref. 30. Copyright of Elsevier.)

Water types/ characteristics	MBR effluent	NF permeate	RO permeate	RO and MBR mixture (2:1)	Tap water	unit	Degree of restriction on use		
							none	slight to moderate	severe
				Salinity					
EC	3.62	2.27	0.04	1.15	0.80	(mS/cm)	<0.7	0.7–3.0	>3.0
TDS	1.88	1.30	0.37	0.57	0.38	(g/L)	<0.45	0.45–2.0	>2.0
				Permeability					
SAR	11.20	25.77	1.67	5.30	0.82		<18	18–26	>26
PAR	0.69	1.43	0.12	0.36	0.03		<5	5–10	>10
Class	1st	3rd	3rd	2nd	1st				
				Specific ion toxicity and nutrients					
Na^+	652.3	533.3	7.8	190.0	36.8	(mg/L)	<69	>69	
K^+	67.9	50.4	0.9	22.2	2.5	(mg/L)			
Ca^{2+}	191.5	28.4	1.6	79.8	110.0	(mg/L)	<400	>400	
Mg^{2+}	39.0	2.4	0.06	10.32	24.6	(mg/L)	<61	>61	
F^-	<0.5	<0.5	<0.5	<0.5	<0.5	(mg/L)		<1.0	
Cl^-	661.0	527.5	4.96	184.67	36.8	(mg/L)	<106.5	>106.5	
Boron	2.6	2.6	1.6	1.8	0.4	(mg/L)	<0.7	0.7–3.0	>3.0

(Continued)

Table 3.6. (*Continued*)

Water types/ characteristics	MBR effluent	NF permeate	RO permeate	RO and MBR mixture (2:1)	Tap water	unit	Degree of restriction on use: none	slight to moderate	severe
PO_4-P	<0.05	<0.05	<0.05	<0.05	<0.05	(mg/L)		<2 (<5)	
NO_3-N	25.7	24.38	1.21	8.22	8.9	(mg/L)	<5	5–30	>30
Total N	36.0	26.78	2.55	9.24	10.1	(mg/L)	<5	5–30	>30
HCO_3^-	9.8	6.30	2.18	3.87	32.3	(mg/L)	<90	90–500	>500
Other parameters									
pH	7.0	7.4	7.4	7.0	6.9			6.0–9.0	
TSS	3	1	1	2	2	(mg/L)	<50	50–100	>100
Turbidity	0.81	0.51	0.23	0.27	0.33	NTU		<0.2	

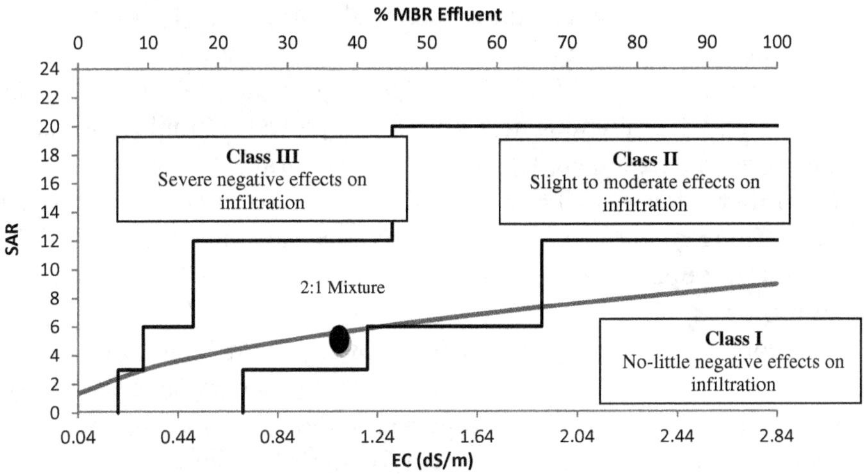

Fig. 3.8. SAR-EC graph for RO permeate-MBR permeate theoretical mixtures. (Reproduced with permission from Ref. 32.)

(alongside with all other parameters related with irrigation) values of such theoretical mixture were calculated for all mixing ratios. By drawing Fig. 3.8 and checking Table 3.6, it was decided that a mixture of RO Permeate-MBR permeate (2:1) would be the best option of irrigation.

This decision was based on finding an optimized water for irrigation, by finding an agreement between its SAR-EC comparison, salinity and ion toxicity. All parameters for this theoretical mixture (2:1) either fits to "No restriction on use" or they are close to that class. Although its soil infiltration class is second class, it is close to the first class. It cannot be directly taken as 50% or more MBR permeate because that is where salinity and ion toxicity would create problems in long-term irrigation although its soil infiltration is optimal. Also, MBR permeate cannot be lower than 20% of mixture since the soil infiltration of solution would be like RO permeate. Therefore, properties were found optimal between 20% and 35% of MBR permeate in the mixture. Final appropriate solution was chosen as RO permeate-MBR permeate with a volume ratio of 2:1 because it was desired to recover more amounts of MBR permeate. Falizi *et al.*[30] showed that not all parameters can be satisfied together. However, it was also noted that RO permeate:MBR effluent (2:1) mixture can satisfy most of the parameters.

In another example, Hacifazlioğlu *et al.*[21] used RO membrane (BW30) and NF membrane (NF270) employed a mini-pilot membrane test system shown in Fig. 3.4 to demineralize the MBR treated industrial wastewater. They showed that the RO permeate was not useable in irrigation due to its very low mineral content while NF270 was found applicable for plants that have tolerance to medium salinity levels.

According to data given in Table 3.3, the MBR permeate can be used for the irrigation of salinity tolerant plants such as barley, cotton, sugar beet, asparagus, safflower and date palm. Depending on its feed salinity, it may also be used for slightly tolerant plants such as oat, rye, sorghum, soya bean, artichoke, red beet and pumpkin. Howerever it cannot be used to irrigate non-tolerant plants. NF permeate can also be used for irrigation of slightly non-tolerant plants such as corn, pea, linen, rice, sugar cane, cabbage, celery, cucumber, lettuce, potato, spinach, tomato, turnip and grape. However, it may cause infiltration problems in long-term irrigation since its second class in SAR-EC classification. It was shown that RO permeate-MBR permeate mixture (2:1) was the best irrigation option among other water types. It can be used for irrigation of all types of plants whether it is tolerant to salinity or non-tolerant. Since most of the fruit plants (e.g. almond, apricot, blackberry, orange, peach, plum, strawberry) and plants like bean, carrot, onion are sensitive to salinity and they can be irrigated by using only these type of mixtures.

Irrigation can also be performed for landscape purposes rather than agricultural purposes. Meadow plants are generally relatively more tolerant to salinity and ion toxicity, therefore irrigation options for NF permeate increases. There are various types of meadow plants raised for landscaping and their salinity tolerance levels are given in Table 3.7.

As indicated in Table 3.7 while MBR permeate can be used for irrigation of Bermuda grass, timothy, harding grass, Sudan grass and wheatgrass; NF permeate and RO-MBR permeates mixture (2:1) can be used also for clover, fruit orchard, sesbania and ordinary grasses.

Table 3.7. Salinity tolerances of meadow plants.[27]

Tolerance level (TDS, mg/L)	Meadow plants
Non-Tolerant (500–1000)	—
Slightly Non-Tolerant (1000–1500)	Clover, fruit orchard, sesbania, ordinary grass
Slightly Tolerant (1500–2000)	Timothy, harding grass, Sudan grass, wheatgrass
Tolerant (>2000)	Bermuda Grass

In order to increase water recovery, Hacifazlioglu *et al.*[33] used BW30-BW30, NF90-NF90 and BW30-NF270 membrane-in-series configurations (where concentrate stream of the first membrane is fed to the second) enabling over 80% recovery when treating MBR-treated industrial wastewaters. They found that the theoretical mixtures of 5–15% of MBR effluent and 85–95% of BW30-BW30 or NF90-NF90 permeate would be less toxic than MBR effluent and more nutritious than NF and RO permeates while not posing critical effect on soil hydraulic properties. Properties of such theoretical mixtures (calculated and optimum percentages) were given in Table 3.8. It was stated that a first class soil-permeability can not be obtained since it requires higher salinity values and it is another problem. However, close to first class can be achieved with such mixtures as given above, these irrigation waters would fit given criteria for evaluation of water for agricultural irrigation.

Heavy metal concentrations of MBR feed are given in Table 3.9. Since all heavy metal concentrations of the feed solution are under detectable limits, heavy metal concentrations were not measured for permeate samples.

The concentrations of some organic micropollutants in water samples including MBR effluent, NF and RO permeates were analyzed and the measured values were summarized in Table 3.10. The concentrations of micropollutants in water samples, except DMP, were found to be very low. However, the RO method was found to be effective in DMP removal, since much lower amounts of DMP were detected in RO permeate than in MBR effluent and NF

Table 3.8. Properties of optimal theoretical mixtures of NF90-NF90 and BW30-BW30 permeates with MBR effluent. (Reproduced with permission from Ref. 33). Copyright of Elsevier.)

Two membrane-in series configuration	Unit	BW30-BW30	NF90-NF90
MBR Effluent (%)		12	9
EC	mS/cm	0.4	0.4
SAR	—	3.9	4.3
TDS	g/L	0.2	0.2
NO_3^-N	mg/L	7.6	11.3
TN	mg/L	8.9	13.4
PO_4-P	mg/L	0.1	0.1
PAR	—	0.4	0.5
HCO_3^-	mg/L	8.9	13.4
B	mg/L	<0.5	0.5–1.0
Cl^-	mg/L	70.3	71.4
Na^+	mg/L	65.8	68.0
Free chlorine	mg/L	<0.1	<0.1
TSS	mg/L	0.2	0.2
pH	—	6.5	6.9
Turbidity	NTU	0.1	0.3

Table 3.9. Concentrations of heavy metals measured in MBR permeate. (Adapted from Ref. 35.)

Heavy metal	Concentration (mg/L)
Cadmium	<0.5
Zinc	<0.1
Lead	<0.1
Copper	<0.5
Nickel	<0.5
Arsenic	<0.1
Manganese	<0.1
Iron	<0.1
Chromium	<0.1
Cobalt	<0.1

Table 3.10. The concentrations of organic micropollutants measured in water samples.* (Adapted from Ref. 35.)

Micropollutant	MBR effluent	NF permeate	RO permeate
Dimethyl phtalate (DMP) (mg/L)	0.2804	0.2002	0.02701
Diethyl phtalate (DEP) (mg/L)	0.0002	0.0001	0.00002
Triclosan (μg/L)	0.1545	<0.05	<0.05
17α-Ethynylestradiol (μg/L)	<0.05	<0.05	<0.05
17β-Estradiol (μg/L)	<0.05	<0.05	<0.05
Carbamazepine (μg/L)	0.0975	<0.01	<0.01
Bisphenol A (BPA) (μg/L)	<0.5	<0.5	<0.5
2-4-dichlorophenol (μg/L)	<0.5	<0.5	<0.5
2-3-4-trichlorophenol (μg/L)	<0.5	<0.5	<0.5

*TM620 NF and TM720D RO membranes were employed.

permeate. Also, some amounts of carbamazepine and triclosan were detected in MBR effluent. On the other hand, neither carbamazepine nor triclosan were detected in NF and RO permeates, so it was understood that NF and RO membranes were effective in removing these chemicals. According to the results reported by Capangpangan *et al.*,[34] membranes provide a physical barrier with high removal capacity for micropollutants removal, but the removal depends on the membrane structures (size, polarity) and properties. NF and RO membranes are more effective than MF and UF in EDC removal.

It was concluded that the concentrations of micropollutants measured in MBR treated wastewater samples are generally low except dimethyl phtalate (DMP), triclosan and carbamazepine. On the other hand, both NF and RO processes were effective to reduce the concentrations of triclosan and carbamazepine below detection limits. RO membranes were more successful than NF membranes to eliminate DMP from MBR effluent.

3.3. Utilization of MBR Effluent for Industrial Reuse after NF and RO Treatments

Water is required in all types of power generation technologies for cooling purposes, steam generation, infrastructure manufacture.[36]

MBR systems are promising treatment methods for reuse of wastewater, both as pre-treatment of NF and RO or with the effluent directly reused for unrestricted irrigation.[37]

MBR effluent has been used perfectly for water reuse in commercial laundries and in food, beverage and cosmetics industries. MBR technology has also the ability to efficiently treat textile wastewater. However, in textile industry the dye molecules with low molecular weight can pass the MBR membranes and advance treatment for MBR effluent is required.[38] Melin *et al.*[39] stated that the reuse of MBR effluent is a hopefull issue. Moreover, MBR treatment plants have been implemented nearly at all of the Europe. However, reuse projects are not very common. Bixio *et al.*[37] warned about the lack of water resources in the future. Even for Europe where there are plenty of water sources compared to other parts of the world, the situation is critical. Integrated membrane systems for wastewater reclamation and reuse can be implemented on a large scale for augmenting water supply. Also, this solution will decrease the impact of human activities on the environment.

Table 3.11 summarized the demand for process water in various industrial applications.

Economic arrangements and water reuse guidelines are needed to achieve high yield for the water reuse applications.[37] The water quality requirements for different purposes are given in Table 3.12.

Table 3.11. Water demand for different sectors. (Adapted from Ref. 40.)

Industry	Water demand
Paper	29 m^3/t paper produced
Brewing	10–15 m^3/m^3beer
Dairy	140 m^3/m^3 milk
Sugar	8–9 m^3/t sugar
Automotive	450 m^3/car for metal production
	760 m^3/car for tire production
Dying	100 m^3/t fabric processed
Soap	2 m^3/t soap produced
Power	60 m^3/MWH for cooling
	3 m^3/MWH for steam

Table 3.12. Required quality of water in reuse applications. (Adapted from Ref. 41.)

General use	Specific use	Quality requirement
Domestic supply	—	• Free from chemical substances and organisms harmful to health • Low aggressiveness and hardness • Aesthetically pleasant
Industrial supply	Water incorporated into the product (foods, drinks, medicine)	• Free from chemical substances and organisms harmful to health • Aesthetically pleasant
	Water that enters into contact with the product	• Variable with the product
	Water that does not enter into contact with the product (refrigeration units, boilers)	• Low hardness • Low aggressiveness
Irrigation	Horticulture, product ingested raw or with skin	• Free from chemical substances and organisms harmful to health • Non-excessive salinity
Aquaculture	Animal breeding	• Free from chemical substances harmful to health of animals, workers and consumer • Free from organisms harmful to health of animals, workers and consumers • Availability of nutrients
	Vegetable growing	Free from chemical substances toxic to vegetables and consumers Availability of nutrients
Recreation and leisure	Primary contact with water (bathing, swimming)	• Free from chemical substances and organisms harmful to health • Free from organisms harmful to health • Low levels of suspended solids and oils and grease
	Secondary contact (fishing)	• Pleasant appearance

(*Continued*)

Table 3.12. (*Continued*)

General use	Specific use	Quality requirement
Energy generation	Hydroelectric power plants	• Low aggressiveness
	Nuclear or thermoelectric power plants (cooling towers)	• Low hardness

In assessing the above goals, Parlar[41] studied the quality of product waters obtained by NF and RO permeates of MBR effluent. For the above goals, TM620 NF and TM720D RO membranes were employed in a pilot-scale system described in Figs. 3.6 and 3.7. The comparison between the qualities NF permeates obtained and the required quality standards defined in the literature[42] for cooling water and steam generation were given in Table 3.13. The NF permeate had only suitable pH values for the production of cooling water and low pressure steam production. Other quality parameters were not suitable for cooling water or steam production.

The qualities of NF permeates collected were also compared with the required quality standards given for a typical textile industry (Table 3.14). Generally, the NF permeates met the required quality for the needs of the textile industry. It was seen that the TDS value of NF permeate were higher than standards, while sulphate parameter was too low with a value of 0.50 mg/L for all NF permeates.

As seen in Table 3.15, NF permeates satisfied some of the required quality parameters for paper industry except high TDS values. TDS parameters were very close to the upper boundary of required quality for brown grades paper. It exceeded the boundary which is 300 mg/L for bleached kraft and also for bleached papers.

In case of RO permeates, they were mostly suitable for cooling water and steam generation as summarized in Table 3.16. But

Table 3.13. Comparison of qualities of NF permeates with required standards for cooling water and steam generation. (Adapted from Ref. 41.)

Parameters	NF 1	NF 2	NF 3	Cooling water	Steam generation: Low pressure (<10 bar)	Steam generation: Medium pressure (10–50 bar)	Steam generation: High pressure (>50 bar)
TDS (mg/L)	979	939	969	500	700	500	200
pH	7.58	7.55	7.61	6.9–9.0	7.0–10	8.2–10	8.2–9.0
COD (mg/L)	8.52	5.33	6.82	75	5	5	1
Ca (mg/L)	22.9	14.7	18.4	50	+	0.4	0.01
Mg (mg/L)	0.50	1.13	1.34	0.5	+	0.25	0.01
Cu (mg/L)	<0.1	<0.1	<0.1	+	0.5	0.05	0.05
Zn (mg/L)	<0.1	<0.1	<0.1	+	+	0.01	0.01
Fe (mg/L)	0.17	0.17	0.14	0.5	1	0.3	0.05

+ Acceptable if the other parameters are fine.

Table 3.14. Comparison of qualities of NF permeates with required quality standards for a typical textile industry. (Adapted from Ref. 41.)

Parameters	NF 1	NF 2	NF 3	Quality required for a typical textile industry
Al (mg/L)	<0.1	<0.1	<0.1	8
Color (Pt-Co)	0	0	0	0–5
Cu (mg/L)	<0.1	<0.1	<0.1	0.01–5
Fe (mg/L)	0.17	0.17	0.14	0.1–0.3
Mn (mg/L)	<0.1	<0.1	0.14	0.01–0.05
pH	7.58	7.55	7.61	6–8
SO_4 (mg/L)	0.50	0.50	0.50	100
TSS (mg/L)	0	0	0	0–5
TDS (mg/L)	978.8	939.0	969.0	100–200
Hardness (mg/L $CaCO_3$)	42.39	38.73	38.73	0–50

according to pH and COD values, they were not suitable for high pressure steam generation purposes. In addition to this, calcium and magnesium concentrations are making them to be suitable for low pressure steam generation or cooling purposes. According to copper, iron and zinc ion concentrations, medium or high pressure steam generation should not be preferred.

Table 3.17 shows the comparison of RO permeates and required quality standards given for a typical textile industry. The qualities of RO permeates were mostly suitable for textile industry as well. Also, as shown in Table 3.18, product water of RO membranes was much more better than required qualities for paper production industry. TDS and alkalinity parameters of the samples were not between the range for specific paper production. Hardness and chloride ion concentrations of permeate samples were lower than the required quality.

The water qualities obtained by integration of MBR with NF or RO were compared with water qualities of MBR + NF + RO

Table 3.15. Comparison of the qualities of NF permeates with required quality standards for paper production. (Adapted from Ref. 41.)

				Brown		White	
Parameter	NF 1	NF 2	NF 3	Ground paper	Kraft	Bleached kraft	Bleached
Turbidity (NTU)	0.05	0.13	0.22	70	35	40	14–56
Color (Pt-Co)	0.00	0.00	0.00	30	5	25	5–25
TDS (mg/L)	979	939	969	250–1000	250–1000	300	75–650
TSS (mg/L)	0	0	0	40	10	10	10–30
Alkalinity (mg/L) $CaCO_3$	31.79	29.05	29.05	75–150	75–150	75	75–125
Hardness (mg/L) $CaCO_3$	42.39	38.73	38.73	100–200	100–200	100	100
Cl (mg/L)	490.7	414.1	434.5	75	75	200	200
Fe (mg/L)	0.17	0.17	0.14	0.3	0.1	0.2	0.1
Mn (mg/L)	<0.1	<0.1	0.14	0.1	0.05	0.1	0.03
SiO_2 (mg/L)	4.20	2.80	3.10	50	20	50	9–20
SO_4 (mg/L)	0.50	0.50	0.50	Trace	—	—	100–300

(—): Not given

Table 3.16. Comparison of the qualities of RO permeates with standards for cooling water and steam generation. (Adapted from Ref. 41.)

Parameters	RO 1	RO 2	RO 3	Cooling water	Steam generation		
					Low pressure (<10 bar)	Medium pressure (10–50 bar)	High pressure (>50 bar)
TDS (mg/L)	21.2	16.0	16.0	500	700	500	200
pH	6.58	6.76	6.53	6.9–9.0	7.0–10	8.2–10	8.2–9.0
COD (mg/L)	5.00	5.00	5.00	75	5	5	1
Ca (mg/L)	1.80	1.32	1.49	50	+	0.4	0.01
Mg (mg/L)	0.50	0.50	0.50	0.5	+	0.25	0.01
Cu (mg/L)	<0.1	<0.1	<0.1	+	0.5	0.05	0.05
Zn (mg/L)	<0.1	<0.1	<0.1	+	+	0.01	0.01
Fe (mg/L)	0.13	0.12	<0.1	0.5	1	0.3	0.05

+ Acceptable if the other parameters are OK.

Table 3.17. Comparison of the qualities of RO permeates with required quality standards for a typical textile industry. (Adapted from Ref. 41.)

Parameters	RO 1	RO 2	RO 3	Quality required for a typical textile industry
Al (mg/L)	<0.1	<0.1	<0.1	8
Color (Pt-Co)	0	0	0	0–5
Cu (mg/L)	<0.1	<0.1	<0.1	0.01–5
Fe (mg/L)	0.13	0.12	<0.1	0.1–0.3
Mn (mg/L)	<0.1	<0.1	<0.1	0.01–0.05
pH	6.58	6.76	6.53	6–8
SO_4(mg/L)	0.50	0.50	0.50	100
TSS (mg/L)	0.67	0.33	0.00	0–5
TDS (mg/L)	21.2	16.0	16.0	100–200
Hardness (mg/L $CaCO_3$)	1.32	1.10	2.48	0–50

integration. According to the Table 3.19, NF permeate was inferior while RO permeate with or without NF pretreatment were satisfying required quality defined for steam production. The RO permeate pretreated with NF had better quality as expected. However, Ca^{2+} instead of Mg^{2+} ions conecntrations in the product water were not suitable for high pressure steam generation.

Table 3.20 shows the effect of NF pretreatment and before RO system following MBR process and required quality standards for a typical textile industry. NF permeate was not suitable due to the high TDS and hardness. On the other hand, integration of NF with RO after MBR gave the best water quality for textile industry. Also, single RO study after MBR was enough for obtaining water requirements by typical textile industry.

As shown in Table 3.21, NF pretreatment before RO system following MBR provided with a better quality of water than RO only for paper industry.

Table 3.18. Comparison of RO permeate produced in membrane configuration studies with required quality standards for paper production, defined by 45 (Adapted from Ref. 41.)

				Brown		White	
Parameters	RO 1	RO 2	RO 3	Ground paper	Kraft	Bleached Kraft	Bleached
Turbidity (NTU)	0.02	0.02	0.13	70	35	40	14–56
Color (Pt-Co)	0.00	0.00	0.00	30	5	25	5–25
TDS (mg/L)	21.2	16.0	16.0	250–1000	250–1000	300	75–650
TSS (mg/L)	0.67	0.33	0	40	10	10	10–30
Alkalinity (mg/L) $CaCO_3$	0.99	0.83	1.86	75–150	75–150	75	75–125
Hardness (mg/L) $CaCO_3$	1.32	1.10	2.48	100–200	100–200	100	100
Cl (mg/L)	7.50	5.27	6.80	75	75	200	200
Fe (mg/L)	0.13	0.12	<0.1	0.3	0.1	0.2	0.1
Mn (mg/L)	<0.1	<0.1	<0.1	0.1	0.05	0.1	0.03
SiO_2 (mg/L)	0.98	1.09	1.11	50	20	50	9–20
SO_4 (mg/L)	0.50	0.50	0.50	Trace	—	—	100–300

Table 3.19. Comparison of NF, NF + RO and single RO permeates with cooling water and steam generation purposes for process water production. (Adapted from Ref. 41.)

					Steam generation		
Parameters	MBR + NF	MBR + NF + RO	MBR + RO	Cooling water	Low pressure (>10 bar)	Medium pressure (10–50 bar)	High pressure (>50 bar)
TDS (mg/L)	1285.5	17.09	29.61	500	700	500	200
pH	7.50	6.88	7.28	6.9–9.0	7.0–10	8.2–10	8.2–9.0
COD (mg/L)	5.00	5.00	5.00	75	5	5	1
Ca (mg/L)	13.77	1.07	3.92	50	+	0.4	0.01
Mg (mg/L)	1.70	0.50	0.70	0.5	+	0.25	0.01
Cu (mg/L)	<0.1	<0.1	<0.1	+	0.5	0.05	0.05
Zn (mg/L)	<0.1	<0.1	<0.1	+	+	0.01	0.01
Fe (mg/L)	<0.1	<0.1	<0.1	0.5	1	0.3	0.05

+ Acceptable if the other parameters are OK.

Table 3.20. Comparison of water qualities obtained by integrated MBR + NF, MBR + NF + RO, MBR + RO systems with required quality standards for a typical textile industry. (Adapted from Ref. 35.)

Parameters	MBR + NF	MBR + NF + RO	MBR + RO	Quality Required for a typical textile industry
Al (mg/L)	<0.1	<0.1	<0.1	8
Color (Pt-Co)	2.67	1.67	2.33	0–5
Cu (mg/L)	<0.1	<0.1	<0.1	0.01–5
Fe (mg/L)	<0.1	<0.1	<0.1	0.1–0.3
Mn (mg/L)	<0.1	<0.1	<0.1	0.01–0.05
pH	7.50	6.88	7.28	6–8
SO_4 (mg/L)	2.00	2.00	2.00	100
TSS (mg/L)	1.00	1.00	1.67	0–5
TDS (mg/L)	1286	17.1	29.6	100–200
Hardness (mg/L $CaCO_3$)	51.16	2.19	2.19	0–50

Table 3.21. Comparison of water qualities obtained by integration of MBR with only NF, NF + RO and only RO with required quality standards for paper production. (Adapted from Ref. 41.)

				Brown		White	
Parameters	MBR + NF	MBR + NF + RO	MBR + RO	Ground paper	Kraft	Bleached Kraft	Bleached
Turbidity (NTU)	0.42	0.22	0.36	70	35	40	14–56
Color (Pt-Co)	2.67	1.67	2.33	30	5	25	5–25
TDS (mg/L)	1286	17.09	29.61	250–1000	250–1000	300	75–650
TSS (mg/L)	1.00	1.00	1.67	40	10	10	10–30
Alkalinity (mg/L) $CaCO_3$	33.4	1.64	1.64	75–150	75–150	75	75–125
Hardness (mg/L) $CaCO_3$	51.2	2.19	2.19	100–200	100–200	100	100
Cl (mg/L)	442	4.42	6.92	75	75	200	200
Fe (mg/L)	<0.1	<0.1	<0.1	0.3	0.1	0.2	0.1
Mn (mg/L)	<0.1	<0.1	<0.1	0.1	0.05	0.1	0.03
SiO_2 (mg/L)	0.05	0.02	0.03	50	20	50	9–20
SO_4 (mg/L)	2.00	2.00	2.00	Trace	—	—	100–300

(—) Not given

3.4. Membrane Fouling and Concentrate Management for RO and NF Processes Used for Treatment of MBR Effluent

Two important problems to arise while working with NF and RO membranes are membrane fouling and concentrate management. Hacifazlioğlu *et al.*[43] performed an extensive research on how much the fouling created by MBR effluent affects NF and RO membranes performances and how the optimal chemical cleaning strategy can be developed. They reported that while the decrease in NF permeate quality was not significant, RO permeate quality showed the effects of membrane fouling very clearly. Table 3.22 included the average properties of MBR, NF and RO permeates measured from the samples obtained in five cycles of 24 h tests with MBR permeate fed to NF 90 and BW 30 membranes.

Table 3.22. Average properties of MBR, NF 90 and BW 30 membranes in a five cycle 24-h experiments. (Adapted from Ref. 32.)

Parameter	Unit	MBR permeate	NF 90 permeate	BW 30 permeate
EC	μS/cm	3960	120.7	228.1
TDS	mg/L	2050	57.1	109.4
NO_3-N	mg/L	8.43	1.04	1.21
Total N	mg/L	6.06	1.44	1.25
PO_4-P	mg/L	<0.1	<0.1	<0.1
HCO_3^-	mg/L	70.6	5.78	0.80
Cl^-	mg/L	940	47.9	21.0
SO_4^{2-}	mg/L	209	1.58	4.12
Ca^{2+}	mg/L	160	4.73	4.97
Mg^{2+}	mg/L	32.0	0.03	0.03
K^+	mg/L	81.1	2.93	2.14
Na^+	mg/L	707	26.5	46.7
TOC	mg/L	6.72	1.46	0.91
Free chlorine	mg/L	<0.5	<0.5	<0.5
TSS	mg/L	1.5	0.50	0.50
pH	—	6.93	6.61	6.83
Turbidity	NTU	1.51	0.63	0.43

Previous studies — also mentioned before in this chapter — used either fresh membranes and close-to-fresh ones. Those studies proved the superiority of BW30 to NF90 when all quality measurements indicated that BW30 rejected all measured ions and other parameters better than NF90. This study experimentally proved the point of countless other researchers that RO membranes are more prone to fouling than NF membranes because of their intrinsic properties.

Hacifazlioglu *et al.*[43] also performed permeability tests with both deionized water and MBR effluent after each 24 h test to develop a mathematical model to calculate water permeability. Those water permeability values can pinpoint on how much the fouling affects the membranes. Permeability tests were performed between 8 and 32 bar with a 2 bar pressure margin while keeping water recovery of around 50%. Alkaline cleaning was conducted by using 2.5–3.0 ppm · h NaOH and 5.3–5.8 ppm · h NaOCl solutions while acid cleaning was conducted by using 650–700 ppm · h citric acid solution. The concentration of citric acid solution was increased to 1350–1400 ppm · h before the final cycle because of the very low permeate flux obtained before cleaning. Two different cleaning periods were tested (15 min alkaline-20 min acid and 30 min alkaline-40 min acid) and the later one was found to be more efficient.

Figures 3.9 and 3.10 respectively show the results of water permeability tests for NF90 and BW30 membranes by using deionized water as the feed solution while Figs. 3.11 and 3.12 respectively show the results of water permeability tests for NF90 and BW30 membranes by using MBR permeate as the feed solution. Slopes of lines are water permeability values as $L/m^2 \cdot h \cdot bar$.

Figure 3.12 shows that the RO membranes are affected from fouling worse than NF membranes and membrane fouling effects were becoming more drastic after each 24 h cycle. Normalized flux ($L/m^2 \cdot h$) and water permeability ($L/m^2 \cdot h \cdot bar$) values were given in Table 3.23.

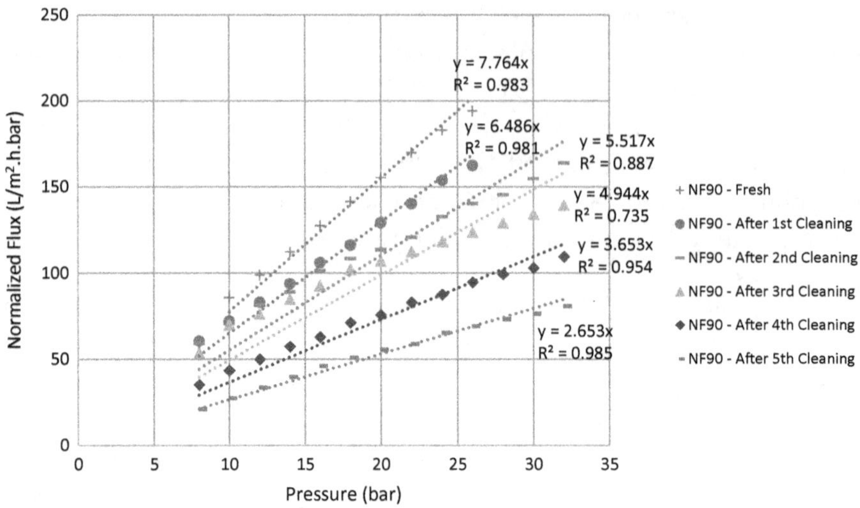

Fig. 3.9. Water permeability test results for NF90 by using deionized water as the feed solution. (Reproduced with permission from Ref. 43. Copyright of Elsevier.)

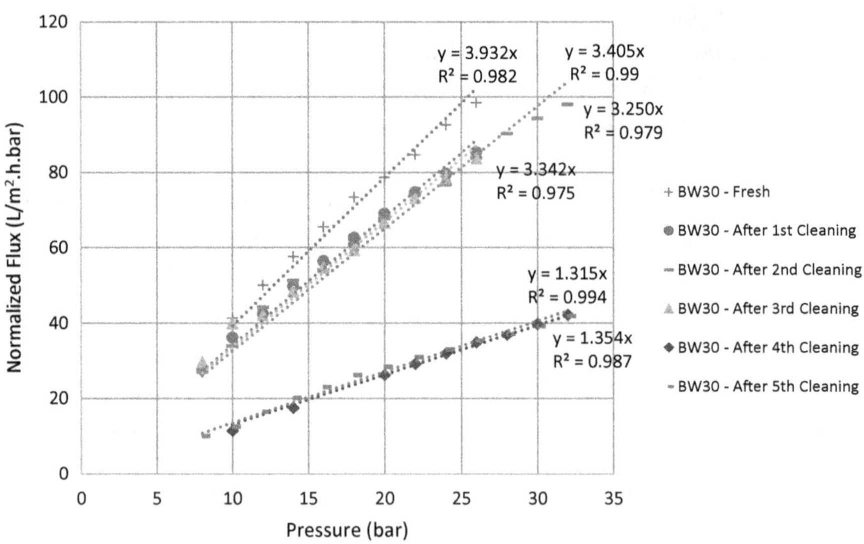

Fig. 3.10. Water permeability test results for BW30 by using deionized water as the feed solution. (Reproduced with permission from Ref. 43. Copyright of Elsevier.)

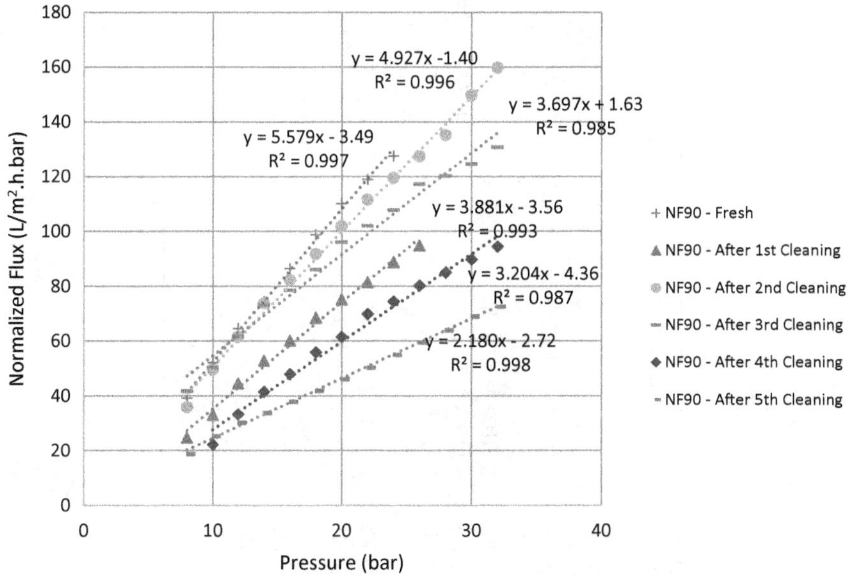

Fig. 3.11. Water permeability test results for NF90 by using MBR permeate as the feed solution. (Reproduced with permission from Ref. 43. Copyright of Elsevier.)

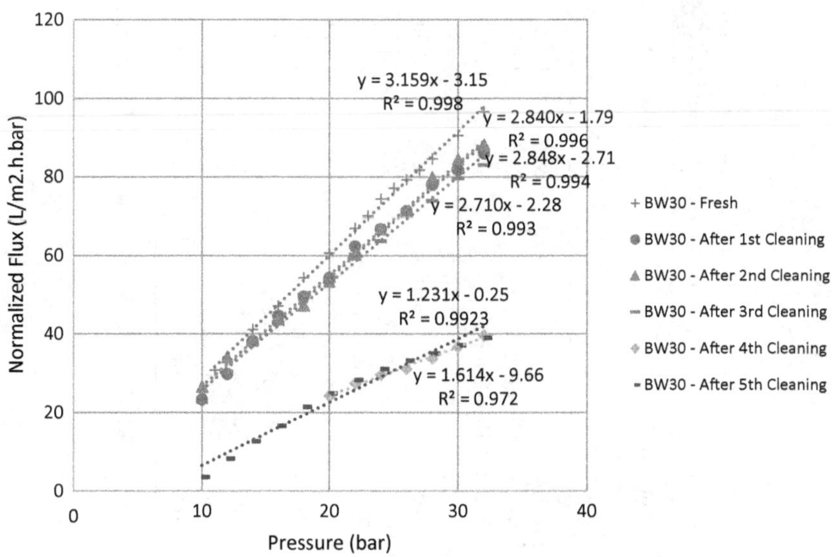

Fig. 3.12. Water permeability test results for BW30 by using MBR permeate as the feed solution. (Reproduced with permission from Ref. 43. Copyright of Elsevier.)

Table 3.23. Normalized flux and water permeability values obtained in five cycles of 24 h operation and cleaning. (Adapted from Ref. 43. Copyright of Elsevier.)

Membrane-Feed water	BW30-DI*		BW30-MBR*		NF90-DI		NF90-MBR	
Parameter	Flux	Perm.	Flux	Perm.	Flux	Perm.	Flux	Perm.
Fresh	78.7	3.932	60.6	3.159	155.4	7.764	110.1	5.579
After 1st Cleaning	68.9	3.405	54.3	2.840	129.1	6.486	99.6	3.881
After 2nd Experiment	67.2	3.257	53.3	2.696	110.9	5.489	75.2	4.202
After 2nd Organic Cleaning	67.7	3.264	53.3	2.739	113.6	5.512	101.2	4.699
After 2nd Inorganic Cleaning	68.1	3.250	54.2	2.848	113.6	5.517	101.9	4.927
After 3rd Experiment	72.7	2.775	47.3	2.762	95.0	4.475	75.3	3.889
After 3rd Organic Cleaning	75.0	2.842	48.2	2.802	95.9	4.520	79.0	3.923
After 3rd Inorganic Cleaning	76.2	3.342	53.3	2.710	107.2	4.944	96.1	3.697
After 4th Experiment	17.6	0.821	15.0	0.933	66.4	3.241	49.4	2.347
After 4th Organic Cleaning	18.1	0.823	15.2	0.993	73.8	3.619	60.5	2.835
After 4th Inorganic Cleaning	26.2	1.315	24.2	1.508	75.6	3.653	61.4	3.204
After 5th Experiment	2.7	0.115	2.1	0.199	24.4	1.209	21.8	1.178
After 5th Cleaning	28.3	1.354	24.9	1.667	55.3	2.653	46.0	2.180

*DI: Deionized water as feed, *MBR: MBR effluent as feed

These detailed values indicated that inorganic fouling is a more important matter than organic fouling in cases where MBR permeate was used as the feed solution. This finding was justified by the mechanism of MBR process where most of the organic content is treated but inorganics can pass through MBR membranes.

Authors decided an optimal cleaning strategy after these experiments such that frequent acid cleaning with a less frequent alkaline cleaning was proposed. They also suggested to use a pre-treatment for MBR permeate with UF membranes. Another study performed[30] with same MBR permeate by using TM720D RO and TM620 NF membranes pinpoints the importance of pre-treatment. Prior to NF and RO processes, sequential treatment steps of sand filter, activated carbon filter and UF were applied to the MBR permeate. Even though the total operation time of membranes were over 200 h, only around 5% of flux drop was observed in NF and RO membranes. It was seen that UF lost more than 60% of its initial flux even with six cleaning in 200 h which suggests that UF membranes should be replaced. However, as UF membranes are much cheaper than NF and RO membranes, it seems a more feasible approach.

It must also be noted that both pre-treatment requirements and cleaning strategies depend on the properties of feed water and membrane type. While the procedure performed by authors can be used to develop a mathematical model for anyone else, it cannot be taken as a directly applicable method.

In addition to membrane fouling issue, another important problem is the concentrate management. Membrane processes are not purification processes but separation processes which means the separated stream includes a cleaner, desired stream and the highly concentrated, undesired stream. These streams are generally not directly dischargeable and requires certain considerations before a membrane process is designed. Although cleaner concentrate streams can be discharged to oceans is possible, as NF and especially RO concentrate streams are more concentrated than MBR permeate, they must be dealt with before discharge. Another approach is to leaving NF/RO concentrates to natural evaporation although it would require a very long time and area if the volume is high.

Valuable compounds can be recovered while the concentrate amount can be reduced via additional processes however these additional steps also increase the cost of the process and hard to scale-up.

Concentrate recirculation and membrane-in-series configurations are viable methods to deal with concentrate streams. In concentration recirculation, some part of the concentrate streams is sent back to the feed tank to reduce the concentrate amount to be discharged. Recirculation cuts the concentrate amount significantly and permeate quality is easy to control, also requires lower investment costs. However, it cannot be used in long continous operations and increases membrane fouling in long batch times. A study performed on this method by using MBR permeate as the feed solution used BW30 (RO) membrane.[44] The authors recirculated 25%, 35%, 45% and 60% of the concentrate back to the feed tank in 4 different experiments. Table 3.24 shows some properties of permeates obtained in each case.

Table 3.24 shows that the permeate quality was unaffected by the increase in recirculation ratio. It must be noted that in longer operations, increasing salinity (conductivity in the feed tank was increased from about 4 mS/cm to 6.8 mS/cm in 3 h) in the feed tank would increase the fouling of membrane and hence, decrease the

Table 3.24. Characterization of average permeates in concentration recirculation. (Reproduced with permission from Ref. 44. Copyright of Elsevier.)

Percent of concentrate sent back	25%	35%	45%	60%
Na^+ (mg/L)	27.9	28.4	8.49	12.3
Ca^{2+} (mg/L)	2.70	1.18	2.06	2.45
K^+ (mg/L)	21.0	23.1	10.4	12.5
Mg^{2+} (mg/L)	<0.5	<0.5	<0.5	<0.5
SO_4^{2-} (mg/L)	<2	<2	<2	<2
Cl^- (mg/L)	21.8	26.3	8.48	19.8
Hardness (mg/L $CaCO_3$)	<2	<2	<2	<2
Conductivity (μS/cm)	179	205	80.5	112
TDS (mg/L)	85.4	97.7	38.9	52.6

Table 3.25. Minimum and maximum properties of feed solutions in concentration recirculation. (Reproduced with permission from Ref. 44. Copyright of Elsevier.)

Percent of concentrate sent back to feed	25%	35%	45%	60%
Na^+ (mg/L)	608–717	608–1019	564–870	599–1020
Ca^{2+} (mg/L)	82.8–107	85.0–127.4	81.9–120	114–161
K^+ (mg/L)	436–577	386–593	392–566	326–519
Mg^{2+} (mg/L)	37.8–54.6	43.6–71.5	41.8–65.2	38.5–65.1
SO_4^{2-} (mg/L)	521–703	822–898	521–804	579–976
Cl^- (mg/L)	762–1068	718–1186	665–1071	772–1240
Hardness (mg/L $CaCO_3$)	374–580	502–726	352–676	445–879
Conductivity (mS/cm)	4.20–5.79	4.27–6.42	3.82–6.04	3.94–6.84
TDS (g/L)	2.18–2.63	2.21–3.42	1.98–3.20	2.05–3.65

permeate quality. Therefore, the operation must be stopped when a certain feed salinity value is reached and new feed must be used to continue. This can be understood better by Table 3.25, showing minimum and maximum values of some parameters of feed solution in each case.

As opposed to this batch type of operation, a continous method is to use membrane-in-series in order to reduce the amount of concentrate. This method can also drastically decrease the amount of concentrate and can produce a high quality permeate. It can be also modified with an interstage pump or permeate throttling to increase the driving force in the consecutive membranes or can be used in a pyramidal design where concentrate stream of more than one membrane would feed the later membranes (with 2:1 feeder-taker ratio) which increases flexibility and ease to control. Only setback of this process is the increased fouling in the later stages which makes more than 3–4 stages not feasible. Hacifazlioglu *et al.*[33] used this method to treat MBR permeates by using NF90-NF90 and BW30-BW30 (RO) pairs. Water recovery in this method was between 81.5–81.6% which is much higher compared to around 50–60% water recovery obtained in single membrane operation. Table 3.26 shows feed and permeate compositions obtained in that study.

Table 3.26. Average feed and permeate compositions obtained in membrane-in-series configuration. (Reproduced with permission from Ref. 33. Copyright of Elsevier.)

Parameter	Unit	Average feed	BW30–BW30 permeate	NF90–NF90 permeate
Conductivity	mS/cm	4.10	0.1	0.2
TDS	g/L	1.40	<0.1	0.1
NO_3-N	mg/L	36.2	3.8	8.7
Total N	mg/L	37.9	5.1	10.9
HCO_3^-	mg/L	15.7	2.7	1.9
Cl^-	mg/L	493	13.4	29.6
Na^+	mg/L	419	21.0	35.2
TSS	mg/L	1.40	0	0

As reported by Hacıfazlıoğlu *et al.*,[33] a high permeate quality was obtained with 81.6% of water recovery. However, performance of later stages must be observed carefully to detect critical fouling levels if high permeate quality is required.

3.5. Conclusions

MBR systems employed in wastewater treatment plants seem to be good alternatives for reuse of treated wastewater as they provided a good quality of product water. However, decreasing the salinity of product water is needed in some cases if the MBR effluent will be recovered and reused for irrigation and/or as process water. For this, advanced membrane treatment processes such as RO and NF should be installed for post-treatment of MBR effluent. These pressure driven membrane processes can produce water of reliably high quality. On the other hand, a proper strategy should be developed for concentrate reduction in NF and RO processes. One of the approaches is applying membrane-in-series configuration. Another way for concentrate management could be concentrate recirculation which allows to reduce the brine generation. In this method, a high system water recovery and concentrate retention was achieved with the additional advantage of a reduced membrane water recovery, which minimizes the fouling effects.

Acknowledgement

The authors would like to thank Professor Yoram Cohen for the kind invitation to write this chapter and their co-workers cited in various references for their great collaborations.

N. Kabay acknowledges especially Ministry of Science, Industry, and Technology, Republic of Turkey (SANTEZ Project Number: 03.STZ.2013-02) and TUBITAK (Project No. 114Y500) for their financial supports to conduct researches on wastewater reclamation and reuse and giving scholarship to our postgraduate students.

We sincerely thank to the people at Wastewater Treatment Plant of ITOB Organized Industrial Zone, Menderes-Izmir, Turkey for their great support to conduct our pilot membrane tests in wastewater treatment plant.

References

1. United Nations, Managing water under uncertainty and risk, The United Nations World Water Development Report 4 (2012).
2. S. Burn, M. Hoang, D. Zarzo, F. Olewniak, E. Campos, B. Bolto, O. Barron, Desalination techniques — A review of the opportunities for desalination in agriculture, *Desalination*, **364** (2015) 2–16.
3. C. A. Quist-Jensen, F. Macedonio, E. Drioli, Membrane technology for water production in agriculture: Desalination and Wastewater Reuse, *Desalination*, **364** (2015) 17–32.
4. M. Rebhun, Desalination of reclaimed wastewater to prevent salinization of soils and groundwater, *Desalination*, **160** (2004) 143–149.
5. D. W. Westcot, R. S. Ayers, *Irrigation with Reclaimed Municipal Wastewater: A Guidance Manual.* Lewis Publishers, Inc., Chelsea, MI, (1985).
6. D. Fatta-Kassinos, I. K. Kalavrouziotis, P. H. Koukoulakis, M. I. Vasquez, The risks associated with wastewater reuse and xenobiotics in the agroecological environment, *Sci. Total Environ.*, **409**(19) (2011) 3555–3563.
7. B. K. Pramanik, F. A. Roddick, L. Fan, S. Jeong, S. Vigneswaran, Assessment of biological activated carbon treatment to control membrane fouling in reverse osmosis of secondary effluent forreuse in irrigation, *Desalination*, **364** (2015) 90–95.
8. Y. Dreizin, Ashkelon seawater desalination project — off-taker's self costs, supplied water costs, total costs and benefits, *Desalination*, **190** (2006) 104–116.

9. C. Garcia, F. Molina, D. Zarzo, 7 year operation of a BWRO plant with raw water from a coastal aquifer for agricultural irrigation, *Desalin. Water Treat.*, **31** (2011) 331–338.
10. M. S. Mohsen, Water strategies and potential of desalination in Jordan, *Desalination*, **203** (2007) 27–46.
11. K. Chon, S. Sarp, S. Lee, J. A. Lopez-Ramirez, J. Cho, Evaluation of a membrane bioreactor and nanofiltration for municipal wastewater reclamation: Trace contaminant control and fouling mitigation, *Desalination*, **272**(1–3) (2011) 128–134.
12. K. Chon, J. Cho, H. K. Shon, K. Chon, Advanced characterization of organic foulants of ultrafiltration and reverse osmosis from water reclamation, *Desalination*, **301** (2012) 59–66.
13. S. Bunani, E. Yörükoğlu, G. Sert, Ü. Yüksel, M. Yüksel, N. Kabay, Application of nanofiltration for reuse of municipal wastewater and quality analysis of product water, *Desalination*, **315** (2013) 33–36.
14. S. Bunani, E. Yörükoğlu, Ü. Yüksel, N. Kabay, M. Yüksel, G. Sert, Application of reverse osmosis for reuse of secondary treated urban wastewater in agricultural irrigation, *Desalination*, **364** (2015) 68–74.
15. S. Shanmuganathan, S. Vigneswaran, T. V. Nguyen, P. Loganathan, J. Kandasamy, Use of nanofiltration and reverse osmosis in reclaiming micro-filtered biologically treated sewage effluent for irrigation, *Desalination*, **364** (2015) 119–125.
16. P. Cartagena, M. El Kaddouri, V. Cases, A. Trapote, D. Prats, Reduction of emerging micropollutants, organic matter, nutrients and salinity from real wastewater by combined MBR-NF/RO treatment, *Sep. Purif. Technol.*, **110** (2013) 132–143.
17. D. De Jager, M. S. Sheldon, W. Edwards, Colour removal from textile wastewater using a pilot-scale dual-stage MBR and subsequent RO system, *Sep. Purif. Technol.*, **135** (2014) 135–144.
18. G. Sert, S. Bunani, N. Kabay, E. Yörükoğlu, Ö. Egemen, M. Yüksel, Performances of some NF and RO membranes for desalination of MBR treated wastewater, *J. Water Process. Eng.*, **16** (2017) 193–198.
19. G. Sert, S. Bunani, N. Kabay, Ö. Egemen, M. Arda, T. Ö. Pek, M. Yüksel, Investigation of mini pilot scale MBR-NF and MBR-RO integrated systems performance — preliminary field tests, *J. Water Process. Eng.*, **12** (2016) 72–77.
20. M. Gündoğdu, Y. A. Jarma, N. Kabay, T. Ö. Pek, M. Yüksel, Integration of MBR with NF/RO Processes for Industrial Wastewater Reclamation and Water Reuse-Effect of membrane type on product water quality, *J. Water Process. Eng.*, **29** (2019) 100574.

21. M. C. Hacifazlioglu, H. R. Tomasini, N. Kabay, L. Bertin, T. O. Pek, M. Kitis, N. Yigit, M. Yuksel, Effect of pressure on desalination of MBR effluents with high salinity by using NF and RO processes for reuse in irrigation, *J. Water Process. Eng.*, **25** (2018) 22–27.
22. F. Pedrero, I. Kalavrouziotis, J. J. Alarcon, P. Koukoulakis, T. Asano, Use of treated municipal wastewater in irrigated agriculture review of some practices in Spain and Greece, *Agric. Water Manag.*, **97**(9) (2010) 1233–1241.
23. D. Norton-Brandao, S. Scherrenberg, J. B. van Lier, Reclamation of used urban waters for irrigation purposes — A review of treatment technologies, *J. Environ. Manage.*, **122** (2013) 85–98.
24. United States Environmental Protection Agency (EPA), Tables 2-7. Recommended limits for constituents in reclaimed water for irrigation, Tables 4-13 Suggested guidelines for water reuse, types of reuse: urban reuse (2004).
25. United Nations Food and Agricultural Organization (FAO), Table 8: Recommended microbiological quality guidelines for wastewater use in agriculture. Table 9: Guidelines for interpretation of water quality for irrigation. Table 27: Water quality and clogging potential in drip irrigation systems (1992).
26. World Health Organization (WHO), Table A 1.1. Water Quality in Irrigation (2016).
27. Ministry of Environment and Urbanization, Republic of Turkey, Technical procedure declaration for wastewater treatment plants, (2010) 91–93.
28. C. J. Smith, J. D. Oster, G. Sposito, Potassium and magnesium in irrigation water quality assessment, *Agric. Water Manag.*, **157** (2015) 59–64.
29. S. Babel, T. A. Kurniawan, Cr(VI) removal from synthetic wastewater using coconut shell charcoal and commercial activated carbon modified with oxidizing agents and/or chitosan, *Chemospere*, **54** (2004) 951–967.
30. N. J. Falizi, M. C. Hacifazlioglu, I. Parlar, N. Kabay, T. O. Pek, M. Yuksel, Evaluation of MBR treated industrial wastewater quality before and after desalination by NF and RO processes for agricultural reuse, *J. Water Process. Eng.*, **22** (2018) 103–108.
31. I. Parlar, M. Hacifazlioglu, N. Kabay, T. Ö. Pek, M. Yüksel, Performance comparison of reverse osmosis (RO) with integrated nanofiltration (NF) and reverse osmosis process for desalination of MBR effluent, *J. Water Process. Eng.*, **29** (2019) 100640.

32. M. C. Hacifazlioğlu, Post-treatment of MBR treated wastewater with high salinity by means of NF/RO processes for wastewater reuse in irrigation-pilot-scale tests at ITOB wastewater treatment plant, MS Thesis, Ege University, Izmir (2018).
33. M. C. Hacıfazlıoğlu, H. R. Tomasini, L. Bertin, T. O. Pek, N. Kabay, Concentrate reduction in NF and RO desalination systems by membrane-in-series configurations — Evaluation of product water for reuse in irrigation, *Desalination*, **466** (2019a) 89–96.
34. M. B. Capangpangan, I. H. Suffet, "Optimization of Drying Methods for Filtered Suspended Solids from Natural Waters for Supercritical Fluid Extraction Analysis of Hydrophobic Organic Compounds", *J. Chromatogr.*, **738**(2) (1996) 253–264.
35. N. J. Falizi, Utilization of wastewater treated by membrane bioreactor (MBR) process in safflower cultivation and investigation of biodiesel production from safflower harvested, PhD Thesis, Ege University, Izmir (2019).
36. J. J. Klemes, Industrial water recycle/reuse, *Curr. Opin. Chem. Eng.*, **1**(3) (2012) 238–245.
37. D. Bixio, C. Thoeye, J. De Koning, D. Joksimovic, D. Savic, T. Wintgens, T. Melin, Wastewater reuse in Europe, *Desalination*, **187** (2006) 89–101.
38. J. Hoinkins, S. A. Deowan, V. Panten, A. Figoli, R. R. Hunag, E. Drioli, Membrane Bioreactor (MBR) technology — A promising approach for industrial water reuse, *Procedia Eng.*, **33** (2012) 234–241.
39. T. Melin, B. Jefferson, D. Bixio, C. Thyoeye, W. D. Wilde, J. D. Koning, J. van der Graff, T. Wintgens, Membrane bioreactor technology for wastewater treatment and reuse, *Desalination*, **187** (2006) 271–282.
40. S. Judd, B. Jefferson, *Membranes for Industrial Wastewater Recovery and Reuse*, Elsevier Science Ltd., Oxford OX5 lGB, UK (2003).
41. I. Parlar, Pilot tests with MBR system for wastewater treatment and production of process water from wastewater with integrated MBR-NF, MBR-RO processes, MS Thesis, Ege University, Izmir, (2018).
42. L. H. Andrade, F. D. S. Mendes, J. C. Espindola, M. C. S. Amara, Nanofiltration as tertiary treatment for the reuse of dairy wastewater treated by membrane bioreactor, Department of Sanitary and Environmental Engineering, Federal University of Minas Gerais, (2014).
43. M. C. Hacifazlioglu, İ. Parlar, T. O. Pek, N. Kabay, Evaluation of chemical cleaning to control fouling on nanofiltration and reverse osmosis membranes after desalination of MBR effluent, *Desalination*, **466** (2019b) 44–51.

44. H. R. Tomasini, M. C. Hacifazlioglu, N. Kabay, L. Bertin, T. O. Pek, M. Yuksel, Concentrate management for integrated MBR-RO process for wastewater reclamation and reuse-preliminary tests, *J. Water Process. Eng.*, **29** (2019) 100455.
45. T. Asano, F. L. Burton, H. L. Leverenz, R. Tsuchihashi, G. Tchobanoglous, *Water Reuse: Issues, Technologies and Applications*, 1st Ed., McGraw-Hill (2007).

CHAPTER 4

Polyelectrolyte Coagulants and Flocculants in Wastewater Treatment: A Fundamental Perspective

Divya Jayaram Iyer, Advait Holkar and Samanvaya Srivastava*

Department of Chemical and Biomolecular Engineering, University of California, Los Angeles, Los Angeles, CA 90095, USA

4.1. Introduction

Water in its clean, usable form, is a dwindling resource.[1] In 2015, WHO estimated that 663 million people across the world did not have access to clean potable water.[2] The global demand for water is projected to further rise by 20–30% by 2050,[3] depriving an even more significant population of access to clean water.[4] Water sanitation and hygiene have therefore been identified as a key public health issue by the United Nations and is the main focus of one of their sustainable development goals.[5] Governing and regulatory bodies across the globe have consistently emphasized on the need to recycle wastewater to improve water reuse while reinforcing increasingly stricter water quality norms, to alleviate the contamination of existing water sources.[6] Development and implementation of effective and energy-efficient technologies for wastewater treatment and recycling have consequently become one of the leading priorities of the US Department of Energy.[7]

*Corresponding author: samsri@ucla.edu

Over the years, research directed towards water purification technologies[8–13] has garnered widespread attention of scientists, given its indispensable role in the survival of various forms of life. Feasibility and industrial scalability of many of these processes are continually improving, and considerable efforts are being made towards charting out a cost-effective and technologically-driven scheme for wastewater treatment.[14]

Wastewater contains many impurities, spanning a multitude of sizes. The impurities can be broadly classified as colloids (5 nm–0.1 mm) and solutes (<1 nm) that are stable in water owing to the combined influence of van der Waals interactions, electrostatic interactions, steric repulsive and thermal forces.[6,15] Colloidal impurities are removed through technologies like ultrafiltration, adsorption and physical straining, while the removal of solutes is typically achieved through modern technologies like nanofiltration and reverse osmosis. The efficacy of these separation processes is significantly improved by coagulation and flocculation of solute and colloidal particles. Therefore, coagulation and flocculation steps have emerged as integral parts of the wastewater treatment process. Coagulation and flocculation aid in removing natural organic matter,[16,17] pathogens[18] and other suspended solids, and also reduce turbidity and improve color. Furthermore, coagulation and flocculation prevent clogging of equipment or fouling of membranes.[19–23] Both these processes are employed as parts of water pre-treatment (to prepare wastewater for effective filtration), post-treatment (to enhance the treated water quality) or as the primary treatment process (to remove a majority of the impurities).[6,12,13,24,25]

Coagulation is the process of destabilization of impurities; flocculation brings about their agglomeration to form large flocs and aggregates.[6,24] These flocs settle and are removed from the wastewater as sludge during further treatment steps involving screening and sedimentation, brought about by grit chambers and clarifiers. Further, treatment involving gravity sand filtration and ultraviolet disinfection is employed to improve water quality in order to meet regulatory or recycling requirements.[14,26,27] Coagulation and flocculation technologies have evolved over the last few decades to cater to the varying water quality in various parts of the world. Yet, further improvements in coagulation and flocculation technology, to maximize treatment efficiency, minimize sludge volumes and enable industrial scale adoption, require a concerted effort to understand the fundamental science behind these processes and the roles of conventional, natural and modern synthetic coagulant and flocculant additives in wastewater treatment. In this chapter, we review the intermolecular interactions that stabilize colloidal impurities in water and elucidate the mechanisms through which polymeric coagulants and flocculants can destabilize and agglomerate colloidal impurities, along with a few examples of the use of polymeric coagulants and flocculants in water treatment processes. This chapter is organized as follows: in Section 4.2, we describe the relevance of colloidal and intermolecular interactions in coagulation and flocculation processes and the factors that influence their effectiveness. In Section 4.3, we discuss the use of natural and synthetic polyelectrolytes as coagulants and flocculants, backed by a fundamental perspective on their interactions with colloids. The chapter concludes with a discussion on recent advances made in this field.

4.2. Coagulation & Flocculation in Water Treatment

The use of naturally-sourced coagulants and flocculants, often referred to as precious 'gems' in folklore, can be traced in ancient literature. Kataka (*Strychnos potatorum*) seeds, roots of lotus, pearls, and rock crystal alum have all been mentioned in various ancient scriptures to possess water-clarifying abilities.[28] Unsurprisingly, Kataka seeds are still used in rural regions of India and southeast Asia. Egyptian civilizations were known to use almonds for coagulation.[29] Although the use of alum as a medical aid finds mention in the Indian medical text 'Susruta' (first century A.D.), the earliest mention of its use as a water clarifying agent appears in a Chinese encyclopedia from the 17th century.[28] In modern times, scientific advances have led to an enhanced understanding of the mechanisms behind coagulation and have contributed to the development of improved coagulants and flocculants. For example, the coagulating action of pearls, used in the ancient times for water clarification, can be ascribed to weakly coagulating Ca^{2+} ions produced upon mixing crushed pearls (comprising 83–92% $CaCO_3$) with weak acids.[28] Over the years, it was ascertained that gemstones with weakly bound trivalent ions like Fe^{3+} and Al^{3+} brought about better clarifying performance than pearls, owing to their propensity to hydrolyze and initiate coagulation.[28] By mid-18th century, aluminum sulfate (alum) was being used in treatment of water in England.[30] Additionally other aluminum and iron-based coagulants like polyaluminum chloride,[10,31–34] aluminum chloride,[35,36] ferric chloride,[37–39] polyferric sulfate,[13,40] amongst others, also find applications as coagulants. Alum remains one of the most widely used coagulants owing to its economic and abundant nature.[34,41] However, it has shortcomings like high sludge production, persistent aluminum levels in treated water[41,42] and fragility of flocs that limit its industrial-scale use.[34]

4.2.1. *Colloidal Interparticle Interactions*

An understanding of interactions that (de)stabilize colloidal particles in suspension is integral to the development of next generation of coagulants and flocculants. Typical mineral and organic colloidal particles possess charges on their surfaces owing to ionizable surface functional groups, lattice imperfections or adsorption of other ions present in their surroundings.[43] Particle stability against aggregation can therefore be quantified as the interplay between repulsive (screened) electrostatic forces and attractive van der Waals forces in the framework of the Derjaguin-Landau-Verwey-Overbeek (DLVO) theory.[15,44]

A charged surface in water attracts oppositely charged ions (counterions) and repels similarly charged ions (co-ions). This leads to the formation of an electrical double layer and a net reduction in the electrical potential of the particle. The formation of double layer around charged colloids has been explained by models like the Gouy-Chapman and the Gouy-Chapman-Stern model, amongst others.[15,45,46] The Gouy-Chapman model assumed the ions to be point charges, which led to the notion that there were no physical barriers to an ion approaching the particle surface.[45] This shortcoming was

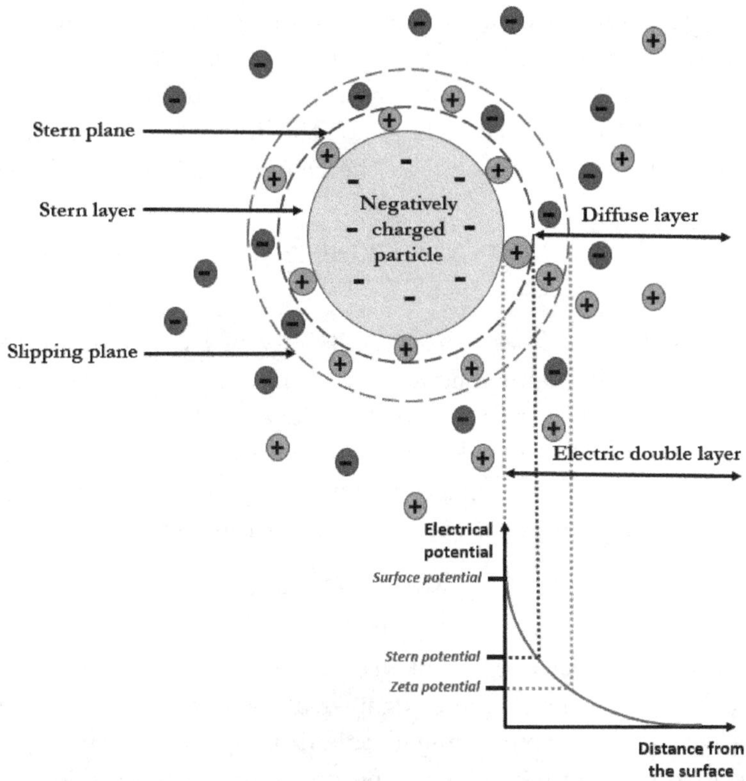

Fig. 4.1. Schematic depicting the electric double layer, consisting of the surface charge, bound counterions (Stern layer) and free ions (diffuse layer) around a negatively charged spherical particle. The Stern plane lies at the interface of the Stern and diffuse layers, while the interface between the latter and dispersion medium is called the slipping plane. (Adapted from Park *et al.*[47]).

accounted for in Stern's modification of the model, which attributes a finite size to the ions, thus stating that the first layer of ions is restricted to approach within a distance δ (of the order of a few Å) from the particle surface, referred to as the Stern layer.[45] Stern also accounted for possible adsorption of counterions on the charged surface, leading to a partial neutralization of the surface charges within the Stern layer. The Stern layer is followed by the diffuse layer, a region in which the concentration of the counterions continually decreases with distance from the surface, as does the electric potential

(Fig. 4.1).[47] The electric potentials corresponding to the surface charge, Stern layer and diffuse layer are called surface, Stern and zeta potentials, respectively. As shown in the illustration in Fig. 4.1, the characteristic distance up to which the diffuse layer and therefore, the electric field, persist is called the Debye length κ^{-1}, defined as:

$$\kappa^{-1} = \left(\sum_i \frac{\rho_{\infty i} e^2 z_i^2}{\epsilon_0 \epsilon k_B T} \right)^{-\frac{1}{2}}$$

Here, ϵ, ϵ_0, k_B and T are the permittivity of free space and of the medium, the Boltzmann constant and temperature, respectively. $\rho_{\infty i}$ and z_i are the concentration and the valency of the ith ionic moieties in solution. The screened electric potential U_e as a function of x, the distance away from the charged surface of a sphere of radius R is thus described in terms of the Debye length as:

$$U_e = \frac{64\pi kTR\gamma^2}{\kappa^2} e^{-2\kappa x}$$

Here, $\gamma = \tanh(ze\psi_o/4kT)$, with ψ_o being the surface potential. The Coulombic repulsion competes against attractive long-range van der Waals interactions comprising dipole-dipole, dipole-induced dipole and London dispersion forces. The interparticle interaction energy (U_{vdW}) between two spherical particles of radii R_1 and R_2 can thus be described as an inverse relation with the distance between the particle surfaces d as:

$$U_{vdW} = -\frac{AR_1R_2}{6(R_1 + R_2)d}$$

Here, A is the Hamaker constant, typically of the order of 10^{-19} J for interactions in condensed systems as seen in vacuum.[15] The net interparticle interaction potential $U(x)$, for similarly sized particles of radius R, can thus be described as:

$$U(x) = U_e + U_{vdW} = \frac{64\pi k_B TR\rho_\infty \gamma^2}{\kappa^2} e^{-2\kappa x} - \frac{AR}{12x}$$

Consequently, the rate of aggregation of particles can be estimated based on a rate constant (k_{agg}), which is related to the barrier potential U_{max}, defined as the maximum value of the net interparticle interaction potential $U(x)$ and viscosity of the suspending medium (η_0) as[48,49]:

$$k_{agg} \propto \left(\frac{k_B T}{\eta_0}\right) \exp\left(-\frac{U_{\text{max}}}{k_B T}\right)$$

Figure 4.2 depicts interaction potentials between two flat surfaces as a function of normalized inter-surface distance with varying surface charge density; interparticle interaction potentials between two spherical particles also follow a similar evolution with surface charge density.[15] Strong repulsive energy barriers between particles with high surface charge density inhibit their agglomeration, while weakly charged particles present long-range interparticle attraction and agglomerate readily in suspension. For typical charged colloids, the

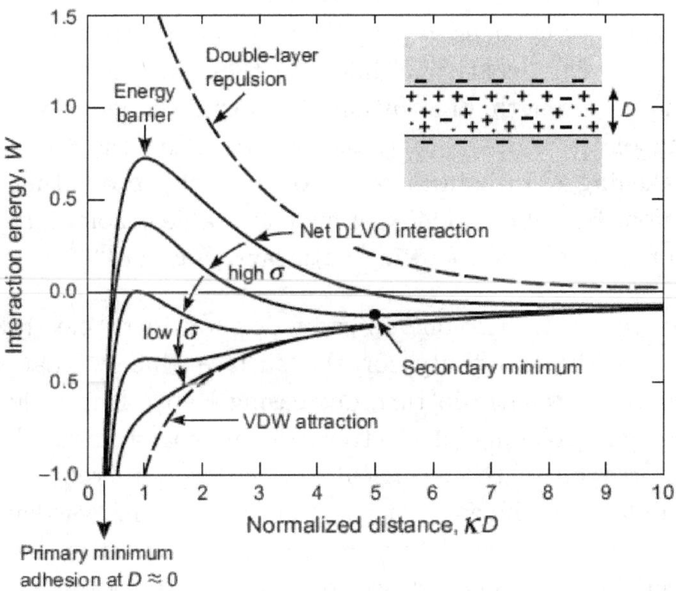

Fig. 4.2. A schematic of the change in interaction energy (W) with distance between the species (D), at varying surface charge density (σ). The double-layer curve represents a stable system which approaches instability due to change in σ, ultimately leading to an unstable system, involving only van der Waals (VDW) attraction. The secondary minimum represents a kinetically stable system. (Adapted from Israelachvili.[15])

repulsive energy barrier is often too high for the particles to overcome readily; the particles either remain at distances corresponding to the secondary minimum or farther away wherein they are completely dispersed thereby forming a stable system.[15]

4.2.2. *Mechanisms for Coagulation and Flocculation*

Coagulation and flocculation of colloidal particles can be induced by weakening repulsive interactions or by introducing additional attractive interparticle interactions. While the former can be brought about by reducing the surface charge density of particles, or by restricting the length scales over which the repulsive interactions are effective (by increasing the solution ionic strength to reduce the Debye length), the latter can be mediated by multivalent macroions or polyelectrolytes that can bridge particles.[33,42,50–52] The van der Waals interactions are relatively unaffected by electrolyte concentration or changes in pH. Repulsive interparticle interactions, however, are significantly modified by these factors.[15] An increase in the ionic strength of solution, for instance, leads to stronger screening of the surface charges, thus reducing the Debye length and weakening electrostatic repulsion, thereby promoting particle aggregation. Destabilization of previously stable colloidal dispersions can occur through the following pathways (Fig. 4.3)[53,54]:

(i) Charge neutralization: Introduction of inert electrolytes or salts, with no affinity for the surface charge, that increase the ionic strength, in turn decreasing Debye length, leading to stronger screening of electrostatic interparticle repulsion and compression of the electrical double layer, thereby promoting coagulation. Since the impurities present in wastewater are generally negatively charged, charge neutralization is brought about by positively charged ions of coagulant species.
(ii) Sweep flocculation: When coagulants and flocculants (particularly aluminum and ferric-based) are dosed beyond pH ranges or amounts conducive to their solubility, hydroxide precipitates form, which then act as traps for impurities.[33] Since the impurities are electrostatically attached to the precipitates,

MECHANISM	DESCRIPTION	ILLUSTRATION
Charge Neutralization	Reduction of Double Layer thickness and charge neutralization	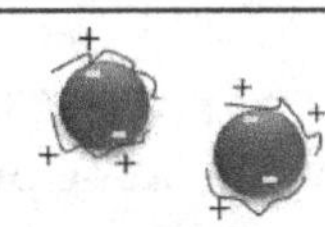
Sweeping Flocculation	Small colloidal pollutants are trapped in the mesh of polymeric precipitates	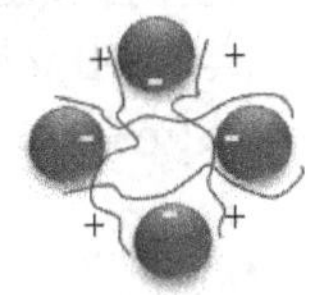
Bridging and Patch Flocculation	Primary flocs are adsorbed and connected via large MW flocculants	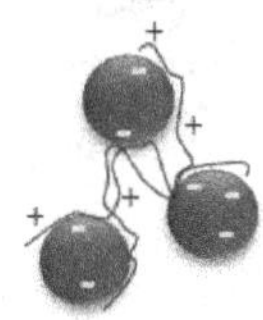

Fig. 4.3. Different mechanisms of coagulation and flocculation. (Adapted from Suopajärvi[53] and Yang *et al.*[54]).

sweep flocculation can be considered a form of charge neutralization.[55]

(iii) Bridging and Patch Flocculation: Both these processes entail adsorption of polymer on particle surface.[56–58] In patch flocculation, low molecular weight polymers are adsorbed on particle surfaces, thereby creating charged patches on particle surfaces. In bridge flocculation, adsorption of high molecular weight polymers on multiple particles results in agglomeration of particles. Patch flocculation is significant for polymers with molecular weights between 10^5 and 2×10^6 g/mol; polymers with molecular weights above that range participate in bridging.[59]

The mechanism through which coagulation and flocculation proceed in wastewater treatment often depends on the nature of the impurities present as well as the coagulant being used. For example, coagulation using natural agents is typically brought about by adsorption of ions leading to charge neutralization. Naturally occurring polyelectrolytes induce bridging by binding particles, causing them to aggregate and thus aiding flocculation.[50] Aluminum and iron salts induce charge neutralization upon adsorption of the multivalent cations on the charged particle surfaces. At high coagulant dosages, sweep flocculation, which involves encapsulation and precipitation of dispersed particles, dominates.[14] For instance, in a system with kaolin impurities, at a pH of 7, a reduction in suspension turbidity at 15 μM aluminum sulfate dosage was attributed to charge neutralization, while sweep flocculation was said to play a role beyond 60 μM dosage, due to hydrolysis and precipitation of metal coagulant species.[33]

Bridging and patch flocculation are particularly important in polyelectrolyte-based systems,[42] with the surface coverage of the colloids by the polyelectrolytes, in turn influenced by polyelectrolyte dosage, being a key factor in influencing the bridging efficiency.[60] Very high coverage can reinstate colloidal stability following charge reversal and a lack of adsorption sites by bridging polyelectrolytes while too low a coverage can lead to insufficient bridging contacts. It is, therefore, important to achieve an optimum dosage.[42] Bridging in polyelectrolyte systems also leads to stronger flocs as compared to metal coagulants, which in turn can result in formation of larger flocs.[61] Flocs formed via patch flocculation have strengths intermediate to those formed by simple charge neutralization and those formed by bridging.[42]

4.2.3. *Factors Influencing Coagulation and Flocculation*

4.2.3.1. *Solution pH*

pH of the solution influences the effectiveness of coagulation and flocculation processes.[24,31,35,62–68] pH variations affect the speciation of surface groups on particulate impurities, thereby affecting suspension stability. Additionally, solubility and performance of coagulants are also influenced by variations in pH.[13,65] pH also affects the process of flocculation, particularly the size and strength of floc.[40] Thus, many coagulants are supplied with a recommended pH range to achieve optimal performance.

Alum, for instance, is known to be very effective within a pH range of 6–8.[36,64–66] Similarly, in the use of polyferric chloride (PFC) to treat water with low dissolved organic carbon (1.86–3.71 mg/L), high alkalinity (245.1–262.7 mg $CaCO_3$/L) and pH of 8.15–8.4, it was observed that adjusting the pH to neutral values was conducive for attaining larger floc size and greater rate of growth as compared to other pH values, at a polyferric chloride dosage of 0.15 mmol/L.[64] When pH was increased from 4 to 6.25, a decrease in the residual turbidity was observed, which increased slightly when the pH rose to 9. This behavior was attributed to restrained hydrolysis of the PFC in acidic conditions, thereby causing charge neutralization to dominate the process.[64,68] The persistent low residual turbidity at alkaline pH was inferred to be the result of adsorption and sweep flocculation.[64] A slightly different trend was observed in case of dissolved organic carbon (DOC) removal by polyferric chloride, with the highest removal (47.34%) being observed at 5.5 pH. DOC removal, too, is known to depend on an interplay of charge neutralization, coagulant and organic matter speciation, in addition to adsorption and sweep flocculation.[16,39] The effective DOC removal at lower pH can be attributed to charge neutralization and ferric-organic complexation[40] while the drop in DOC removal, despite good hydrolysis of PFC at higher pH, can be attributed to interference of OH^- ions, thereby impeding coagulant-organic matter interaction.[64] This effect was seen to be an interplay of two phenomena: at lower pH values, formation of microflocs is bolstered by collision and aggregation of particles caused due to charge neutralization being at the core of the coagulation process, whereas, at higher pH, PFC is prone to hydrolysis, with sweep coagulation and adsorption dominating the process.[64]

The effect of pH on coagulation and flocculation efficacy was highlighted by a comparison of the performances of polyaluminum chloride (PAC), aluminum chloride ($AlCl_3$), and aluminum sulfate ($Al_2(SO_4)_3$) in the treatment of humic acid-kaolin containing wastewater in different initial pH ranges (acidic, neutral and alkaline).[36] Turbidity decreased by more than 90% upon addition of the three coagulants at a dosage of 10mg/L at pH less than 6 but showed little improvement at higher pH in PAC & $AlCl_3$ based systems. The optimum pH range for $AlCl_3$ and $Al_2(SO_4)_3$ was established to be 6–7, while the pH range was broader for PAC (5–8).[36] This behavior was attributed to the fact that at lower pH (<5), the three Al-based coagulants hydrolyzed to form positive charged species like $Al(OH)^{2+}$, $Al(OH)_2^+$, $Al_2(OH)_2^{4+}$, and $Al_3(OH)_4^{5+}$, which induced charge neutralization of the negatively charged humic acid-kaolin particles, in addition to allowing their adsorption on the hydrolysates. At higher pH values (>8), the hydrolyzed species were negatively charged and thus ineffective as coagulants.[36]

The charge density of weak polyelectrolytes is dependent on the degree of ionization of the ionizable groups, and therefore, on the solution pH. Ester group hydrolysis and subsequent loss of positive charge in cationic polyacrylamides are known to be dependent on pH.[69] Similarly, charge density of anionic polyelectrolytes that commonly possess weakly acidic (low pK_a) carboxylic acid groups is also dependent on pH.[42] Typical polycationic coagulants like polydiallyldimethyl ammonium chloride (PDADMAC), however, comprise strongly basic quaternary ammonium groups that remain ionized across a broad range of pH (high pK_a) and are therefore, not very sensitive to pH changes.[42]

4.2.3.2. *Dosage*

In commercial setups, cost is a great motivation for optimizing coagulant dosage. Coagulant dosage needs to be optimized to ensure effective coagulation and flocculation. Suboptimal coagulant dosage undermines the performance of downstream treatment processes involving membrane filtration. At the same time, excess coagulant dosage can lead to reversal of charge, causing re-stabilization of the colloidal impurities, thereby decreasing the coagulation and flocculation efficiency.[65] Furthermore, while addition of coagulant prevents fouling of membranes,[19–21,70] dosage beyond an optimum risks passage of coagulant into the permeate stream.[71] Since the dosage can vary significantly with the quality of feed water, the need for real-time adjustment of coagulant dosage is critical in water treatment setups.[71]

Blankert *et al.*[72] reported the use of an in-line coagulation system functioning through real-time dosage adjustment in dead-end ultrafiltration (UF) systems, by empirically correlating the UF filtration resistance post backwash (UF PB) to the filtered volume on the basis of pilot tests. The difference between the observed and empirically determined UF PB resistance values was then fed to a proportional-integral controller to enable dosage adjustment. However, because this approach was largely empirical, obtaining optimal dosage values remained a challenge. This drawback was allayed by the self-adaptive approach demonstrated by Gao *et al.*[70] wherein optimum coagulant dosage was determined based on real-time tracking of UF PB resistance post every cycle of dosage adjustment (Fig. 4.4).[71]

It must be noted that coagulants that operate via sweep floc-coagulation, particularly aluminum and ferric salts that precipitate as hydroxides at high dosages instead of exhibiting charge reversal, are effective even at excessive dosages, but suffer from poor flocculation efficiency, and hence require introduction of co-coagulants and flocculants.[33]

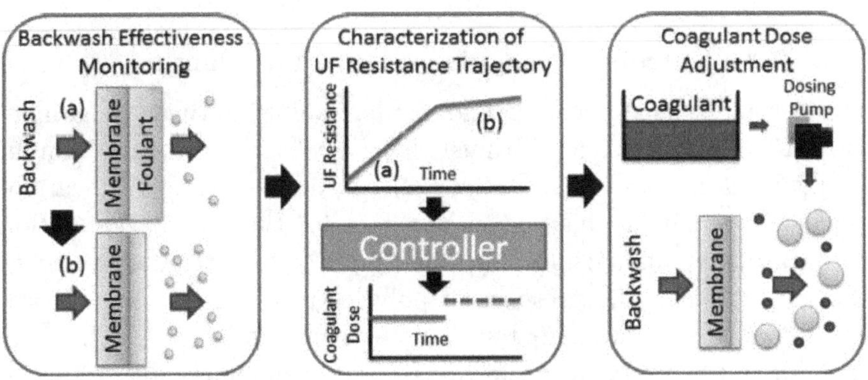

Fig. 4.4. A quantification of the backwash effectiveness and membrane fouling, by tracking the ultrafiltration resistance trajectory in real-time, is optimal to adapt and control coagulant dosage. Adapted from Gao *et al.*[71]

4.2.3.3. *Temperature*

Temperature of operation is a vital parameter because it not only affects water viscosity and chemistry and stability of coagulants/flocculants but also influences the rate of flocculation and the size and strength of the flocs, thereby impacting coagulation and flocculation efficiency. Additionally, temperature also influences the stability of impurities, thereby affecting the turbidity of wastewater, thus altering the intensity of coagulation and flocculation required.[73] Alum has been noted to become less effective at lower temperatures as compared to polyaluminum chloride because of reduced hydrolysis of the former.[13] At the same time, flocs formed from aluminum-based coagulants (alum and polyaluminum chloride) were reported to be more sensitive to temperature changes as compared to ferric sulfate flocs; this was attributed to formation of smaller, lower density flocs.[74] Higher than optimal temperatures can also adversely affect floc formation. Performance of alum in terms of total suspended solids removal from palm oil mill effluents dropped at temperatures greater than 20 °C, owing to an increase in kinetic energy of particles at higher temperatures impeding floc formation.[75] Furthermore, it was suggested that convection currents due to heating could also prevent settling of particles.[76,77]

4.3. Polyelectrolyte Coagulants and Flocculants

For centuries, cost-effective and biodegradable natural coagulants such as Moringa oleifera (drumstick), *Strychnos potatorum*, tannin, opuntia, chitin and sago have been used for water purification. Various mechanisms have been proposed for the coagulating action of natural coagulants; most of these hypotheses attribute the coagulating action to the presence of polyelectrolytes and charged proteins that adsorb on colloid surfaces and induce flocculation.[50,78,79] Yet, their usage is limited owing to low availability, difficulties in scalability, large sludge volumes, poor performance at lower temperatures and in adverse pH conditions, and a lack of suitability for industrial wastewater treatment. Scalability of natural coagulants, with production requiring multi-step processing, is one of the major hindrances that limit their widespread use.[79–82] These drawbacks are alleviated using synthetic polyelectrolytes (PEs), which can provide effective coagulation and flocculation even in extreme temperature and pH conditions; and can be designed to cater to the specific impurities present in wastewater stream from industrial effluents.

Polymers can interact with colloids in solution, leading to coagulation and flocculation of impurities.[83] In solution, simultaneous interaction of polymers with multiple colloidal particles also results in formation multi-colloid flocs, which can be further separated by sedimentation or membrane filtration. Typical impurities in natural water systems have an anionic surface charge[84] and hence cationic polyelectrolytes, or polycations, such as poly (diallyldimethyl ammonium chloride) (PDADMAC) and poly(dimethylamino)ethyl methacrylate (PDMAEMA) are primarily used in their treatment. Polycations possess several advantages over conventional coagulants and flocculants (like alum) — denser sludges are formed, lower dosing is required and reduced levels of mineral residue left in the purified water, leading to an overall treatment cost reduction by 25–30%.[42] Similarly, anionic PEs such as anionic polyacrylamides (APAM) are used for treatment of water with cationic impurities such as clay, and as scale inhibitors.[42,60] They are also utilized as polymer aids to cationic salts to form larger aggregate flocs which settle faster.[42] Substantial reduction in dosage requirements (40–60%) of alum is seen when APAM is used as a polymer aid.[85] Non-ionic polymers such as polyethylene oxide are also used as polymer aids as they are generally less toxic compared to polycations and are preferred in cases when biodegradability is a priority. Figure 4.5 summarizes the different polymers used for water treatment.[14]

PE characteristics, including charge density, degree of ionization, size and persistence length, all affect coagulation and flocculation efficiency. In the following subsections, we discuss the effect of these parameters on the structure as well as the phase behavior of polyelectrolyte-colloid systems from a fundamental perspective to enable the rational design of polyelectrolytes tailored towards specific

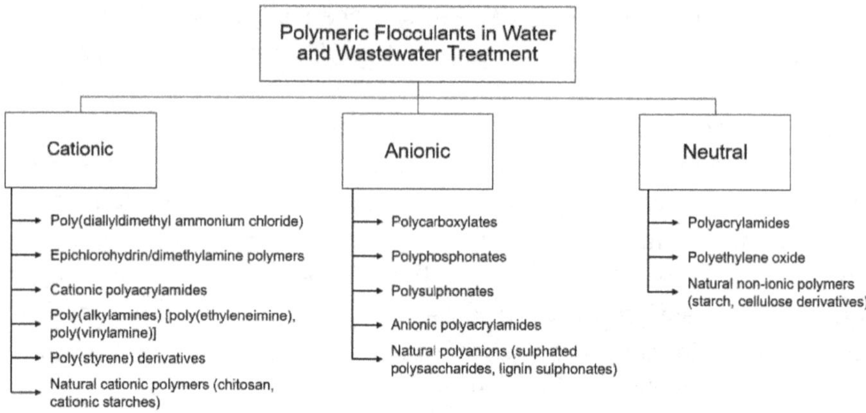

Fig. 4.5. Classification of some polymeric flocculants used in wastewater treatment. Adapted from Renault *et al.*[14]

requirements. Subsequently, we discuss representative examples of natural and synthetic PEs used for water treatment applications.

4.3.1. *Polyelectrolyte-colloid interactions: A fundamental perspective*

4.3.1.1. *Structure of polyelectrolyte-colloid mixtures*

Interactions between oppositely charged polyelectrolytes (PEs) and colloidal particles lead to PE adsorption on the particle surfaces. Adsorption is driven by gain in free energy from the electrostatic interaction between the colloid surface and PE monomers. Also, charged entities in aqueous media are surrounded by a cloud of oppositely charged ions, known as condensed counterions. When a PE monomer adsorbs on a colloidal surface, the counterions are released into the solution, which leads to a gain in their translational entropy along with a decrease in the PE conformational entropy. The net change in free energy from these contributions dictate the strength of interaction between PE chain and colloidal particles[86,87] Correspondingly, PE chains can adsorb in different conformations on the surface of the colloidal particles, forming trains, loops and tails, wherein parts of the PE chain are adsorbed completely on the surface, adsorbed partially with the unadsorbed segments forming a

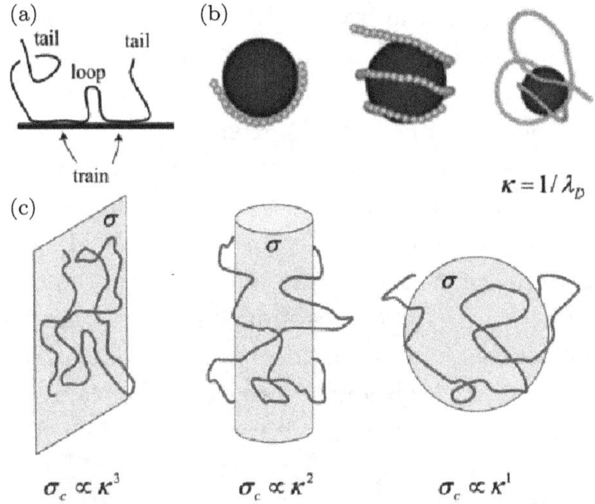

Fig. 4.6. (a) A schematic representing the tail, loop and train configuration of polymer adsorption on surfaces. Adapted from Netz and Andelman.[88] (b) Illustration of U-shaped, solenoid and rosette structures of PE-colloid complex from left to right. Adapted from Ulrich *et al.*[89] (c) Dependence of the critical surface charge of colloid on the inverse Debye length for planar, cylindrical and spherical surfaces. Adapted from Cherstvy and Winkler.[86]

loop, and adsorbed partially with the chain end extending away from the absorbing surface, respectively (Fig. 4.6(a)).[88] PE adsorption on the colloidal particles lead to a wrapping of the PE chains around the oppositely charged colloids and the formation of various structures such as U-shaped, solenoidal and rosette-like PE-colloid complexes (Fig. 4.6(b)),[89] eventually leading to multi-particle flocs following neutralization and bridging.

Solution ionic strength influences the adsorption of PE on colloid surfaces significantly. An increase in the ionic strength reduces the electrostatic interactions between the PE and the colloid due to enhanced electrostatic screening (a decrease in Debye length κ^{-1}). Lower gain in entropy from the release of condensed counterions is also observed, which further reduces the overall decrease in free energy and thus hampers PE adsorption.[87,90] Simulations have shown the existence of a critical surface charge σ_c corresponding to a PE charge density below which PE chains do not adsorb on a charged surface. σ_c depends on the inverse Debye screening length κ[86] and on the curvature of the adsorbing surface: at low ionic strengths ($\kappa a \ll 1$), $\sigma_c \propto \kappa^3$ for planar surfaces, $\sigma_c \propto \kappa^2$ for cylinders and $\sigma_c \propto \kappa^1$ for spheres (Fig. 4.6(c)). Hence, at low ionic strengths, spherical particles such as silica particles require a higher surface charge for PE adsorption and subsequent flocculation as compared to particles with lower curvature, such as platelet-like clay particles. At the same time, when the radius of curvature a is comparable to κ^{-1}, that is $\kappa a \sim 1$, σ_c becomes independent of the curvature and follows $\sigma_c \propto \kappa^3$ corresponding to adsorption on planar surfaces for all $a > 1/\kappa$. Therefore, at high ionic strengths or for large particles, the shape of the particle does not significantly influence the σ_c required for PE adsorption.

PE persistence length also plays a critical role in dictating the PE-colloid interactions and the floc structure. Simulations have shown that rigid PEs wrap around the particle to a lower extent with fewer monomers interacting with the colloid surface, leading to a weaker adsorption as compared to flexible chains which can conform with relative ease around the particle surfaces (Fig. 4.7(a)).[91–93] In contrast, flexible adsorbed PEs have folded structures at the monomer length scale where the PE trains are not flush with the colloidal surface leading to a lower surface coverage. Consequently, semi-flexible chains are shown to have the optimum monomer adsorption on colloid surfaces.[93,94]

Individual PE chains can also bridge multiple particles together by adsorbing on their surfaces leading to particle aggregation and enabling flocculation. Open structures of such complexes with larger interparticle distances are obtained for rigid chains as opposed to compact complexes formed with flexible chains. Higher electrostatic screening of colloidal surface charges with flexible chains, due to higher surface coverage, allows particles to come closer due to reduced interparticle repulsion. Rigid PE chains screen the charges on the colloids lesser, leading to interparticle repulsion in the complexes and thus it is energetically more favorable for particles to be further apart, forming more open structures. Moreover, at higher colloid concentrations, binding of colloids to charge neutral complexes is stronger for

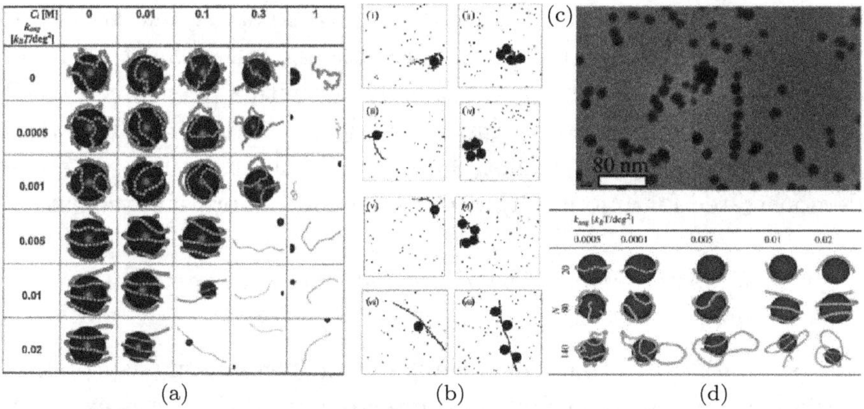

Fig. 4.7. (a) Various PE-NP complex structures obtained from simulations with increasing persistence length (from top to bottom) and increasing ion concentrations (increasing left to right). Adapted from Stoll and Chodanowski.[93] (b) Panels showing simulations of a single PE chain interacting with 1 particle (in the left column) and 4 particles (in the right column), with increasing polyelectrolyte flexibility (top to bottom). Adapted from Jonsson and Linse.[92] (c) Cryo-TEM image of rod-like complex with high silica nanoparticle and low chitosan concentrations. Adapted from Shi *et al.*[95] (d) PE-NP complex structures at different PE lengths (increasing from top to bottom) and persistence lengths (increasing from left to right). Adapted from Ulrich *et al.*[89]

rigid chains as compared to flexible ones.[92] This is because rigid chain complexes have more open, extended conformations due to their open structures, which allows un-complexed particles more access to monomers. The effect of chain stiffness on the complexes also becomes increasingly prominent with increasing length of the chains; longer chains can form larger loops as well as have longer tails, leading to lower surface coverage (Fig. 4.7(d)).[89]

Linear arrangement of colloids along the backbone of a rigid PE chain forming rod-like complex structures has also been predicted by simulations[91,92] and demonstrated using small angle scattering (X-ray and neutron) and cryo-electron microscopy experiments in systems with high PE and low colloid concentrations. Experiments with chitosan (cationic PE) and anionic silica nanoparticle systems showed multiparticle strings in electron microscopy images (Fig. 4.7(b)) along with evidence of rod-like structures in X-ray scattering measurements.[95] In contrast, these string-like structures were not observed in polylysine-silica complexes. It must be noted that chitosan is a relatively stiff biopolymer with a persistence length $l_p = 9$ nm, as compared to that of polylysine with $l_p = 1$ nm which is for typical polymer chains. Similar rod-like structures were also demonstrated for anionic PE hyaluronan and the protein lysozyme.[96]

4.3.1.2. *Phase behavior of polyelectrolyte-colloidal particle mixtures*

Phase separation of multi-particle flocs that form upon adsorption of PE chains on colloid surfaces and inter-particle bridging is exploited for removal of particulate impurities during water purification. An in-depth understanding of the different factors that influence the phase behavior of PE-colloid flocs is therefore critical for design of optimal PE flocculants for water treatment systems. We note that the PE-colloid flocs begin to scatter light as they grow to several microns in size. Thus, typical flocculation processes proceed with an increase in the turbidity of the solution. Turbidimetry, therefore, has been the experimental technique of choice to study phase behavior of polyelectrolyte-colloid mixtures due to its simplicity portability and low costs.[42,91,96–98]

Experimental phase diagrams of PE-colloid mixtures were constructed by Shi *et al.* by monitoring the turbidity and X-ray scattering signatures of aqueous mixtures of polycation chitosan and anionic silica nanoparticles (Fig. 4.8(a)).[95] The mixtures were shown to form monophasic regions when an excess of either components was present in the mixtures. A biphasic region was reported in the region containing comparable concentrations of chitosan and silica particles. Small angle scattering experiments on the solutions and complexes in these concentration regimes revealed the varying structures of the complexes — branched polyelectrolyte-particle complexes with protruding short strands of polyelectrolyte in the excess polyelectrolyte region, globular complexes in the biphasic region, and single polyelectrolyte chains with multiple complexed nanoparticles in the silica rich solutions. Similar phase behavior was

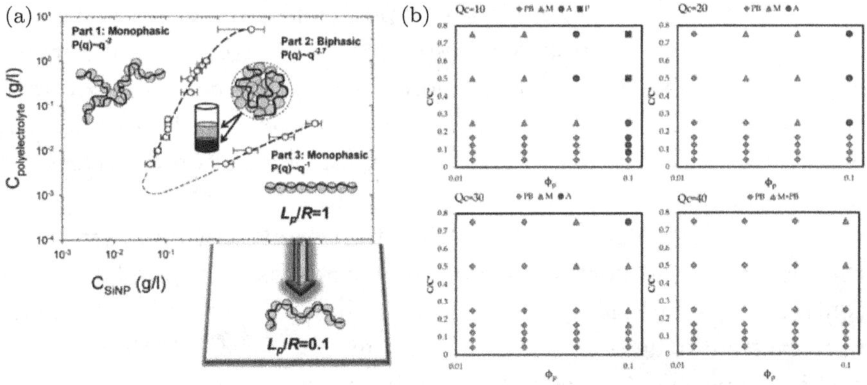

Fig. 4.8. (a) A general phase diagram for turbidity in PE-NP complexes with respect to concentration of PE and NPs. Adapted from Shi *et al.*[95] (b) Polymer bridged (PB), monomer dominated aggregates (*M*), aggregates (A) and percolated aggregates (*P*) identified in simulations of PE-NP complexes for different PE and NP concentrations and NP charge (Qc). Adapted from Pandav *et al.*[87]

also reported for the complexes of more flexible polylysine chains with silica nanoparticles.

Further classification of the structure of the PE-colloid complexes were projected onto the phase diagrams through mean field self-consistent field theory (SCFT) calculations that considered multibody interactions (Fig. 4.8(b)).[87] Complexes were classified as percolated (P) complexes which span the entire volume of the solution, PE-colloid aggregates (A), monomer dominated small aggregates (M) and polymer bridged (PB) aggregates. Phase diagrams with varying PE concentration (C) normalized by its critical overlap concentration (C^*) and increasing particle volume fraction (ϕ_p) for particle surface charges of $Q_C = 10, 20, 30$ and 40 have been calculated. Increasing the concentrations of either the PE or the colloid was reported to bring particles closer and lead to formation of larger aggregates. At low Q_C, particle aggregates formed largely due to PE induced depletion. On increasing particle charge, the PE and colloids interacted electrostatically, and polymer bridged (PB) aggregates were prevalent over larger concentration ranges. Moreover, higher concentration of PE was required to transition from PB to M and PB to A type complexes for all particle concentrations; this was attributed to the stronger interparticle electrostatic repulsion requiring more PE adsorption surface coverage to screen these charges bring the particles closer. Increasing PE charge was shown to increase PB aggregate formation as well as an increase in the size of the clusters formed.

Further insights into PE-colloid interactions can be drawn from the rich body of work on PE-protein complexes inspired by their central role in myriad applications including protein purification and preservation, drug delivery and gene therapy.[90,91,99–103] Proteins are colloidal macromolecules with sizes in the nanometer range, with a pH dependent heterogenous surface charge comprising both positive and negatively charged moieties which can electrostatically interact with PEs in aqueous media. At pH values significantly lower or higher than the protein's isoelectric point, their surfaces can be considered to be quasi-homogenous with either only positive or only negative charges expressed, respectively. Therefore, direct analogies between homogenously charged colloidal particles and proteins and their interaction to PEs can be drawn.

The model system of protein lysozyme and polyanion polystyrene sulfonate (PSS) was studied extensively by Gummel and coworkers by small-angle neutron scattering, which revealed the spatial arrangement of lysozyme as well as the PSS chain conformation in the PE-protein complexes. The quasi-homogeneous positive charge of lysozyme at pH = 3, the pH at which these experiments were conducted, allowed an analogy to be drawn between the phase behavior of these systems with homogenous PE-colloid mixtures. PSS-lysozyme complexes were generally categorized as either gel-like complexes, where the PE chains were distributed throughout the solvent and were physically crosslinked by protein particles, and a liquid-like regime with spherical globules of PE-protein complexes (Fig. 4.9).[104] PSS chains in the two regimes were found to exhibit worm-like behavior in the gel-like complex in contrast to being in a collapsed state in the dense globular complexes. The transition between these two regimes was observed at the C^* of the PE chains. However, this C^* was higher in the PE-colloid complexes

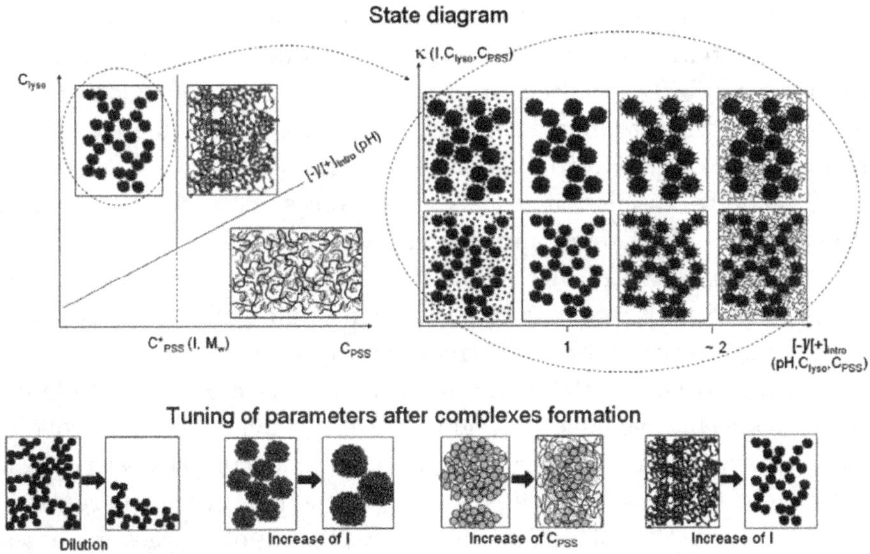

Fig. 4.9. Different structures of PE-protein complexes comprising polystyrene sulfonate and lysozyme, respectively, and their dependence on PE and protein concentrations, ionic strength and PE length. Adapted from Gummel *et al.*[104]

than in pure PE solutions, and was attributed to the screening of intrachain electrostatic repulsion upon complexation of the PSS chains leading to smaller persistence lengths (l_p = 5 nm for PSS in solution and l_p = 2 nm for PSS in PSS-lysozyme complexes). Furthermore, an increase in PE chain length, which decreases C^*, was shown to decrease the overlap concentration in PE-colloid complexes.[105] Change in ionic strength also affect C^*; increasing ionic strength decreased PE persistence length by screening monomer — monomer repulsions, reducing C^* for PE chains in PE-colloid complexes.[106]

Increase in the size of globules was seen on decreasing Debye length due to a decrease in electrostatic screening. Moreover, the charge ratio inside the globules was observed to be always ~1; that is the cluster core is neutral.[107] This implies that their system was in fact in equilibrium and not kinetically trapped with reaction limited colloidal aggregation as the mechanism for aggregation and complex formation. Excess positive or negative charge was also induced on the surface of the globules by a decrease or increase in the PE/protein charge ratio (Fig. 4.9).[108] Addition of PE to already formed complexes resulted in a phase transition from globular to gelled state whereas increasing ionic strength resulted in a transition from gelled complexes to globular complexes (Fig. 4.9).[109]

The above discussion reveals the nuances presented by charged polymer-colloid systems and the effect of several factors influencing complex architecture and phase behavior. Approaching the phenomenon of PE-colloid complexation from such a fundamental standpoint is crucial for water treatment applications as it allows for the rational design of PEs and associated filtration processes tailored towards specific systems. Changing concentrations of PE and colloid lead to varying structures ranging from rod-like complexes to dense, globular complexes. Ionic strength of water is a crucial parameter which dictates PE-colloid interaction. Screening of charges increases on increasing ionic strength and higher charge densities of PE or colloids are required for complexation. The persistence length of the coagulant PE is another critical characteristic of the PE which affects the size as well as density of flocs. PEs with larger persistence length lead to complexes with a more open structure as well as larger flocs by forming PE networks crosslinked by colloids. Similarly, PEs with a larger chain size also result in larger network formation.

4.3.2. *Biopolymer Flocculants*

4.3.2.1. *Chitosan-based flocculants*

The widespread use of chitosan as a flocculant and coagulant is backed by its economic feasibility, abundance, ease of handling and biodegradable nature.[14,81] Chitosan, a biopolymer, consisting of D-glucosamine and N-acetyl-D-glucosamine, is typically synthesized by deacetylation of chitin by a strong base like NaOH.[14,110] The cationic nature of chitosan in acidic solutions, coupled with its long chain lengths, render it suitable for coagulation and flocculation of mineral[111] as well as organic impurities[112] by the mechanisms discussed in sub-section 4.2.2.[14] The coagulation and flocculation performance of chitosan is determined by its molecular weight, extent of acetylation as well as operational parameters like dosage (excess dosage can reinstate stability) and pH of solution (different flocculation mechanisms dominate at different values of pH), all of which need to be well-monitored and controlled.[14,112]

In order to enhance its coagulation and flocculation performance, chitosan is often modified with other chemical compounds; grafting and etherification are two prevalent modification strategies.[36] The molecular architecture (and therefore, flocculation mechanism) of the chitosan can be controlled by grafting different monomers like acrylamide (non-ionic),[113] N-vinyl-2-pyrrolidone (cationic)[114] and methyacrylic acid (anionic),[115] among others.[36] For example, Wang *et al.* studied the effect of grafting acrylamide to chitosan, on its flocculating ability in kaolin suspensions, and various factors affecting it. A synergistic effect was observed, wherein the grafted copolymer displayed superior performance over a broader range of pH and lower dosage, as compared to pristine chitosan. This was attributed to the enhanced adherence to impurities, brought about by the flexible polyacrylamide component.[113] Similarly, etherification of chitosan can be brought about by its reaction with compounds like 3-chloro-2-hydroxypropyl trimethylammonium chloride (CTA) (cationic)[116] or organic acids[117] and sulfates (anionic), thereby influencing the molecular weight and flocculation mechanism.[36]

4.3.2.2. *Starch-based flocculants*

Due to their low cost, environmental benignity and abundance,[118–121] starch (majorly in the cationic form) and its derivatives are used as flocculants for wastewater consisting of biomass,[122] humic acid,[121] kaolin[119,123,124] and other impurities.[125] Various modifications via blending, esterification, etherification, oxidation, cross-linking and graft polymerization have been attempted to further improve the water solubility and flocculation efficacy of starch based flocculants.[126,127] For example, grafting starch molecules with quaternary ammonium groups have been shown to increase their flocculation performance in silica suspensions,[125] and also render antimicrobial properties.[119]

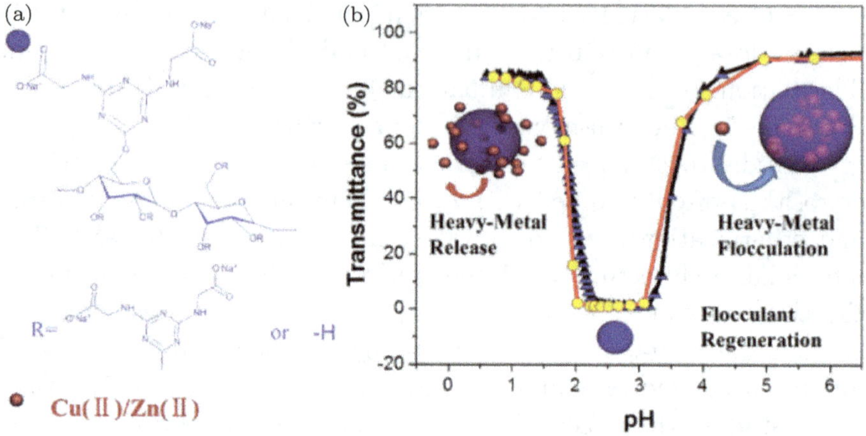

Fig. 4.10. (a) Molecular structure of the pH-responsive amphoteric starch (PRAS). (b) Schematic representation of the cyclic flocculation/regeneration process along with the observed turbidity in the solution as a function of solution pH. Adapted from Wu *et al.*[116]

Recently, Wu *et al.* demonstrated the removal and recovery of heavy metals (zinc and copper) from water by employing pH responsive amphoteric starch (PRAS) formed by etherifying starch with 2-chloro-4,6-diglycino-1,3,5-triazine (CDT) (Fig. 4.10(a)).[128] The pH responsiveness was brought about by (de)protonation of carboxyl and amine groups at different pH values (1–8) (Fig. 4.10(b)). It was also noted that PRAS remained effective (>90% removal of both metals) post three cycles of use, thereby suggesting effective elution and reusability.[128] Similarly, Shi *et al.* used (2,4-bis(dimethylamino)-[1,3,5]-triazine-6-yl)-starch, another etherified form of starch, for effective removal of anionic dyes from water.[129]

4.3.3. *Incorporation and Performance of Polyelectrolytes in Wastewater Treatment*

As discussed in the preceding sections, PEs interact with colloidal impurities to form aggregate flocs and induce phase separation of these particles and are used in various fields such as dyestuff treatment,[37] oil sand tailing sedimentation[130] and protein purification.[131] These PEs act as coagulants and flocculants are typically added in the preliminary treatment of wastewater. Long term usage of polycations PDADMAC and epichlorhydrin-dimethylamine (epi-DMA) in Ugemi, South Africa has shown a decrease in operation costs by over 30%. This is attributed to lower PE dosages (< 1/10) and sludge production as compared to Al and Fe salts.[132] It should be noted, however, that appropriate dosage of PE is required as resuspension of particulates due to surface charge reversal on flocs on excess PE addition is seen (Fig. 4.11(a)).[132] Addition of cationic polyelectrolytes with metal salts has also been shown to increase the efficiency of solid particle removal without an increase in the volume of sludge produced.[133] For instance, Fig. 4.11(b) shows improvement

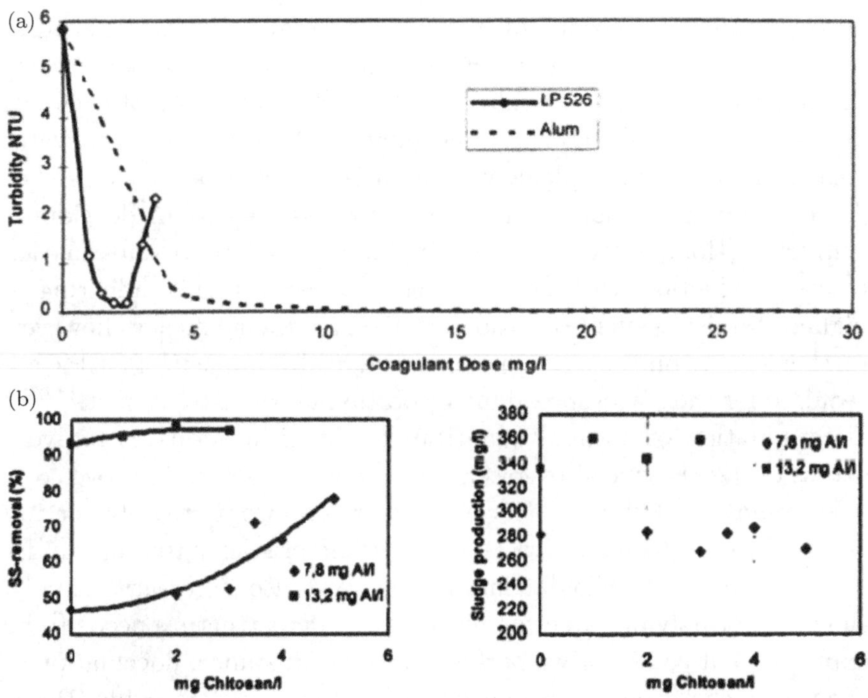

Fig. 4.11. (a) Change in turbidity of wastewater depicting re-stabilization using LP 526 (blended PDADMAC and polyamine-based coagulant) as compared to alum. Adapted from Nozaic *et al.*[132] (b) Effect on addition of chitosan with given concentration of Aluminum on suspended solids (SS) removal and sludge production. (Adapted from Odegaard.[133])

of suspended solids (SS) removal on addition of chitosan at lower Al dosage, while maintaining a constant sludge production. Moreover, settling speed of flocs is affected by addition of anionic flocculants, with increase in settling speeds with increasing PE dosage.[133] Metal oxide particles (such as Fe_2O_3, Al_2O_3 and MnO_2), which serve as adsorption sites for natural organic matter by ligand exchange and hydrogen bonding, can also be used with polycations and such systems showed improved color removal compared to polycation alone.[134]

Furthermore, chemically enhanced primary treatment (CEPT) strategies have been developed that focus on the usage of a combination of metallic (iron and aluminum) salts and polyelectrolytes (such as PDADMAC and anionic polyacrylamides) for the removal of pollutants. A recent review by Shewa and Dagnew surmises that appropriately selected CEPT additives, based on wastewater characteristics, show removal of >70% of SS and >40% of chemical oxygen demand (COD) and phosphates.[135] CEPT in wastewater treatment has been implemented in cities in developed nations (Los Angeles, Orange County and San Diego) as well as in developing countries (Hong Kong, Mexico City and Sao Paulo) and has led to costs reduction with increased plant capacities while adhering to higher standards of purification.[136] Despite their efficacy, however, further studies on the environmental and health impacts of CEPT are required for their widespread incorporation in water treatment.[132,135] Determination of optimal quantities of chemical additives to avoid excess dosage is crucial to lower this impact as well as improve cost-effectiveness.[135] Guidelines for choosing the appropriate polymer and salt additives specific to wastewater characteristics based on fundamental studies of colloidal interactions are also necessary. Finally, the effect of polymer architecture on floc microstructure needs to be explored as it can improve performance. For instance, flocculation of negatively charged clay suspensions with polypropylene oxide (PPO) grafted PAM showed increased dewaterability as well as higher solid content as compared to linear PAM chains.[130]

4.4. Summary and Future Perspectives

Polymeric coagulants and flocculants are gaining increasing popularity in diverse water treatment settings owing to their scalable production, widespread availability at low costs and water treatment efficacy. Yet, several issues associated with their use persist. Synthetic polyelectrolytes leave behind unreacted residual monomers such as acrylamide, ethyleneimine, and trimethylolmelamine. Other residual chemicals from polyelectrolyte synthesis may also make their way into water systems such as epihydrochlorin, formaldehyde and dimethylamine. Unregulated dosing of polyelectrolytes in water treatment may lead to excess polyelectrolytes persisting in the treated water, posing detrimental environmental and health effects. For instance, acrylamide is proven to be neurotoxic and to cause gastrointestinal ailments,[83] while cationic polyelectrolytes are toxic, particularly to aquatic organisms.[134] Thus, major research thrust is required, with a particular emphasis on improving the design of polyelectrolyte molecules that can reduce the polyelectrolyte dosing and produce denser sludge and controlled mixing of the coagulants and flocculants to enhance flocculation effectiveness before feeding the treated water to the settling tank. At the same time, a real-time control on the polyelectrolyte dosage is required to ensure optimal impurity removal while minimizing polyelectrolyte residue in the treated water. Development of cheap and robust methods to detect polyelectrolyte concentrations in the untreated and treated waters will enable continuous optimization of polyelectrolyte dosage, thus improving the safety and reliability of water treatment with polyelectrolytes coagulants and flocculants.

References

1. D. Hinrichsen, H. Tacio, Finding the source: The coming freshwater crisis is already here, *The Linkages between Population and Water* (*The Wilson Center*), (2002) 1–26.
2. WHO, and UNICEF. Key Facts from JMP 2015 Report (2015).

3. A. Boretti, L. Rosa, Reassessing the projections of the World Water Development Report, *Nat. Partn. Journals Clean Water*, **2** (2019) 1–15.
4. P. Burek, Y. Satoh, G. Fischer, M. T. Kahil, A. Scherzer, S. Tramberend, L. F. Nava, Y. Wada, S. Eisner, M. Flörke, *et al.*, Water Futures and Solution — Fast Track Initiative (Final Report). IIASA Work. (2015) Pap. 11.
5. UNICEF, and WHO. WASH in the 2030 Agenda (2016).
6. J. Bratby, Introduction, *Coagulation & Flocculation in Water and Wastewater Treatment*, Ed. 3 IWA Publishing (2016) 1–8.
7. D. Bauer, M. Philbrick, B. Vallario, The Water-Energy Nexus: Challenges and Opportunities (United States of America, Department of Energy) (2014).
8. M. N. Chong, B. Jin, C. W. K. Chow, C. Saint, Recent developments in photocatalytic water treatment technology: A review, *Water Res.*, **44** (2010) 2997–3027.
9. G. Graziano, Water purification: No-filter filtration, *Nat. Rev. Chem.*, **1** (2017) 0049.
10. P. A. Moussas, N. D. Tzoupanos, A. I. Zouboulis, Advances in coagulation/flocculation field: Al- and Fe-based composite coagulation reagents, *Desalin. Water Treat.*, **33** (2011) 140–146.
11. A. Nqombolo, A. Mpupa, R. M. Moutloali, P. N. Nomngongo, Wastewater Treatment Using Membrane Technology, *Wastewater and Water Quality*, IntechOpen (2018) 29–40.
12. A. Szyguła, E. Guibal, M. A. Palacín, M. Ruiz, A. M. Sastre, Removal of an anionic dye (Acid Blue 92) by coagulation-flocculation using chitosan, *J. Environ. Manage.*, **90** (2009) 2979–2986.
13. C. Y. Teh, P. M. Budiman, K. P. Y. Shak, T. Y. Wu, Recent Advancement of Coagulation-Flocculation and Its Application in Wastewater Treatment, *Ind. Eng. Chem. Res.*, **55** (2016) 4363–4389.
14. F. Renault, B. Sancey, P. M. Badot, G. Crini, Chitosan for coagulation/flocculation processes — An eco-friendly approach, *Eur. Polym. J.*, **45** (2009) 1337–1348.
15. J. N. Israelachvili, Intermolecular & Surface forces (Academic Press) (2011).
16. A. Matilainen, M. Vepsäläinen, M. Sillanpää, Natural organic matter removal by coagulation during drinking water treatment: A review, *Adv. Colloid Interface Sci.*, **159** (2010) 189–197.
17. M. Sillanpää, M. C. Ncibi, A. Matilainen, M. Vepsäläinen, Removal of natural organic matter in drinking water treatment by coagulation: A comprehensive review, *Chemosphere*, **190** (2018) 54–71.

18. E. Zamudio, M. Vaca, L. Torres, M. N. Rojas-Valencia, Pathogen Contaminant's Removal by Coagulation-flocculation using a Galactomannan-type Biopolymer, (2012), 1–4.
19. C. Guigui, J. C. Roucha, L. Durand-Bourlierb, V. Bonnelyeb, P. Aptel, Impact of coagulation conditions on the in-line coagulation/UF process for drinking water production, *Desalination*, **147** (2002) 95–100.
20. N. Y. Jang, Y. Watanabe, S. Minegishi, Performance of ultrafiltration membrane process combined with coagulation/sedimentation, *Water Sci. Technol.*, **51** (2005) 209–219.
21. K. Kimura, Y. Hane, Y. Watanabe, G. Amy, N. Ohkuma, Irreversible membrane fouling during ultrafiltration of surface water, *Water Res.*, **38** (2004) 3431–3441.
22. E. Marañón, L. Castrillón, Y. Fernández-Nava, A. Fernández-Méndez, A. Fernández-Sánchez, Coagulation-flocculation as a pretreatment process at a landfill leachate nitrification-denitrification plant, *J. Hazard. Mater.*, **156** (2008) 538–544.
23. M. R. Wiesner, M. M. Clark, J. Mallevialle, Membrane filtration of coagulated suspensions, *J. Environ. Eng.*, **115** (1989) 20–40.
24. K. K. Au, (Amos), S. M. Alpert, D. J. Pernitsky, Particle and Natural Organic Matter Removal in Drinking Water, in *Operational Control of Coagulation and Filtration Processes* (American Water Works Association), (2010) 1–16.
25. L. D. Xavier, L. Yokoyama, V. R. de Oliveira, G. T. Ribeiro, O. Araújo, The role of coagulation-flocculation in the pretreatment of reverse osmosis in power plant, *J. Sustain. Dev. Energy, Water Environ. Syst.*, **8** (2020) 118–131.
26. United States Environmental Protection Agency, Wastewater Treatment Works... The Basics, *US EPA Office of Water*, 4204 (1998), 1–6.
27. F. Spellman, Handbook of Water and Wastewater Treatment Plant Operations (CRC Press) (2013).
28. S. A. A. Jahn, From clarifying pearls and gems to water coagulation with alum history, surviving practices, and technical assessment, *Anthropos*, **94** (1999) 419–430.
29. N. B. Prakash, V. Sockan, P. Jayakaran, Waste Water Treatment by Coagulation and Flocculation, *Certif. Int. J. Eng. Sci. Innov. Technol.*, **9001** (2014) 2319–5967.
30. O. Sahu, P. Chaudhari, Review on Chemical treatment of Industrial Waste Water, *J. Appl. Sci. Environ. Manag.*, **17** (2013) 241–257.
31. A. L. Ahmad, S. Sumathi, B. H. Hameed, Coagulation of residue oil and suspended solid in palm oil mill effluent by chitosan, alum and PAC, *Chem. Eng. J.*, **118** (2006) 99–105.

32. J. Bratby, Treatment with metal coagulants, *Coagulation and Flocculation in Water and Wastewater Treatment*, Ed. 3 IWA Publishing (2016) 81–245.
33. J. Duan, J. Gregory, Coagulation by hydrolysing metal salts, *Adv. Colloid Interface Sci.*, **100–102** (2003) 475–502.
34. P. Gebbie, Using Polyaluminium Coagulants in Water Treatment. In 64th Annual Water Industry Engineers and Operators' Conference, (2001), pp. 39–47.
35. P. K. Chaudhari, B. Majumdar, R. Choudhary, D. K. Yadav, S. Chand, Treatment of paper and pulp mill effluent by coagulation, *Environ. Technol.*, **31** (2010) 357–363.
36. Z. L. Yang, B. Y. Gao, Q. Y. Yue, Y. Wang, Effect of pH on the coagulation performance of Al-based coagulants and residual aluminum speciation during the treatment of humic acid-kaolin synthetic water, *J. Hazard. Mater.*, **178** (2010) 596–603.
37. M. Riera-Torres, C. Gutiérrez-Bouzán, M. Crespi, Combination of coagulation-flocculation and nanofiltration techniques for dye removal and water reuse in textile effluents, *Desalination*, **252** (2010) 53–59.
38. A. C. Rodrigues, M. Boroski, N. S. Shimada, J. C. Garcia, J. Nozaki, N. Hioka, Treatment of paper pulp and paper mill wastewater by coagulation-flocculation followed by heterogeneous photocatalysis, *J. Photochem. Photobiol. A Chem.*, **194** (2008) 1–10.
39. Y. X. Zhao, B. Y. Gao, H. K. Shon, B. C. Cao, J. H. Kim, Coagulation characteristics of titanium (Ti) salt coagulant compared with aluminum (Al) and iron (Fe) salts, *J. Hazard. Mater.*, **185** (2011) 1536–1542.
40. B. Cao, B. Gao, C. Xu, Y. Fu, X. Liu, Effects of pH on coagulation behavior and floc properties in Yellow River water treatment using ferric based coagulants, *Chinese Sci. Bull.*, **55** (2010) 1382–1387.
41. A. H. Jagaba, S. Abubakar, I. M. Lawal, A. A. A. Latiff, I. Umaru, Wastewater Treatment Using Alum, the Combinations of Alum-Ferric Chloride, Alum-Chitosan, Alum-Zeolite and Alum-Moringa Oleifera as Adsorbent and Coagulant, *Int. J. Eng. Manag.*, **2** (2018) 67–75.
42. B. Bolto, J. Gregory, Organic polyelectrolytes in water treatment, *Water Res.*, **41** (2007) 2301–2324.
43. J. Bratby, Colloids and Interfaces, *Coagulation & Flocculation in Water and Wastewater Treatment*, Ed. 3 IWA Publishing (2016) 9–32.
44. M. Kobayashi, F. Juillerat, P. Galletto, P. Bowen, M. Borkovec, Aggregation and charging of colloidal silica particles: Effect of particle size, *Langmuir*, **21** (2005) 5761–5769.

45. M. Elimelech, J. Gregory, X. Jia, R. A. Williams, Electrical properties of interfaces. In Particle Deposition & Aggregation: Measurement, Modelling and Simulation, (World Scientific), (1995) pp. 9–32.
46. T. Radeva, Physical Chemistry of Polyelectrolytes (Ed. T. Radeva, Marcel Dekker) (2001).
47. S. J. Park, M. K. Seo, Intermolecular Force. In Interface Science and Composites, (2011) pp. 1–57.
48. C. O. Metin, R. T. Bonnecaze, L. W. Lake, C. R. Miranda, Q. P. Nguyen, Aggregation kinetics and shear rheology of aqueous silica suspensions, *Appl. Nanosci.*, **4** (2014) 169–178.
49. H. Reerink, J. T. G. Overbeek, The rate of coagulation as a measure of the stability of silver iodide sols, *Discuss. Faraday Soc.*, **18** (1954) 74–84.
50. C. Y. Yin, Emerging usage of plant-based coagulants for water and wastewater treatmen, *Process Biochem.*, **45** (2010) 1437–1444.
51. N. D. Tzoupanos, A. I. Zouboulis, Coagulation-Flocculation Processes in Water/Wastewater Treatment?: The Application of New Generation of Chemical Reagents, *6th IASME/WSEAS Int. Conf. HEAT Transf. Therm. Eng. Environ.*, (2008) 309–317.
52. R. Krishna, O. Sahu, Reduction of COD and Color by Polymeric Coagulant (Chitosan), *J. Polym. Biopolym. Phys. Chem.*, **1** (2013) 22–25.
53. T. Suopajärvi, Functionalized nanocelluloses in wastewater treatment applications, *Dissertation: University of Oulu*, (2015) 1–80.
54. R. Yang, H. Li, M. Huang, H. Yang, A. Li, A review on chitosan-based flocculants and their applications in water treatment, *Water Res.*, **95** (2016) 59–89.
55. D. Ghernaout, B. Ghernaout, Sweep flocculation as a second form of charge neutralisation — A review, *Desalin. Water Treat.* **44** (2012) 15–28.
56. J. Gregory, Rates of flocculation of latex particles by cationic polymers, *J. Colloid Interface Sci.*, **42** (1973) 448–456.
57. T. W. Healy, V. K. La Mer, The adsorption-flocculation reactions of a polymer with an aqueous colloidal dispersion, *J. Phys. Chem.*, **66** (1962) 1835–1838.
58. R. H. Smellie, V. K. La Mer, Flocculation, subsidence and filtration of phosphate slimes. VI. A quantitative theory of filtration of flocculated suspensions, *J. Colloid Sci.*, **13** (1958) 589–599.
59. M. Hubbe, Mini-Encyclopedia of Papermaking Wet-End Chemistry (2020).

60. V. K. La Mer, Filtration of colloidal dispersions flocculated by anionic and cationic polyelectrolytes, *Discuss. Faraday Soc.*, **42** (1966) 248–254.
61. Z. Zhu, Theory on Orthokinetic Flocculation of Cohesive Sediment: A Review, *J. Geosci. Environ. Prot.*, **02** (2014) 13–23.
62. R. J. Banu, K. U. Do, I. T. Yeom, Phosphorus removal in low alkalinity secondary effluent using alum, *Int. J. Environ. Sci. Technol.*, **5** (2008) 93–98.
63. J. Beltrán-Heredia, J. Sánchez Martín, Azo dye removal by Moringa oleifera seed extract coagulation, *Color. Technol.*, **124** (2008) 310–317.
64. B. Cao, B. Gao, X. Liu, M. Wang, Z. Yang, Q. Yue, The impact of pH on floc structure characteristic of polyferric chloride in a low DOC and high alkalinity surface water treatment, *Water Res.*, **45** (2011) 6181–6188.
65. Q. H. Malik, Performance of alum and assorted coagulants in turbidity removal of muddy water, *Appl. Water Sci.*, **8** (2018) 1–4.
66. V. Saritha, N. Srinivas, N. V. Srikanth Vuppala, Analysis and optimization of coagulation and flocculation process, *Appl. Water Sci.*, **7** (2017) 451–460.
67. S. Verma, B. Prasad, I. M. Mishra, Pretreatment of petrochemical wastewater by coagulation and flocculation and the sludge characteristics, *J. Hazard. Mater.*, **178** (2010) 1055–1064.
68. X. Zhan, B. Gao, Q. Yue, Y. Wang, B. Cao, Coagulation behavior of polyferric chloride for removing NOM from surface water with low concentration of organic matter and its effect on chlorine decay model, *Sep. Purif. Technol.*, **75** (2010) 61–68.
69. R. Aksberg, L. Wågberg, Hydrolysis of cationic polyacrylamides, *J. Appl. Polym. Sci.*, **38** (1989) 297–304.
70. S. A. A. Tabatabai, S. I. Gaulinger, M. D. Kennedy, G. L. Amy, J. C. Schippers, Optimization of inline coagulation in integrated membrane systems: A study of $FeCl_3$ Desalin, *Water Treat.*, **10** (2009) 121–127.
71. L. X. Gao, H. Gu, A. Rahardianto, P. D. Christofides, Y. Cohen, Self-adaptive cycle-to-cycle control of in-line coagulant dosing in ultrafiltration for pre-treatment of reverse osmosis feed water, *Desalination*, **401** (2017) 22–31.
72. B. Blankert, B. H. L. Betlem, B. Roffel, Development of a control system for in-line coagulation in an ultrafiltration process, *J. Memb. Sci.*, **301** (2007) 39–45.
73. H. Xu, F. Xiao, D. Wang, C. Ye, Survey of treatment process in water treatment plant and the characteristics of flocs formed by two new coagulants, *Colloids Surfaces A Physicochem. Eng. Asp.*, **456** (2014) 211–221.

74. C. S. B. Fitzpatrick, E. Fradin, J. Gregory, Temperature effects on flocculation, using different coagulants, *Water Sci. Technol.*, **50** (2004) 171–175.
75. C. Phalakornkule, J. Mangmeemak, K. Intrachod, B. Nuntakumjorn, Pretreatment of palm oil mill effluent by electrocoagulation and coagulation, *ScienceAsia*, **36** (2010) 142–149.
76. N. G. Marriott, Essentials of food sanitation (Chapman & Hall) (1997).
77. K. P. Y. Shak, T. Y. Wu, Coagulation-flocculation treatment of high-strength agro-industrial wastewater using natural Cassia obtusifolia seed gum: Treatment efficiencies and flocs characterization, *Chem. Eng. J.*, **256** (2014) 293–305.
78. S. M. Miller, E. J. Fugate, V. O. Craver, J. A. Smith, J. B. Zimmerman, Toward understanding the efficacy and mechanism of Opuntia spp. as a natural coagulant for potential application in water treatment, *Environ. Sci. Technol.*, **42** (2008) 4274–4279.
79. T. Okuda, A. U. Baes, W. Nishijima, M. Okada, Coagulation mechanism of salt solution-extracted active component in Moringa oleifera seeds, *Water Res.*, **35**(3) (2001) 830–834.
80. K. A. Ghebremichael, K. R. Gunaratna, H. Henriksson, H. Brumer, G. Dalhammar, A simple purification and activity assay of the coagulant protein from Moringa oleifera seed, *Water Res.*, **39** (2005) 2338–2344.
81. T. Okuda, A. U. B. Baes, W. Nishijima, M. Okada, Improvement of extraction method of coagulation active components from Moringa oleifera seed, *Water Res.*, **33** (1999) 3373–3378.
82. T. Okuda, A. U. Baes, W. Nishijima, M. Okada, Isolation and characterization of coagulant extracted from Moringa oleifera seed by salt solution, *Water Res.*, **35**(2) (2001) 405–410.
83. J. Bratby, Coagulants, *Coagulation & Flocculation in Water and Wastewater Treatment*, Ed. 3 IWA Publishing (2016) 34–80.
84. B. A. Bolt, Soluble Polymers in Water Purification, *Prog. Polym. Sci.*, **20** (1995) 987–1041.
85. Hyde, and Hall, Water Treatment Processes and Practices (Ed. T. Hall & R. Hyde; WRc Swindon, Swindon, Wilts) (1992).
86. A. G. Cherstvy, R. G. Winkler, Polyelectrolyte adsorption onto oppositely charged interfaces: Unified approach for plane, cylinder, and sphere, *Phys. Chem. Chem. Phys.*, **13** (2011) 11686–11693.
87. G. Pandav, V. Pryamitsyn, J. Errington, V. Ganesan, Multibody Interactions, Phase Behavior, and Clustering in Nanoparticle-Polyelectrolyte Mixtures, *J. Phys. Chem. B*, **119** (2015) 14536–14550.

88. R. R. Netz, D. Andelman, Neutral and charged polymers at interfaces, *Phys. Rep.*, **380** (2003) 1–95.
89. S. Ulrich, M. Seijo, S. Stoll, The many facets of polyelectrolytes and oppositely charged macroions complex formation, *Curr. Opin. Colloid Interface Sci.*, **11** (2006) 268–272.
90. X. Xu, S. Angioletti-Uberti, Y. Lu, J. Dzubiella, M. Ballauff, Interaction of Proteins with Polyelectrolytes: Comparison of Theory to Experiment, *Langmuir*, **35** (2019) 5373–5391.
91. C. L. Cooper, A. Goulding, A. B. Kayitmazer, S. Ulrich, S. Stoll, S. Turksen, S. Yusa, A. Kumar, P. L. Dubin, Effects of polyelectrolyte chain stiffness, charge mobility, and charge sequences on binding to proteins and micelles, *Biomacromolecules*, **7** (2006) 1025–1035.
92. M. Jonsson, P. Linse, Polyelectrolyte-macroion complexation. I. Effect of linear charge density, chain length, and macroion charge, *J. Chem. Phys.*, **115** (2001) 3406–3418.
93. S. Stoll, P. Chodanowski, Polyelectrolyte adsorption on an oppositely charged spherical particle. Chain rigidity effects, *Macromolecules*, **35** (2002) 9556–9562.
94. A. Akinchina, P. Linse, Monte Carlo simulations of polyion-macroion complexes: 1. Equal absolute polyion and macroion charges, *Macromolecules*, **35** (2002) 5183–5193.
95. L. Shi, F. Carn, F. Boue, G. Mosser, E. Buhler, Control over the electrostatic self-assembly of nanoparticle semiflexible biopolyelectrolyte complexes, *Soft Matter*, **9** (2013) 5004–5015.
96. I. Morfin, E. Buhler, F. Cousin, I. Grillo, F. Boué, Rodlike complexes of a polyelectrolyte (hyaluronan) and a protein (lysozyme) observed by SANS, *Biomacromolecules*, **12** (2011) 859–870.
97. E. Kizilay, A. B. Kayitmazer, P. L. Dubin, Complexation and coacervation of polyelectrolytes with oppositely charged colloids, *Adv. Colloid Interface Sci.*, **167** (2011) 24–37.
98. L. Shi, F. Carn, F. Boué, E. Buhler, Role of the ratio of biopolyelectrolyte persistence length to nanoparticle size in the structural tuning of electrostatic complexes, *Phys. Rev. E*, **94** (2016) 1–12.
99. F. Comert, P. L. Dubin, Liquid-liquid and liquid-solid phase separation in protein-polyelectrolyte systems, *Adv. Colloid Interface Sci.*, **239** (2017) 213–217.
100. C. L. Cooper, P. L. Dubin, A. B. Kayitmazer, S. Turksen, Polyelectrolyte-protein complexes, *Curr. Opin. Colloid Interface Sci.*, **10** (2005) 52–78.
101. S. Gao, A. Holkar, S. Srivastava, Protein-Polyelectrolyte Complexes and Micellar Assemblies, *Polymers (Basel)*, **11**(1097) (2019), 1–34.

102. A. B. Kayitmazer, D. Seeman, B. B. Minsky, P. L. Dubin, Y. Xu, Protein-polyelectrolyte interactions, *Soft Matter*, **9** (2013) 2553–2583.
103. A. B. Kayitmazer, A. F. Koksal, E. Kilic Iyilik, Complex coacervation of hyaluronic acid and chitosan: Effects of pH, ionic strength, charge density, chain length and the charge ratio, *Soft Matter*, **11** (2015) 8605–8612.
104. J. Gummel, F. Cousin, F. Boué, Structure transition in PSS/lysozyme complexes: A chain-conformation-driven process, as directly seen by small angle neutron scattering, *Macromolecules*, **41** (2008) 2898–2907.
105. F. Cousin, J. Gummel, D. Ung, F. Boué, Polyelectrolyte-protein complexes: Structure and conformation of each specie revealed by SANS, *Langmuir*, **21** (2005) 9675–9688.
106. F. Cousin, J. Gummel, S. Combet, F. Boué, The model Lysozyme-PSSNa system for electrostatic complexation: Similarities and differences with complex coacervation, *Adv. Colloid Interface Sci.*, **167** (2011) 71–84.
107. J. Gummel, F. Boue, B. Deme, F. Cousin, Charge Stoichiometry Inside Polyelectrolyte-Protein Complexes: A Direct SANS Measurement for the PSSNa-Lysozyme System, *J. Phys. Chem. B.*, **110** (2006) 24837–24846.
108. J. Gummel, F. Boué, D. Clemens, F. Cousin, Finite size and inner structure controlled by electrostatic screening in globular complexes of proteins and polyelectrolytes, *Soft Matter*, **4** (2008) 1653–1664.
109. F. Cousin, J. Gummel, D. Clemens, I. Grillo, F. Boué, Multiple scale reorganization of electrostatic complexes of poly(styrenesulfonate) and lysozyme, *Langmuir*, **26** (2010) 7078–7085.
110. K. Kurita, Chitin and chitosan: Functional biopolymers from marine crustaceans, *Mar. Biotechnol.*, **8** (2006) 203–226.
111. J. Roussy, M. Van Vooren, E. Guibal, Chitosan for the coagulation and flocculation of mineral colloids, *J. Dispers. Sci. Technol.*, **25** (2004) 663–677.
112. E. Guibal, J. Roussy, Coagulation and flocculation of dye-containing solutions using a biopolymer (Chitosan), *React. Funct. Polym.*, **67** (2007) 33–42.
113. J. P. Wang, Y. Z. Chen, S. J. Zhang, H. Q. Yu, A chitosan-based flocculant prepared with gamma-irradiation-induced grafting, *Bioresour. Technol.*, **99** (2008) 3397–3402.
114. A. Srivastava, D. K. Mishra, K. Behari, Graft copolymerization of N-vinyl-2-pyrrolidone onto chitosan: Synthesis, characterization and study of physicochemical properties, *Carbohydr. Polym.*, **80** (2010) 790–798.

115. D. K. Mishra, J. Tripathy, A. Srivastava, P. K. Pandey, K. Behari, Synthesis and characterization of chitosan-g-methacrylic acid and studies of its additional physicochemical properties, such as swelling, metal-ion sorption, and flocculation behavior, *J. Appl. Polym. Sci.*, **113** (2009) 2429–2439.
116. S. A. Ali, S. Pal, R. P. Singh, Flocculation performance of modified chitosan in an aqueous suspension, *J. Appl. Polym. Sci.*, **118** (2010) 2592–2600.
117. Z.-S. Cai, C.-S. Yang, X.-M. Zhu, Preparation of quaternized carboxymethyl chitosan and its capacity to flocculate COD from printing wastewater, *J. Appl. Polym. Sci.*, **118** (2010) 299–305.
118. M. Huang, Y. Wang, J. Cai, J. Bai, H. Yang, A. Li, Preparation of dual-function starch-based flocculants for the simultaneous removal of turbidity and inhibition of Escherichia coli in water, *Water Res.*, **98** (2016) 128–137.
119. M. Huang, Z. Liu, A. Li, H. Yang, Dual functionality of a graft starch flocculant: Flocculation and antibacterial performance, *J. Environ. Manage.*, **196** (2017) 63–71.
120. G. Sen, R. Kumar, S. Ghosh, S. Pal, A novel polymeric flocculant based on polyacrylamide grafted carboxymethylstarch, *Carbohydr. Polym.*, **77** (2009) 822–831.
121. H. Wu, Z. Liu, H. Yang, A. Li, Evaluation of chain architectures and charge properties of various starch-based flocculants for flocculation of humic acid from water, *Water Res.*, **96** (2016) 126–135.
122. R. Anthony, R. Sims, Cationic starch for microalgae and total phosphorus removal from wastewater, *J. Appl. Polym. Sci.*, **130** (2013) 2572–2578.
123. Y. Chen, S. Liu, G. Wang, Flocculation properties and adsorption kinetics of cationic starches in kaolin suspensions, *J. Appl. Polym. Sci.*, **105** (2007) 2841–2849.
124. J. P. Wang, S. J. Yuan, Y. Wang, H. Q. Yu, Synthesis, characterization and application of a novel starch-based flocculant with high flocculation and dewatering properties, *Water Res.*, **47** (2013) 2643–2648.
125. S. Pal, D. Mal, R. P. Singh, Cationic starch: An effective flocculating agent, *Carbohydr. Polym.*, **59** (2005) 417–423.
126. M. Haroon, L. Wang, H. Yu, N. M. Abbasi, M. Zain-Ul-Abdin, Saleem, R. U. Khan, R. S. Ullah, Q. Chen, J. Wu, Chemical modification of starch and its application as an adsorbent material, *RSC Adv.*, **6** (2016) 78264–78285.
127. Zia-ud-Din, H. Xiong, P. Fei, Physical and chemical modification of starches: A review, *Crit. Rev. Food Sci. Nutr.*, **57** (2017) 2691–2705.

128. L. Wu, X. Zhang, L. Chen, H. Zhang, C. Li, Y. Lv, Y. Xu, X. Jia, Y. Shi, X. Guo, Amphoteric starch derivatives as reusable flocculant for heavy-metal removal, *RSC Adv.*, **8** (2018) 1274–1280.
129. Y. Shi, B. Ju, S. Zhang, Flocculation behavior of a new recyclable flocculant based on pH responsive tertiary amine starch ether, *Carbohydr. Polym.*, **88** (2012) 132–138.
130. S. P. Gumfekar, V. Vajihinejad, J. B. P. Soares, Advanced Polymer Flocculants for Solid–Liquid Separation in Oil Sands Tailings, *Macromol. Rapid Commun.*, **40** (2019) 1–18.
131. X. Du, P. L. Dubin, D. A. Hoagland, L. Sun, Protein-selective coacervation with hyaluronic acid, *Biomacromolecules*, **15** (2014) 726–734.
132. D. J. Nozaic, S. D. Freese, P. Thompson, Longterm experience in the use of polymeric coagulants at Umgeni Water, *Water Sci. Technol. Water Supply*, **1** (2001) 43–50.
133. H. Odegaard, Optimized Particle Separation in Primary Step Of Wastewater Treatment, **37** (1998) 43–53.
134. B. Bolto, D. Dixon, R. Eldridge, S. King, Cationic polymer and clay or metal oxide combinations for natural organic matter removal, *Water Res.*, **35** (2001) 2669–2676.
135. W. A. Shewa, M. Dagnew, Revisiting chemically enhanced primary treatment of wastewater: A review, *Sustain.*, **12** (2020) 1–19.
136. D. R. F. Harleman, S. Murcott, The role of physical-chemical wastewater treatment in the mega-cities of the developing world, *Water Science and Technology*, (1999) 75–80.

CHAPTER 5

Pressure Retarded Osmosis: Modelling, Mirages and Prospects

Robert W. Field[a,*] and Jun Jie Wu[b]

[a] *Faculty of Engineering and Environment, Northumbria University, UK*
[b] *Department of Engineering, Durham University, UK*

5.1. Background

Great interest is being shown in processes with the potential to produce clean sustainable energy. Unlike solar and wind energy, which are both dependent on availability and the weather, the Blue Energy processes based upon the controlled mixing of seawater and fresh water have the advantage that they can be operated continuously.[1] One process for producing renewable energy from salinity gradients is Pressure Retarded Osmosis (PRO) and this will be our focus; the other is Reverse Electrodialysis (RED). With RED, the salinity gradient between seawater and river water is exploited via a stack of alternating cation and anion exchange membranes, and due to the chemical potential difference between salt and fresh water a voltage is generated across each membrane. It has received significant research attention.[2–5] A 50kW pilot plant on the Afsluitdijk causeway in the Netherlands was opened in late 2014.[6] The aim of this pilot was to assess the technical feasibility of RED under real-life conditions using

*Corresponding author: Emeritus Professor, Department of Engineering Science, University of Oxford, robert.field@eng.ox.ac.uk

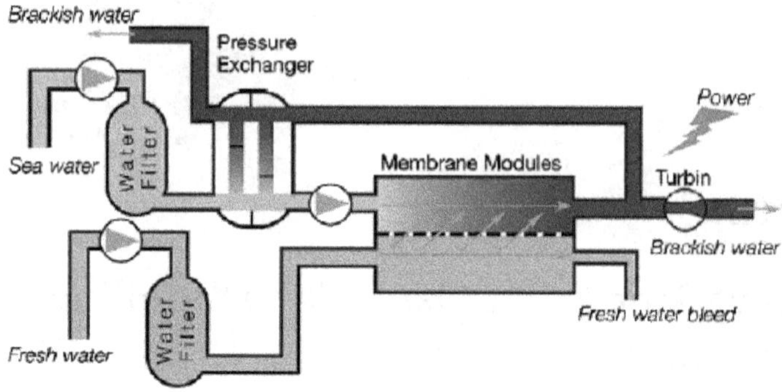

Fig. 5.1. Schematic diagram of the first prototype PRO installation opened in Norway on November 24, 2009 and designed and operated by Statkraft (Oslo, Norway). Extracted from Wang *et al.*[35]

fresh Ijsselmeer water, which is on one side of the causeway and salt water from the Wadden Sea which is on the other side. There is mention in the literature of plans for a full-scale 200 MW plant but there seems to be no evidence of it having materialised. With RED, electricity is generated directly from the membrane stack, and it has been suggested as a potential pre-treatment or better "pre-desalination" option to a traditional reverse osmosis process. However herein the almost exclusive focus is upon an analysis of PRO. This is a process that uses the osmotic pressure difference between a saline stream and fresh water to produce an osmotic flow, which together with an appropriate back pressure, is used to drive a hydro turbine as shown schematically in Fig. 5.1. The pressure exchanger is a key component of the overall process scheme. See Fig. 5.6 of Field and Wu.[7] Although the efficiency of these devices has now increased to *circa* 96–97%[8] the flowrate of the draw is an order of magnitude greater than the bleed to the hydro turbine and the net output from a PRO plant will be considerably smaller than the output of the hydro turbine itself.[7] The inter-relationship between Forward Osmosis, PRO and reverse osmosis (RO) is illuminated schematically in Fig. 5.2.

The pioneers of reverse osmosis recognised the conceptual simplicity of PRO[9] but concern about costs has always been a problem.[10–12] It was estimated that the osmotic technology would soon produce around 1,600 TWh a year globally, which is about

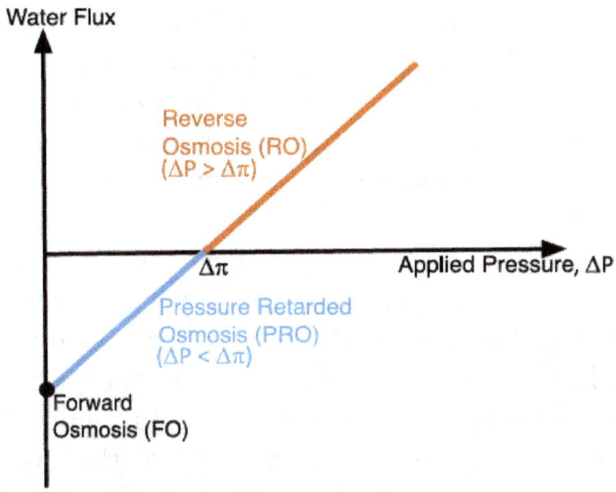

Fig. 5.2. Schematic diagram comparing reverse osmosis, forward osmosis and pressure retarded osmosis (PRO) for an ideal semipermeable membrane. The optimal point of operation for PRO is around half $\Delta\pi$. Extracted from Field and Wu.[7]

13 times the annual hydropower generation of Norway. The potential for Norway itself was put at 12 TWh.[13] However unlike reverse osmosis for seawater desalination, which has developed from being a relatively minor local means of supplementing potable water supplies into a globally significant technology (becoming absolutely crucial for the well being of a number of countries), the development of PRO has stalled. Indeed, in the twentieth century the concept of PRO remained just that — a concept — whereas RO rose to become by the year 2000 the main desalination technology of the world (see Fig. 5.1 of Ref. 8).

The idea of PRO was revived in the early twenty-first century because of the need to develop renewable energy options. This led to step changes in membrane development and osmotically driven processes more generally. PRO itself has received significant attention over the past 15 years with the first pilot osmotic power plant being opened by the Norwegian energy company Statkraft in 2009. It was reported the following year that the Statkraft prototype was designed to generate 10 kW of power and that the company planned to build a full-scale 25 MW osmotic power plant by 2015.[14] However, the pilot plant closed in 2014 due to the low power density (power per unit of membrane area) of the prototypes and the questionable economic feasibility of the process. Indeed at its peak, the facility had produced power of just 2–4 kW before being closed. Despite the company's efforts, they concluded that within the current market outlook the technology could not be sufficiently developed to become competitive "within the foreseeable future".[15]

It is interesting to re-read, with the benefit of hindsight the review titled, "Pressure retarded osmosis: From the vision of Sidney Loeb to the first prototype installation".[14] It was noted that at current energy prices and achievable power densities, and for an expected membrane life of 5 years, the revenue per square metre of membrane would only cover the cost of the membranes assuming that they were purchased at a price at the lowest end of the range of the current estimated bulk cost of RO membranes. Around the same time, and before the pilot plant closed, scepticism was also expressed elsewhere.[7]

Today, as discussed in a Section 5.3, it is generally accepted that if PRO is to be commercially viable then it will be necessary to use resources with a higher salinity than seawater, for example brine from a reverse osmosis desalination plant.[16,17] While this would tend to increase the osmosis and permit a higher backpressure, both of which lead to higher power densities, there is the inhibiting effect of a relatively high salt concentration within the support layer, due to the flux driven phenomenon known as internal concentration polarisation (ICP) which will be explained in the next section. This acts to reduce the difference in salinity across the active layer of the membrane and thereby dampen the positive effects of increased draw salinity.[18] Ideally a doubling of draw salinity would lead to a four-fold increase in power density but the mass transfer boundary layers, both external and ICP, have a negative impact. In accord with examples given later, it is generally accepted that increases in power density due to an increase in draw salinity will be more modest.[7,18]

Following an introduction in Sections 5.2.1–5.2.3 to the various elements needed for the modelling of a PRO process, including the Spiegler-Kedem-Katchalsky (SKK) model,[19] Section 5.2.4 will introduce Yaroshchuk's recent model,[20] which is an application of the SKK model to pressure retarded osmosis (PRO). This analysis held out the tantalising prospect of a major breakthrough but for reasons given herein and elsewhere there is unfortunately no reason to believe that the theoretically predicted high power densities can be realised in practice.

5.2. Modelling Elements

5.2.1. *Concentration polarisation in PRO*

Normally authors start with an equation for describing water transport through the active layer such as equation (8) which is below. In its most basic form this would model the highly idealised system illustrated in Fig. 5.3(a). Even if allowance is made for the

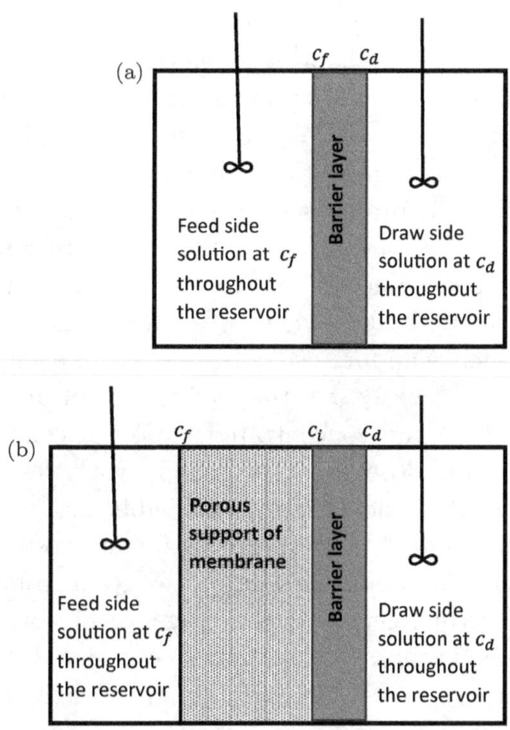

Fig. 5.3. (a) Unsupported membrane. Idealised mass transfer conditions in the feed and draw reservoirs; both taken to be perfectly stirred. (b) Supported membrane. Idealised mass transfer conditions in the feed and draw reservoirs; both taken to be perfectly stirred. Extracted from Field and Wu.[1]

mass transfer resistance of the support layer, which corresponds to modelling the system shown in Fig. 5.3(b), important mass transfer resistances are being neglected. This is particularly true of Forward Osmosis (FO) and PRO because there are also two external mass transfer resistances and one of these will involved so-called dilutive concentration polarisation. In this section the concept of concentration polarisation is introduced, and the relative importance of concentrative and dilutive concentration polarisation emphasised.

Concentration polarization refers to the emergence of concentration gradients at a membrane/solution interface resulting from selective transfer through the membrane. The link between this natural consequence of permselectivity and the attenuation of driving forces across the active layer of a membrane has been examined for many years but the asymmetrical impact of this natural process depending upon whether the membrane was preferential selective to, or rejecting of the minor component was not effectively explained until the seminal work of Wijmans *et al.*[21] Without doubt the term "concentration polarisation" first emerged in the membrane arena in the field of pressure driven membrane processes such as UF. With such processes one could identify a region of particular high concentration adjacent to the membrane and one of lower concentration in the bulk so there was a resonance between its use in membrane science and its use outside science; a solute was indeed concentrated. In the early era, the 'gel' theory of ultrafiltration was popular[22] and here the concentrated region is considered to be a gel or a high level close to a gel. Today with the recognition of dilutive concentration polarization, probably first in dialysis, then in pervaporation and more recently in forward osmosis, there has been a generalisation of the meaning of concentration polarization. Nowadays the term refers to the emergence of concentration gradients at a membrane/solution interface resulting from selective transfer of some species through the membrane, be it a minor species preferentially transferred (in which case there is dilutive concentration polarization) or if a minor species is preferentially retained it is concentrated within the mass transfer boundary layer.

Whilst both are a natural consequence of permselectivity the link with the everyday use of the term polarized has been abandoned in order to have a unity of meaning within the field of membrane science.

In their analysis Wijmans *et al.*[21] clearly explained why boundary layer effects are much more severe in pervaporation (where the minor components preferentially permeate the membrane) than in ultrafiltration, where minor components are rejected. With regard to osmotic processes, our more recent work[18] showed for the first time that one of the unique features of forward osmosis (FO) is that due to the reverse salt flux there is a maximum boundary layer Peclet number *i.e.* for a given mass transfer coefficient there is a maximum possible flux irrespective of the strength of the osmotic driving force.

The classical expression for concentration polarisation as derived elsewhere (e.g.[23]) is:

$$C_m = (C_b - C_p)\exp(Pe) + C_p \tag{1}$$

where Pe is the boundary layer Peclet number and is defined as J/k_{bl}. J is the volumetric flux through the membrane and k_{bl} is the boundary layer mass transfer coefficient.

It is now emphasized that dilutive and concentrative concentration polarisation have a common basis as first shown in the 1990's.[21] The resulting general expression is valid for compounds that are enriched in the permeate as well as for compounds that are depleted in the permeate. An intrinsic enrichment factor was defined:

$$E_0 = C_p/C_m \tag{2}$$

where C_p is the concentration of the minor component in the permeate and C_m is the concentration of that component at the membrane surface. The intrinsic enrichment factor is large and positive for pervaporation where the minor component is enriched in the permeate but for reverse osmosis (RO) and ultrafiltration (UF) there is rejection of the minor component so E_0 is typically in the range 0.001 to 0.1 (i.e 90% to 99.9% rejection). The expression for the concentration polarization modulus (defined as C_m/C_b where C_b is the concentration of the component of interest in the bulk is a useful measure of the extent of concentration polarization. This ratio is equal

$$\frac{C_m}{C_b} = \frac{\exp(Pe)}{1 + E_0[\exp(Pe) - 1]} \tag{3}$$

where Pe is defined as J/k_{bl}, J being the volumetric flux through the membrane and k_{bl} being the mass transfer coefficient describing diffusive transport in the boundary layer. The variation of the extent of concentration polarization with Pe and E_0 is shown in Fig. 5.4. Figure 5.5 clearly demonstrates why concentration polarization effects are more severe in pervaporation than in UF and RO.

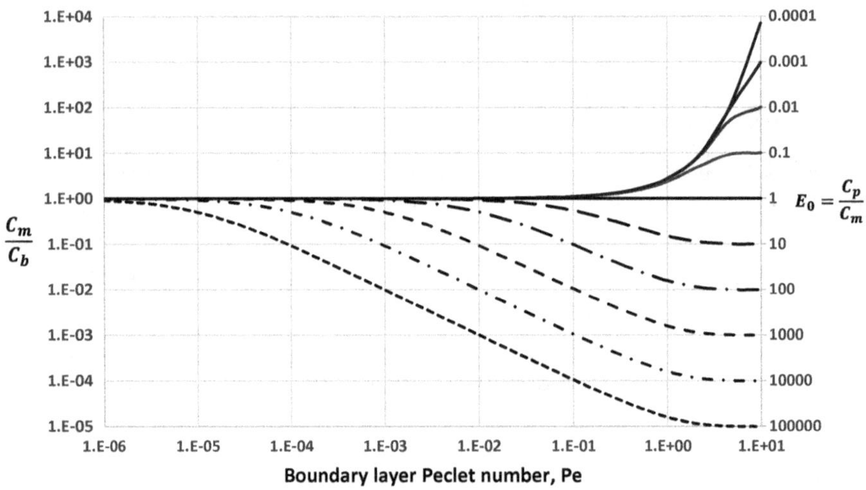

Fig. 5.4. Concentration polarization modulus, C_m/C_b, as a function of Boundary layer Peclet number $Pe(= J/k_{bl})$ for a wide range of values of the intrinsic enrichment factor E_0. Lines calculated from Eq. (3). The solid lines are for components that are rejected (i.e. the values of E_0 are less that unity) and there is concentration of these component in the concentration boundary layer. The dashed lines are for the components that are enriched in the permeate ($E_0 > 1$); there is dilution of these components in the boundary layer. Adapted from Wijmans *et al.*[21]

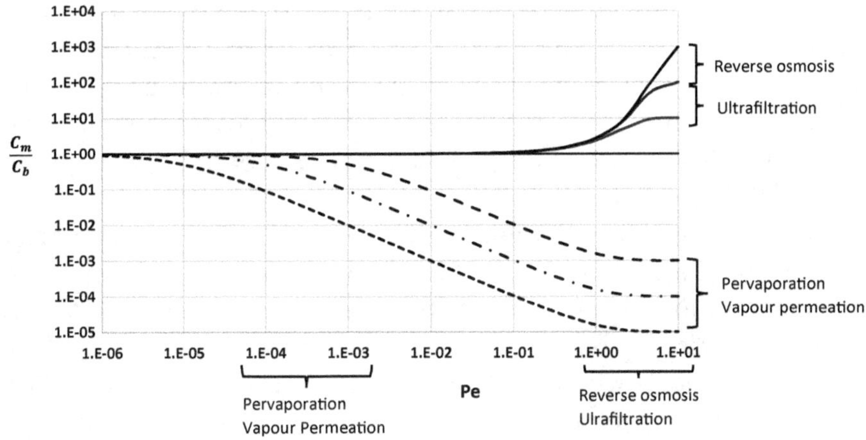

Fig. 5.5. Illustration of the differing impact of concentration polarization depending upon the membrane process involved. For RO and UF the effect of concentration polarisation is concentrative; for pervaporation and vapour permeation (e.g. VOC recovery) it is dilutive and the effect is significant even at low values of the boundary layer Peclet number. Adapted from Wijmans *et al.*[21]

In pervaporation, the concentration of solute at the membrane surface is often one-tenth or less of the concentration in the bulk solution because of the huge dilutive concentration polarization effect and therefore the driving force across the membrane is greatly reduced compared with its nominal value. These observations are very pertinent to FO and PRO because both processes naturally have both concentrative and dilutive concentration polarisation. The draw solution experiences dilutive concentration polarisation, whilst on the feed side solutes are concentrated. In PRO, the barrier layer (which is often referred to as the active layer) faces the saline solution (draw solution) whilst the feed solution (which is often envisaged to be filtered river water) flows past the support layer and solutes will be concentrated at the interface between the feed stream and the support and further within the support layer. The source of solutes within the support layer are two-fold. Firstly, some are convected in with that fraction of water which is drawn from the feed into the draw solution, and secondly as empirically observed there is diffusion across the barrier layer of some of the solutes in the draw solution. This solute flux is counter to that of the solvent and is often labelled as "reverse salt diffusion". In PRO, both processes contribute to the high concentration of solutes at the interface between the barrier layer and the support. The resulting concentration profile within the support layer is referred to as *internal concentration polarisation.*

The reverse salt diffusion i.e. diffusion of salt from the draw solution to the feed solution which is in the opposite direction to the main volumetric flux, is to be found in both FO and PRO. Flux equations are required for both the water and the salt. The pair of equations will be introduced in Section 5.2.3 but suffice to say, at this stage, they include two mass transfer coefficients.[24] Adapting the symbols used previously to reflect the focus upon PRO, the mass transfer coefficients are k_{cd} and k_{sup} where the mass transfer coefficient designated as k_{cd} is for the draw solution that is invariably facing the active layer and k_{sup} is the combined mass transfer coefficient for the support layer itself and the adjacent external mass transfer layer. The mass transfer coefficient for the feed solution on the side of the support layer consists of two parts. The main part relates to the support itself and can be written as D/S where D is the diffusivity of salt in water (it has a moderate dependency upon concentration[25]) and S is the structural parameter, defined as the product of the support layer thickness and tortuosity over its porosity.[26] The two equations are:

$$S = t \cdot \tau / \varepsilon \tag{4}$$

Thus

$$\frac{1}{k_{sup}} = \frac{1}{k_{cf}} + \frac{S}{D} \tag{5}$$

where k_{cf} is for the feed channel adjacent to the support.

The relatively small value of D/S compared with k_{cf} indicates that the support layer dramatically affects mass transport of solutes on that side of the membrane. The term internal concentration polarization (ICP) was coined to describe this phenomenon. Unlike external concentration polarisation (ECP) it is not ameliorated by crossflow. A relatively open support (high porosity) with low thickness and tortuosity will reduce the effective thickness of the support, S.

To illustrate the importance of the mass transfer coefficients the relevant expression for FO is now given for the orientation where the active layer faces the draw solution i.e. the same orientation as PRO.

$$J_w = A\left[\pi_{ds}\exp\left(-\frac{J_w}{k_{cd}}\right) - \pi_f\exp\left(\frac{J_w}{k_{\text{sup}}}\right)\right] + B\left[\exp\left(-\frac{J_w}{k_{cd}}\right) - \exp\left(\frac{J_w}{k_{\text{sup}}}\right)\right] \tag{6}$$

where 'A' is the parameter characterising water permeability (m/s per Pa); 'B' characterises salt flux (m/s) and other symbols have already been introduced. This equation is fully consistent with those found elsewhere, *e.g.*,[27] but it is presented in a manner which avoids the use of an opaque overall mass transfer coefficient. In moving from Eq. (6) to Eq. (7) the opportunity has been taken to invert the terms in the second square bracket so that both terms in square brackets are positive. This format emphasizes that reverse salt diffusion reduces water flux. In terms of Peclet numbers, $Pe_{\text{draw}} = \frac{J_w}{k_{ca}}$ and $Pe_{\text{feed}} = \frac{J_w}{k_{\text{sup}}}$ and thus the equation can be expressed as:

$$J_w = A[\pi_{ds}\exp(-Pe_{\text{draw}}) - \pi_f\exp(Pe_{\text{feed}})] - B[\exp(Pe_{\text{feed}}) - \exp(-Pe_{\text{draw}})] \tag{7}$$

If the flux is $3\,\mu\text{m/s}$ and $k_{cd} = 10^{-5}$ m/s and $k_{\text{sup}} = 3 \times 10^{-6}$ m/s, $Pe_{\text{draw}} = 0.3$ and $Pe_{\text{feed}} = 1$. Combining these illustrative values with Eq. (7) will give one an indication of the important role of concentration polarisation; the Peclet numbers are of a magnitude such that the nominal driving term π_{ds} is significantly reduced and the effective osmotic backpressure is significantly increased above its nominal value of π_f. Concentration moduli on both sides are important and due allowance must be made.

As a final point it is noted that the expressions taken from Ref. 18 and used herein are more comprehensive than those presented in

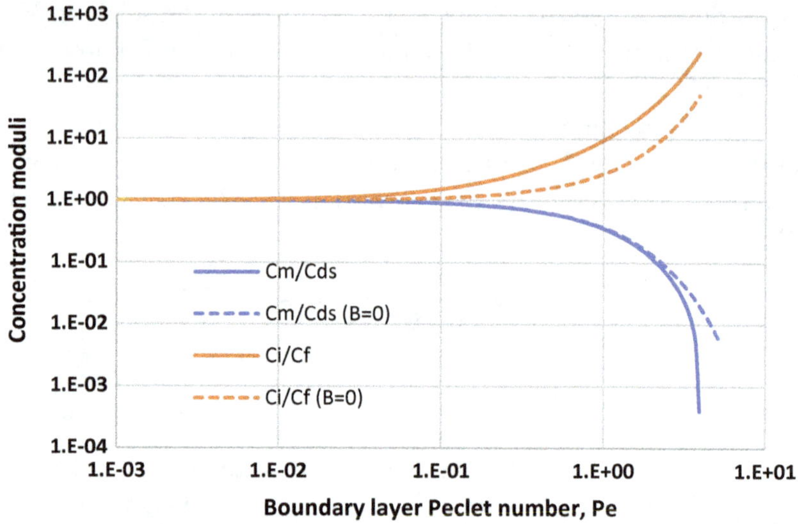

Fig. 5.6. Variation of the concentration polarization moduli with boundary layer Peclet number. Solid lines drawn for $A = 2\times10^{-12}$ m/s per Pa; $B = 2\times10^{-7}$ m/s; $S = 350\,\mu$m; $\pi_{ds} = 5$ MPa; $\pi_f = 0.025$ MPa. Dashed lines are for same parameters except $B = 0$.

earlier works e.g. Ref. 28 because allowance is made for the salt leakage, and all of the mass transfer coefficients. As shown in Fig. 5.6 the influence of reverse salt diffusion is significant at realistic values of Pe.

5.2.2. *Support layer*

The structure of an asymmetric membranes consists of two parts. There is the selective/barrier layer that performs the actual separation function and it is supported on a porous layer that provides mechanical integrity. In RO such support layers have little influence upon performance and besides providing the necessary mechanical support, they need to be inexpensive, inert and easy to handle during manufacture. As discussed elsewhere[29] the emergence of processes such as FO and PRO challenged the membrane community to re-evaluate conventional wisdom regarding the influence of a support layer. The importance of the support layer will be appreciated if the value of D/S introduced in the previous section is calculated and compared with estimated values of the mass transfer coefficients. Given that the inhibiting of diffusive transport is important the new challenge was to have a support providing adequate strength but with as low an S value as possible.

In Section 5.2.4 a recent original theoretical analysis will be examined because it suggested that an overlooked 'breakthrough' mode of operating FO and PRO membranes would exist if membranes could be manufactured such that they had properties consistent with a Spiegler-Kedem-Katchalsky model[19] provided the reflection coefficient was around 0.99; in other words with a minor deviations from ideal semi-permeability. This is mentioned here because in the envisaged 'breakthrough' mode the support layer thickness was supposed to be unimportant.

5.2.3. *Modelling of the barrier layer and overall equation*

The standard two equations for the active layer, based upon a solution-diffusion approach, are:

$$J_w = A[\pi_{d,surface} - \pi_{f,interface} - \Delta P] \tag{8}$$

$$J_s = B[c_{d,surface} - c_{f,interface}] \tag{9}$$

where for PRO mode, the subscript 'surface' refers to the external surface of the active layer, 'interface' to the surface abutting the support layer, 'd' refers to the draw solution and 'f' to the feed. Here ΔP is the hydraulic pressure difference across the active layer.

With due allowance for ECP and ICP, Eq. (6) can be derived for FO, where, by definition, $\Delta P = 0$. The equivalent equation for PRO is:

$$J_w = A\left[\pi_{ds}\exp\left(-\frac{J_w}{k_{cd}}\right) - \pi_f\exp\left(\frac{J_w}{k_{\text{sup}}}\right) - \Delta P\right] + B\left(1 + \frac{A\Delta P}{J_w}\right)\left[\exp\left(-\frac{J_w}{k_{cd}}\right) - \exp\left(\frac{J_w}{k_{\text{sup}}}\right)\right] \tag{10}$$

In early modelling, one or more of the mass transfer resistances were ignored, *e.g.*[26,30] but more recently due allowance has been made for all three relevant resistances, e.g. Refs. 18, and 27. Equation (10) above can be considered to be the accepted state-of-the-art equation for the lumped parameter approach to the modelling of PRO.

Now in reverse osmosis, the Spiegler-Kedem-Katchalsky (SKK) model is a well-established three-parameter model[19] and one can ask whether it could with advantage be applied to osmotically driven processes. The SKK model has rarely been used for FO and comparisons with the standard model are rare. Attarde *et al.* did compare the fit of their data both with the standard two-parameter solution-diffusion (S-D) model and with the SKK model; little difference was found.[31]

Now as Forward Osmosis (FO) is simply osmosis, it might be thought that the SKK model could be applied without reservation to FO and by extension to pressure retarded osmosis (PRO) as in Ref. 20. This is considered in the next section.

5.2.4. *Modelling PRO using the SKK model*

A recent thought provoking theoretical paper applied the SKK model to PRO and deduced if one allowed for minor deviations from ideal semi-permeability then under certain circumstances PRO could operate in an overlooked mode of "breakthrough" osmosis which would yield relatively large rates of osmosis because ICP would be circumvented due to the absence of reverse solute diffusion. There was the consequential prediction that one could envisage power densities that would easily exceed by one order of magnitude those achieved in the conventional mode. A second claim was that much more robust support layers could be used without thicker supports having a detrimental effect upon flux. The mathematics of the derivation are sound but doubts about the boundary conditions used (specifically the assumption of perfect stirring on the draw side) and questions regarding the validity of the underlying SKK model were raised.[1,32] The latter could prove to be of wider significance as mentioned below.

The basis of Yaroshchuk's innovative theoretical analysis has been reviewed elsewhere[1,32] where consideration was also given as to whether the envisaged process of "breakthrough" could be extended beyond the purely analytical. Even though there was no experimental confirmation, it was clearly suggested[20] that "breakthrough" could involve solute flux sign reversal (i.e., a transition from counter-current to co-current fluxes) and that the "dramatic change in the behaviour is ultimately caused by the change in the direction of solute flow through the membrane". This specific possibility was rebutted. The prospect of tailoring membranes to create a step change in output was a tantalising prospect but ultimately it was, in our opinion, a mirage.

Claims of a major scientific advance should ideally be experimentally verified, or as a minimum, demonstrated through calculation using practical values. Elsewhere the analysis has been extended to include the draw side mass transfer boundary layer because as fluxes increase it is unreasonable to ignore it as confirmed in the above section on concentration polarisation. Overall, it was shown that a degree of 'leakiness' (i.e. a reflection coefficient σ close to but not equal to unity) brings no discernible benefits. In these evaluations the analytical dimensionless analysis was converted into an engineering science analysis based on the use of practical values for the parameters. This showed that even if the model were to be taken at face value then for practical salinities of the draw solution the predicted outcomes indicate that the impact arising from the inclusion of a "leakiness" parameter is minor.

In investigating the possibility and practicality of employing tailored PRO membranes to generate very high power densities, the foundations of the SKK model were found to be questionable. This was an unexpected finding.[1] FO and PRO are the simplest of settings for the application of the Spiegler-Kedem-Katchalsky model. That this model has been shown to yield irrational results indicates that some detail in the SKK model is wrong. Now the equation for the salt flux includes the term $(1 - \sigma)cJ_v$ where c is the reference (virtual) solute concentration (as given in Ref. 33 and used in Ref. 20), σ is the solute reflection coefficient, and J_v is the solvent flux. In reverse osmosis (RO), this term can be linked to a pressure contribution because the solvent flux J_v is strongly influenced by feedside pressure in RO. Indeed in an alternative model, the solution-diffusion-imperfection model,[34] the term $(1-\sigma)cJ_v$ is explicitly replaced by the term $L_i c_d \Delta P$ and there are no absurdities when this model is applied to FO. This suggests that the fundamental fault with the SKK model, when applied to FO, lays with the term $(1-\sigma)cJ_v$ where σ is assumed to be independent of both the solute concentration in the barrier layer and the hydraulic pressure difference across it. Notwithstanding the utility that the SKK model has demonstrated over a number of decades, it seems to have a fundamental shortcoming.[1,32] Others too have observed that whilst irreversible thermodynamic arguments were used to derive the solute and solvent transport equations of the SKK model the membrane itself is treated as a "black box".[35]

As discussed elsewhere, others have overestimated the potential of PRO. For example very high velocities have been used to mitigate the effects of ECP and achieve relatively high fluxes but the high velocities inside the hollow fibres are such that the energy required for pumping exceeds that which would be produced via a PRO process. It is vital to evaluate the overall energy balance.[7]

5.2.5. *State of the art model for PRO*

Drawbacks of the semi-empirical model approach that is embodied in Eq. (10) have been discussed elsewhere where the focus wa upon better characterisation of the support layer.[29] There might also in future be a desire to move to a three-parameter model for the selective layer. Now for RO, the solution-diffusion-imperfection model (SDI)[34] was formulated to address some of the shortfalls in the basic solution-diffusion model.[35] The key equations for the barrier layer, when applying it to PRO, would be:

$$J_v = A(\Delta P - \Delta\pi) + L_i \Delta P \tag{11}$$

$$J_s = B(c_d - c_i) + L_i c_d \Delta P \tag{12}$$

where L_i is the permeability of imperfections, ΔP is the trans-membrane pressure difference, $\Delta\pi$ is the effective osmotic pressure difference across the barrier layer and other symbols have their normal meaning with 'A' being the water permeance and 'B' the solute permeance through the barrier layer. Clearly for PRO this model suggests an augmentation of the solute flux (compared with the S-D model) in the direction away from the pressurised draw solution; see Eq. (12). For FO operation the pressure difference (ΔP) would be zero and the SDI model reverts to the basic solution-diffusion model. This model would be an appropriate starting point for the modelling of the barrier layer if three as opposed to two parameters were thought to be desirable. However, this has not been developed herein because, as noted earlier, Eq. (10) above can be considered the accepted state-of-the-art equation for the modelling of PRO. The active layer is characterised by the classic 'A' and 'B' parameters.

$$J_w = A\left[\pi_{ds}\exp\left(-\frac{J_w}{k_{\text{draw}}}\right) - \pi_f\exp\left(\frac{J_w}{k_{\text{sup}}}\right) - \Delta P\right] + B\left(1+\frac{A\Delta P}{J_w}\right)\left[\exp\left(-\frac{J_w}{k_{\text{draw}}}\right) - \exp\left(\frac{J_w}{k_{\text{sup}}}\right)\right] \quad (13)$$

An informative timeline of the development of semi-empirical models for PRO is elsewhere.[29] With enhanced permeability of the barrier layer and greatly improved support layers it is no longer appropriate to use the earlier models of Loeb[9] or Lee.[11] For example, ECP must now be included because new forward osmosis membrane developed this century, such as those from Hydration Technology Innovations (HTI), have substantially improved water flux which means that old assumptions and approximations are no longer valid. The equation in Ref. 12 is a special case of Eq. (10); therein Loeb assumed a salt free river water (i.e. $\pi_f = 0$) and perfect mass transfer for the draw solution i.e. $k_{\text{draw}} \to \infty$. The importance of not overlooking these two factors is illustrated in the next section. McCutcheon and Elimelech[28] were the first group of researchers to develop a model incorporating the effect of ECP on the selective layer side of an FO membrane but their model assumed that no draw solute crossed the membrane. It is a special case and simplification of Eq. (10) with $B = 0$. Several models have been summarised in

Ref. 29; these are essentially the same as Eq. (10) but are typically presented in a more convoluted form,

5.3. Prospects for PRO

The April 8th 2010 issue of National Geographic featured three future technologies for sea water desalination, namely FO, carbon nanotubes and biomimetic systems such as aquaporins. The FO system featured the Yale work with their innovative draw solution[36] which is basically aqueous ammonium bicarbonate. A potential 30% cut in energy requirements was mentioned, a figure which has not been realised. Research from the same laboratory have also highlighted FO processes that either avoid draw solution regeneration[37] or involve an integration of FO with RO for a non-potable application.[38] The former considered other regeneration-free FO applications under development including the use of fertilizers as draw agents and algal harvesting/dewatering at sea but over the last decade these have not, to our knowledge, led to any commercial developments. As for the latter, the scheme proposed an alternative to a conventional two-pass reverse osmosis process for agricultural irrigation (the stringent boron and chloride standards making RO alone more energy intensive than RO as used for potable water production). Now today the integration of PRO with FO has been proposed by the "Mega-ton Water System" which was a funded project under the Japanese national "Funding Program for World-Leading Innovative R&D on Science and Technology" (FIRST Program).[8] They noted that the technology was near to being ready for adoption in commercial plants. The South Korean SeaHERO Program is now in its second phase of funding and it too is also seeking to improve and integrate RO and PRO.

The prototype plant developed in the FIRST Program achieved 13.3 W/m^2 of the maximum membrane power density per surface area using 10-inch membrane modules. On the other hand, the lab-scale plant showed 17.1 W/m^2 with a five-inch module, as shown in figure 16 in Ref. 8. That larger modules give lower output is not unexpected. One can envisage poorer feed distribution and smaller overall mass transfer coefficients. However, an output around

13 W/m^2 may prove to be sufficient to encourage further commercial development. Whether this is the predicted output from the hydro turbine or from the overall PRO system is not clear, but our calculation suggest that it is more likely to be the former.

The current reality is that this may prove to be the only area of PRO that is moving towards commercial exploitation. In addition to the combination of high saline + waste water, there is the classic one of sea-water and river water or to be more precise filtered sea water and filter river water. Around 15 years ago, Loeb presented an analysis of the potential of PRO using the Mississippi as the fresh water source.[12] The process design calculations of that work has been extended to include all relevant mass transfer boundary layers. Details including membrane parameters are given in Table 5.1. The chosen values are based in part upon unpublished work on the feasibility of implementing PRO systems in the North of Mexico; the unpublished work of one of us (JJW) has assessed data for the rivers named Panuco, Soto La Marina and Mayo. Data on their salinities as a distance from the sea was used and allowance for feed side salinity was shown to be important.

Clearly PRO is very dependent upon the efficiency of the pressure exchanger that recovers energy from part of the outlet brine and transfers it to the incoming brine. This point was not included in earlier analyses. Table 5.2 indicates our output and reinforces the assessment that the only area of PRO that may find a commercial

Table 5.1. Basis for generation of flux and power data.

Parameter	Hydraulic permeability	Solute permeability	Structural parameter	Draw side mass transfer coefficient
Symbol (units)	A (μm s^{-1} MPa^{-1})	B (μm s^{-1})	S (μm)	k_{bl}(μm s^{-1})
Value	9.03	0.0637	325	20
Source/ comment	[12], similar to TFC	[12], similar to TFC	Based on value in Table 1 of [24]	About 3× greater than value found for current modules [24]

Table 5.2. Illustrative outputs for two potential applications.

Application	Osmotic draw, π_{ds} (bar)	Osmotic pressure of feed (bar)	Estimated flux (μm s^{-1})	Specific power from hydro turbine (W/m^2)	System efficiency %	System output (W/m^2)
sea water — river water	25	0.1	5.4	6.5	67	4.4
Brine-waste water	45	1*	5.75	12.7	67	8.4

*based on NEWater brine

niche is the one involving a water factory where one has high strength brine that is an output from RO and a low salinity waste water stream from waste water treatment. The latter is fed to the support side of the FO membrane but there is still the potential for fouling of the support which may prove difficult to clean. Time will tell whether this process is commercial.

5.4. Concluding Remarks

With PRO and FO it is empirically observed that the diffusion of the solute is counter to that of the solvent i.e. there is so-called "reverse salt diffusion". However, with the Spiegler-Kedem-Katchalsky (SKK) model there is an allowance for a minor deviation from ideal semi-permeability and the ratio of the solute flux and solvent flux is no longer constant. Now it has been recently suggested, in a purely theoretical paper, that if allowance is made for minor deviations from ideal semi-permeability then operation in an overlooked mode of "breakthrough" osmosis would be possible and importantly it would yield relatively large rates of osmosis. A consequential prediction was that in "breakthrough mode", Pressure Retarded Osmosis (PRO) would generate very high-power densities. Whilst we find no reason to believe this is possible (and indeed have given references to two papers arguing otherwise[1,32]) the fact that the paramount model for RO does not capture the performance of osmotically driven processes clearly suggests that some re-thinking of the basis of the Spielger-Kedem Katchalsky model is required.

ICP and ECP have been shown to be important and only one application of PRO seems to be meriting commercial consideration currently. This sole area envisages high strength brine as the draw and a low salinity wastewater stream as the feed.

References

1. R. W. Field, J. J. Wu, On the understanding and feasibility of "Breakthrough" Osmosis, *Open Access Scientific Reports*, 2019
2. A. Daniilidis, D. A. Vermaas, R. Herber, K. Nijmeijer, Experimentally obtainable energy from mixing river water, seawater or brines with reverse electrodialysis, *Renew. Energy*, **64** (2014) 123–131.

3. E. Güler, W. van Baak, M. Saakes, K. Nijmeijer, Monovalent-ion-selective membranes for reverse electrodialysis, *J. Membr. Sci.*, **455** (2014) 254–270.
4. J. G. Hong, *et al.* Potential ion exchange membranes and system performance in reverse electrodialysis for power generation: A review, *J. Membr. Sci.*, **486** (2015) 71–88.
5. M. Tedesco, *et al.* Performance of the first reverse electrodialysis pilot plant for power production from saline waters and concentrated brines, *J. Membr. Sci.*, **500** (2015) 33–45.
6. https://www.dutchwatersector.com/news-events/news/12388-dutch-king-opens-world-s-first-red-power-plant-driven-on-fresh-salt-water-mixing.html Accessed September 6th 2018.
7. R. W. Field, J. J. Wu, Mass transfer limitations in forward osmosis: Are some potential applications overhyped? *Desalination*, **318** (2013) 118–124.
8. M. Kurihara, H. Takeuch, SWRO-PRO System in "Mega-ton Water System" for Energy Reduction and Low Environmental Impact, *Water*, **10** (2018) 48; doi: 10.3390/w10010048
9. S. Loeb, Production of energy from concentrated brines by pressure retarded osmosis. I. Preliminary technical and economic correlations, *J. Membr. Sci.*, **1** (1976) 49–63.
10. S. Loeb, R. S. Norman, Osmotic power plants, *Science*, **189**, 654–655 (1975).
11. K. L. Lee, R. Baker, H. Lonsdale, Membranes for power generation by pressure-retarded osmosis, *J. Membr. Sci*, **8** (1981) 141–171.
12. S. Loeb, Large-scale power production by pressure-retarded osmosis using river water and sea water passing through spiral wound modules, *Desalination*, **143** (2002) 115–122.
13. https://www.power-technology.com/projects/statkraft-osmotic/ Accessed January 30th 2020.
14. A. Achilli, A. E. Childress, Pressure retarded osmosis: From the vision of Sidney Loeb to the first prototype installation — Review, *Desalination*, **261** (2010) 205–211.
15. http://www.powermag.com/statkraft-shelves-osmotic-power-project/ Accessed 27th October 2017.
16. A. P. Straub, A. Deshmukh, M. Elimelech, Pressure-retarded osmosis for power generation from salinity gradients: is it viable? *Energy Environ. Sci.*, **9** (2016) 31–48.
17. S. Lin, A. P. Straub, M. Elimelech, Thermodynamic limits of extractable energy by pressure retarded osmosis, *Energy Environ. Sci.*, **7** (2014) 276–2714.

18. R. W. Field, J. J. Wu, On boundary layers and the attenuation of driving forces in forward osmosis and other membrane processes, *Desalination*, **429** (2018) 167–174.
19. K. S. Spiegler, O. Kedem, Thermodynamics of hyperfiltration (reverse osmosis): criteria for efficient membranes, *Desalination*, **1** (1966) 311–326.
20. A. Yaroshchuk, "Breakthrough" osmosis and unusually high power densities in Pressure-Retarded Osmosis in non-ideally semi-permeable supported membranes, *Sci. Rep.*, **7** (2017) 45168; doi: 10.1038/srep45168.
21. J. G. Wijmans, A. L. Athayde, R. Daniels, J. H. Ly, H. D. Kamaruddin, I. Pinnau, The role of boundary layers in the removal of volatile organic compounds from water by pervaporation, *J. Membr. Sci.*, **109** (1996) 135–146.
22. R. W. Field, P. Aimar, Ideal limiting fluxes in ultrafiltration: comparison of various theoretical relationships, *J. Membr. Sci.*, **80**(1) (1993) 107–115.
23. S. S. Vasan, R. W. Field, On maintaining the consistency between the film model and the profile of the concentration polarisation layer, *J. Membr. Sci.*, **279** (2006) 434–438.
24. R. W. Field, F. A. Siddiqui, P. Ang, J. J. Wu, Analysis of the influence of module construction upon forward osmosis performance, *Desalination*, **431** (2018) 151–156.
25. V. Vitagliano, P. A. Lyons, Diffusion coefficients for aqueous solutions of sodium chloride and barium chloride, *J. Am. Chem. Soc.*, **78**(18) (1956) 4538–4542; doi: 10.1021/ja01599a008.
26. S. Loeb, L. Titelman, E. Korngold, J. Freiman, Effect of porous support fabric on osmosis through a Loeb–Sourirajan type asymmetric membrane, *J. Membr. Sci.* **129** (1997) 243–249.
27. Q. She, R. Wang, A. G. Fane, C. Y. Tang, Membrane fouling in osmotically driven membrane processes: A review, *J. Membr. Sci.*, **499** (2016) 201–233.
28. J. R. McCutcheon, M. Elimelech, Influence of concentrative and dilutive internal concentration polarization on flux behavior in forward osmosis, *J. Membr. Sci.*, **284**(1–2) (2006) 237–247.
29. S. S. Manickam, J. R. McCutcheon, Understanding mass transfer through asymmetric membranes during forward osmosis: A historical perspective and critical review on measuring structural parameter with semi-empirical models and characterization approaches, *Desalination*, **421** (2017) 110–126.
30. C. Y. Tang, Q. She, W. C. L. Lay, R. Wang, A. G. Fane, Coupled effects of internal concentration polarization and fouling on flux behavior of

forward osmosis membranes during humic acid filtration, *J. Membr. Sci.*, **354** (2010) 123–133.
31. D. Attarde, M. Jain, S. K. Gupta, Modeling of a forward osmosis and a pressure-retarded osmosis spiral wound module using the Spiegler-Kedem model and experimental validation, *Separation and Purification Technology*, **164** (2016) 182–197.
32. J. J. Wu, On the application of the Spiegler-Kedem model to forward osmosis, *BMC Chem. Eng.*, **1** (2019) 15. https://doi.org/10.1186/s42480-019-0014-8.
33. A. E. Yaroshchuk, Osmosis and reverse osmosis in fine-porous charged diaphragms and membranes, *Adv. Colloid Interface Sci.*, **60** (1995) 1–93.
34. A. E. Yaroshchuk, Solution-Diffusion-Imperfection model revised, *J. Membr. Sci.*, **101** (1995) 83–87.
35. J. Wang, *et al.* A critical review of transport through osmotic membranes, *J. Membr. Sci.*, **454** (2014) 516–537.
36. J. R. McCutcheon, R. L. McGinnis, M. A. Elimelech, Novel ammonia–carbon dioxide forward (direct) osmosis desalination process, *Desalination*, **174** (2005) 1–11.
37. L. A. Hoover, W. A. Phillip, A. Tiraferri, N. Y. Yip, M. Elimelech, Forward with osmosis: emerging applications for greater sustainability, *Environ. Sci. Technol.*, **45** (2011) 9824–9830.
38. D. L. Shaffer, N. Y. Yip, J. Gilron, M. Elimelech, Seawater desalination for agriculture by integrated forward and reverse osmosis: improved product water quality for potentially less energy, *J. Membr. Sci.*, 415–416 (2012) 1–8.

CHAPTER 6

Modeling Ion Transport in Electrodialysis of Concentrated Solutions

Soraya Honarparvar[a], Chau-Chyun Chen[a], and Danny Reible[a,b,*]

[a] *Department of Chemical Engineering, Texas Tech University, Lubbock, TX 79409-3121, USA*
[b] *Department of Civil, Environmental, and Construction Engineering, Texas Tech University, Lubbock, TX 79409-1023, USA*

6.1. Introduction

Fresh water scarcity in many areas, due to population, industrial, and agricultural growth and drought, has increased demand for saline water desalination.[1–3] Electrodialysis (ED) is an electro-membrane desalination technique which utilizes an electric power source to separate ions from water via selective transport through ion exchange membranes (IEMs).[4,5] ED has been applied for brackish and seawater desalination, deacidification in the food industry, industrial effluent wastewater treatment, production of the boiler feed water, a pre-treatment technology for reverse osmosis, and in some cases, as a pre-evaporation step for salt recovery from seawater.[6–8] The efficiencies of ED at high salt concentrations[9] have also led to consideration of ED for the partial desalination of highly saline oil and gas flowback and produced water.[8] Use of ED in such an application, however, generates concern for the effects of non-ideal solution behavior at high salt concentrations.

*Corresponding author: danny.reible@ttu.edu

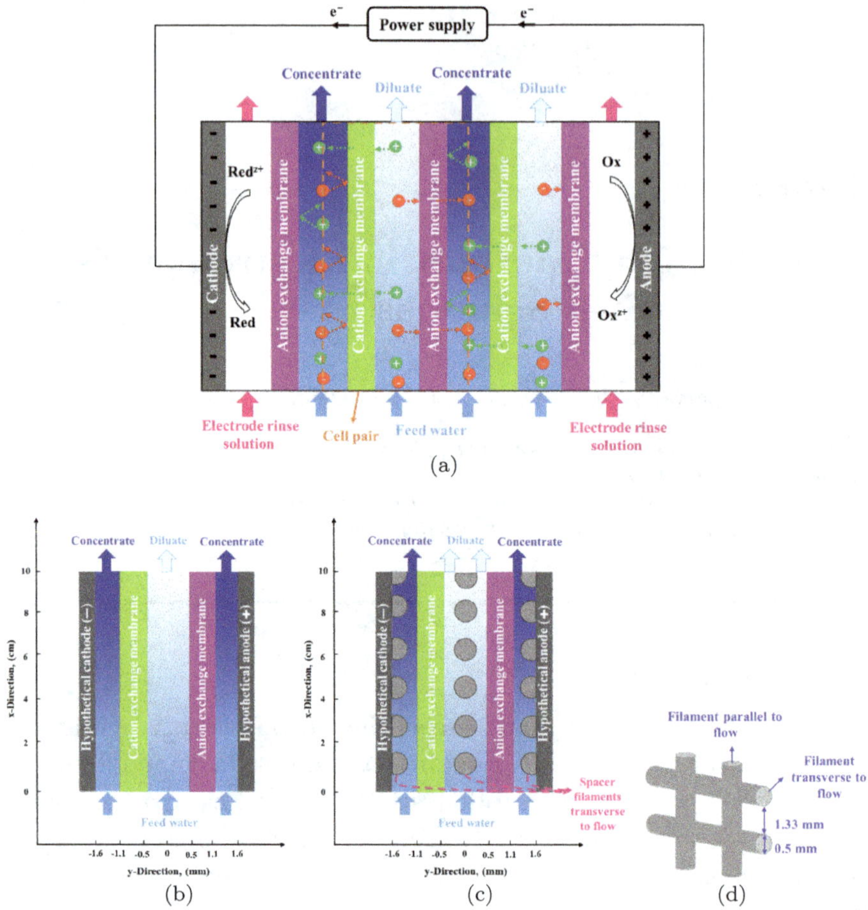

Fig. 6.1. (a) A schematic diagram of an ED stack, (b) a schematic of the spacer-free cell, (c) a schematic of the spacer-filled cell, and (d) a schematic of the spacer mesh.

Figure 6.1(a) demonstrates a schematic diagram of an ED stack. As feed water flows in channels between cation and anion exchange membranes (CEMs and AEMs), alternately placed between two electrodes, the imposed electric field drives ions toward opposite-signed electrodes. Counterions permeate through membranes (IEMs) while co-ions are hindered due to selective permeability of ion exchange IEMs. This results in decrease of ionic concentration in diluate channel and increase in adjacent concentrate compartments.

As shown in Fig. 6.1(a), an ED stack is made of repeating cell that contains a diluate channel, a concentrate channel, an AEM, and a CEM.

The higher transport of counterions inside IEMs compared to that in the solution, results in formation of concentration gradients in the boundary layers of the diluate and concentrate channels. This phenomenon is known as concentration polarization and can affect ion removal efficiency via increasing the potential drops at membrane-solution interfaces and raising ionic back diffusion from concentrate to diluate channel. Spacers are generally used inside channels to keep membranes separated and minimize concentration polarization.[10] These spacers can have various mesh size, filament diameter size, voidage, hydrodynamic angle, and material.[11] Spacer geometry and design influence the flow pattern and pressure drops in channels.[12,13]

Accurate modeling of ion transport in an ED cell helps optimize the operational conditions and reach maximum separation efficiency. Numerous studies have developed theoretical and empirical models for transport phenomena in ED. These models are generally based on the Nernst-Planck (NP) equation.[4,5,14–24] Using the NP equation for electrolyte solutions with equal valences of cations and anions leads to a convective-diffusion type equation, which is the basis of several models for ED.[4,25–27] Other modeling approaches include the use of Stefan-Maxwell equation,[28,29] non-equilibrium thermodynamics,[30–32] and semi-empirical models.[33–35]

Non-ideal solution behavior affects ion transport in ED. However, only a few studies have taken these impacts into account.[15,18] In application of ED for partial desalination of highly saline waters such as oil and gas produced water,[7,8] these non-ideal solution effects may become important. Numerous activity coefficient models developed for electrolyte solutions[36,37] can be used to model the non-ideality of bulk solution in channels. However, the non-ideal solution behavior inside IEMs has rarely been studied. Kamcev *et al.*[38] reviewed the experimental and theoretical studies estimating the ionic activity coefficients in IEMs. They further employed the Manning's condensation theory developed for polyelectrolyte solution to calculate the ionic activity coefficient in IEMs. Yu *et al.*[39] developed the polyelectrolyte nonrandom two-liquid (polyelectrolyte NRTL) activity coefficient model for polyelectrolyte solutions by which they successfully calculated the ionic activity coefficients in IEMs. These studies did not, however, evaluate the effects of the non-ideal solution behavior on ion transport in desalination systems.

In this study, a 2-dimensional ion transport model is developed to investigate the effects of solution non-ideality on ion transport in ED. The model accounts for co-ion transport through IEMs and eliminate the simplifying assumption of ideal membranes. The polyelectrolyte NRTL activity coefficient model is employed to calculate the ionic activity coefficients inside channels and IEMs. Both spacer-free and spacer-filled ED cells are modeled to investigate spacer effects on flow and ion transport. The investigation of these phenomena is used to evaluate their importance in subsequent applications of the developed model to examine selectivity or optimal operating conditions under various applications of ED. The model can also be used to explore and optimize conditions within an ED cell where no experimental data exist.

6.2. Modeling Framework

Figures 6.1(b) and 6.1(c), respectively, demonstrate the spacer-free and spacer-filled ED cells under study. Each cell includes a diluate channel (the middle domain), two half-concentrate channels, a CEM, an AEM, and a hypothetical cathode and an anode. The hypothetical cathode and anode represent the potential placed across a single cell by the overall potential difference placed across the entire multi-celled ED system. The baseline parameters used in the model including the membrane and channel dimensions, fixed charge concentration and water volume fractions of IEMs, diffusion coefficients of ions in bulk solution, ionic concentrations in feed water, inlet flow velocity, and temperature are summarized in Table 6.1. The reported values by Kamcev *et al.*[38,40] of physical and thermodynamic model parameters for CR61-CMP and AR103-QDP (Ionics Inc., USA) membranes are

Table 6.1. ED cell properties.

Parameter	Value	Unit
Channel width (W_{Ch})	1	[mm]
Channel length (L)	10	[cm]
Membrane width (W_M)	0.6	[mm]
Flow inlet velocity (V_0)	0.06	[m/s]
Inlet concentration (c_{in})	500–1500	[mol/m^3]
Spacer filament diameter (D_f)	0.5	[mm]
Temperature	298.15	[K]
Na^+ Diffusion coefficient in channels	1.33×10^{-9}	[m^2/s]
Cl^- Diffusion coefficient in channels	2.03×10^{-9}	[m^2/s]
Concentration of fixed charges on CEM[38,40]	3,210	[mol/m^3 (adsorbed water)]
Concentration of fixed charges on AEM[38,40]	3,580	[mol/m^3 (adsorbed water)]
Pure water volume fraction of CEM, $\phi_{w,CEM}$[38,40]	0.5 ± 0.014	[L(water)/L(swollen polymer)]
Pure water volume fraction of AEM, $\phi_{w,AEM}$[38,40]	0.4 ± 0.018	[L(water)/L(swollen polymer)]
ξ_{CEM}[38,40]	1.83	—
ξ_{AEM}[38,40]	2.21	—

used as model input parameters for CEM and AEM, respectively. As shown in Figs. 6.1(c) and 6.1(d), the submerged spacer configuration is used for the spacer-filled cell and the diameter of spacer filament and the mesh size are, respectively, set to 0.5 mm and 1.33 mm (associated with channel porosity of 85%) in the model.

NaCl is assumed to completely dissociate to its ions while water remains undissociated in the system. ED is typically operated under such potentials (0.5–1.5 V) that water dissociation is minimized. The flow is assumed to be incompressible and isothermal. Osmosis water transport through the membranes is neglected. The feed flows co-currently in the positive x-direction in channels while the electric field is in y-direction perpendicular to the flow streams. The fundamental equations applied for electrochemical systems including the NP equation, the Teorell diffusion, and the Navier-Stokes (NS) equations are adopted from literature[14,41–45] to model ion transport.

6.2.1. *Modeling transport in channels*

Molar balance at steady-state results in

$$\nabla \cdot N_i = 0 \tag{1}$$

where N_i is the molar flux of species i. Applying the NP equation with Teorell diffusion results in Eq. (2). for a non-ideal solution.

$$N_i = -M_i c_i \nabla \mu_i - z_i M_i F c_i \nabla \emptyset_s + c_i V \tag{2}$$

where M_i, c_i, μ_i, z_i are the mobility, the molar concentration, the chemical potential, and the charge number of species i (+1 for Na^+ and −1 for Cl^-), respectively; F denotes Faraday's constant (9.6×10^4 C/mol); $\emptyset_s$ is the potential in channels; and V is the flow velocity. In the NP equation, the first term represents the diffusion flux, the second term shows the electromigration flux, and the last term stands for the convection flux of species i.

The mobility of species i is calculated based on its diffusion coefficient using the Nernst-Einstein relation as

$$M_i = \frac{D_i}{RT} \tag{3}$$

here, D_i, R, and T are the diffusion coefficient of species i, the universal gas constant (8.314 J/mol.K), and the temperature, respectively. The chemical potential of components i is expressed as

$$\mu_i = \mu_{i,0} + RT \ln(a_i) \tag{4}$$

where $\mu_{i,0}$ denotes the chemical potential of species i at reference state and a_i is its activity defined as

$$a_i = c_i \gamma_i^c \tag{5}$$

where γ_i^c is the molar activity coefficient of component i. The substitution of Eqs. (3)–(5). into Eq. (2). leads to

$$N_i = -D_i \left(1 + \frac{\partial \ln \gamma_i^c}{\partial \ln c_i}\right) \nabla c_i - \frac{D_i}{RT} z_i F c_i \nabla \emptyset_s + c_i V \tag{6}$$

Substituting Eq. (6) into Eq. (1) results in

$$V \cdot \nabla c_i = D_i \nabla \cdot \left(\left(1 + \frac{\partial \ln \gamma_i^c}{\partial \ln c_i} \right) \nabla c_i \right) + \frac{D_i}{RT} z_i F \nabla \cdot (c_i \nabla \emptyset_s) \quad (7)$$

For a system containing N different ions, Eq. (7). provides a set of N equations and N + 1 unknowns, i.e., concentrations of N species and the potential field. The additional required relationship to solve this set of equations is provided by the electroneutrality assumption shown in Eq. (8). Note that the electroneutrality assumption is valid when the Debye length of the solution is negligible relative to the cell dimensions (for the concentration range of this study, Debye length is of the order of a nanometer).[46]

$$\sum_i z_i c_i = 0 \quad (8)$$

The current density, I, in channels is calculated as

$$I = F \sum z_i N_i$$
$$= F \sum z_i \left(-D_i \left(1 + \frac{\partial \ln \gamma_i^c}{\partial \ln c_i} \right) \nabla c_i - \frac{D_i}{RT} z_i F c_i \nabla \emptyset_s + c_i V \right) \quad (9)$$

Under the assumption of electroneutrality, the advection term is zero, hence, Eq. (9) becomes

$$I = F \sum_i z_i \left(-D_i \left(1 + \frac{\partial \ln \gamma_i^c}{\partial \ln c_i} \right) \nabla c_i - \frac{D_i}{RT} z_i F c_i \nabla \emptyset_s \right) \quad (10)$$

Here, the polyelectrolyte NRTL model is used to calculate the ionic activity coefficients inside channels and IEMs.[39] The detailed descriptions of the polyelectrolyte NRTL model is presented elsewhere[39] and a brief summary of the terms and parameters is provided here. According to the polyelectrolyte NRTL model, the ionic activity coefficients in the polyelectrolyte solution is calculated via Eq. (11).

$$\ln \gamma_i = \ln \gamma_i^M + \ln \gamma_i^{PDH} + \ln \gamma_i^{lc} \quad (11)$$

where γ_i^M is the Manning's term described by Manning's condensation theory and accounts for the long range electrostatic interactions between polyions (polymer containing charged functional groups) and free-ions; γ_i^{PDH} denotes the Pitzer-Debye-Hückel term which describes the electrostatic interactions between ions including the free-ions (co-ions and un-condensed counterions) and non-neutralized charged groups of polyions; γ_i^{lc} is the local term calculated from the NRTL model which represents the interactions between each species in their immediate neighborhood that includes the molecule-molecule, the ion-ion, and the ion-molecule interactions (note that ions include both free-ions and charged segments of the polyions); and i denotes cations, anions, and molecules in the solution.

In the absence of polyions in solution, e.g., in channels, the polyelectrolyte NRTL model provides the identical framework as electrolyte NRTL (eNRTL) model.[47] Within the concentration range of this study, the non-ideal behavior of the solution in the channels is mainly due to the electrostatic interactions between point ions. Hence, the Pitzer-Debye-Hückel term is dominant in calculating the species activity coefficients. Note that the contribution of local interactions becomes significant at higher ionic concentrations. Thus, the rational ionic activity coefficients (mole fraction basic) in channels is calculated as

$$\ln \gamma_i^x = \ln \gamma_i^{PDH} = -A_\phi \left\{ \left(\frac{2zi^2}{\rho} \right) \ln \left[1 + \rho I_x^{1/2} \right] + \frac{z_i^2 I_x^{1/2} - 2I_x^{3/2}}{1 + \rho I_x^{1/2}} \right\} \tag{12}$$

$$A_\phi = \frac{1}{3} \left(\frac{2\pi N_A}{\nu} \right)^{1/2} \left(\frac{Q_e^2}{\epsilon K_B T} \right)^{3/2} \tag{13}$$

$$I_x = \frac{1}{2} \sum_i z_i^2 x_i \tag{14}$$

$$\epsilon = A + B \left(\frac{1}{T} - \frac{1}{C} \right) \tag{15}$$

here, γ_i^x is the rational activity coefficient of species i; A_ϕ is the Debye-Hückel parameter; ρ denotes the closest approach parameter; I_x is the mole fraction-based ionic strength of the solution; N_A is the Avogadro's number; ν denotes the molar volume of water (cm^3/mol); Q_e is the electron charge (4.80298×10^{-10} E.S.U); ϵ is the dielectric constant of water; K_B is the Boltzman constant (1.3805×10^{-16});

x_i is the mole fraction of species i; A, B, and C are the correlation parameters of ϵ with respect to temperature and their values for water are 78.54, 31989.4, and 298.15, respectively; i refers to the cations and the anions in the system. At 298.15 K, A_φ and ρ are 3.04 and 14.9, respectively. The rational activity coefficients are converted to molarity-based activity coefficients using Eq. (16).[36]

$$\gamma_i^c = \frac{d_0}{d + 0.001c(\nu w_A - w_B)}\gamma_i^x \tag{16}$$

where d_0 denotes the solvent density; d is the solution density; c is the salt concentration (for NaCl in water $c_{\mathrm{NaCl}} = c_{\mathrm{Na+}} = c_{\mathrm{Cl-}}$); ν is the number of ions produced in the solution from dissociation of the electrolyte (two for NaCl); w_A is the solvent molecular weight; and w_B is the electrolyte molecular weight. The mole fraction of each component inside the solution is calculated as

$$x_i = \frac{c_i}{\nu c + \frac{(1000d - cw_B)}{W_A}} \tag{17}$$

Using the activity coefficient model, the non-ideal thermodynamic factor in the diffusion term in Eq. (10) becomes

$$\left(1 + \frac{\partial \ln \gamma_i^c}{\partial \ln c_i}\right) = \left(1 + \left(-A_\varphi c_i \left(\left(\frac{I_x^{-1/2}}{1+\rho I_x^{1/2}}\right) + \frac{1}{2}\left(\frac{I_x^{-1/2}}{(1+\rho I_x^{1/2})^2}\right)\right.\right.\right.$$
$$\left.\left. - \left(\frac{3I_x^{1/2} + 2\rho I_x}{(1+\rho I_x^{1/2})^2}\right)\right)\left(\frac{\frac{1000d}{W_A}}{\left(\nu c + \frac{(1000d - cw_B)}{W_A}\right)^2}\right)$$
$$\left. + \frac{0.001c_i(\nu w_A - w_B)}{d + 0.001c(\nu w_A - w_B)}\right) \tag{18}$$

6.2.2. *Modeling transport in membranes*

Here, the osmotic transport of solvent through membranes is neglected. Hence, diffusion and electromigration are the only transport mechanisms in IEMs. The molar balance equation in IEMs at steady-state is given as

$$D_{i,m}\nabla\cdot\left(\left(1+\frac{\partial \ln \gamma_{i,m}^{c}}{\partial \ln c_{i,m}}\right)\nabla c_{i,m}\right)+\frac{D_{i,m}}{RT}z_iF\nabla\cdot(c_{i,m}\nabla\O_m)=0 \quad (19)$$

where $D_{i,m}$ denotes the diffusion coefficient of species i inside IEMs; $\gamma_{i,m}^{c}$ is the molar activity coefficient of ions in the IEMs; $c_{i,m}$ is the concentration of species i in the IEMs; $\O_m$ is the potential field in the membrane domains; and i denotes the free ions (counterions or co-ions) in the IEMs.

Tortuosity effects due to the inflexible chain network of polymers in IEMs influence the ions diffusion coefficients inside membranes.[40] To account for these effects, Mackie and Meares[15] proposed a relationship between ion diffusion coefficient in aqueous solution and in IEMs presented as Eq. (20)

$$\frac{D_{i,m}}{D_i}=\left(\frac{\varphi_w}{2-\varphi_w}\right)^2 \quad (20)$$

where, φ_w is the volume of water uptake. Here, Eq. (20). is used to calculate ionic diffusion coefficients inside membranes using the reported values of φ_w for each membrane.[40] Note that electrostatic effects of membrane fixed charges on ionic diffusion coefficients[48,49] are neglected due to their minor contribution relative to those of the tortuosity.[40]

Electroneutrality approximation in IEMs accounts for the concentration of fixed charges as described in Eq. (21).

$$\sum_i z_i c_{i,m} + z_F c_F = 0 \quad (21)$$

here, c_F denotes the concentration of fixed charges on the membranes and z_F is the associated charge number (+1 for the AEM and −1 for the CEM). The current density is calculated via Eq. (22).

$$I_m = F\sum_i z_i\left(-D_{i,m}\left(1+\frac{\partial \ln \gamma_{i,m}^{c}}{\partial \ln c_{i,m}}\right)\nabla c_{i,m}-\frac{D_{i,m}}{RT}z_iFc_{i,m}\nabla\O_m\right) \quad (22)$$

IEMs contain free ions (co-ions and counterions), charged-containing polymers (with fixed charge groups) treated as polyions here, water molecules, and non-charged polymers. The ion exchange capacity (IEC) of IEMs represents the number of active charged functional groups per gram of dry polymer. For membranes with high IEC, the electrostatic interactions between polyions (the polymer with fixed charge groups) and free ions contribute significantly to solution non-ideality in membranes. Thus, the Manning's term becomes dominant in calculating ions activity coefficients and the contribution of the local and Pitzer-Debye-Hückel terms in polyelectrolyte NRTL model are negligible. Due to the high IEC of IEMs modeled here, the ionic activity coefficients are estimated using only the Manning's term.

$$\ln \gamma_{i,m}^{c} = \ln \gamma_{i,m}^{M,c} \tag{23}$$

where $\gamma_{i,m}^{c}$ is the molarity-based activity coefficient of species i in IEMs, $\gamma_{i,m}^{M,c}$ is the Manning's term of the polyelectrolyte NRTL model, and i represents co-ions or counterions inside membranes. The Manning's condensation theory has a single parameter, ξ, defined as the dimensionless linear charge density of polyions. The critical value of ξ parameter in polyelectrolyte solutions is defined as

$$\xi_{\text{crit}} = \frac{1}{|z_i z_p|} \tag{24}$$

here, z_p is the charge number of each charged segment of the polyions. In polyelectrolyte solution with ξ value greater than ξ_{crit}, counterions condensation occurs. The Manning's terms for molar activity coefficients of co-ions and counterions in IEMs[50] are described as

$$\ln \gamma_{\text{counterion}}^{M,c} = \ln \frac{X/\xi + 1}{X + 1} - \frac{X/\xi}{2(X/\xi + 2)}, \quad \xi > \xi_{\text{crit}} \tag{25}$$

$$\ln \gamma_{\text{co-ion}}^{M,c} = -\frac{X/\xi}{2\,(X/\xi + 2)}, \quad \xi > \xi_{\text{crit}} \tag{26}$$

$$\ln \gamma^{M,c}_{\text{counterion or co-ion}} = -\frac{\xi X}{2(X+2)}, \quad \xi < \xi_{\text{crit}} \tag{27}$$

$$X = \frac{c_P}{c_{\text{co-ion}}} \tag{28}$$

where c_P is the concentration of charged segments of the polyion.

For IEMs, c_p is equivalent to the concentration of fixed charges (c_F) which is calculated from IEC and φ_w,[38] and z_p is same as the charge number of fixed charge group (z_F). Kamcev *et al.*[38] used the membranes chemical structure, IEC, molecular mass of charged and uncharged monomers, and the dielectric constant of the water-swollen membrane to estimate the ξ for membranes used in this model.

The activities of species i in the IEMs, $a_{i,m}$, is described as

$$a_{i,m} = c_{i,m}\gamma^{M,c}_{i,m} \tag{29}$$

Using Eqs. (25)–(28), the non-ideal thermodynamic factor of diffusion fluxes in IEMs are expressed as

For $\xi > \xi_{\text{crit}}$

$$\left(1 + \frac{\partial \ln \gamma^{m,c}_{\text{counterion}}}{\partial \ln c_{\text{counterion}}}\right) = 1 + \frac{c_{\text{counterion}}}{c_{\text{co-ion}}}\left(\frac{\left(1-\frac{1}{\xi}\right)X}{(X+1)\left(1+\frac{X}{\xi}\right)} + \frac{\frac{x}{\xi}}{\left(\frac{X}{\xi}+2\right)^2}\right) \tag{30}$$

$$\left(1 + \frac{\partial \ln \gamma^{m,c}_{\text{co-ion}}}{\partial \ln c_{\text{co-ion}}}\right) = 1 + \frac{\xi X}{(X+2\xi)^2} \tag{31}$$

For $\xi < \xi_{\text{crit}}$

$$\left(1 + \frac{\partial \ln \gamma^{m,c}_{\text{counterion}}}{\partial \ln c_{\text{counterion}}}\right) = 1 + \frac{c_{\text{counterion}}}{c_{\text{co-ion}}}\left(\frac{\xi X}{(X+2)^2}\right) \tag{32}$$

$$\left(1 + \frac{\partial \ln \gamma^{m,c}_{\text{co-ion}}}{\partial \ln c_{\text{co-ion}}}\right) = 1 + \frac{\xi X}{(X+2)^2} \tag{33}$$

6.2.3. *The ideal solution model*

The ideal solution model serves as the limiting case in this study for comparison. Note that the ionic activity coefficients in the ideal solution are unity, thus

$$\left(1 + \frac{\partial \ln \gamma_i^c}{\partial \ln c_i}\right) = \left(1 + \frac{\partial \ln \gamma_{i,m}^c}{\partial \ln c_{i,m}}\right) = 1 \tag{34}$$

$$a_i = c_i \tag{35}$$

$$a_{i,m} = c_{i,m} \tag{36}$$

Substituting Eqs. (34)–(36) in the framework of the non-ideal solution model provides the outlines of the ideal solution model.

6.2.4. *Modeling flow*

6.2.4.1. *Spacer-free channels*

For the flow in our base case which is a spacer-free channel with the average inlet velocity of 0.06 m/s, the Reynolds number is 67, which is within the range of laminar flow.[10] Hence, the NS equations are solved to obtain the velocity and pressure profiles in channels. The IEM-solution interfaces are treated as no-slip boundaries. Symmetric boundary conditions are assumed for the rightmost and leftmost boundaries. At channels inlets, the normal inflow velocity profiles are imposed while the outlet boundaries are set to zero pressure and suppressing backflow conditions.

6.2.4.2. *Spacer-filled channels*

In a spacer-filled channel, the Reynolds number is calculated using the hydraulic diameter of the channel described as

$$d_h = \frac{4\epsilon}{\frac{2}{W_{ch}} + (1-\epsilon)S_{SpF}} \tag{37}$$

here, w_{ch} denotes the channel thickness; ϵ is the channel porosity; and S_{SpF} is the spacer filament specific surface area. The specific surface area of a cylindrical filament is calculated as

$$S_{SpF} = \frac{S_s}{V_s} = \frac{4}{d_f} \tag{38}$$

where S_S, V_S, and d_f are the surface area, volume, and diameter of the spacer filament, respectively. For the spacer-filled channel with d_f/w_{ch} of 0.5 and porosity of 0.85, the calculated Reynolds number is 72. Flow under these conditions is also expected to remain laminar.[10,51] Thus, the NS equations with the laminar flow assumption are employed to achieve velocity and pressure profiles. No-slip boundary condition is assumed on the surface of the spacer filament. The remaining boundaries are treated under the same conditions used for spacer-free channels.

6.2.5. *Boundary conditions*

Following boundary conditions are employed to solve the set of equations. The ion concentrations are set to zero at the membrane inlet, while they are fixed to the initial NaCl concentration at channel inlet. Due to symmetry, the concentration of species i at the left and right most boundaries are assumed to be equal. At steady-state, the continuity of ionic fluxes and current densities exist at the membrane-solution interfaces as described in Eqs. (39) and (40)

$$\hat{n} \cdot N_i = \hat{n} \cdot N_{i,m} \tag{39}$$

$$\hat{n} \cdot I = \hat{n} \cdot I_m \tag{40}$$

$\hat{n}$ is the unit vector perpendicular to membrane surfaces. Donnan equilibrium is assumed at the membrane-solution interfaces which results in

$$\emptyset_s - \emptyset_m = \frac{-RT}{Fz_i} \ln\left(\frac{a_i}{a_{i,m}}\right) \tag{41}$$

Hence, the total potential drop of the cell, $\Delta\emptyset_{\text{cell}}$, is calculated as

$$\Delta\emptyset_{\text{cell}} = \Delta\emptyset^c + \Delta\emptyset^m + \Delta\emptyset^D \tag{42}$$

where $\Delta\emptyset^c$ is the summation of ohmic potential drops in diluate and concentrate channels; $\Delta\emptyset^D$ is the summation of Donnan potential drops at the CEM-solution and AEM-solution interfaces; and $\Delta\varnothing^M$ is the summation of ohmic potential drops in CEM and AEM. Note that through Eq. (41), the boundary condition at the membrane involves the ratio of the activities and thus a ratio of the activity coefficients inside and outside the membrane. This may moderate non-ideal solution effects even though the activity coefficients are significantly different from unity for concentrated brine solutions.

6.2.6. *Numerical method and mesh structure*

The set of equations with defined boundary conditions are numerically solved using the finite element method with COMSOL Multiphysics 5.4 software with *electrochemistry* module. *Laminar Flow* interface with linear discretization of velocity and pressure and *Tertiary Current Distribution, Nernst-Planck* interface with quadratic discretization of concentration and potential are chosen to calculate the velocity, concentration, and potential distributions in the cell.

In the spacer-free, the mesh structure is simply a rectangular mesh with a higher density of elements near the boundaries and more widely spaced in the middle of the channels and membranes. A hybrid mesh structure is opted for the spacer-filled cell. Triangular elements are employed for regions outside the boundaries of the spacers and membranes which is well-suited to the system geometry. For the boundaries adjacent to the IEMs-solution interfaces and spacer filaments surfaces where sharp gradients of concentration or velocity exist, rectangular elements with relatively small size are used. For both cells, five mesh refinement steps were taken until results were determined to be mesh-independent. Figures 6.2(a) and 6.2(b) determine the mesh dependency of NaCl concentration, calculated by ideal solution model, at the IEM-solution interface of diluate channel for the cell potential of 1 V, cell length of 10 cm, and feed water concentration of 500 mol/m^3 in the spacer-free and spacer-filled cells, respectively. To assure the mesh independent results, the number of grid elements were set to 160,000 and 167,289 for the spacer-free and spacer-filled cells, respectively.

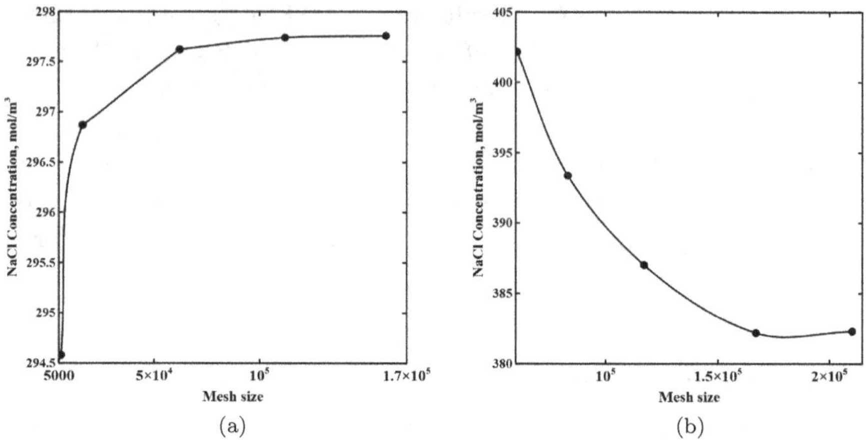

Fig. 6.2. The calculated NaCl concentration based on the ideal solution model at CEM-solution interface at the outlet of the diluate channel versus the mesh size at $x \approx 10\,\mathrm{cm}$, $V_{\mathrm{cell}} = 1\,\mathrm{V}$, and $c_{\mathrm{in}} = 500\,\mathrm{mol/m^3}$, (a) spacer-free cell and (b) spacer-filled cell.

6.2.7. *Comparison of the model results with the experimental data*

Comparison of the detailed model to experimental data is difficult primarily due to limited reporting of model parameters related to the IEMs and spacer characteristics. Here, we compare the calculated limiting current densities to experimental data, reported by Lee *et al.*[52] The limiting current density occurs when the concentration of ions at the membrane surface approaches zero. To reach a sound comparison of the modeling results with the experimental data, the model input parameters are selected according to reported values in the experimental studies of Lee *et al.*[52,53] A summary of these parameters are reported in Table 6.2. However, the concentration of fixed charges on the membranes and the associated water volume fractions as well as the physical dimensions of the spacers were not identified in the experimental study.[52,53] In our simulation, we use the values that are typically reported for the water volume fraction and fixed charge concentrations of the IEMs as well as the porosity and filament diameter of the commercial spacers. Those parameters are summarized in Table 6.2. Also, due to the very low NaCl

Table 6.2. ED cell properties for comparison with experimental data.

Parameter	Value	Unit
Channel width (W_{Ch})	6	[mm]
Channel length (L)	5	[cm]
CEM width (W_{CEM})	0.17	[mm]
AEM width (W_{AEM})	0.15	[mm]
Spacer filament diameter (D_f)	2	[mm]
Flow inlet velocity (V_0)	0.017–0.039	[m/s]
Inlet concentration (c_{in})	100	[mol/m^3]
Temperature	298.15	[K]
Na^+ Diffusion coefficient in channels	1.33×10^{-9}	[m^2/s]
Cl^- Diffusion coefficient in channels	2.03×10^{-9}	[m^2/s]
Concentration of fixed charges on CEM	3,210	[mol/m^3(adsorbed water)]
Concentration of fixed charges on AEM	3,580	[mol/m^3(adsorbed water)]
Pure water volume fraction of CEM, $\varphi_{w,\,CEM}$	0.5 ± 0.014	[L(water)/L(swollen polymer)]
Pure water volume fraction of AEM, $\varphi_{w,\,AEM}$	0.4 ± 0.018	[L(water)/L(swollen polymer)]

concentration of the feed water in the experiment (100 mol/m^3), we use the ideal solution model framework for the limiting current calculations. Testing at such a low concentration will still allow confirmation of the basic model structure.

Figure 6.3 demonstrates a comparison between the calculated limiting current densities in the spacer-free and spacer-filled cells and the reported experimental data[52] for NaCl concentration of 100 mol/m^3 and inlet flow velocity at 0.017–0.039 m/s. Two different mesh sizes are assumed for the spacer screens in the cell (resulting in channel porosity of 0.85 and 0.95) to investigate the effects of spacer dimensions on the calculated results. The model calculated and experimental limiting currents show an identical trend with velocity and are similar. The spacer-free cell, however, does not accurately reflect the experimental conditions since spacers are generally used in the channel to assure cross-channel mixing and to keep membranes separated. The high deviation between the experimental values and the calculated results for the spacer-free cell is due the elimination of

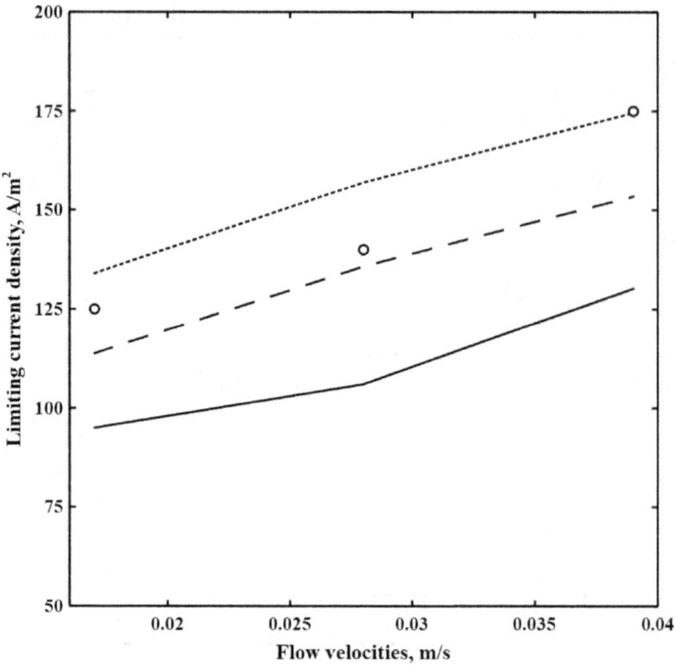

Fig. 6.3. The comparison between the calculated limiting current densities and reported experimental data[52] for NaCl concentration of 100 mol/m^3, $L = 5$ cm, $W_{Ch} = 6$ mm, $W_{CEM} = 0.17$ mm, $W_{AEM} = 0.15$ mm, and $D_f = 2$ mm. (——) Spacer-free cell, (– – –) spacer-filled cell with channel porosity of 0.95, (............) spacer-filled cell with channel porosity of 0.85, and (○) experimental measurements.[52]

such hydrodynamic effects on ion transport. The calculated limiting current densities for the channel with porosity of 0.95 are closer to the experimental values at lower inlet flow velocities. Since laminar flow may not exist at higher velocity, the deviation between the modeling results and experimental data at such condition can be caused by the laminar flow assumption in the modeling framework. For the spacer-filled cell with lower porosity, the enhanced cross-channel mixing results in better agreement between the calculated and experimental limiting current densities at higher velocity. The observed current densities may also be influenced by non-ideal flow and dispersion associated with processes such as entrance or exit effects or dead spots within the channel that were not simulated. Dispersion or a modified spacer structure could be fit to the experimental data but without the experimental observations of the missing parameters, this fitting would not provide further confidence in the model's ability to predict ion transport.

6.3. Results and Discussion

6.3.1. *Ideal solution–spacer-free cell*

The general concentration and potential distributions in the cell under baseline conditions without accounting for the impacts of the non-ideal solution behavior and spacers in channels are presented in Figs. 6.4 and 6.5, respectively. Note that the simulations are based upon a fairly short cell (L = 10 cm, Table 6.1) and therefore the results near the exit of the cell still represent developing profiles and not intended to indicate concentration profiles at the exit of a typical commercial application. Once the current reaches the limiting current density, however, the ion removal rate remains constant and the behavior in the cell is controlled by that removal rate.

Figure 6.4 shows the Na^+ concentration distribution across the cell at $x \approx 10\,\text{cm}$, V_{cell} of 1 V, and c_{in} of $500\,\text{mol/m}^3$. Due to electroneutrality, Na^+ and Cl^- concentration distributions in channels are identical, while the concentration of counterions inside IEMs (Na^+ in CEM and Cl^- in AEM) is the sum of the concentration of membrane fixed charges and co-ions (Cl^- in CEM and Na^+ in AEM). Higher transport of counterions inside the membranes compared to that in the solution results in depletion and increase of ion concentration in the boundary layers of diluate and concentrate channels, respectively. These boundary layers grow along the cell and eventually become fully developed. Due to the smaller diffusion coefficient and mobility of Na^+ ions compared to those of Cl^- ions, the diffusion and electromigration fluxes of Na^+ ions are relatively low. This results in greater concentration polarizations in boundary layers adjacent to the CEM than the ones next to the AEM.

In the existence of an electric field, co-ions move in the opposite direction of counterions and migrate back from concentrate channel to the diluate channel, conveying the equivalent amount of

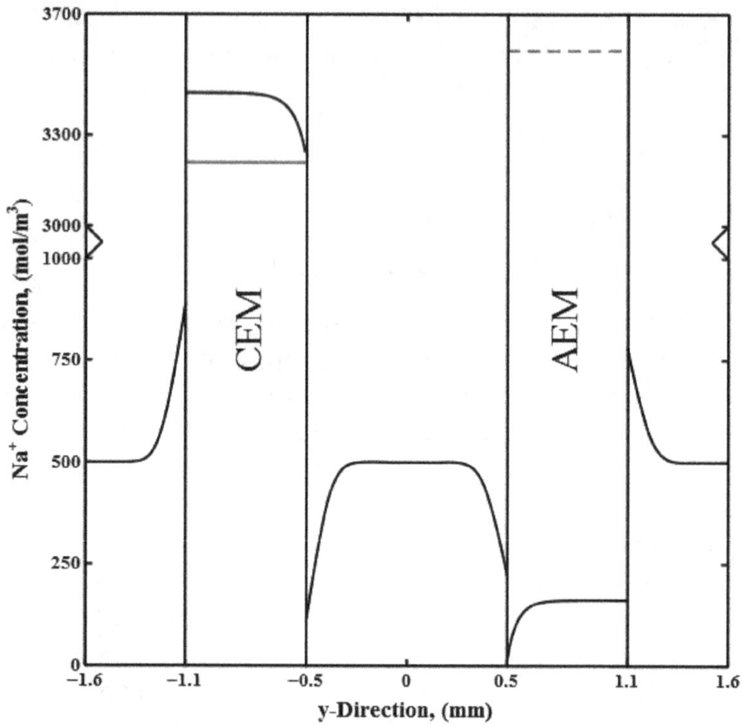

Fig. 6.4. Na^+ concentration distribution across the cell at $x \approx 10\,\text{cm}$, $V_{\text{cell}} = 1\,\text{V}$, and $c_{\text{in}} = 500\,\text{mol/m}^3$ are shown for the ideal solution model in the spacer-free cell (——); (————) the concentration of fixed charges on the CEM; (– – – – –) the concentration of fixed charges on the AEM.

counterions. Donnan potential drops at the solution-IEM interface oppose co-ions sorption to the membranes in the concentrate channel, while enhancing co-ions desorption to the solution in the diluate channel. Higher co-ions concentrations and lower Donnan potential drops at the concentrate channel-IEM interface result in a greater fraction of co-ions transport to the membranes.

Figure 6.5 shows the potential distribution across the cell at $x \approx 10\,\text{cm}$, V_{cell} of $1\,\text{V}$, and c_{in} of $500\,\text{mol/m}^3$. Outside the boundary layers, the uniform ion concentration distribution results in linear potential drops, while non-linear potential changes occur within these regions. Donnan equilibrium raises a discontinuity and a step change in potential at the IEMs-solution interfaces.

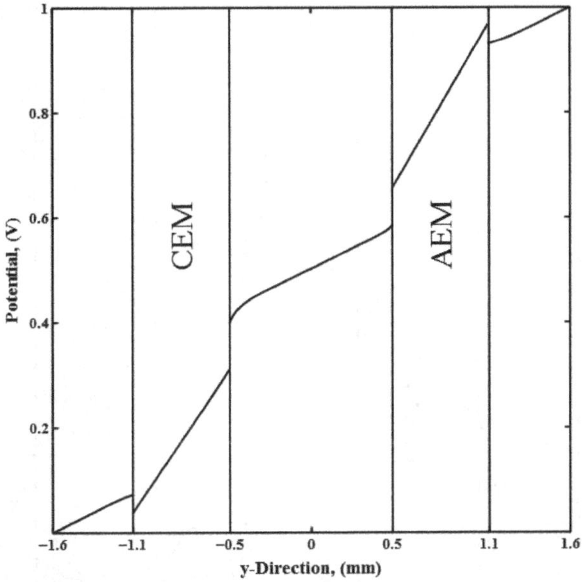

Fig. 6.5. The potential distribution across the cell at $x \approx 10\,\text{cm}$, $V_{\text{cell}} = 1\,\text{V}$, and $c_{\text{in}} = 500\,\text{mol/m}^3$ for the ideal solution model in the spacer-free cell (——).

6.3.2. *Non-ideal solution impacts*

The non-ideal solution model deviates from ideal solution model due to substitution of activity for concentration in calculating the Donnan potential drops at the IEM-solution interfaces and also accounting for non-ideal solution thermodynamic effects on diffusion fluxes. Figures 6.6 and 6.7 show the deviations between activities and diffusion coefficients calculated with ideal and non-ideal solution models, respectively. The associated impacts on the concentration distribution and consequently, the current density in the cell are shown in Figs. 6.8 and 6.9, respectively. The reported results are in the spacer-free cell.

Figure 6.6 shows the comparison between ideal and non-ideal solution models for Na^+ activities across the channel at V_{cell} of $1\,\text{V}$, $x \approx 10\,\text{cm}$, and c_{in} of $500\,\text{mol/m}^3$. Note that in the ideal solution model, activity is equivalent to concentration. The high concentration of counterions inside the membranes leads to a greater deviation between the ideal and non-ideal solution models within the

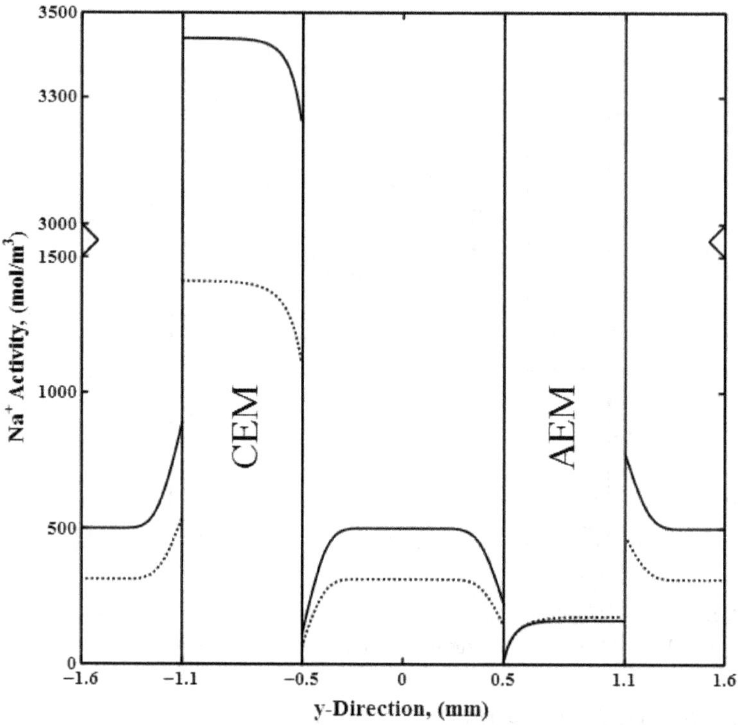

Fig. 6.6. Na^+ activity distribution across the cell at $x \approx 10\,\text{cm}$, $V_{\text{cell}} = 1\,\text{V}$, and $c_{\text{in}} = 500\,\text{mol/m}^3$ for the ideal solution model (_____) and the non-ideal solution model (. . . .) in the spacer-free cell.

membranes. For co-ions, the discrepancy is negligible due to the small absorbed concentration. In boundary layers of diluate channel, the ionic concentrations decrease and the activity coefficients approach unity, resulting in less deviations between ideal and non-ideal solution models results.

Figure 6.7 compares the ideal and non-ideal solution models results for the Na^+ diffusion coefficient across the cell at V_{cell} of 1 V, $x \approx 10\,\text{cm}$, and c_{in} of $500\,\text{mol/m}^3$. For the non-ideal solution model, the ionic diffusion coefficients in channels are smaller than those of the ideal solutions while in membranes, they are larger. Again, the deviations are modest at $500\,\text{mol/m}^3$ ionic strength although they become more important at $1500\,\text{mol/m}^3$.

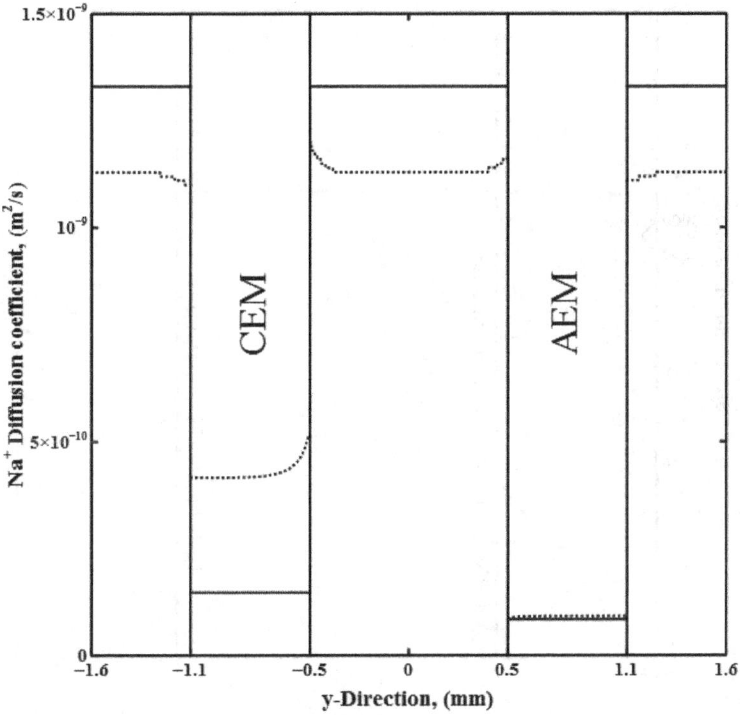

Fig. 6.7. Na^+ diffusion coefficient across the cell at $x \approx 10\,\text{cm}$, $V_{\text{cell}} = 1\,\text{V}$, and $c_{\text{in}} = 500\,\text{mol/m}^3$ for the ideal solution model (——) and the non-ideal solution model (..............) in the spacer-free cell.

Figure 6.8 depicts the Na^+ concentration distribution at V_{cell} of 1 V, $x \approx 10\,\text{cm}$, and c_{in} of 500 and $1500\,\text{mol/m}^3$ inside the CEM and in its vicinities in channels for both ideal and non-ideal solution models. As expected, the effect of non-ideality is more significant as the inlet concentration increases from 500 to $1{,}500\,\text{mol/m}^3$. The co-ion activity coefficient decreases with concentration in boundary layers of the concentrate channel, whereas it increases inside IEMs. Thus, the calculated concentrations using the non-ideal solution model inside IEMs at higher inlet concentrations are smaller than those of the ideal solution model. In no case, however, does the calculated concentration differ by more than 10% by assuming an ideal solution, even at $1500\,\text{mol/m}^3$. As noted previously, this is likely the result of the appearance of the ratio of ionic activities at the

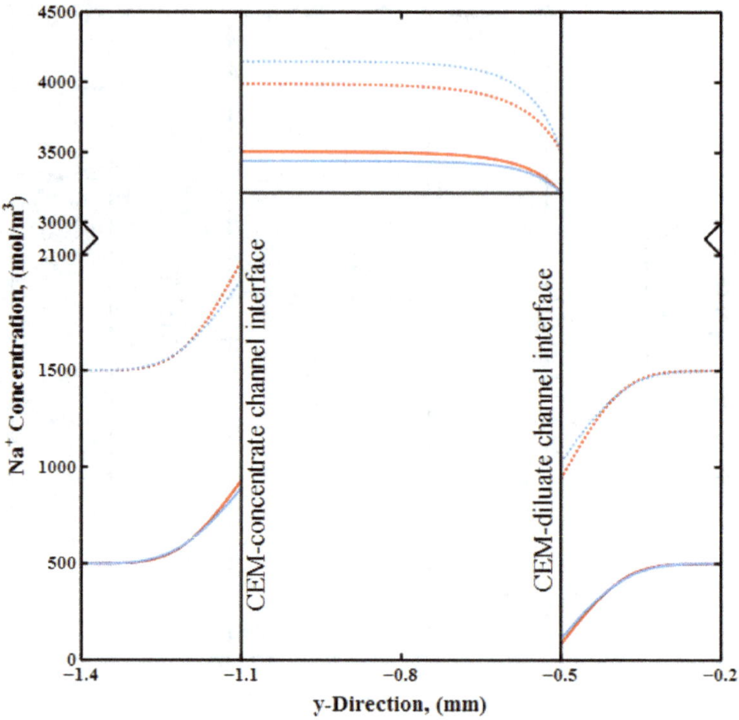

Fig. 6.8. Na^+ concentration distribution inside the CEM and in its outside vicinities in channels at $x \approx 10\,\text{cm}$ and $V_{\text{cell}} = 1\,\text{V}$ are shown for the ideal solution model, in blue, and for the non-ideal solution model, in red, in the spacer-free cell. (——) and (——) at $c_{\text{in}} = 500\,\text{mol/m}^3$, (……) and (……) at $c_{\text{in}} = 1500\,\text{mol/m}^3$; (——) fixed charge concentrations on the CEM.

membrane-solution interface which limits the deviations from ideality even though individual activity coefficients may differ substantially from unity.

Figure 6.9 compares the current density calculated with the ideal and non-ideal solution models in x-direction along the CEM in the diluate channel. The decrease in ionic concentrations along the cell reduces the current density. The smaller Donnan potential drops calculated with the non-ideal solution model results in higher calculated current densities compared to those of the ideal solution model. Again, the deviations by assuming ideal solution are less than 10% from those assuming non-ideal solutions.

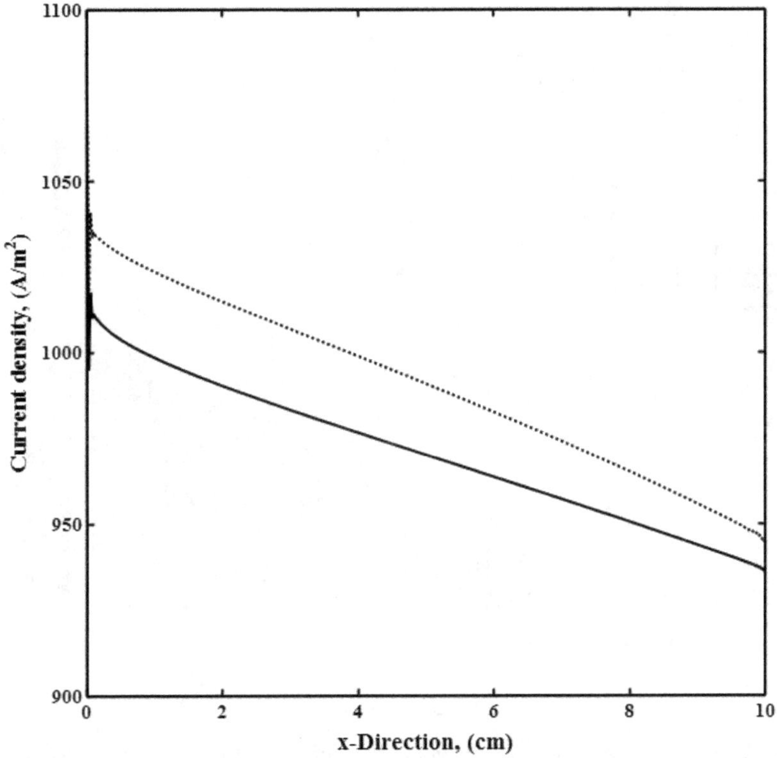

Fig. 6.9. The current density along the CEM in the diluate channel of the spacer free cell at $V_{\text{cell}} = 1\,\text{V}$ and $c_{\text{in}} = 500\,\text{mol/m}^3$ for the ideal solution model (______) and the non-ideal solution model (.............).

Overall, the non-ideal solution effects on ion transport are minor for salt concentrations up to $1500\,\text{mol/m}^3$ (three times higher than the salinity of seawater).

6.3.3. *Spacer impacts*

The impacts of the spacer filaments on flow pattern in channels are discussed in Fig. 6.10 and the comparison between concentration distribution and current density with and without spacer in the cell are provided in Figs. 6.11 and 6.12, respectively. In these simulations, an ideal solution is assumed as a result of the modest deviations in ion transport behavior due to the non-ideal solution effects.

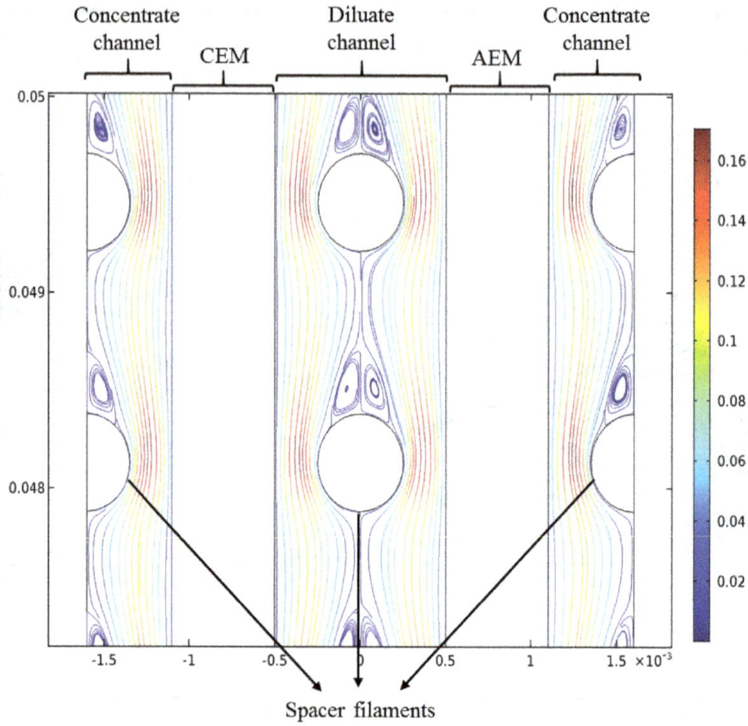

Fig. 6.10. A schematic of an enlarged section from the spacer-filled cell representing the flow streamlines in the cell for inlet flow velocity of 0.06 m/s. The color expression indicates the velocity magnitude (m/s) of the flow.

Figure 6.10 shows the flow streamlines in the spacer-filled cells. In spacer-free channels, the flow streamlines are simply parallel reflecting a simple unidirectional flow with a parabolic velocity profile in channels with maximum velocity of 0.09 m/s (50% larger than the average) at the center and zero velocity at the IEMs-solution interfaces. In spacer-filled channels, the maximum velocity of 0.17 m/s (almost triple the average flow) occurs at regions between each spacer filament and membranes surfaces. At the low Reynolds number of this study, the downstream filament has no impact on the recirculation region formed behind the upstream filament. The primary effect of the spacer filaments in channels is greater cross-channel mixing and smaller boundary layer thickness. The increase

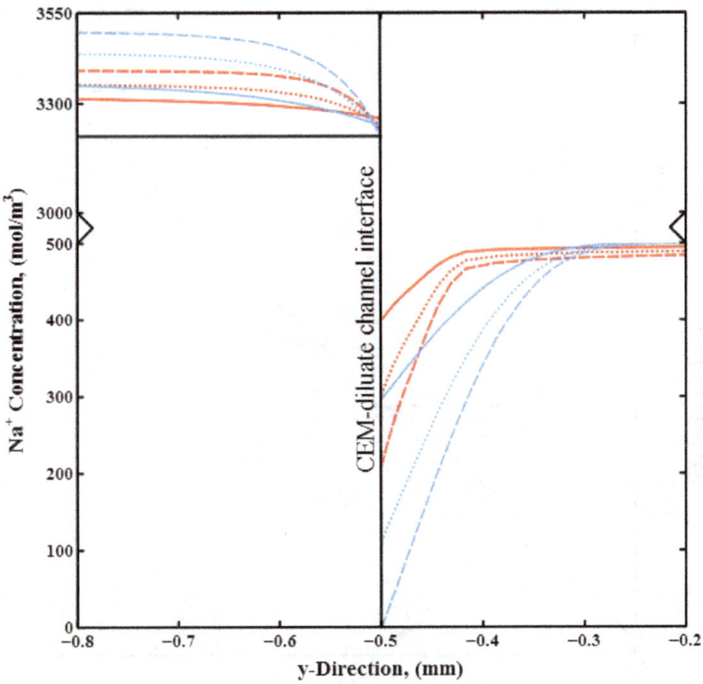

Fig. 6.11. Na^+ concentration distribution in the vicinity of the CEM-diluate channel interface at $x \approx 10\,\text{cm}$ and $c_{\text{in}} = 500\,\text{mol/m}^3$, shown for the ideal solution model in the spacer-free cell, in blue, and in the spacer-filled cell, in red, at various cell potentials. (——) and (——) at $V_{\text{cell}} = 0.5\,\text{V}$, (⋯⋯) and (⋯⋯) at $V_{\text{cell}} = 1\,\text{V}$, (- - - -) and (— — —) at $V_{\text{cell}} = 1.5\,\text{V}$; (——) fixed charge groups concentrations in the CEM.

in pressure drops from 260 Pa in the spacer-free channel to 1000 Pa in the spacer-filled channel indicates that the use of spacers, while decreasing the concentration polarization in the cell, increases the required pumping energy.

Figure 6.11 compares the Na^+ concentration distribution in the spacer-free and spacer-filled cells at $x \approx 10\,\text{cm}$, c_{in} of $500\,\text{mol/m}^3$, and V_{cell} of 0.5, 1, and 1.5 V in the vicinity of the diluate channel-CEM interface. Increasing the cell potential enhances the electromigration and the amount of ion transport through the membranes. In the spacer-free cell, the thickness of boundary layers and ion depletion increase at higher potentials. In the spacer-filled cell, however, the

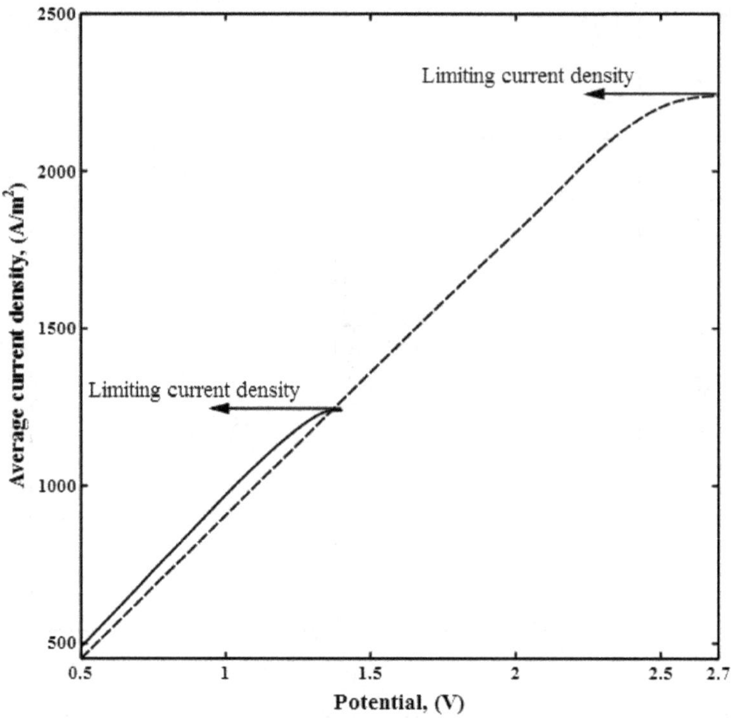

Fig. 6.12. The average current density along the CEM versus cell potentials plotted up to the limiting current density for the ideal solution model in the spacer-free cell (_____) and in the spacer-filled cell (_ _ _) at $c_{\text{in}} = 500\,\text{mol/m}^3$.

greater electromigration leads to decrease in the concentration across the whole channel width. The enhanced cross-channel mixing caused by spacer increases the ion concentrations at the IEMs-solution interface compared to those of the spacer-free cell.

Figure 6.12 shows the comparison between the average current density along the CEM in the diluate channel of the spacer-free and spacer-filled cells versus the imposed cell potential at c_{in} of $500\,\text{mol/m}^3$. Increasing the cell potential enhances the transport of counterions through the membranes, resulting in a decrease of ions in the boundary layers of the diluate channel. The limiting current density is reached in the system once the counterions concentrations at the membrane-solution interface approach zero. In general, the current-potential curves contain three regions as follows: a linear ohmic region (under-limiting current) followed by a plateau (limiting current region) before a linear increase (over-limiting current) which reaches an asymptotic value. In the ohmic region, which is the focus of the current effort, the electroneutrality assumption is valid. The decrease in ionic concentration in the boundary layers of the diluate channel increases the Debye length of the solution. At limiting and over-limiting current regimes, the electroneutrality assumption does not hold and Poisson equation should be used instead.[54,55] Beyond the limiting current density, further increasing the cell potential results in water dissociation and pH changes thus, leading to salt precipitation and damage to the membranes. As shown in Fig. 6.12, due to the enhanced mixing caused by spacer filaments in the spacer-filled channels, the required potential to reach limiting current density is high relative to the spacer-free cell. However, below the limiting current values, the use of spacers while increases the pressure drop in the cell does not improve the produced current density. For highly concentrated bulk solutions which is the focus of the current study, the required potentials to reach the limiting current density (maximum ion depletion at the membrane-solution interface) increases[52] and is beyond the typical values under which ED system is normally operated. The combination of these factors suggest that a spacer filled ED cell will exhibit no advantages over a cell with minimal spacer (simply to separate the membranes) for high salinity of feed water.

6.4. Conclusion

We develop a 2-dimensional steady-state ion transport model for an ED cell that accounts for co-ion transport through IEMs. The main focus of the study is to evaluate the effects of non-ideal solution and cell hydrodynamics on ion transport in ED of concentrated solution operated in the ohmic regime. The polyelectrolyte NRTL activity coefficient model is employed to calculate the ion activity coefficient in channels and membranes.

The developed ideal and non-ideal solution models are capable of describing the concentration and potential distributions in ED.

The non-ideal solution model takes into account the thermodynamic effects of highly saline solution on diffusion fluxes and calculates Donnan potential drops more accurately, thus providing a sound framework for modeling ED of concentrated feed water. The comparison between the results of ideal and non-ideal solution models exhibits slight deviations which increase with increasing feed water concentrations. The calculated Donnan potential drops of the non-ideal solution model are smaller compared to those of the ideal solution model, leading to a higher estimated current density in the cell. The developed ideal and non-ideal solution models can be used to optimize the cell design and operational condition for ED to reach high ion removal efficiency.

The comparison between results of the spacer-free and spacer-filled cells indicates that use of spacers reduces the thickness of the boundary layers and concentration polarization in channels. Thus, the limiting current density occurs at higher potentials in spacer-filled cells but this provides no real advantage in highly concentrated solutions which achieve limiting currents at higher than normally applied potentials. Below the limiting current value, using spacers increases the required pumping energy without any significant improvement of ion transport.

Acknowledgment

The authors gratefully acknowledge the financial support of the Donovan Maddox Distinguished Engineering Chaired Professorship and the Jack Maddox Distinguished Engineering Chaired Professorship in Sustainable Energy sponsored by the J. F. Maddox Foundation.

References

1. K. Reddy, N. Ghaffour, Overview of the cost of desalinated water and costing methodologies, *Desalination*, **20**(1–3) (2007) 340–353.
2. N. Ghaffour, T. M. Missimer, G. L. Amy, Technical review and evaluation of the economics of water desalination: Current and future challenges for better water supply sustainability, *Desalination*, **309** (2013) 197–207.

3. I. C. Karagiannis, P. G. Soldatos, Water desalination cost literature: review and assessment, *Desalination*, **223**(1–3) (2008) 448–456.
4. R. F. Probstein, *Physicochemical Hydrodynamics: An Introduction.* John Wiley & Sons, 2005.
5. M. M. Benjamin, D. F. Lawler, *Water Quality Engineering: Physical/Chemical Treatment Processes.* John Wiley & Sons, 2013.
6. W. Ho, K. Sirkar, *Membrane Handbook.* Springer Science & Business Media, 2012.
7. A. S. Yoshino, *Desalination of Produced Water with Electrodialysis: Ion Exchange Membrane Fouling, System Performance, and Fate of Organic Constituents.* Colorado School of Mines, 2017.
8. T. Sirivedhin, J. McCue, L. Dallbauman, Reclaiming produced water for beneficial use: salt removal by electrodialysis, *J. Membr. Sci.*, **243**(1–2) (2004) 335–343.
9. R. K. McGovern, A. M. Weiner, L. Sun, C. G. Chambers, S. M. Zubair, On the cost of electrodialysis for the desalination of high salinity feeds, *Appl. Energy*, **136** (2014) 649–661.
10. J. Schwinge, D. Wiley, D. Fletcher, Simulation of the flow around spacer filaments between narrow channel walls. 1. Hydrodynamics, *Ind. Eng. Chem. Res.*, **41**(12) (2002) 2977–2987.
11. J. D. D. Ocampo-Díaz, B. Valdez-Salaz, M. Shorr, M. Sauceda, N. Rosas-González, Review of corrosion and scaling problems in Cerro Prieto geothermal field over 31 years of commercial operations, in *Proceedings of World Geothermal Congress, International Geothermal Association (IGA), Antalya, Turkey,* 2005.
12. A. R. Da Costa, A. G. Fane, Net-type spacers: effect of configuration on fluid flow path and ultrafiltration flux, *Ind. Eng. Chem. Res.*, **33**(7) (1994) 1845–1851.
13. G. Belfort, G. A. Guter, An experimental study of electrodialysis hydrodynamics, *Desalination*, **10**(3) (1972) 221–262.
14. T. Teorell, Transport processes and electrical phenomena in ionic membranes, *Prog. Biophys. Mol. Biol.*, **3** (1953) 305–369.
15. J. Mackie, P. Meares, The diffusion of electrolytes in a cation-exchange resin membrane. I. Theoretical, in *Proceedings of the Royal Society of London A: Mathematical, Physical and Engineering Sciences,* **232**(1191) (1955) 498–509: The Royal Society.
16. A. G. Guzmán-Garcia, P. N. Pintauro, M. W. Verbrugge, R. F. Hill, Development of a space-charge transport model for ion-exchange membranes, *AIChE J.*, **36**(7) (1990) 1061–1074.
17. P. Moon, G. Sandí, D. Stevens, R. Kizilel, Computational modeling of ionic transport in continuous and batch electrodialysis, *Sep. Sci. Technol.*, **39**(11) (2004) 2531–2555.

18. J. A. S. J. M. Ortiz, E. Exposito, F. Gallud, V. Garcia-Garcia, V. Montiel, A. Aldaz, Brackish water desalination by electrodialysis: Batch recirculation operation modeling, *J. Membr. Sci.*, **252**(1) (2005) 65–75.
19. V. Volgin, A. Davydov, Ionic transport through ion-exchange and bipolar membranes, *J. Membr. Sci.*, **259**(1) (2005) 110–121.
20. A. Ribeiro, J. Rodriguez-Maroto, E. Mateus, E. Velizarova, L. M. Ottosen, Modeling of electrodialytic and dialytic removal of Cr, Cu and As from CCA-treated wood chips, *Chemosphere*, **66**(9) (2007) 1716–1726.
21. Y. Kim, W. S. Walker, D. F. Lawler, Electrodialysis with spacers: Effects of variation and correlation of boundary layer thickness, *Desalination*, **274**(1) (2011) 54–63.
22. M. Tedesco, H. Hamelers, P. Biesheuvel, Nernst-Planck transport theory for (reverse) electrodialysis: I. Effect of co-ion transport through the membranes, *J. Membr. Sci.*, **510** (2016) 370–381.
23. A. Moya, Numerical simulation of ionic transport processes through bilayer ion-exchange membranes in reverse electrodialysis stacks, *J. Membr. Sci.*, **524** (2017) 400–408.
24. M. Tedesco, H. Hamelers, P. Biesheuvel, Nernst-Planck transport theory for (reverse) electrodialysis: II. Effect of water transport through ion-exchange membranes, *J. Membr. Sci.*, **531** (2017) 172–182.
25. V. Shaposhnik, V. Kuzminykh, O. Grigorchuk, V. Vasil'eva, Analytical model of laminar flow electrodialysis with ion-exchange membranes, *J. Membr. Sci.*, **133**(1) (1997) 27–37.
26. V. Nikonenko, A. Istoshin, M. K. Urtenov, V. Zabolotsky, C. Larchet, J. Benzaria, Analysis of electrodialysis water desalination costs by convective-diffusion model, *Desalination*, **126**(1) (1999) 207–211.
27. A. Pismenskiy, V. Nikonenko, M. Urtenov, G. Pourcelly, Mathematical modelling of gravitational convection in electrodialysis processes, *Desalination*, **192**(1–3) (2006) 374–379.
28. P. N. Pintauro, D. N. Bennion, Mass transport of electrolytes in membranes. 1. Development of mathematical transport model, *Ind. Eng. Chem. Fundam.*, **23**(2) (1984) 230–234.
29. G. Kraaijeveld, V. Sumberova, S. Kuindersma, H. Wesselingh, Modelling electrodialysis using the Maxwell-Stefan description, *Chem. Eng. J. Biochem. Eng. J.*, **57**(2) (1995) 163–176.
30. A. Staverman, Non-equilibrium thermodynamics of membrane processes, *Transactions of the Faraday Society*, **48** (1952) 176–185.
31. B. Baranowski, Non-equilibrium thermodynamics as applied to membrane transport, *J. Membr. Sci.*, **57**(2–3) (1991) 119–159.

32. V. Nikonenko, V. Zabolotsky, C. Larchet, B. Auclair, G. Pourcelly, Mathematical description of ion transport in membrane systems, *Desalination*, **147**(1) (2002) 369–374.
33. Y. Tanaka, Mass transport and energy consumption in ion-exchange membrane electrodialysis of seawater, *J. Membr. Sci.*, **215**(1) (2003) 265–279.
34. Y. Tanaka, Concentration polarization in ion-exchange membrane electrodialysis — the events arising in a flowing solution in a desalting cell, *J. Membr. Sci.*, **216**(1) (2003) 149–164.
35. Y. Tanaka, M. Reig, S. Casas, C. Aladjem, J. L. Cortina, Computer simulation of ion-exchange membrane electrodialysis for salt concentration and reduction of RO discharged brine for salt production and marine environment conservation, *Desalination*, **367** (2015) 76–89.
36. J. F. Zemaitis Jr, D. M. Clark, M. Rafal, N. C. Scrivner, *Handbook of Aqueous Electrolyte Thermodynamics: Theory & Application.* John Wiley & Sons, 2010.
37. S. H. Saravi, S. Honarparvar, C.-C. Chen, Modeling aqueous electrolyte systems, *Chem. Eng. Prog.*, **111**(3) (2015) 65–75.
38. J. Kamcev, D. R. Paul, B. D. Freeman, Ion activity coefficients in ion exchange polymers: applicability of Manning's counterion condensation theory, *Macromolecules*, **48**(21) (2015) 8011–8024.
39. Y. L. Y. Yu, N. Hossain, C.-C. Chen, Nonrandom two-liquid activity coefficient model for aqueous polyelectrolyte solutions, *Fluid Phase Equilibria*, **497** (2019) 1–9.
40. J. Kamcev, D. R. Paul, G. S. Manning, B. D. Freeman, Predicting salt permeability coefficients in highly swollen, highly charged ion exchange membranes, *ACS Appl. Mater. Interfaces*, **9**(4) (2017) 4044–4056.
41. R. F. Probstein, *Physicochemical Hydrodynamics: An Introduction.* Butterworth-Heinemann, 2013.
42. R. B. Bird, W. E. Stewart, E. N. Lightfoot, *Transport Phenomena.* John Wiley & Sons, 2007.
43. J. Koryta, J. Dvorak, A. M. Bond, *Principles of Electrochemistry.* Wiley, New York, ed: Elsevier, 1987.
44. M. H. Friedman, *Principles and Models of Biological Transport.* Springer Science & Business Media, 2008.
45. J. Newman, K. E. Thomas-Alyea, *Electrochemical Systems.* John Wiley & Sons, 2012.
46. E. J. Dickinson, J. G. Limon-Petersen, R. G. Compton, The electroneutrality approximation in electrochemistry, *J. Solid State Electrochem.*, **15**(7–8) (2011) 1335–1345.

47. Y. Song, C.-C. Chen, Symmetric electrolyte nonrandom two-liquid activity coefficient model, *Ind. Eng. Chem. Res.*, **48**(16) (2009) 7788–7797.
48. G. S. Manning, Nonconvective ionic flow in fixed-charge systems, *J. Chem. Phys.*, **46**(6) (1967) 2324–2333.
49. G. S. Manning, Limiting laws and counterion condensation in polyelectrolyte solutions II. Self-diffusion of the small ions, *J. Chem. Phys.*, **51**(3) (1969) 934–938.
50. N. H. Y. Yu, C.-C. Chen, Modeling of polyelectrolytes system with electrolyte nonrandom two-liquid model, *AIChE Journal*, (2018).
51. J. Schwinge, D. Wiley, D. Fletcher, Simulation of unsteady flow and vortex shedding for narrow spacer-filled channels, *Ind. Eng. Chem. Res.*, **42**(20) (2003) 4962–4977.
52. H.-J. Lee, H. Strathmann, S.-H. Moon, Determination of the limiting current density in electrodialysis desalination as an empirical function of linear velocity, *Desalination*, **190**(1–3) (2006) 43–50.
53. H.-J. Lee, F. Sarfert, H. Strathmann, S.-H. Moon, Designing of an electrodialysis desalination plant, *Desalination*, **142**(3) (2002) 267–286.
54. N. D. P. V. Nikonenko, E. I. Belova, P. Sistat, P. Huguet, G. Pourcelly, C. Larchet, Intensive current transfer in membrane systems: Modelling, mechanisms and application in electrodialysis, *Adv. Colloid Interface Sci.*, **160**(1–2) (2010) 101–123.
55. M. A.-K. Urtenov, E. V. Kirillova, N. M. Seidova, V. V. Nikonenko, Decoupling of the Nernst–planck and poisson Equations. Application to a Membrane System at overlimiting Currents, *J. Chem. Phys. B*, **111**(51) (2007) 14208–14222.

CHAPTER 7

Membrane Distillation Pilot Units for Seawater Desalination

F. Macedonio[a,*], A. Criscuoli[a,*], and E. Drioli[a,b,*]

[a] *Institute on Membrane Technology, National Research Council of Italy (CNR-ITM), Rende, Italy*
[b] *University of Calabria — Department of Environmental and Chemical Engineering, Rende, Italy*

7.1. Introduction

The term desalination refers to any of several processes that remove excess salts and other minerals from water. Water is desalinated in order to be converted to fresh water suitable for human consumption or irrigation in regions where the availability of fresh water is limited. This process can also produce different types of salts as by-products. The principal desalination technologies can be classified by their separation mechanism into thermal and membrane desalination methods. Thermal desalination processes (such as Multi-stage flash (MSF), Multi-effect distillation (MED), Vapour compression (VC)) separate salt from water by evaporation and condensation, whereas membrane processes (such as reverse osmosis (RO), nanofiltration (NF), electrodialysis (ED)) use semi-permeable membranes and pressure or an electrical field to separate salts from water.

*Corresponding authors: f.macedonio@itm.cnr.it; a.criscuoli@itm.cnr.it; enrico.drioli@unical.it

Thermal desalination has been used for hundreds of years. At present, RO is the the most widespread. The key to extensive interest and implementation of seawater RO plants has been a significant reduction in capital and operation/maintenance costs over the past 30 years. Several factors have helped in reducing RO energy consumption and costs, including improvements in membrane materials and technology (higher flux, higher salt rejection, lower hydrostatic pressure required, lower materials cost) and the use of pressure recovery devices. Despite the great success of RO desalination, some aspects need to be further developed, such as the increasing of water recovery and the minimization of brine disposal problems. For each liter of desalted water, around a liter of brine is also produced and needs to be disposed. A solution for improvement of desalination process is offered by membrane distillation (MD). The latter is a thermally driven separation technique, using a hydrophobic microporous membrane. The hydrophobicity of the membrane retains the liquid feed solution, whereas the microporous membrane structure allows permeation of vapors. The lower vapor pressure on the permeate side can be imposed by a variety of methods, resulting in four basic configurations: in direct contact membrane distillation (DCMD), a cold permeate is in direct contact with the membrane. In air-gap membrane distillation (AGMD) an air gap is introduced between the membrane and a cold surface on the permeate side in order to reduce the heat loss of DCMD due to conduction. A cold sweep gas on the permeate side provides the driving force in sweeping gas membrane distillation (SGMD), removing the water vapor, which is condensed in an external condenser. In vacuum membrane distillation (VMD), a vacuum is applied on the permeate side. Recently, other configurations arose, such as:

- permeate/liquid gap membrane distillation (PGMD/LGMD), where the gap in AGMD is completely filled with permeate;
- material gap membrane distillation (MGMD) where the air gap in AGMD is filled with different materials like sponge (polyurethane) and PP mesh;

- vacuum-multi effect membrane distillation (V-MEMD) that integrates the concept of multi effect distillation into the VMD.

With respect to RO, MD presents a lower electrical energy consumption, because it works at atmospheric pressure, and it is much less affected by osmotic pressure limitations, being the driving force a difference of vapor pressure, so that it is able to efficiently treat high concentrated streams. There are, however, some drawbacks that have to be overcome for its industrial implementation, like the need of developing membranes and module designs with specific properties, of evaluating its performance in long-run tests and of reducing the thermal energy consumption. In this respect, many researches have been carried out and are still in progress, and promising improvements have been reached. This chapter aims at providing an update of progresses made in MD for desalination during last decades. An overview of the MD projects in the field and of MD pilots tested for both seawater and RO brine treatment is presented and discussed.

7.2. Membrane Distillation Projects

An important contribution to membrane distillation development was given by the various research projects funded, during the years, by various Organizations and, in particular, by the European Commission (EC). Table 7.1 summarizes main research projects on MD in the water sector (i.e., desalination, wastewater treatment, etc.).

Some of the funded projects focused on MD for industrial water management, such as INSPIREWater and MATChING:

- The EU project INSPIREWATER aimed to reduce water, raw material and energy consumption in the process industry. Special focus was on the steel and chemical industries, with the long-term goal of reducing fresh water consumption by 40 to 80 percent, combining reliable existing treatment processes and new technologies (such as forward osmosis and membrane distillation).

Table 7.1. Projects related to MD technology (modified from Ref. 1).

Sector	Title of the project	Note	Reference
Desalination	PV and thermally driven small-scale, stand-alone desalination systems with very low maintenance needs	Acronym: SMADES — Project ID: ICA3-CT-2002-10025 Funded under: FP5-INCO 2. From 2003-01-01 to 2005-12-31	https://cordis.europa.eu/project/rcn/67217/factsheet/en
Desalination	Development of stand-alone, solar thermally driven and PV-supplied desalination system based on innovative membrane distillation	Acronym: MEMDIS — Project ID: 13533 Funded under: FP5-EESD. From 2003-03-19 to 2008-03-18	https://cordis.europa.eu/project/rcn/86935/factsheet/en
Water production	Integrated Reverse Osmosis and Membrane Distillation for Brackish Water Desalination	The project is funded by WERC: A Consortium for Environmental Education and Technology Development at NMSU (New Mexico State University).	
Water production	Pilot-Scale Studies for Direct Contact Membrane Distillation-Based Desalination Process.	Agreement No. 04-FC-81-1037 From: October 2004 To: June 2007 Project funded by Desalination and Water Purification and Development Program, Bureau of Reclamation, Denver, Colorado.	https://www.usbr.gov/research/dwpr/reportpdfs/report134.pdf

(*Continued*)

Table 7.1. (*Continued*)

Sector	Title of the project	Note	Reference
Desalination	A study of the state of the art, commercial potential, and prospects for advancement of desalination by membrane distillation	Project funded by MEDRC	
Desalination	Design of novel membranes for desalination by direct contact membrane distillation	Project funded by MEDRC	
Water	Innovative Water Treatment	Acronym: INNOWA Project funded by Asia ProEco Call Period: May 2005–May 2007	http://www.innowa.org/
Water production	Seawater desalination by innovative solar-powered membrane-distillation system	Acronym: MEDESOL Project funded during the FP6 of the European Commission Period: 2006-10-01/2010-05-31	https://cordis.europa.eu/project/rcn/81390/factsheet/it
Desalination	Membrane-based Desalination: An Integrated Approach	Acronym: MEDINA Project funded by the European Commission under the FP6 programme Period: 15/October/2006-14/January/2010	https://cordis.europa.eu/project/rcn/81392/factsheet/en

(*Continued*)

Table 7.1. (*Continued*)

Sector	Title of the project	Note	Reference
Water production	Direct Contact Membrane Distillation of Brackish and Contaminated Water Sources for Sourcing Potable Water	Project ID: 2008DE130B Start Date: 3/01/2008. End Date: 2/28/2009 Project funded by: USGS	https://water.usgs.gov/wrri/grant-details.php?ProjectID=2008DE130B&Type=Annual
Desalination	Promotion of Renewable Energy for Water production through Desalination	Acronym: PRODES Period: 1/October/2008-1/October/2010 Project co-financed by the Intelligent Energy for Europe programme (contract number IEE/07/781/SI2.499059)	https://ec.europa.eu/energy/intelligent/projects/en/projects/prodes
Water treatment	Joint project: memsys clearwater Pte Ltd (memsys) — Nanyang Technological University (NTU) partnership to enhance water desalination technology	Period: eight-month joint project. Status: The project ended in August 2011. Project funded by Singapore's Environment and Water Industry Programme Office (EWI)	https://www.wateronline.com/doc/memsys-ntu-partnership-to-enhance-water-0001
Wastewater	Converting waste or solar heat to treated water using membrane distillation	Project Number: 2016. Project funded by WQRA. Status: Complete	Membrane Technology. Volume 2011, Issue 2, February 2011, Pages 1, 16

(*Continued*)

Table 7.1. (*Continued*)

Sector	Title of the project	Note	Reference
Desalination	Design Of Novel Membranes For Desalination By Direct Contact Membrane Distillation	Project Number: 04-AS-006 Project funded by MEDRC	
Water Supply, Treatment, Water Use	Direct Contact Membrane Distillation — "Addition of Fine Particles Mitigates Flux Decline in Direct Contact Membrane Distillation"	Project ID: 2009DE157B Project funded by U.S. geological Survey Federal Funds Start Date: 3/01/2009. End Date: 2/28/2010	https://water.usgs.gov/wrri/grant-details.php?ProjectID=2009DE157B&Type=Annual
Desalination	MEmbrane DIstillation in Remote AreaS	Acronym: MEDIRAS Project funded by the European Commission under the FP7 programme Period: 1st September 2008–31st August 2011	https://cordis.europa.eu/project/rcn/90319/factsheet/de
3R strategy: Reduce-Replace-Redesign	Zero Carbon Resorts	A SWITCH-Asia project funded by European Union Duration: 11/2009–4/2014	https://zerocarbonresorts.eu/
Desalination and juice concentration	Direct-Contact Membrane Distillation, Osmotic Distillation and Membrane Crystallization	Project funded by King Abdulaziz City for Science and Technology (KACST) Beginning and duration: November 2011–36 months	https://www.itm.cnr.it/index.php/en/component/content/article?id=31&Itemid=579

(*Continued*)

Table 7.1. (*Continued*)

Sector	Title of the project	Note	Reference
Desalination	Nanotechnological Application in WAter DESalination	Acronym: NAWADES Project type: FP7-ENV-2012-two-stage — Collaborative Project; ENV.2012.6.3-1 : Innovative resource efficient technologies, processes and services Funding Board: European Community — Project Reference: 308439 From: 01/10/2012 — Duration: 48 months	https://cordis.europa.eu/project/rcn/105525/factsheet/es
Desalination	Desalination by Solar Powered Membrane Distillation: Material and Process Optimization	Acronym: SolMeD — Budget: 250.000€ Action Aristeia II (Funded by the European Social Fund (ESF) and National Resources) 2007–2013. Duration : 18 months	http://www.solar.demokritos.gr/projects_en.html
Industrial water and desalination	Economically and Ecologically Efficient Water Management in the European Chemical Industry	Acronym: E4WATER. Project ID: 280756 Funded under: FP7-NMP — From 2012-05-01 to 2016-04-30	https://cordis.europa.eu/project/rcn/103308/factsheet/en

(*Continued*)

Table 7.1. (*Continued*)

Sector	Title of the project	Note	Reference
Wastewater and Drinking Water	Fully Integrated Membrane Bioreactor System for Wastewater Treatment in Remote Applications	Funded by SERDP (Strategic Environmental Research and Development Program). From 7/6/2012 to 7/6/2016.	https://www.serdp-estcp.org/Program-Areas/Environmental-Restoration/Wastewater-and-Drinking-Water/ER-2237
Water	Transfer of INNOvative techniques for sustainable WAter use in FERtigated crops	Acronym: FERTINNOWA. Project ID: 689687 Funded under: H2020 — From 2016-01-01 to 2018-12-31	https://www.fertinnowa.com/project/
Water	Resource recovery from industrial waste water by cutting edge membrane technologies	Project acronym: ReWaCEM Project starting date: 01 October 2016 — Duration in months: 36 Call (part) identifier: H2020-SPIRE-2016 Grant agreement no.: 723729	https://www.rewacem.eu/profile.html
Water water management in the process industry	Innovative Solutions in the Process Industry for next generation Resource Efficient Water management	Acronym: INSPIREWater. Project ID: 723702 Funded under: H2020-EU.2.1.5.3. — Sustainable, resource-efficient and low-carbon technologies in energy-intensive process industries From 2016-10-01 to 2020-03-31, ongoing project	https://cordis.europa.eu/project/rcn/205618/factsheet/en

(*Continued*)

Table 7.1. (*Continued*)

Sector	Title of the project	Note	Reference
Desalination	Global Membrane distillation, Valuable resource recovery, Pressure retarded osmosis Project	Acronym: Global MVP and/or GMVP. From 2013–To 2018. Project funded by Ministry of Land, Infrastructure and Transport (MOLIT) and Korea Agency for Infrastructure Technology Advancement (KAIA).	www.globalmvp.org
Water and power	Materials Technologies for Performance Improvement of Cooling Systems in Power Plants	Acronym: MATCHING (H2020). Project ID: 686031 Funded under: H2020-EU.2.1.3. — INDUSTRIAL LEADERSHIP — Leadership in enabling and industrial technologies — Advanced materials From: 2016-03-01 to 2019-08-31.	http://matching-project.eu/

- -MD was tested, at lab and pilot scale, in MATChING project for reducing the cooling water demand in the power sector.

However, the most part of the projects focused on seawater and/or brackish water desalination and many of them on solar powered MD:

- One of SMADES objectives was to develop appropriate small-scale, stand-alone desalination systems with low maintenance needs based on membrane distillation. The systems were in particular addressed to people living in remote areas (away from the public grid) via the utilization of improved MD modules with internal heat recovery function coupled with solar thermal collectors.
- Stand-alone MD desalination systems were developed in MEMDIS project, too. The systems integrate solar thermal and PV energy. The desalination energy was supplied entirely by solar thermal solar collectors and the electrical auxiliary energy was supplied by a PV system.
- MEDESOL project focused on the development of an environmentally friendly improved-cost desalination technology to fresh water supply in arid and semi-arid regions in EU and Third Countries based on solar MD;
- The main objective of MEDIRAS project was the development and demonstration of cost effective and very reliable MD solar driven desalination systems for water scarcity affected regions with high insolation.
- The coupling of the MD unit with solar energy was also investigated in MEDINA^2 project, both theoretically (through the modeling of different configurations involving a salinity gradient solar pond and a solar thermal collector) and experimentally (through a semi-industrial unit with VMD units coupled to solar collectors built in Tunisia).

Moreover, in various projects (such as MEDINA, E4WATER and GMVP), membrane distillation was investigated together with its extension, membrane crystallization (MCr), for improving the productivity of water treatment systems while recovering some of the valuable salts present in the highly concentrated streams of the plants.

In the case of E4Water,[3] pilot experiments were performed on a small pilot system, using spiral wound modules with PE membranes of about $12\,m^2$. The MD process was operated at a temperature of 70°C with a temperature difference across the membrane of 10°C. The pilot ran for more than three months, testing different parameters and set-ups with the conclusion that MD is a promising technology, especially for treating difficult process water streams, but not economically feasible yet on the required scale. In MEDINA and Global MVP projects, as it will be described in the following sections, the feasibility of MD/MCr operations for crystals recovery from RO brine streams was demonstrated.

7.3. Commercial MD Modules for Pilot Scale Applications

Despite the huge number of researches and studies on membrane distillation reported in the literature, few are the modules developed and available for large scale applications of the technique. The existing commercial modules are mainly based on the use of flat membranes in spiral wound and plate and frame configurations. Some attempts were made by KmX corportation (Canada) to produce hollow fiber membrane-based modules, but not many information about their properties and performance is provided.[4] A successful effort was done by Econity whose hollow fiber membrane modules have been recently used in the SWRO-MD pilot plant built at the institute of Fisheries Science in Pukyong National University (located in the Korean South-Eastern port city of Busan) in the framework of the GMVP project.

Referring to the membrane distillation configurations, the Air Gap Membrane Distillation, the Permeate/Liquid Gap Membrane Distillation and the Vacuum-driven Multi-Effect Membrane Distillation are those mostly considered for commercial modules design and development. In the following text, main properties of the commercial membrane distillation modules are presented. It has to be mentioned that one of the first modules for pilot/large scale applications of MD was that produced by TNO (Netherlands Organisation for Applied Scientific Research), who developed the Memstill system based on the AGMD configuration with internal heat recovery.[5,6] The Memstill technology was licenced to Aquastill and Keppel Seghers, but it was no longer exploited in recent years. Therefore, no detailed information on Memstill modules is provided in this section.

7.3.1. *Scarab development AB (Sweden)*

The Swedish Company[7] produces plate and frame modules working in AGMD configuration. In this case, the hot feed is in contact with one side of the membrane, while an air gap is created at the other membrane side, inserting a condensing foil which is in direct contact with a cooling stream. During the operation, the water vapour moves from the hot feed-membrane interface towards the condensing foil passing through the membrane micropores and the air gap. Then, it condenses and can be recovered as liquid (see Fig. 7.1).

A typical module has a membrane area of $2.8\,m^2$ and consists of 10 plastic cassettes, each one containing two parallel flat sheet membranes and two condensing walls (air gap width of 1 mm).

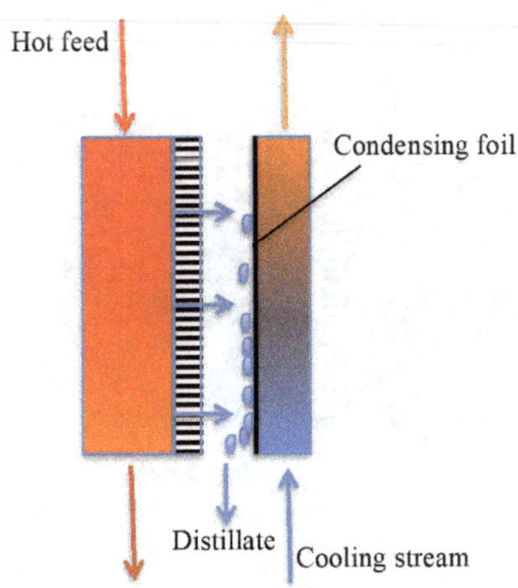

Fig. 7.1. Scheme of the AGMD configuration used in SCARAB modules.

Membranes used are spun-bounded polytetrafluoroethylene (PTFE) with a pore size of 0.2 μm, a porosity of 80% and a thickness of 200 μm.[8]

7.3.2. *Solar spring GmbH (Germany)*

The AGMD configuration is also chosen by Solar Spring GmbH (Germany).[9] However, in this case, spiral wound modules are produced with internal heat recovery: the feed to be treated is sent at ambient temperature (inlet temperature) inside the module where it is pre-heated thanks to the heat of condensation of the water vapour. The temperature difference across the membrane is ensured by supplying external heat to the hot feed stream that enters at the other side of the module at the so-called evaporator temperature (see Fig. 7.2).

The membrane is in PTFE with a pore size of 0.2 μm, a porosity of 80% and a thickness of 70 μm, and is supported on polypropylene (PP) having a porosity of 50% and a thickness of 280 μm. The module membrane area typically ranges between 7 and 12 m^2.[10,11] Solar Spring also produces another type of spiral wound membrane module where the air gap is replaced by a permeate liquid gap (PGMD).

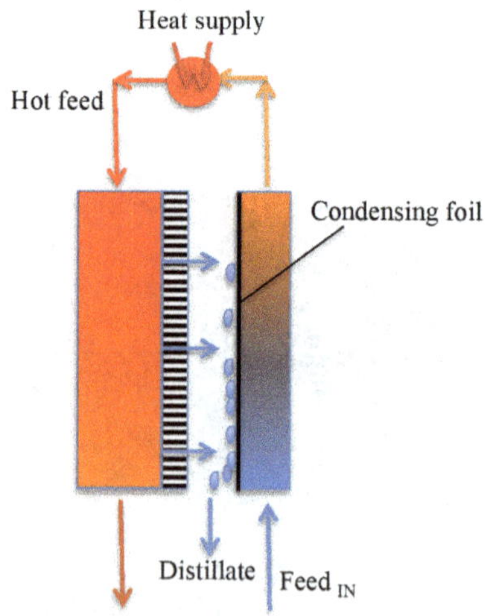

Fig. 7.2. AGMD configuration with internal heat recovery used in Solar Spring modules.

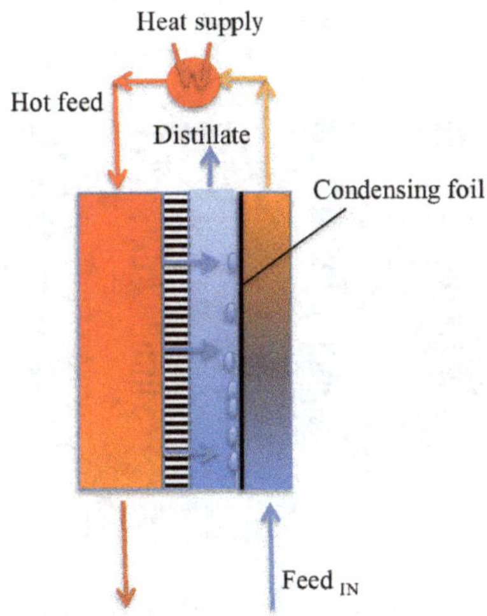

Fig. 7.3. PGMD configuration with internal heat recovery used in Solar Spring modules.

The typical membrane area varies from 5 to 14 m^2.[12] Figure 7.3 shows how the developed module works in the PGMD configuration.

7.3.3. *Aquastill BV (The Netherlands)*

Aquastill BV,[13] as Solar Spring, produces spiral wound membrane modules working mainly in AGMD with internal heat recovery. In this case, the membrane is not supported and is made of low-density polyethylene (LDPE) with a pore size of 0.3 μm, a porosity of 85% and a thickness of 76 μm. The module membrane area ranges between 7.2 and 24 m^2.[14] Recently, LGMD spiral wound modules have also been produced.

7.3.4. *Memsys GmbH (Germany)*

Memsys Company[15] produces VMEMD plate and frame modules equipped with PTFE membranes (pore size, 0.2 μm, porosity 70–75%, thickness 20 μm, 200 μm if including the supporting layer).[16] The module is equipped with an evaporator (steam raiser) and a condenser. In between there are the different distillation stages (effects). The water vapour production and condensation occur in each effect that works at lower temperature and pressure with respect

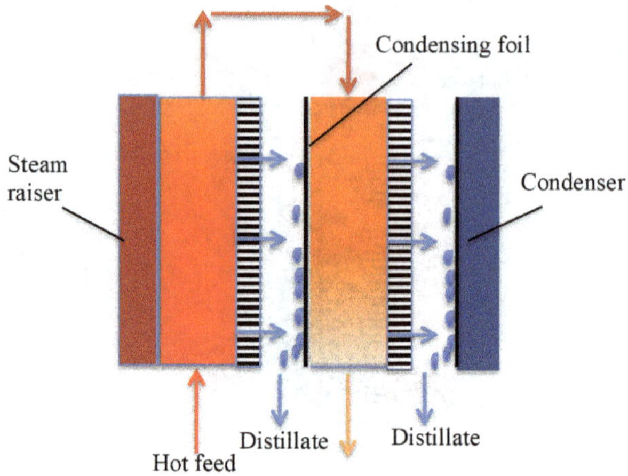

Fig. 7.4. Scheme of the VMEMD configuration for a module with two evaporation effects.

to the previous one (see Fig. 7.4). Membrane areas bigger than 100 m^2 can be obtained by adding more effects.

7.3.5. *Econity, INC (Korea))*

Econity's hollow fiber membrane[17] is made by stretching method with pores on the membrane surface structured in slit and ellipse in order to induce to a less-fouling function. In the framework of the GMVP project two different hollow fiber membrane modules[18] were utilized:

- a small one for the MD prototype: membrane material PVDF, effective membrane area 7.6 m^2, capacity 1 m^3/d (with 2 modules) and flux 6LMH when operated in DCMD configuration at feed temperature 80°C and temperature difference ΔT across the membrane of 20°C;
- a larger one used in the MD pilot (capacity 10 m^3/d) and MD demo pilot (400 m^3/d): area 10 m^2, pore size 0.1 μm, module material ABS, length 400 mm, configuration VMD.

7.4. Membrane Distillation Pilot Plants

A list of pilot MD systems used in desalination can be found in Table 7.2.[1,28] Only pilots that were led by commercial entities were

Table 7.2. List of pilot MD systems and their details (modified from Refs. 1 and 28). Reprinted with permission from Elsevier.

Project year	Executing company	Plant location	Capacity [m^3/day]	Energy source	Application	Membrane configuration	Membrane material	MD configuration	Notes	Ref.
1999	SCARAB	El Paso, USA	—	Thermal heat from solar ponds	Brackish water desalination	—	—	AGMD	Pilot plant	19,20
2004	SolarSpring	Gran Canaria, Spain	0.12	Solar themal + PV	Seawater desalination	Spiral wound	PTFE	AGMD/ PGMD	Pilot plant	21,22
2006	Keppel Seghers	Johor, Singapore	0.5–1	—	Seawater desalination	Spiral wound	PTFE	AGMD	Demonstration plant	23–25
2006	Keppel Seghers	Rotterdam, Netherlands	0.5–1	2 bar steam from coal fired power plant	Brackish water desalination	Spiral wound	PTFE	AGMD	Demonstration plant	24,25
2008	Keppel Seghers	Rotterdam, Netherlands	0.5–1	Waste heat from incineration plant	Seawater desalination	Spiral wound	PTFE	AGMD	Demonstration plant	24,25
2010	SolarSpring	Pantelleria Island, Italy	3.69	Solar thermal + waste heat from diesel generators	Seawater desalination	Spiral wound	PTFE	AGMD/ PGMD	Pilot plant	22

(*Continued*)

Table 7.2. (*Continued*)

Project year	Executing company	Plant location	Capacity [m^3/day]	Energy source	Application	Membrane configuration	Membrane material	MD configuration	Notes	Ref.
2010	SolarSpring	Amarika-Village, Namibia	5	Solar themal + PV	Brackish water desalination	Spiral wound	PTFE	AGMD/ PGMD	Pilot plant	22
2010	SolarSpring	Medinine, Tunisia	0.15	Solar themal + PV	Seawater desalination	Spiral wound	PTFE	AGMD/ PGMD	Pilot plant	26
2011	SolarSpring	Gran Canaria, Spain	1.4	Solar themal + PV	Seawater desalination	Spiral wound	PTFE	AGMD/ PGMD	Pilot plant	26
2011	Memsys Clearwater Pte. Ltd	Marina Barrage, Singapore	1	Solar themal + PV	Seawater desalination	Flat Sheet	PTFE	VMEMD	Pilot plant	27
2011	Memsys Clearwater Pte. Ltd	Tjuntjunjarra, Australia	1	—	Brackish water desalination	Flat Sheet	PTFE	VMEMD	Pilot plant	28
2011	Keppel Seghers	Jurong Island, Singapore	100	Heat from tempered water	Seawater desalination	Spiral wound	PTFE	AGMD	Demonstration plant	23

(*Continued*)

Project year	Executing company	Plant location	Capacity [m^3/day]	Energy source	Application	Membrane configuration	Membrane material	MD configuration	Notes	Ref.
2011	Scarab Development AB	Plataforma Solar de Almería (PSA), Spain	—	Solar	Marine salt water solutions of 1 and 35 g/l	Flat sheet	Spun-bounded PTFE	AGMD	Pilot plant	33
2012	Aquaver, Memsys Clearwater Pte. Ltd	Senoko Power Plant, Singapore	50	Waste heat from the boiler blow down from the refinery	Seawater desalination	Flat Sheet	PTFE	VMEMD	Demonstration	28
2013	Aquaver Memsys	Kamperland, Zeeland, Netherlands	1.5	Waste heat from diesel generator	Seawater desalination	Flat Sheet	PTFE	VMEMD	Pilot plant	28
2014	Aquaver, Memsys Clearwater Pte. Ltd	Kunfunadhoo Island, Maldives	100	Waste heat from existing diesel generators	Seawater desalination	Flat Sheet	PTFE	VMEMD	Commercial installation	28
2014	Aquaver, Memsys Clearwater Pte. Ltd	Gulhi Island, Maldives	10	Waste heat from diesel generator	Brackish water desalination	Flat Sheet	PTFE	VMEMD	Commercial Installation	28

(*Continued*)

Project year	Executing company	Plant location	Capacity [m^3/day]	Energy source	Application	Membrane configuration	Membrane material	MD configuration	Notes	Ref.
2016	Abengoa Water Memsys	Ghantoot, United Arab Emirates	60	Waste heat from diesel generator	SWRO brine concentration	Flat Sheet	PTFE	VMEMD	Demonstration plant	28
2013–2018	Econity (GMVP project)	South Korea	400	Partial energy requirement will be met by using MD brine as draw solution for PRO energy generation (5.0W/m^3)	Seawater and RO brine	Hollow fiber	PVDF	DCMD in the prototype/VMD in the pilot	Pilot demonstration plant	29–31

included. The table provides information such as plant capacity, membrane material and configuration, MD configuration, energy source and application.

As it can be observed, MD configurations more widespread in the pilots listed in Table 7.2 are AGMD and VMEMD due to the presence, as executing companies, of either SCARAB/Solar Spring/Keppel Seghers or Aquaver/Memsys. For what concerns the membrane configuration, Memsys-based pilots used flat sheet MD modules, whereas SolarSpring and Keppel Seghers used spiral wound modules. In all cases, the membranes were in PTFE. Completely different is the situation for the Global MVP project. It was financially supported by the Korean Ministry of Land, Infrastructure, and Transportation through the Korea Agency for Infrastructure Technology Advancement. Objective of the project was to utilize a RO-MD-PRO hybrid system to reduce the brine volume by 30% (through RO-MD) and to recover osmotic energy (through RO-PRO). The built MD plant utilized hollow fiber membrane modules in PVDF and DCMD configuration in the MD prototype whilst VMD configuration was adopted in the pilot. The latter consists of two units. Each unit has the capacity of $200\,m^3/d$ for a total water production of $400\,m^3/d$. Unit 1 has 120 hollow fiber membrane modules and unit 2 reduced the module number to 104 by improving the heat transfer efficiency (in both units each membrane module has a membrane area of $10\,m^2$). Average flux was 7 LMH for the unit 1, and 8 LMH for unit 2. Moreover, pilot plants realized in GMVP project as well the industrial installation of Abengoa Water in Ghantoot, concerned the application of MD to RO brine. Main results on the use of Memsys MD system in Ghantoot for the production of fresh water from RO brine are presented in the Memsys website.[15] A distillate production of $2.5\,m^3/h$, with a conductivity lower than $20\,\mu m/cm$ was obtained, at 47% of recovery rate. The specific thermal and electrical energy consumptions were 160 and $3\,kWh/m^3$, respectively.

In addition to the systems shown in Table 7.2, various other MD pilots have been developed in several research projects and not only, both for seawater desalination and RO brine concentration, as discussed in the following sections. Below it is reported Table 7.3 summarizing main characteristics of some of the MD systems used for these investigations.

7.4.1. *Membrane Distillation pilots for seawater treatment*

In the framework of the SMADES project,[32] a stand-alone solar-driven membrane desalination unit was designed and manufactured. The plant was installed at the Jordan University of Science and Technology (JUST) campus and prepared for long-term testing in the weather conditions of Irbid, Jordan. The plant integrates solar thermal and PV energy. The desalination system is composed of a spiral-wound AGMD module with internal heat recovery function. The membrane was in PTFE with an effective area of $10\,m^2$. The main advantages of the plant were: no need for external electricity, no additional water for cooling, and no impact on the environment. The plant has been continuously operated producing as high as 120 L/d with an approximate distillate conductivity of $5\,\mu S/cm$. The thermal energy required by the process was in the range of $200–300\,kWh/m^3$.

A solar AGMD desalination pilot plant coupled with a static solar collector's field was built in the context of the MEDESOL project.[33] Each employed MD module had a total membrane surface area of $2.8\,m^2$. The MD system contemplated the use of a multi-stage layout to minimize energy consumption. The modules showed specific distillate flux values up to a maximum of $6.5\,l/hm^2$ at a very high temperature difference between the membrane sides (65°C) when 1 g/l salt solution was used as feed. However, during the tests leakage and wetting problems were registered (in particular in the long-term operation due to suspected membrane damage).

A single-pass, pilot-scale AGMD process was analyzed by Duong *et al.*[34] The system contained a spiral-wound membrane module (Aquastill, Sittard, The Netherlands), designed specifically to recover the latent heat of condensation, and with microporous low-density polyethylene (LDPE) membrane (nominal pore size of $0.3\,\mu m$, thickness of $76\,\mu m$, membrane surface area of $7.2\,m^2$ and porosity of 85%). At the optimum operating conditions identified in the cited study,[34]

Table 7.3. Main characteristics of some of the MD systems applied in seawater and RO brine pilot investigations.

Module	Manufacturer	MD configuration	Membrane area (m^2)
SC	Scarab AB	Plate and frame AGMD	2.8
PT5	Keppel Seghers	Plate and frame LGMD (3 in series)	3 (* 3)
Oryx 150	Solar Spring	Spiral-wound PGMD	10
WTS-40A	Memsys/Aquaver	Plate and frame V-MEMD	5.76
memDist 4 — 6.4	Memsys	Plate and frame V-MEMD	6.4
AS7, AS24	Aquastill	Spiral-wound AGMD and V-AGMD	7.2/24

a stable AGMD operation for seawater desalination was achieved with specific thermal and electrical energy consumption of 90 and 0.13 kWh/m^3, respectively.

Air-gap, permeate-gap and vacuum MD were compared by Zaragoza *et al.*[35] utilizing commercial prototypes with flat-plate or spiral-wound membranes. In particular, they evaluated the commercially available or pre-commercial MD prototypes coupled with solar thermal named SC, PT5, Oryx 150 and WTS-40A, which consists in 6 effects, each one with a membrane area of 0.96 m^2 (see Table 7.3).

Regarding the productivity, the obtained distillated fluxes were in the same order of magnitude for the AGMD, LGMD and V-MEMD technologies, varying with operational factors like the feed flow rate and the temperature of the operation. Regarding the quality of distillate, acceptable values (distillate conductivity lower than 6 μS/cm) were obtained for all modules except in the cases when leaking occurred (for the Scarab modules and the Keppel Seghers' PT5 prototype). In terms of energy efficiency, there was a clear effect of the heat recovery in the Keppel Seghers' PT5 module but especially in the case of the Solar Spring's Oryx 150, which reached values of STEC as low as 210 kW h/m^3 (a factor of 5 lower than the minimum values obtained for the AGMD Scarab modules). Therefore, the spiral-wound Oryx 150 resulted as the most energy-efficient module of all the single-effect technologies tested. Considerable heat recovery was also demonstrated for the VMEMD technology represented by the Memsys/Aquaver WTS-40A prototype, with STEC values constantly lower than 400 kWh/m^3 and even decreasing below 200 kWh/m^3, while having a concentration factor one order of magnitude larger than in the single-effect technologies. However, this improvement was associated to a larger electric consumption due to the vacuum system. Energy consumption and heat recovery of MD modules were studied by Ruiz-Aguirre *et al.*,[14] too. They analyzed the different performance of the two multi-channel spiral-wound air-gap membrane AS7 and AS24 distillation modules at commercial-scale for seawater desalination. They found that, in the longest module, the heat recovery through the membrane was increased, reducing the permeate production and the energy consumption, while in the shortest module the opposite effect was taking place.

In the framework of the research project SolMed, the Memsys Vacuum Multi-Effect Membrane Distillation (V-MEMD) system was designed, installed and tested in an outdoor test site at NCSR "DEMOKRITOS".[36] Experimental results of the four effects V-MEMD showed that the distillate production increased from 0 to 50 L/h at temperatures of 50 and 80°C, respectively. Moreover, it was found that the distillate production increased with the reduction of the feed flow rate (due to the uncontrolled cooling water temperature that increased at higher feed flow rate). The energy efficiency of the system was calculated from the experimental data and showed a Specific Thermal Energy Consumption in the range of 300–700 kWh/m^3, Gained Output Ratio (GOR) was recorded to vary between 1–2.2 and Performance Ratio (PR) was calculated to be in the range 0.4–0.9 kg/Mj.

A solar compact MD demonstration plant of 100 L/d was installed in the facilities of the Instituto Tecnológico de Canarias (ITC) in Playa de Pozo Izquierdo (Gran Canary Island-Spain).[37] The unit was designed and installed at the end of 2004 as part of a co-funded FP6 DGTREN research project called 'Development of stand-alone, solar thermally driven and PV-supplied desalination systems based on innovative membrane distillation' (MEMDIS) and continued to be tested within the FP7 EU co-funded project called

'MEmbrane DIstillation in Remote AreaS — MEDIRAS'. The solar MD system was developed by the MEMDIS project coordinator — Fraunhofer ISE (Institute for Solar Energy Systems). The effective surface of the MD module was $8.5\,m^2$–$10\,m^2$ and operated in PGMD configuration. Specific thermal energy consumption of the system was in the range of 140–350 kWh/m^3 and the distillate water presented a conductivity range of 20–200 μS/cm. The level of autonomy made it an easy to operate and very low maintenance system.

Fraunhofer ISE in collaboration with MAHLE Industriefiltration GmbH[38] developed, constructed and characterized also a new multi-channel spiral-wound DCMD MD module with a membrane surface area of $27.5\,m^2$. The modules, operated with a nominal mass flow rate of $2400\,kgh^{-1}$, an inlet temperature of 80°C and a condenser inlet temperature of 27°C, registered a water production rate of $126\,kgh^{-1}$ at a thermal efficiency of 66%.

A detailed analysis for upscaling of MD module and plant was carried out by Hagedorn *et al.*[39] They tested spiral wound modules with different channel lengths in permeate gap MD (PGMD) and direct contact MD (DCMD) to evaluate the most suitable module configuration and design. The used modules were produced by the Fraunhofer Institute for Solar Energy Systems (ISE) and contained hydrophobic microporous ePTFE (expanded polytetrafluoroethylene) membrane. After pre-investigations of the MD process at lab scale, a pilot plant was designed, constructed and operated for 9 months continuously with seawater and under real conditions before constructing the pre-industrial unit. The pilot plant, installed on the island Helgoland in the North Sea, had a capacity of $2\,m^3$ freshwater per day. In the pre-industrial plant it was decided to go along with DCMD modules with a relatively short channel length. Its production capacity was of $5\,m^3$/d. Moreover, waste heat was used and a heat recovery and recycle concept was implemented and tested successfully. It was proven that long-term operation with real seawater was possible without significant performance decline.

Another DCMD pilot plant, using membrane surface areas between $1.30\,m^2$ and $6.6\,m^2$, was operated in the framework of the project "Pilot-Scale Studies for Direct Contact Membrane

Distillation-Based Desalination Process".[40] The membranes did not allow any salt leakage and achieved an average flux between 15 to 33 kg/m^{2}h when exposed to hot feed at 90–93°C.

7.4.2. *Membrane Distillation pilots for RO brine treatment*

7.4.2.1. *AGMD and PGMD/LGMD modules*

The application of spiral wound AGMD modules produced by Aquastill to the concentration of coal seam gas (CSG) RO brine was reported by Duong *et al.*[41] In particular, a 7.2 m^2 module was fed with the RO brine coming from a pilot UF/RO system working on the CGS produced water of a gas well in Gloucester, New South Wales (Australia). The RO unit was able to achieve 75% of water recovery, producing a brine with a conductivity of 21800 μS/cm. The MD unit was tested at different flow rates (350–450 L/h) and temperatures (50–60°C) and the permeate production ranged between 8 L/h and 16 L/h. Experiments in continous mode carried out at 55°C and 450 L/h led to stable permeate production (around 10 L/h), Gain Output Ratio-GOR (2.5) and Specific Thermal Energy Consumption — STEC (250 kWh/m^3). The rejection was around 98% with a distillate conductivity of 500 μS/cm. The MD water recovery was 80% and the water recovery of the integrated membrane system (UF-RO-MD) was 95%. From the achieved results, it was calculated that 1 ha of flat-plate solar collector was sufficient to ensure the thermal energy demand of the MD unit. The possibility of scaling risk in long-term operations was also highlighted.

Experiments on feed at different salinities (0–105 g/L) were carried out by Winter *et al.*[42] with a 10 m^2 PGMD module produced by SolarSpring. Tests were carried out at 80°C with feed flow rates varying between 200 and 500 L/h. The increase of the feed salinity reduced the distillate production and increased the specific energy consumption. For example, at 500 L/h, already when moving from 0 to 35 g/L (representing water and seawater composition, respectively) the distillate flow rate decreased from 25 to 20 kg/h and the specific energy consumption increased from 200 to 240 kWh/m^3.

Further reduction of the distillate flow rate and increase of the thermal energy demand were registered for feed compositions representing typical RO brines: 18 kg/h and 250 kWh/m^3 were measured for the 50 g/L feed while values of 16 kg/h and 300 kWh/m^3 were obtained for the 75 g/L feed. For the 105 g/L feed both values worsened, being 14 kg/h and about 335 kWh/m^3, respectively. Nevertheless, the distillate quality was always very high, being the distillate conductivity below 3 μS/cm for the investigated range of salinities. Gil *et al.*[43] used Response Surface Methodology (RSM) and Artificial Neural Networks (ANN) to optimize the performance of the module in terms of productivity and specific energy consumption when treating RO brines. Fixing the evaporator temperature at 80°C, it was found that, for high distillate productions, higher feed flow rates and lower inlet temperatures should always be used. However, to improve the thermal efficiency, lower feed flow rates and higher inlet temperatures have to be employed for salinity of 70 g/L, while higher feed flow rates and inlet temperatures should be used for feed salinity of 105 g/L. The performance of the 10 m^2 PGMD module was compared, in terms of quality of the produced distillate, with those of two AGMD Aquastill modules (7.2 and 24 m^2) by Ruiz-Aguirre *et al.*[44] who carried out experiments on synthetic salty solutions (0.6 M–2.4 M). All modules were tested in the same range of evaporator temperature (60–80°C) and inlet temperature (20–30°C), while the feed flow rate and salinity ranges were chosen in accordance to the Companies recommendations. Moreover, the Aquastill modules were also tested in the vacuum-enhanced air-gap (V-AGMD) configuration, by applying vacuum (150–250 mbar abs) in the air gap. The PGMD module showed constant and very high rejection values for all the investigated range of salinities. However, intermittent operation generated a poor permeate quality when the process was restarted (probably due to the fact that liquid feed passes through possible membrane pinholes while no production of permeate occurs). When the module was kept in contact with demineralized water after cleaning, permeate had a very high quality from the beginning. The AGMD and V-AGMD modules led to a good distillate quality only for the 0.6 M feed, with the exception of the

7.2 m^2 module working in AGMD configuration, for which a good removal efficiency was registered also for 1.2M feed. Higher the feed concentrations lower the salt rejection obtained with the Aquastill modules, with a worse performance of the 24 m^2 module with respect to the 7.2 m^2 one. Moreover, the application of vacuum at the gap side further reduced the rejection values, probably due to the higher leakage of liquid through pinholes. To mitigate the membrane wetting phenomenon, that is one of the main drawbacks of MD tests on highly concentrated brines, Schwantes *et al.*[11] investigated a modified 8.34 m^2 spiral wound AGMD SolarSpring module, for which it was possible to drain the air gap channel by low pressure air blowing. In this way, stagnant distillate at the bottom side, that could enhance the wetting risk, was removed. Experiments were carried out at 80°C and 300 L/h with feed salinities up to 240 g/L. Distillate production, GOR and thermal efficiency decreased with salt concentration, reaching values of about 5 kg/h, 0.7 and 25%, respectively, at 240 g/L. Concerning the distillate conductivity, a significant improvement of the permeate quality was registered by using the air blowing, especially for feed conductivities higher than 200 mS/cm.

7.4.2.2. *VMEMD modules*

A VMEMD module with only one effect (0.16 m^2) was studied by Naidu *et al.*[45] for the concentration of a 1M feed. Two feed flow rates (50 and 100 L/h) and three feed temperatures (around 310, 315 and 320 K) were considered for the experiments. Higher permeate fluxes were obtained by working at high feed flow rates and temperatures. The highest initial measured value was 16 kg/hm^2 at 100 L/h, 320 K and 1 M, decreasing to 14 kg/hm^2 at the end of the experiment (where a concentration factor of 3 was reached). Salt rejection was of 99.5% for all the tests, with a permeate conductivity of 0.3 mS/cm. Mendez[46] reported about the treatment of synthetic seawater at different concentrations (from 17 g/l to 175 g/L) by an Aquaver unit (Aquaver Company used Memsys modules for applications development). A permeate production of 60 L/h with a TDS content always lower than 2 mg/l was obtained, indicating the good performance of the module also at high salt content. The Memsys Company also provides some data on the performance of the Memsys modules for treating RO brines on its website.[15] For example, the 6.4 m^2 module, equipped with 4 distillation stages, has a flux of 7 kg/hm^2 when working on a 70 g/L feed at 80°C and at 45% recovery ratio. The thermal and electrical energies involved are 16 kW and 1 kW (max value), respectively.

7.5. Membrane Crystallization in Desalination

In addition to MD, its extension membrane assisted crystallization (MCr) offers the possibility to process the desalination brine streams for producing more desalted water and high-quality salts, thus approaching the concept of "zero-liquid-discharge" and "total raw materials utilization". MCr is able to promote crystals nucleation and growth in a well-controlled pathway, starting from undersaturated solutions. In a membrane crystallizer the membrane matrix acts as a selective gate for solvent evaporation, modulating the final degree and the rate for the generation of the supersaturation. Hence, the possibility to act on the trans-membrane flow rate, by changing the driving force of the process, allows to tailor the final properties of the crystals produced both in terms of structure (polymorphism) and morphology (habit, shape, size, and size distribution). The experimental evidences that can be found in literature,[47] mainly at lab-scale, validate the effectiveness of MCr as an advanced method for performing well-behaved crystallization processes. Figure 7.5 shows pictures of salts crystals recovered by membrane crystallization tests.

The studies carried out by Drioli and co-workers[48–52] showed that the introduction of a MCr unit on NF and RO retentate streams of an integrated membrane-based desalination system constituted by MF/NF/RO increases plant recovery factor so much to reach 92.8%, higher than that of a RO unit (about 45%) and much higher than that of a typical MSF (about 10–20%). Moreover, it has been experimentally shown that the presence of organic compound (i.e., humic acid) in the retentate inhibits crystals growth rate.[51] This

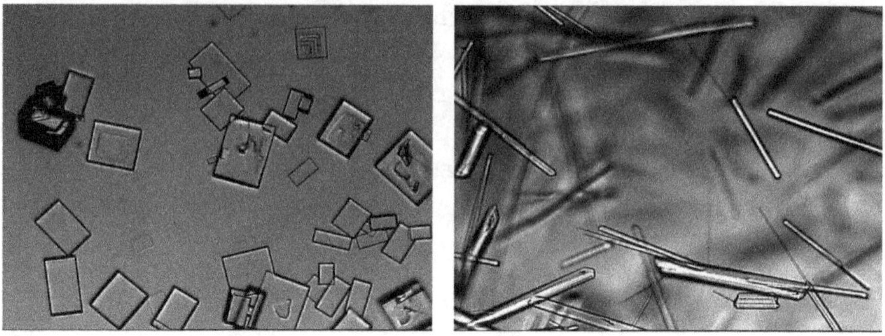

(a) (b)

Fig. 7.5. Crystalline habit of some salts recovered from brine through MCr. Picture from optical microscope: (a) NaCl (magnification: × 20), (b) $MgSO_4^*7H_2O$ (magnification: × 10).

proved the necessity to optimize the RO pre-treatment steps, in order not only to reduce the RO membrane fouling but also to control the crystallization kinetics that are linked with the nature and the amount of the foreign species existing in the highly concentrated brines emerging from the NF and RO stages. In some studies on MCr,[53] a rapid decrease in trans-membrane flux has been observed due to crystals deposition on the membrane which reduced the membrane permeability. This problem can be minimized through a right design of the process and a proper control of the operative conditions.

7.6. Future Research and Main Conclusions

In the last years membrane operations achieved a key role in water desalination. Despite the enormous success of membrane desalination technology and, in particular, of reverse osmosis, improvements are still required in terms of desalted water cost, higher productivity and brine disposal problem. Cost effective and environmentally sensitive concentrate management is today recognized as a significant hurdle to extensive implementation of desalination technologies. Development of innovative technologies, such as membrane distillation, provides unprecedented opportunities for the treatment and reuse of brine streams of desalination plants, reducing brine disposal problem.

National and international projects have also been funded for accelerating the development of this membrane operation. Commercial MD modules for pilot scale applications have been designed and promising results have been registered by applying them to both seawater and RO brine treatment. In most cases MD pilots have been coupled to solar units, so that to supply the thermal energy requested through renewable sources. The improvement of membrane properties, in order to keep their efficiency also in long-run tests with high concentrated feeds, will be an important target to reach for an effective management of the RO brine by MD. Improvements in membrane module designs and energy recovery systems will also help in this sense. Membrane distillation can also be used up to salts crystallization. To date, the potential of membrane crystallization in desalination has been explored mainly at lab-scale, but it is expected that the development in membranes and module designs will open to applications also at pilot-scale. Membrane systems with MD/MCr units offer, therefore, new opportunities in the design, rationalization, and optimization of water desalination cycles. These systems are opening a new era in desalination processes, the 3rd generation desalination installations.

References

1. F. Macedonio, E. Drioli (2019). Membrane distillation development. In *Sustainable Water and Wastewater Processing* (pp. 133–159). Elsevier.
2. E. Drioli, A. Criscuoli, F. Macedonio, (Eds.) *Membrane-Based Desalination: An Integrated Approach (Medina)*. Book Series: European Water Research Series. IWA Publishing, 2011.
3. https://cordis.europa.eu/project/rcn/103308/reporting/en
4. G. Zaragoza, J. A. Andrés-Manas, A. Ruiz-Aguirre, Commercial scale membrane distillation for solar desalination, *npj Clean Water*, **1** (2018) 20; doi:10.1038/s41545-018-0020-z
5. G. W. Meindersma, C. M. Guijt, A. B. de Haan, Desalination and water recycling by air gap membrane distillation, *Desalination*, **187** (2006) 291–301.
6. A. E. Jansen, J. W. Assink, J. H. Hanemaaijer, J. Van Medevoort, E. van Sonsbeek, Development and pilot testing of full-scale membrane distillation modules for deployment of waste heat, *Desalination*, **323** (2013) 55–65.

7. www.scarab.se
8. E. Guillén-Burrieza, J. Blanco, G. Zaragoza, D.-C. Alarcon, P. Palenzuela, M. Ibarra, W. Gernjak, Experimental analysis of an air gap membrane distillation solar desalination pilot system, *J. Membr. Sci.*, **379** (2011) 386–396.
9. www.solarspring.de
10. J. Koschikowski, M. Wieghaus, M. Rommel, V. Subiela Ortin, B. Penate Suarez, J. R. Betancort Rodriguez, Experimental investigations on solar driven stand-alone membrane distillation systems for remote areas, *Desalination*, **248** (2009) 125–131.
11. R. Schwantes, L. Bauer, K. Chavan, D. Ducker, C. Felsmann, J. Pfafferott, Air gap membrane distillation for hypersaline brine concentration: Operational analysis of a full-scale module-New strategies for wetting mitigation, *Desalination*, **444** (2018) 13–25.
12. D. Winter, J. Koschikowski, M. Wieghaus, Desalination using membrane distillation: Experimental studies on full scale spiral wound modules, *J. Membr. Sci.*, **375** (2011) 104–112.
13. www.aquastill.nl
14. A. Ruiz-Aguirre, J. A. Andrés-Manas, J. M. Fernandez-Sevilla, G. Zaragoza, Experimental characterization and optimization of multichannel spiral wound air gap membrane distillation modules for seawater desalination, *Sep. Purif. Technol.*, **205** (2018) 212–222.
15. www.memsys.eu
16. E. Sh. Mohamed, P. Boutikos, E. Mathioulakis, V. Belessiotis, Experimental evaluation of the performance and energy efficiency of a vacuum multi-effect membrane distillation system, *Desalination*, **408** (2017) 70–80.
17. http://econity.com/en/business/products.html
18. S.-H. Kim, Seawater desalination and its brine solution. Busan Global Water Forum, 2017/9/6 (retrieved from http://en.bwf.kr/data/files/B1_SeungHyun_Kim.pdf. Last access on October 31, 2019).
19. H. Lu, J. C. Walton, A. H. P. Swift, Desalination coupled with salinity-gradient solar ponds, *Desalination*, **136** (2001) 13–23
20. R. B. Saffarini, E. K. Summers, H. A. Arafat, J. H. LienhardV, Technical evaluation of stand-alone solar powered membrane distillation systems, *Desalination*, **286** (2012) 332–341
21. J. Koschikowski, M. Wieghaus, M. Rommel, V. S. Ortin, B. P. Suarez, J. R. Betancort Rodríguez, Experimental investigations on solar driven stand-alone membrane distillation systems for remote areas, *Desalination*, **248** (2009) 125–131.

22. K. Zhani, K. Zarzoum, H. Ben Bacha, J. Koschikowski, D. Pfeifle, Autonomous solar powered membrane distillation systems: state of the art, *Desalin. Water Treat.*, **57** (2016) 23038–23051.
23. C. Dotremont, B. Kregersman, R. Sih, K. C. Lai, K. Koh, H. Seah, Seawater desalination with memstill technology — a sustainable solution for the industry, *Water Pract. Technol.*, **5** (2010).
24. A. Jansen, J. Hanemaaijer, J. Assink, E. van Sonsbeek, C. Dotremont, J. Medevoort Van, Pilot plants prove feasibility of a new desalination technique, *Asian Water* (2010) 22–26.
25. L. M. Camacho, L. Dumée, J. Zhang, J. de Li, M. Duke, J. Gomez, *et al.*, Advances in membrane distillation for water desalination and purification applications, *Water (Switzerland)*, **5** (2013) 94–196.
26. Solar Spring, Oryx150 system purifying brackish water using 100% solar powered MD technology, (n.d.). http://www.solarspring.de/fileadmin/templates/solarspring/Downloads/Reference_Oryx_Tunisia_I.pdf (last accessed October 5, 2017).
27. Memsys secures a second pilot study, Global Water Intelligence (GWI), 2011. https://www.globalwaterintel.com/global-water-intelligence-magazine/12/1/general/memsys-secures-a-second-pilot-study (last accessed October 5, 2017).
28. N. Thomas, M. O. Mavukkandy, S. Loutatidou, H. A. Arafat, Membrane distillation research & implementation: Lessons from the past five decades, *Sep. Purif. Technol.*, **189** (2017) 108–127.
29. Global MVP, (n.d.). www.globalmvp.org (last accessed October 5, 2017).
30. E. Drioli, A. Ali, F. Macedonio, Membrane operations for process intensification in desalination, *Appl. Sci.*, **7** (2017) 100.
31. S.-H. Kim, Challenges in Future Desalination Brine Management, 2015. http://www.desaltech2015.com/assets/presenters/Kim_Seung-Hyun.pdf (accessed October 05, 2017).
32. F. Banat, N. Jwaied, M. Rommel, J. Koschikowski, M. Wieghaus, Desalination by a "compact SMADES" autonomous solarpowered membrane distillation unit, *Desalination*, **217**(1–3) (2007) 29–37.
33. E. Guillén-Burrieza, J. Blanco, G. Zaragoza, D. C. Alarcón, P. Palenzuela, M. Ibarra, W. Gernjak, Experimental analysis of an air gap membrane distillation solar desalination pilot system, *J. Membr. Sci.*, **379**(1–2) (2011) 386–396.
34. H. C. Duong, P. Cooper, B. Nelemans, T. Y. Cath, L. D. Nghiem, Evaluating energy consumption of air gap membrane distillation for seawater desalination at pilot scale level, *Sep. Purif. Technol.*, **166** (2016) 55–62.

35. G. Zaragoza, A. Ruiz-Aguirre, E. Guillén-Burrieza, Efficiency in the use of solar thermal energy of small membrane desalination systems for decentralized water production, *Appl. Energ.*, **130** (2014) 491–499.
36. E. S. Mohamed, P. Boutikos, E. Mathioulakis, V. Belessiotis, Experimental evaluation of the performance and energy efficiency of a Vacuum Multi-Effect Membrane Distillation system, *Desalination*, **408** (2017) 70–80.
37. R. G. Raluy, R. Schwantes, V. J. Subiela, B. Peñate, G. Melián, J. R. Betancort, Operational experience of a solar membrane distillation demonstration plant in Pozo Izquierdo-Gran Canaria Island (Spain), *Desalination*, **290** (2012) 1–13.
38. D. Winter, J. Koschikowski, F. Gross, D. Maucher, D. Düver, M. Jositz, . . . , A. Hagedorn, Comparative analysis of full-scale membrane distillation contactors-methods and modules, *J. Membr. Sci.*, **524** (2017) 758–771.
39. A. Hagedorn, G. Fieg, D. Winter, J. Koschikowski, T. Mann, Methodical design and operation of membrane distillation plants for desalination, *Chem. Eng. Res. Des.*, **125** (2017) 265–281.
40. https://www.usbr.gov/research/dwpr/reportpdfs/report134.pdf
41. H. C. Duong, A. R. Chivas, B. Nelemans, M. Duke, S. Gray, T. Y. Cath, L. D. Nghiem, Treatment of RO brine from CSG produced water by spiral-wound air gap membrane distillation — A pilot study, *Desalination*, **366** (2015) 121–129.
42. D. Winter, J. Koschikowski, M. Wieghaus, Desalination using membrane distillation: Experimental studies on full scale spiral wound modules, *J. Membr. Sci.*, **375** (2011) 104–112.
43. J. D. Gil, A. Ruiz-Aguirre, L. Roca, G. Zaragoza, M- Berenguel, Prediction models to analyse the performance of a commercial-scale membrane distillation unit for desalting brines from RO plants, *Desalination*, **445** (2018) 15–28.
44. A. Ruiz-Aguirre, J. A. Andrés-Manas, G. Zaragoza, Evaluation of permeate quality in pilot scale membrane distillation systems, *Membranes*, **9** (2019) 69, doi:10.3390/membranes9060069.
45. G. Naidu, Y. Choi, S. Jeong, T. M. Hwang, S. Vigneswaran, Experiments and modeling of a vacuum membrane distillation for high saline water, *J. Ind. Eng. Chem.*, **20** (2014) 2174–2183.
46. E. Mendez, Sustainable desalination: Membrane distillation delivers *greener* clean water, *Filtration + Separation*, September/October 2012, 26–28.
47. E. Drioli, G. Di Profio, E. Curcio, *Membrane-Assisted Crystallization Technology, Advances in Chemical and Process Engineering — Vol. 2.* Imperial College Press, 2015.

48. F. Macedonio, E. Curcio, E. Drioli, Integrated membrane systems for seawater desalination: energetic and exergetic analysis, economic evaluation, experimental study, *Desalination*, **203** (2007) 260–276.
49. F. Macedonio, E. Drioli, Pressure-driven membrane operations and membrane distillation technology integration for water purification, *Desalination*, **223** (2008) 396–409.
50. F. Macedonio, E. Drioli, E. Curcio, G. Di Profio, Experimental and economical evaluation of a membrane crystallizer plant, *Desalination and Water Treatment*, **9** (2009) 49–53.
51. F. Macedonio, E. Drioli, Hydrophobic Membranes For Salts Recovery from Desalination Plants, *Desalination and Water Treatment*, **18** (2010) 224–234.
52. F. Macedonio, A. Politano, E. Drioli, A. Gugliuzza, Bi 2 Se 3-assisted membrane crystallization. *Materials Horizons*, **5**(5) (2018), 912–919.
53. C. M. Tun, A. G. Fane, J. T. Matheickal, R. Sheikholeslami, Membrane distillation crystallization of concentrated salts — flux and crystal formation, *Journal of Membrane Science*, **257** (2005) 144–155.

CHAPTER 8

Energy Efficiency Metrics in Membrane Distillation

David M. Warsinger*, Sina Nejati, and Hamid Fattahi Juybari

Birck Nanotechnology Center, School of Mechanical Engineering, Purdue University, West Lafayette, IN, USA

8.1. Introduction

New technologies continue to grow and emerge to address emerging challenges to water and energy resources caused by population growth and climate change.[1,2] In water-stressed regions, desalination technologies to extract freshwater from seawater or brackish water are well-established, but efforts continue to make them more efficient and applicable to more water source types.

Membrane distillation (MD) is an emerging thermal desalination technology in which feed water evaporates at a membrane that permits water vapor to pass while preventing any liquid water to penetrate. The vapor is then condensed on the other side of the system producing pure water (Fig. 8.1).[3] MD offers unique advantages over other technologies including the ability to scale down the system to smaller sizes, being fouling resistant, the ability to work at low temperatures, and potentially having competitive efficiencies compared to other thermal desalination technologies like multi-effect distillation (MED) and multi-stage flash (MSF). Deshmukh *et al.*[4]

*Corresponding author: david.warsinger@gmail.com

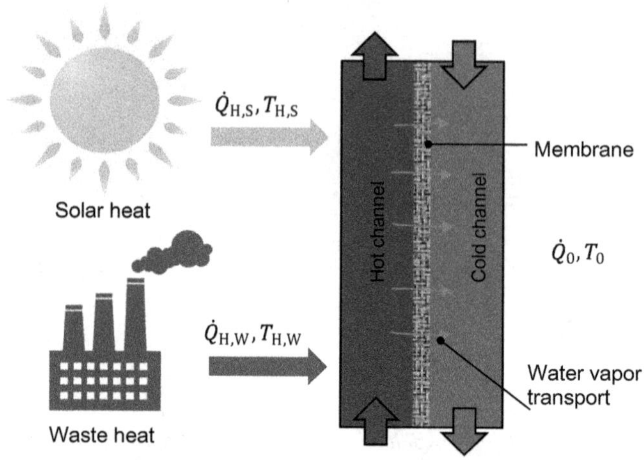

Fig. 8.1. Principle of MD process utilizing solar and waste heat as two energy sources.

showed that in large scales (>1000 m^3 day^{-1} water production), MD is less efficient compared to MED and MSF. However, this trend is reversed for small scale systems. MD systems can operate effectively at lower temperatures than most other industrial processes (<80°C).[5,6] So, MD could use solar and waste heat (Fig. 8.1),[7] and it would be a low carbon desalination technology.[8]

These advantages for low temperatures and small sizes have led MD applications to substantially consider the two underused power sources of waste heat and renewable solar energy. Researchers have also proposed other energy sources with higher amounts of available work, such as Joule heating. However, the energy source chosen can drastically impact the overall thermodynamic efficiency, which is often ignored.[4]

MD systems need further efficiency improvements to truly takeover the niches for thermal technologies, but the best implemented systems[9] remain well below the performance of the best theoretical systems. These systems are complex, as their performance depends on both system-level inputs such as the temperatures and total energy flows,[10] as well as micro-scale transport processes impacted by local properties such as membrane conductivity. In order to advance MD systems to their potential, a better understanding and implementation of MD efficiency is needed.

This chapter aims to unify the various metrics for MD efficiency. Here, we organize, categorize, and include the equations to convert between types of efficiencies, and explain the best uses of each type to maximize performance. To do so, this work chapter divides MD metrics into two primary categories: local and system level metrics. Additionally, it provides modeling results of representative figures that conveys how the metrics are inter-related.

8.2. MD Energy Efficiency Metrics

In this section we present the main performance metrics for MD clearly and concisely, while conveying when to use each one. The sections are organized by thermodynamic category.

In this section, the main metrics are categorized into local and system level groups:

Local:

- Flux
- Thermal efficiency

System level:

- First law efficiency: gained output ratio (GOR), evaporation efficiency, specific energy consumption (SEC)
- Heat transfer effectiveness
- Number of transfer units (NTU)
- Second law efficiency

8.2.1. *Flux*

The permeate flux J of membrane distillation is one of the most significant metrics for the comparison of any membrane technology.[10] Simply, it is the flow rate of permeate through the membrane, defined as:

$$J \equiv \frac{\dot{m}_{\mathrm{p}}}{A} \tag{1}$$

For MD, flux is driven by the vapor pressure difference, often simply described as:

$$J = B_{\mathrm{m}} \Delta P_{\mathrm{vap}} \tag{2}$$

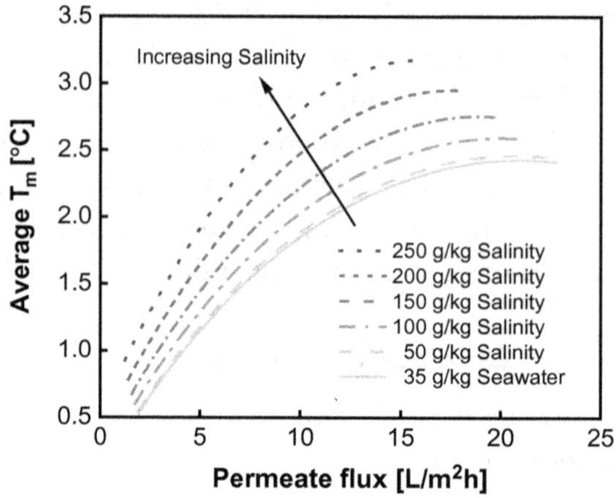

Fig. 8.2. Average temperature difference across the membrane with permeate flux, showing impact of driving temperature across the membrane ΔT_{m} and salinity. Conditions varying module length $L = 0.1\,\mathrm{m}$ to $12\,\mathrm{m}$ at salinities ranging from seawater, 35 g/kg, to 250 g/kg in an AGMD system. Membrane permeability coefficient is set to $B = 2 \times 10^{-10}\,\mathrm{kg/m\,Pa\,s}$, $T_{\mathrm{c,i}} = 25°\mathrm{C}$, $T_{\mathrm{f,i}} = 85°\mathrm{C}$, $\delta_{\mathrm{m}} = 0.2\,\mathrm{mm}$ and $d_{\mathrm{gap}} = 1\,\mathrm{mm}$ represent the parameters used for the computations.

where $\dot{m}_{\mathrm{p}}$ is the mass flow rate of permeate, A is the membrane area, B_{m} is membrane permeability, and ΔP_{vap} is the vapor pressure difference across the membrane. This term ΔP_{vap} is affected by temperature and salinity, but not with absolute pressure.

Figure 8.2 shows the impact of key parameters on membrane flux. The driving force for permeate flux, ΔP_{vap}, is a direct thermodynamic function of the temperature difference across the membrane, ΔT_{m}. However, increasing salinity reduces the vapor pressure on the feed side.

As membrane area is a key capital cost in MD, flux is an essential and easy-to-measure and almost universally used parameter for membrane distillation.

8.2.2. *First-law efficiency: GOR, evaporation efficiency, SEC*

The performance of thermal desalination systems is often described as a function of energy input, regardless of energy type: such a metric is a "First-law efficiency," as it considers only energy, similar to the first law of thermodynamics. Of these, the principle metric typically used to describe these is the GOR. This dimensionless parameter is a ratio of enthalpy of evaporation for the separated water to actual input heat for separation, or:

$$\mathrm{GOR} \equiv \frac{\dot{m}_{\mathrm{p}} h_{\mathrm{fg}}}{\dot{Q}_H} \tag{3}$$

where h_{fg} is the water enthalpy of vaporization (2257 kJ/kg), and $\dot{Q}_{\mathrm{H}}$ is the total heat energy input.

GOR shows how many times the enthalpy of evaporation is reused within the system. Essentially, it directly reflects the energy efficiency and cost of energy used within the system. As a result, the higher the GOR in MD system, the better, although practical design costs often lead to trade-offs with flux which will be discussed later.

Evaporative systems without heat recovery, such as some lab-scale MD systems, or technologies like most steam evaporators and solar stills, often have a GOR between 0 and 1. For example,

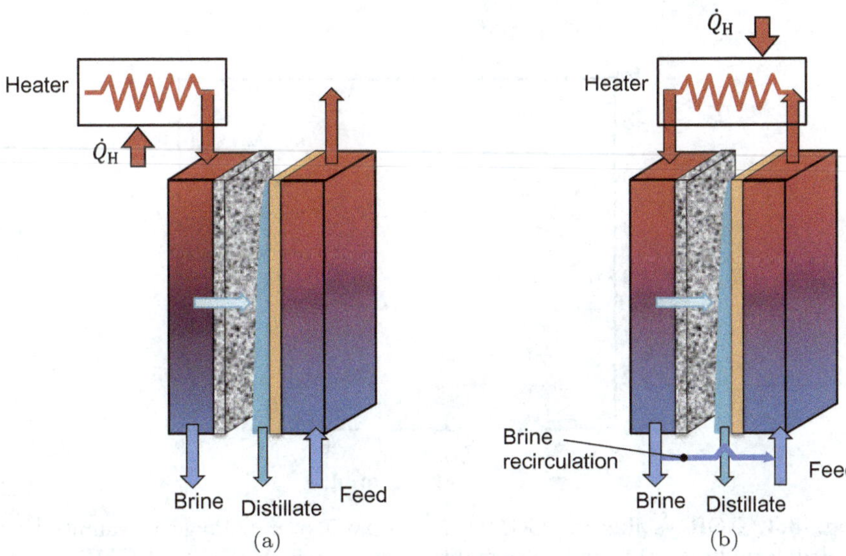

Fig. 8.3. The schematic of (a) separated hot and cold feed at MD (traditional MD setup, GOR $<$ 1) and (b) preheating feed inlet in MD and brine recirculation, necessary for reasonable efficiency (GOR $>$ 1).

Fig. 8.3(a) shows an MD system without heat recovery of the cooling stream, limiting the GOR below 1. Therefore, practical MD systems achieve GOR values greater than 1, by recovering the latent heat of the cooling and brine streams via preheating; Fig. 8.3(b). In the preheating loop, most of the brine heat could be reused by increasing the module length (membrane area), or by multi-stage design. The configuration shown has internal heat recovery (within the membrane module), which is seen in the MD configurations known as air gap (AGMD), permeate gap (PGMD), and conductive gap (CGMD). Recovery of latent heat can also be done with an external heat exchanger,[11] which is seen in the MD systems known as direct contact (DCMD) and vacuum (VMD).[12–17]

Notably, there tends to be a maximum GOR for a given MD operating conditions, based on the membrane area made available (which in turns alters flux). This can be seen as GOR vs flux peaks (Fig. 8.4). The primary limitation causing these peaks is the boiling point elevation (BPE), a term which describes vapor pressure reductions from higher salinity in the feed side. For a given salinity,

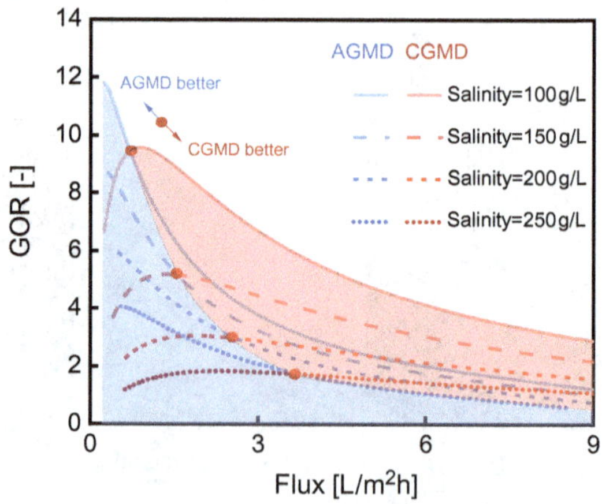

Fig. 8.4. GOR vs flux for CGMD and AGMD over a range of salinity.[13,18] Red dots indicate the trade-off conditions at which AGMD or CGMD begins to dominate with the superior GOR vs flux frontier. Membrane permeability coefficient is set to $B = 2 \times 10^{-10}\,\mathrm{kg/m\,Pa\,s}$, $T_{c,i} = 25°C$, $T_{f,i} = 85°C$, $\delta_m = 0.2\,\mathrm{mm}$ and $d_{gap} = 1\,\mathrm{mm}$ represent the parameters used for the computations.

to have enough vapor pressure to drive flux, there needs to be a higher temperature on the feed side; this ΔT is the BPE. As a result, the maximum possible GOR for a given system is a strong function of salinity. Overall, this comparison is a great way to go about optimizing performance.

However, the first-law metrics like GOR have a serious drawback; they do not consider the available work (exergy) that a heat stream may have. For a thermal process, this available work increases as temperature difference increases. GOR, therefore, is not fully reliable when comparing systems of substantially different operating temperatures or energy input types. As such, GOR is referred to as a *First Law Efficiency* metric, as it considers only the first law of thermodynamics and neglects the *Second Law of Thermodynamics*, which includes entropy. Because of this significant limitation, these first law metrics are largely useful in comparing systems with the same energy sources.

Notably, there are several other first law metrics that can be directly converted to GOR.

However, some reporting of solar evaporation efficiency ignores components of $\dot{Q}_{\text{vap}}$ or $\dot{Q}_{\text{total}}$, they may not match.

For related systems, especially those using solar power, an *evaporation efficiency* is often given: these numbers are usually identical to the thermal efficiency, where $\dot{Q}_{\text{H}}$ is a solar energy input. The three metrics converge in systems with negligible heat recovery, such as in small coupon modules or those without a flowing preheating stream. In those cases, $\dot{Q}_{\text{vap}} = \dot{Q}_{\text{total}}$, and the GOR and thermal efficiency converge. However, for real MD systems with heat recovery, the metrics may not correlate with one another. Usually a full model of an MD module is needed. Such an approach can allow for GOR calcualted from small systems. Those figures are typically below 100% (a GOR of 1) since they usually lack the heat recovery.

Another closely related performance metric is the SEC, or specific energy consumption, which is the energy input used to produce 1 m^3 of distillate (i.e. ratio of energy supplied to the volume of produced fresh water). The SEC varies from about 20 to 9000 kWh/m[4,19] depending on the type and size of the MD system, feed processed water, energy source, energy recovery systems, etc.

The specific energy consumption (SEC)[20] can be calculated with:

$$SEC = \frac{\dot{Q}_H}{\dot{m}_p} = \frac{h_{fg}}{GOR} \tag{4}$$

While these first law metrics are simply related by a constant, they are the reciprocal of one another, so for SEC, lower values are better, unlike the case of GOR.

Usually, the SEC is much higher for small laboratory MD systems compared to larger pilot plants with greater membrane areas.

8.2.3. *Thermal efficiency*

While system-level metrics like GOR are helpful in system design, they are not as easily applied to design choices that are local rather than system-level, such as material and channel depth choices. For such local considerations, the MD thermal efficiency η_{th} is the standard metric and helps with understanding tradeoffs with heat and mass transfer. Researchers find this metric particularly useful as it can bridge lab-scale results to meaningful implications for large systems. The MD thermal efficiency describes how much of the heat transfer between the feed and preheating channel is used to enact desalination, as a fraction of total heat transfer, which includes conduction losses. More specifically, thermal efficiency is defined as the fraction of energy transferred across the membrane as pure vapor transport[21]:

$$\eta_{\text{th}} \equiv \frac{\dot{Q}_{\text{vap}}}{\dot{Q}_{\text{total}}} \tag{5}$$

where $\dot{Q}_{\text{total}}$ is composed of $\dot{Q}_{\text{vap}}$ and $\dot{Q}_{\text{cond}}$. Heat transfer by conduction across the membrane ($\dot{Q}_{\text{cond}}$) is a loss mechanism; Ideally η_{th} approaches 1, meaning that all the energy supplied to the process should be used for the evaporation ($\dot{Q}_{\text{vap}}$) that causes desalination.[21]

The thermal efficiency can be calculated easily from lab-scale experimental data, as $\dot{Q}_{\text{vap}}$ is simply a function of the permeate flow rate and enthalpy of evaporation, while the denominator can be described as a function of total enthalpy removed by the feed stream, which can be described by the heat capacity c_p:

$$\eta_{\text{th}} \approx \frac{\dot{m}_{\text{p}} h_{\text{fg}}}{c_p(\dot{m}_{\text{f,i}} T_{\text{f,i}} - \dot{m}_{\text{f,o}} T_{\text{f,o}})} \tag{6}$$

where the temperatures are taken at the inlet and outlet of some portion of the feed side, as shown in Fig. 8.5.

The thermal efficiency is effectively a measurement of desired thermal transport local to some membrane multilayer sandwich that includes the membrane and major channels. It is thus an excellent parameter for optimizing material choices and layer thicknesses. To improve it, one must increase the resistance to conduction, and/or decrease the resistance to mass transport. Therefore, typically the thermal efficiency can be increased by reducing the thermal conductivity of the membrane, increasing membrane thickness, or by using an air gap. Alternatively, design changes to aid mass transport through the membrane (or air gap) such as a more porous or permeable membranes, can also increase it.

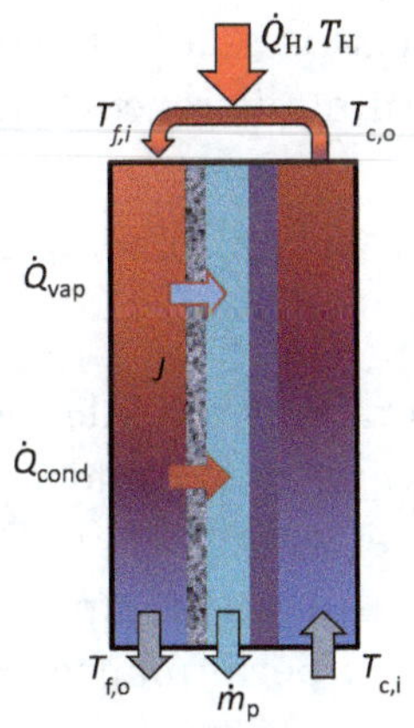

Fig. 8.5. Heat input and heat recovery in the module, showing key parameters for calculated the thermal efficiency of membrane distillation, η_{th}. The heat flow rates shown are those that define η_{th}, and the feed temperatures shown can be used to easily calculate the overall thermal efficiency η_{th}, or the heat transfer effectiveness ε for the system.

However, as it does not consider system-level performance, η_{th} must be used in conjunction with other system-level efficiency metrics, to give any information on optimizing performance. As explained in the following sections, it can be used as a model input to help calculate the other metrics, a use that more researchers need to understand and implement.

8.2.4. *Heat recovery parameter (ε-NTU)*

While other performance parameters indicate total energy use (GOR) or local heat transport (η_{th}), performance parameters that describe operating conditions and system sizing are needed. Two such parameters can be applied from the primary analysis approach for heat exchangers,[10] as MD itself is a heat exchanger with a membrane in the middle. These parameters are the heat transfer effectiveness ε, and the Number of Transfer Units, or NTU; the approach using these parameters is referred to as ε-NTU.

The heat transfer effectiveness ε[22] measures how effectively heat is transferred across the MD module from feed to the cold side, as a fraction of 1, where 1 would be a maximum enthalpy increase for the cold stream. The parameter ε is therefore defined as an enthalpy ratio, but since specific heat is relatively constant in MD, ε can also be expressed in temperature differences only:

$$\varepsilon \equiv \frac{\dot{Q}}{\dot{Q}_{\text{max}}} = \frac{h_{\text{c,o}} - h_{\text{c,i}}}{h_{\text{f,i}} - h_{\text{c,i}}} = \frac{T_{\text{c,o}} - T_{\text{c,i}}}{T_{\text{f,i}} - T_{\text{c,i}}} \tag{7}$$

where $\dot{Q}$ is heat transfer between inlet and outlet streams, h is enthalpy, and the subscripts c, f, i and o represent the cold, feed, inlet, and outlet respectively, as shown in Fig. 8.5. The heat transfer effectiveness ε approaches 1 as the temperature difference across the membrane approaches zero; performance approaches this limit with larger areas (and lower fluxes) and reduced heat transfer resistance in the channels, but boiling point elevation (a result of higher feed salinity) provides an upper limit on values of ε.

This effectiveness (ε) heat exchanger metric is typically used in conjunction with its partner metric, the Number of Transfer Units

(NTU). NTU is a dimensionless system size metric that includes heat transfer. NTU is related to J as they both describes system size, but this new metric is dimensionless which makes it easier to transfer insights across designs. The NTU of an MD module is defined as:

$$\mathrm{NTU} \equiv \frac{UA}{\dot{m}c_p} = \frac{T_{\mathrm{c,o}} - T_{\mathrm{c,i}}}{T_{\mathrm{f,i}} - T_{\mathrm{c,o}}} \tag{8}$$

where, U is the overall heat transfer coefficient across the MD module between cold and hot streams measured in $\mathrm{W/m^2\,K}$, and c_p is the specific heat. The concept of NTU originally appeared in the heat exchanger literature for heat exchanger sizing and rating. Larger values of NTU indicate larger system size and better heat transfer performance, whereas smaller J values usually indicate a larger system size only but can be caused by inferior heat transfer.

Like in heat exchanger analysis, the overall heat transfer coefficient times area, UA, is equal to the inverse of the thermal resistance. Therefore, it can be readily calculated using the resistances in MD, including boundary layer resistance in the cold and hot streams, the effective resistance of the membrane, and the heat conduction resistance of the gap. As a result, the overall heat transfer coefficient U can be expressed as:

$$\frac{1}{U} = \frac{1}{h_{\mathrm{f}}} + \frac{1}{h_{\mathrm{eff,m}}} + \frac{d_{\mathrm{gap}}}{k_{\mathrm{gap}}} + \frac{1}{h_{\mathrm{c}}} \tag{9}$$

where h_{f} is the heat transfer coefficient in the feed, h_{c} is the heat transfer coefficient in the cold channel and $h_{\mathrm{eff,m}}$ is the effective heat transfer coefficient of the membrane. With this term, the ε-NTU framework can be used to directly relate MD performance gains with thermal enhancements, including flow rates, material choices, and channel dimensions.

Figure 8.6 shows the GOR of MD at various salinity levels as a function of system NTU. Beyond a certain system size, the GOR decreases. There exists a value of (NTU*) beyond which the GOR starts declining. At larger system sizes, the flux is also lower since the overall driving temperature difference is lower. This means operating

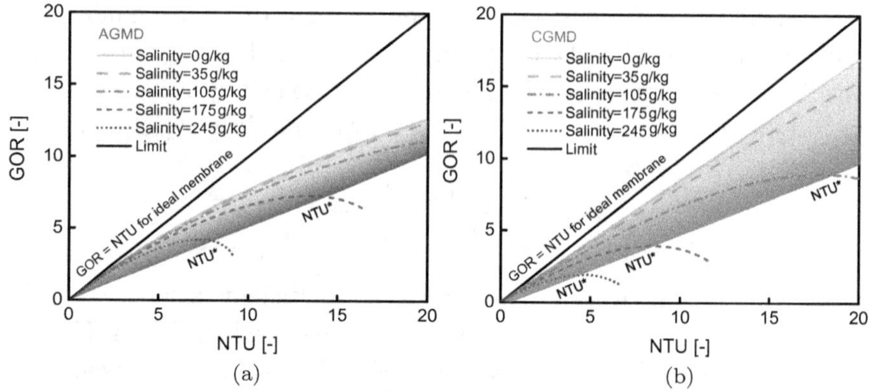

Fig. 8.6. GOR as a function of the heat exchanger dimensionless area parameter NTU (number of transfer units), for (a) AGMD and (b) CGMD at varied salinity. Shaded regions show realistic performance, with the darker bottom indicating the highest GOR (peak performance). These peak values represent dimensionless areas (NTU*) for best performance for each technology, with values to the left of the peak (lighter shading) saving on area cost, and values to the right of each peak are no-go design areas. Shorter module lengths (smaller NTU) have to be used when treating high salinity water with MD.[18] The parameters used for the computations were set to a membrane permeability coefficient is set to $B = 2{\times}10^{-10}$ kg/m Pa s, $T_{c,i} = 25°C$, $T_{f,i} = 85°C$, $\delta_m = 0.2$ mm and $d_{gap} = 1$ mm.

beyond (NTU*) leads to both a lower GOR and lower flux and hence should be avoided.

The dotted lines show that the simplified HX model is valid for high salinity as well. The existence of (NTU*) is indicated by the results from the simplified HX model as well. As a result, we attempt to use this model to determine the (NTU*) to choose safe operating conditions and system design.

For both MD and classic heat exchanger analysis, ε-NTU provides a fast and elegant system-level performance metric. Values of ε in MD modules approaching 1 indicates excellent heat recovery. However, in practice, high ε may mean a low ΔT across the membrane, and thus a small driving force (ΔP_{vap}) and smaller flux (J). Therefore, in optimizing design, comparing ε to a performance metric that considers system size (A, NTU, and J) is useful. In practice, computing NTU values can allow researchers to analyze how given material designs and channel geometries (which result in a given heat transfer effectiveness U) would perform in larger or smaller systems.

8.2.5. *Second law efficiency*

The second law of thermodynamics allows for the most fundamental and universal analysis of energy efficiency. In such analysis, it is often convenient to compare exergy rather than energy.

The concept of exergy, often called available work, is defined as the possible work we can extract from an energy source (in this case, heat) that interacts with the environment's end state. For desalination processes where this exergy difference is an energy barrier to overcome, it is often called the work of separation, $\dot{W}_{\text{sep}}$. The actual work used by the process is denoted by $\dot{W}_{\text{used}}$, and the two together define the second law efficiency for a separation process;

$$\eta_{\text{II}} \equiv \frac{\dot{W}_{\text{sep}}}{\dot{W}_{\text{used}}} = \frac{\text{Exergy}}{\dot{W}_{\text{used}}} \tag{10}$$

Exergy analysis in recent years is a term called second-law efficiency. Second law efficiency, in contrast to the metrics such as first law efficiency (e.g. GOR), is a universal tool to describe the thermodynamic efficiency in any desalination processes, capable of comparing different desalination technologies with different energy sources and heat temperatures.

In the second law efficiency analysis, the heat source (Fig. 8.7) can be described as a function of the enthalpies h_{fg} of the permeate; this describes the first law of thermodynamics.

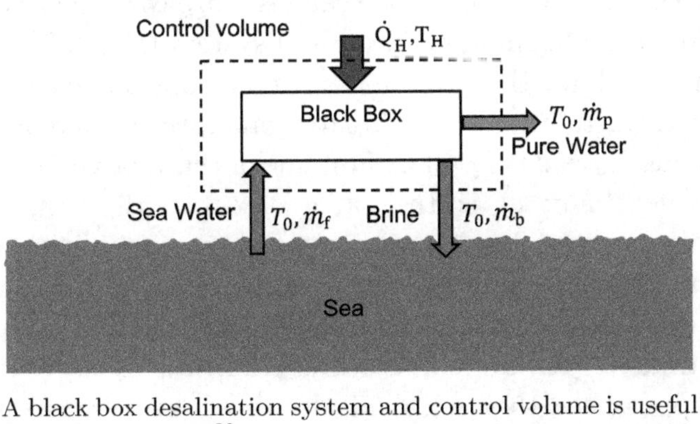

Fig. 8.7. A black box desalination system and control volume is useful for second law efficiency (η_{II}) analysis.[23] η_{II} is agnostic to the process itself and can be calculated with just the input and output streams. For desalination, the minimum least work between the streams is a function of the salinity and recovery ratio of pure water from seawater.

Here, the first law of thermodynamics is written as

$$\dot{Q}_{\mathrm{H}} - \dot{Q}_0 = (\dot{m}h)_{\mathrm{p}} + (\dot{m}h)_{\mathrm{b}} - (\dot{m}h)_{\mathrm{sw}} \tag{11}$$

where $\dot{Q}_{\mathrm{H}}$ is the input energy at T_{H}, $\dot{Q}_0$ is the heat rejected to the environment at T_0, and the second law of thermodynamics is written as:

$$\frac{\dot{Q}_{\mathrm{H}}}{T_{\mathrm{H}}} - \frac{\dot{Q}_0}{T_0} = (\dot{m}S)_{\mathrm{p}} + (\dot{m}S)_b - (\dot{m}S)_{\mathrm{sw}} + \dot{S}_{\mathrm{gen}} \tag{12}$$

The relative magnitude of entropy generation $\dot{S}_{\mathrm{gen}}$ is useful to understand the magnitude of efficiency losses. Using these laws, the work for a separation process like desalination can be described in a useful form that relates the work of separation, normalized by permeate production $\dot{m}_{\mathrm{p}}$, with the least heat for a process, and the flow described by the Gibbs free energy g of the liquid streams;

$$\frac{\dot{W}_{\mathrm{sep}}}{\dot{m}_{\mathrm{p}}} \equiv \left(1 - \frac{T_0}{T_{\mathrm{H}}}\right) \frac{\dot{Q}_{\mathrm{H,sep}}}{\dot{m}_{\mathrm{p}}} = (g_{\mathrm{p}} - g_{\mathrm{b}}) - \frac{1}{RR}(g_{\mathrm{sw}} - g_{\mathrm{b}}) + T_0 \frac{\dot{S}_{\mathrm{gen}}}{\dot{m}_{\mathrm{p}}} \tag{13}$$

where $RR(=\dot{m}_{\mathrm{p}}/m_{\mathrm{f}})$ is the recovery ratio, and the last term describes exergy destroyed by the process.

To provide further clarity, one can visualize these thermodynamic flows through the membrane in an MD system (Fig. 8.8).

Figure 8.9 shows the relation of heat, entropy and exergy transfer in membrane distillation according to temperature difference between cold (T_{c}) and hot side (T_{f}). The first law of thermodynamics dictates that the heat flux $\dot{Q}$ is the same across both sides. However, entropy increases as heat transfers across a temperature gradient- this is the primary cause of exergy destruction, also known as the loss of available work, in membrane distillation. Every occurrence of heat transfer across temperature gradients causes such losses: as MD approaches theoretical ideals, these temperature gradients approach zero, and the $\dot{Q}$ from conduction across the membrane also approaches zero.

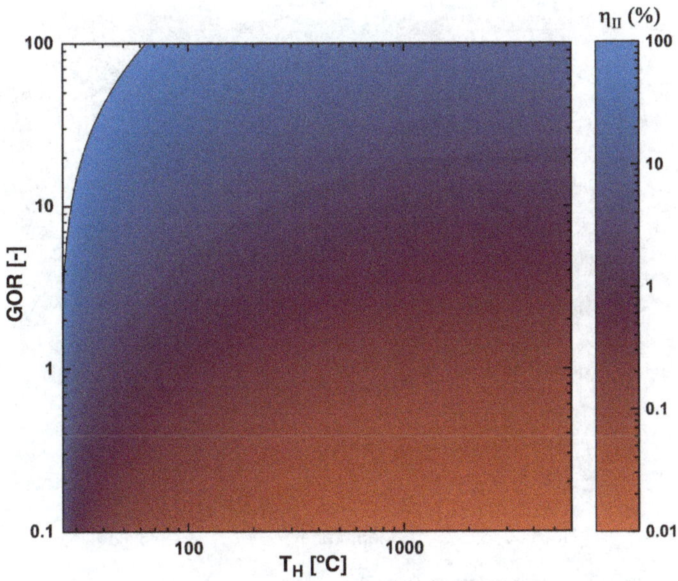

Fig. 8.8. Conversion plot for seawater desalination between second law efficiency η_{II} and first law efficiency (GOR) calculated for the case of an environmental temperature at 25°C and at $RR = 0$; and $\dot{W}_{\text{least}}/\dot{m}_{\text{p}} = 2.6\,\text{kJ/kg}$ and salinity 35 g/kg. Results are derived from $\eta_{II} = GOR \times \frac{\dot{W}_{\text{least}}}{\dot{m}_{\text{p}} h_{\text{fg}}} \times \frac{1}{1-\frac{T_0}{T_H}}$

8.2.6. *Relating performance parameters to one another*

The following subsection goes over equations that can be used to relate the efficiency metrics to one another. More thorough discussion and derivations can be found in previous work by the authors and collaborators.[10]

The second law efficiency can be related to the first law metrics (GOR etc) by using Eq. 10 (which relates work to heat through the temperatures) and the definitions for both efficiencies.

The first law metrics (GOR etc.) can also be related to the thermal efficiency η_{th} and the framework ε-NTU. The standard MD configurations with internal heat recovery (AGMD, PGMD, CGMD, and similar systems) provide straightforward conversions for relating GOR in terms of η_{th} and ε as follows:

$$\text{GOR} = \eta_{\text{th}} \times \frac{\varepsilon}{1-\varepsilon} \tag{14}$$

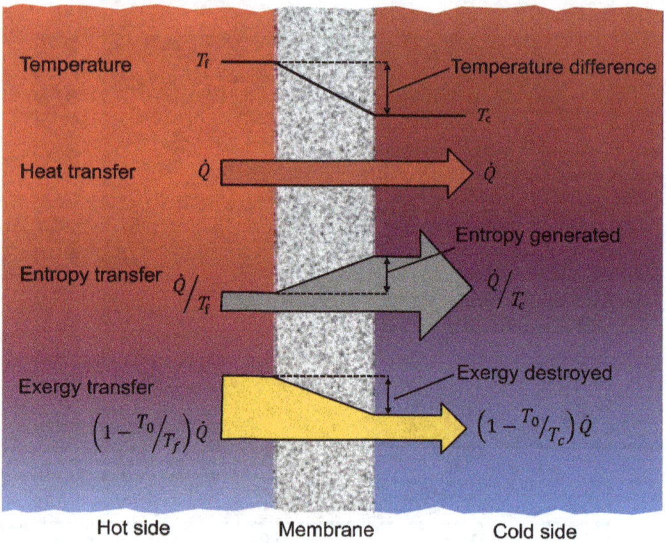

Fig. 8.9. Relation between energy, exergy, and entropy transfer across the membrane in MD.

ε can then be evaluated[24] assuming a perfect counter-flow heat exchanger as:

$$\varepsilon = \frac{NTU}{1 + NTU} \tag{15}$$

With this, the GOR expression derived earlier becomes:

$$GOR = \eta_{\mathrm{th}} \times NTU \tag{16}$$

η_{th} can be evaluated as the fraction of the heat transferred through the mass transfer resistance as:

$$\begin{aligned}\eta_{\mathrm{th}} &= \frac{Bh_{\mathrm{fg}}\Delta p_{\mathrm{m}}}{Bh_{\mathrm{fg}}\Delta p_{\mathrm{m}} + \frac{k_{\mathrm{m}}\Delta T_{\mathrm{m}}}{\delta_{\mathrm{m}}}} \\ &= \frac{1}{1 + \frac{k_m}{\delta_{\mathrm{m}} Bh_{\mathrm{fg}}} \frac{\Delta T_{\mathrm{m}}}{\Delta p_{\mathrm{m}}}} \\ &= \frac{1}{1 + \frac{k_{\mathrm{m}}}{\delta_{\mathrm{m}}} \frac{1}{h_{\mathrm{m}}^{\mathrm{mass}}}}\end{aligned} \tag{17}$$

where $h_{\mathrm{m}}^{\mathrm{mass}}$ is heat transfer coefficient corresponding to vapor transfer across the membrane.

Inspired by the ε-NTU method, η_{th} and ε are rewritten in terms of the transport resistances within the MD system. A single stage membrane distillation module resembles a counter-flow heat exchanger with hot brine and pure product transferring energy into the cooler feed, thereby preheating it. Using similarities with heat exchangers, simplified effectiveness-MTU (mass transfer units) models for reverse osmosis[25] and pressure retarded osmosis systems[26] have also been developed. The overall heat transfer coefficient has been introduced and used to evaluate DCMD flux several times in the literature.[27] Other researchers[28] have used metrics similar to NTU, such as the particular membrane area multiplied by a transfer coefficient.

8.2.7. *Efficiency limits*

With many efficiency metrics, it is critical to convey a clear approach for best optimizing MD systems. Simply put, the capital costs (CapEx) per MD (membrane area) directly relate to flux J, so flux should be maximized. Meanwhile, the operating costs (OpEx) are often readily available in terms of energy input costs, which simply relate to first-law metrics such as GOR. Therefore, GOR vs flux curves can describe the primary tradeoff in MD systems, while other metrics can be used to optimize systems to maximize these curves. Notably, if comparisons are needed to other sources of energy, or energy inputs of different temperature, 2nd law analysis approaches become far more appropriate, with the downside of a slight increase in complexity.

The next line of questioning is just how high each metric can get, and under what conditions this would occur. The fractional metrics (ε, η_{th}) obviously can't approach one, but they can get quite close in well-designed systems. Second law efficiency η_{th} is capped at 100%, but thermal desalination systems very rarely get above 15%, due to significant exergy destruction, especially from heat transfer across temperature gradients. Interestingly, for an ideal membrane, GOR = NTU, but for achievable systems, there is usually a maximum NTU beyond which GOR will decrease. This means

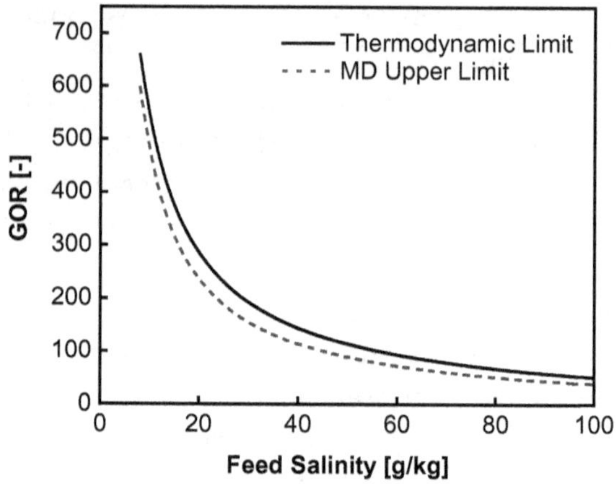

Fig. 8.10. Theoretical limits of MD system energy efficiency given perfect membranes ($\eta_{th} = 1$), perfectly insulating membrane, infinite thermal conductivity in the channels, and optimized design.[10,18]

that beyond a certain membrane area, more membrane area degrades performance, so NTU should never exceed this value, or conversely, the other area parameter, flux J, should never be below the value yielding maximum GOR.

As seen in Fig. 8.10, $GOR_{max,MD}$ is bounded by $GOR_{max,thermodynamic}$. One reason for lower $GOR_{max,MD}$ is that the boiling point elevation varies along the module length. As a result, vapor flux is driven by a non-zero driving force, generating entropy elsewhere. Both η_{th} and ε are lower for real MD systems since the membrane is not a perfect insulator and the area of the system is finite. As a result, real GOR values are at least an order of magnitude lower than the maximum possible GOR, leading to a second law efficiency of less than 10% as observed by Mistry *et al.*[29] A relevant take-away is that current MD systems have significant room for improvement to attain excellent efficiencies, given optimal systems design and further membrane innovations.

For practical systems that can be achieved with real materials, the maximum GOR possible can be modeled, and compared with flux (Fig. 8.11).[13]

As seen in Fig. 8.11, the maximum achievable GOR can be plotted versus flux, and is obtained with optimized membrane thickness. Notably, these efficiencies while achievable, are a significant

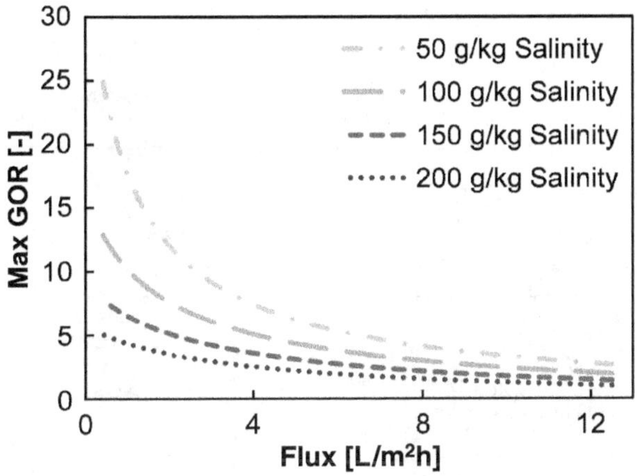

Fig. 8.11. Maximum achievable GOR by using the ideal membrane thickness at each salinity and flux for CGMD.[18]

improvement upon those seen in existing applications, as current products do not optimize membrane thickness for given salinities.

8.2.8. *Performance metric summary*

As a summary, these six parameters provide a comprehensive analysis of MD performance and are all distinctly useful for system design. While flux (J) and overall efficiency (GOR) describe the overall system-level size and energy needs respectively, thermal efficiency η_{th} allows for module-level design especially for the air gap and membrane. On the other hand, and effectiveness (ε) gives insight on the degree of energy recovery and associated temperature differences (ΔT). Through a variety of equations, these parameters can be related to one another, and are detailed in the Section 8.2.6.[10]

8.3. Designing MD Configuration with MD Metrics

The cost of MD is a function of both system capital expenditure (CapEx) and operating expenditure (OpEx), which can be studied from a tradeoff of two metrics for MD performance. First, CapEx is proportional to the system size, and hence inversely related to system productivity, measured by pure water flux. For designing MD systems, the metrics can be summarized and rated for the utility in calculating OpEx, System Side, and Material optimization

Table 8.1. MD metrics and qualitative importance of metrics in design of MD.

Metric		Symbol [unit]	Equation	Average value [lab scale]	Limits	Short definition	Importance in: OpEx	Importance in: System size	Importance in: Material	Refs.
Flux	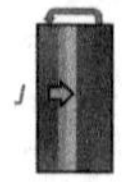	J [L/m²h]	$J \equiv \frac{\dot{m}_p}{A}$ or $J = B_m \Delta P_{vap}$	3-100	NA	Normallized pure water flow rate	★	★★★★★	★★★★	13,14, 30–36
Gained Output Ratio		GOR [-]	$GOR \equiv \frac{\dot{m}_p h_{fg}}{\dot{Q}_H}$	4-13 [<1]	0.1-13 (<860 theoretically)	Energy reuse multiplier	★★★★★	★	★	10,13–18, 23,37–39
Thermal Efficiency	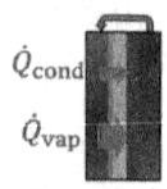	η_{th} [-]	$\eta_{th} = \frac{\dot{Q}_{vap}}{\dot{Q}_{vap} + \dot{Q}_{cond}}$	0.6-0.94	<1	Fraction of heat used for desalination	★★★★		★★★★★	13–15,23, 32,35,38
Heat Transfer Effectiveness	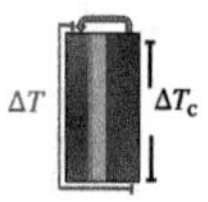	ε [-]	$\varepsilon = \frac{T_{c,o} - T_{c,i}}{T_{f,i} - T_{c,i}}$	0.7-0.93 [0-0.1]	<1	Fraction of maximum possible heat transfer between feed inlet & outlet	★★★★	★★★★★	★★	10,12, 13,18,35
Number of Transfer Units	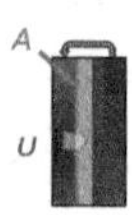	NTU [-]	$NTU = \frac{UA}{\dot{m}c_p}$	5-50 [0-0.1]	NA	Dimensionless area and heat transfer parameter	★	★★★★★	★★★★★	10,12,13, 18,35
Second Law Efficiency	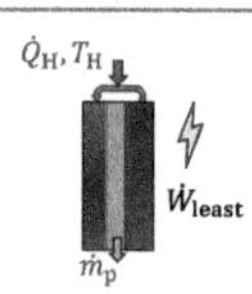	η_{II} [-]	$\eta_{II} = \frac{\dot{Q}_{least}^{min}}{\dot{Q}_H}$	1-12% [0%]	<100%	Universal thermodynamic efficiency	★★★★★	★★★	★★★	17,29, 40–42

(Table 8.1). This summary table gives the units, typical values, limits, and concise definitions for each metric. Also a pictoral representation of each metric is provided, relating them to the key variables and more detailed system diagrams.

The second main metric relates to OpEx, which is mainly representing the cost of energy required to drive the process. In MD, most of the energy supplied is in the form of heat energy, since the pumping power requirement is low in most well-designed MD systems. As a result, OpEx is inversely proportional to the energy efficiency of the system, measured as GOR, thermal efficiency, and second law efficiency.

Acknowledgements

The authors would like to acknowledge Porous Materials Incorporated for helping fund the work. Additionally, this past work draws on experience previously funded by Chevron (FKA CW793339). The authors thank Harsharaj Parmar for assistance in modeling used for figures, Jaichander Swaminathan for helping initially develop many of the efficiency concepts explained here and creating several figures we cite from his work, and John H. Lienhard V for his mentorship on these topics.

Nomenclature

Symbols

A	Membrane area, m^2
B_m	Membrane permeability, $kg/m^2Pa\ s$
B	Membrane coefficient, kg/m Pa s
BPE	Boiling Point Elevation, °C
c_p	Specific heat capacity at constant pressure, J/kg K
d	Depth or thickness, m
h	Heat transfer coefficient, W/m^2K
$h_{eff,m}$	heat transfer coefficient corresponding to vapor transfer across the membrane, W/m^2K
h_{fg}	Enthalpy of vaporization, J/kg
g	Specific Gibbs free energy kJ/kg
HX	Heat Exchanger
J	Permeate flux, $kg/m^2\ s$
k	Thermal conductivity, W/m K
L	Length of module, m

$\dot{m}$	Mass flow rate, kg/s
$\dot{Q}$	Heat transfer rate, W
$\dot{q}$	Heat flux, W/m2
p_{vap}	Vapor Pressure, Pa
S	Entropy J/K
$\dot{S}_{\text{gen}}$	Entropy generation rate W/K
s	Specific entropy J/kg K
SEC	specific energy consumption, kWh/m^3
T	Temperature,°C
T_0	Ambient temperature,°C
U	Overall heat transfer coefficient, W/m^2 K
$\dot{W}$	Work transfer rate [W]
$\dot{W}_{\text{sep}}$	Work of separation [W]
$\dot{W}_{\text{used}}$	Actual work used
w	Module width, m
Greek	
Δ	Change in a variable
ΔP_{vap}	Vapor pressure difference, Pa
δ_{m}	Membrane thickness, m
η_{th}	Thermal efficiency
η_{II}	second law efficiency
ε	Effectiveness
Subscripts	
0	Environment, or global dead state
b	Brine
cond	Conduction
eff	Effective
f	Feed
c	Cold
gen	Generated
H	Heat
i	Inlet
m	Membrane
max	Maximum
o	outlet
p	Permeate (product)

S	Solar heat source
sep	Separation
sw	Sea water
th	thermal
v	Vapor
vap	vapor
W	Waste heat source

Acronyms

AGMD	Air gap membrane distillation
CGMD	Conductive gap membrane distillation
DCMD	Direct contact membrane distillation
GOR	Gained output ratio
MD	Membrane distillation
MR	Mass flow ratio
NTU	Number of transfer units
RR	Recovery ratio
PGMD	Permeate gap membrane distillation
VMD	Vacuum membrane distillation

References

1. A. Murena, L. Borea, T. Zarra, J. Boguniewicz-Zablocka, V. Belgiorno, V. Naddeo, Water–Energy Nexus: Evaluation of the Environmental Impact on the National and International Scenarios, in: V. Naddeo, M. Balakrishnan, K.-H. Choo (Eds.), *Frontiers in Water-Energy-Nexus — Nature Based Solutions, Advanced Technologies and Best Practices for Environmental Sustainability*, Springer International Publishing, 2020, pp. 33–35.
2. R. Damania, The economics of water scarcity and variability, *Oxford Rev. Econ. Policy*, **36** (2020) 24–44. doi:10.1093/oxrep/grz027.
3. A. Alkhudhiri, N. Darwish, N. Hilal, Membrane distillation: A comprehensive review, *Desalination*, **287** (2012) 2–18. doi:10.1016/J.DESAL.2011.08.027.
4. A. Deshmukh, C. Boo, V. Karanikola, S. Lin, A. P. Straub, T. Tong, D. M. Warsinger, M. Elimelech, Membrane distillation at the water-energy nexus: Limits, opportunities, and challenges, *Energy Environ. Sci.*, **11** (2018) 1177–1196. doi:10.1039/c8ee00291f.
5. K. W. Lawson, D. R. Lloyd, Membrane distillation, *J. Memb. Sci.* **124**, 1–25 (1997).

6. D. M. Warsinger, J. Swaminathan, E. Guillen-Burrieza, H. A. Arafat, V. J. H. Lienhard, Scaling and fouling in membrane distillation for desalination applications: A review. Desalination 356, 294–313 (2014).
7. R. Schwantes, A. Cipollina, F. Gross, J. Koschikowski, D. Pfeifle, M. Rolletschek, V. Subiela, Membrane distillation: Solar and waste heat driven demonstration plants for desalination, *Desalination*, **323** (2013) 93–106. doi:10.1016/j.desal.2013.04.011.
8. C. Lienhard, G. P. Thiel, D. M. Warsinger, L. D. Banchik, Low Carbon Desalination: Status and Research, Development, and Demonstration Needs, Report of a workshop conducted at the Massachusetts Institute of Technology in association with the Global Clean Water, *Desalination Alliance*, 2016. http://hdl.handle.net/1721.1/105755.
9. K. S. Pitzer, J. C. Peiper, R. H. Busey, Thermodynamic properties of aqueous sodium chloride solutions, *J. Phys. Chem. Ref. Data.*, **13** (1984) 1–102. doi:10.1063/1.555709.
10. J. Swaminathan, H. W. Chung, D. M. Warsinger, J. H. Lienhard V, Membrane distillation model based on heat exchanger theory and configuration comparison, *Appl. Energy*, **184** (2016) 491–505. doi:10.1016/J.APENERGY.2016.09.090.
11. J. Swaminathan, H. W. Chung, D. M. Warsinger, J. H. Lienhard, Simple method for balancing direct contact membrane distillation, *Desalination*, **383** (2016) 53–59. doi:10.1016/j.desal.2016.01.014.
12. H. Y. Wu, M. Tay, R. W. Field, Novel method for the design and assessment of direct contact membrane distillation modules, *J. Membr. Sci.*, **513** (2016) 260–269. doi:10.1016/J.MEMSCI.2016.04.009.
13. J. Swaminathan, H. W. Chung, D. M. Warsinger, J. H. Lienhard V, Energy efficiency of membrane distillation up to high salinity: Evaluating critical system size and optimal membrane thickness, *Appl. Energy.*, **211** (2018) 715–734. doi:10.1016/J.APENERGY.2017.11.043.
14. J. A. Andrés-Mañas, A. Ruiz-Aguirre, F. G. Acién, G. Zaragoza, Performance increase of membrane distillation pilot scale modules operating in vacuum-enhanced air-gap configuration, *Desalination*, **475** (2020) 114202. doi:10.1016/J.DESAL.2019.114202.
15. D. M. Warsinger, J. Swaminathan, L. L. Morales, J. H. Lienhard V, Comprehensive condensation flow regimes in air gap membrane distillation: Visualization and energy efficiency, *J. Membr. Sci.*, **555** (2018) 517–528. doi:10.1016/J.MEMSCI.2018.03.053.
16. E. K. Summers, H. A. Arafat, J. H. Lienhard, Energy efficiency comparison of single-stage membrane distillation (MD) desalination cycles in different configurations, *Desalination*, **290** (2012) 54–66. doi:10.1016/J.DESAL.2012.01.004.

17. H. W. Chung, J. Swaminathan, D. M. Warsinger, J. H. Lienhard V, Multistage vacuum membrane distillation (MSVMD) systems for high salinity applications, *J. Membr. Sci.*, **497** (2016) 128–141. doi:10.1016/J.MEMSCI.2015.09.009.
18. D. M. Warsinger, G. P. Thiel, J. H. Lienhard V, The Membrane Distillation Efficiency-Flux Frontier, Chevron Rep. CW1251339 (FKA CW793339). (2015).
19. M. Mustakeem, A. Qamar, A. Alpatova, N. Ghaffour, Dead-end membrane distillation with localized interfacial heating for sustainable and energy-efficient desalination. *Water Res.* **189**, 116584 (2021).
20. D. Winter, J. Koschikowski, M. Wieghaus, Desalination using membrane distillation: Experimental studies on full scale spiral wound modules, *J. Membr. Sci.*, **375** (2011) 104–112. doi:10.1016/J.MEMSCI.2011.03.030.
21. J. Swaminathan, H. W. Chung, D. M. Warsinger, V. J. H. Lienhard, Membrane distillation model based on heat exchanger theory and module comparison, *Appl. Energy*, 184, 491–505 (2015).
22. W. Kays, A. London, *Compact Heat Exchangers: A Summary of Basic Heat Transfer and Flow Friction Design Data*, McGraw-Hill, 1958.
23. D. Warsinger, K. Mistry, K. Nayar, H. Chung, J. Lienhard V, Entropy generation of desalination powered by variable temperature waste heat, *Entropy*, **17** (2015) 7530–7566. doi:10.3390/e17117530.
24. J. H. Lienhard IV, J. H. Lienhard V, *A Heat Transfer Textbook*, Dover Publications, Inc, 2011. http://web.mit.edu/lienhard/www/ahtt.html.
25. L. D. Banchik, M. H. Sharqawy, J. H. Lienhard, Effectiveness-mass transfer units (ε-MTU) model of a reverse osmosis membrane mass exchanger, *J. Membr. Sci.*, **458** (2014) 189–198. doi:10.1016/J.MEMSCI.2014.01.039.
26. M. H. Sharqawy, L. D. Banchik, J. H. Lienhard, Effectiveness–mass transfer units (ε–MTU) model of an ideal pressure retarded osmosis membrane mass exchanger, *J. Membr. Sci.*, **445** (2013) 211–219. doi:10.1016/J.MEMSCI.2013.06.027.
27. S. Al-Obaidani, E. Curcio, F. Macedonio, G. Di Profio, H. Al-Hinai, E. Drioli, Potential of membrane distillation in seawater desalination: Thermal efficiency, sensitivity study and cost estimation, *J. Membr. Sci.*, **323** (2008) 85–98. doi:10.1016/J.MEMSCI.2008.06.006.
28. S. Lin, N. Y. Yip, M. Elimelech, Direct contact membrane distillation with heat recovery: Thermodynamic insights from module scale modeling, *J. Membr. Sci.*, **453** (2014) 498–515. doi:10.1016/J.MEMSCI.2013.11.016.
29. K. H. Mistry, R. K. McGovern, G. P. Thiel, E. K. Summers, S. M. Zubair, J. H. Lienhard V, Entropy generation analysis of desalination technologies, *Entropy*, **13** (2011) 1829–1864. doi:10.3390/e13101829.

30. A. E. Khalifa, Flux enhanced water gap membrane distillation process-circulation of gap water, *Sep. Purif. Technol.*, **231** (2020). doi:10.1016/j.seppur.2019.115938.
31. S. M. Alawad, A. E. Khalifa, Analysis of water gap membrane distillation process for water desalination, *Desalination*, **470** (2019). doi:10.1016/j.desal.2019.114088.
32. R. Schwantes, L. Bauer, K. Chavan, D. Dücker, C. Felsmann, J. Pfafferott, Air gap membrane distillation for hypersaline brine concentration: Operational analysis of a full-scale module–New strategies for wetting mitigation, *Desalination*, **444** (2018) 13–25. doi:10.1016/j.desal.2018.06.012.
33. R. Miladi, N. Frikha, A. Kheiri, S. Gabsi, Energetic performance analysis of seawater desalination with a solar membrane distillation, *Energy Convers. Manag.*, **185** (2019) 143–154. doi:10.1016/j.enconman.2019.02.011.
34. L. Gao, J. Zhang, S. Gray, J. De Li, Influence of PGMD module design on the water productivity and energy efficiency in desalination, *Desalination*, **452** (2019) 29–39. doi:10.1016/j.desal.2018.10.005.
35. J. Swaminathan, J. H. Lienhard, Design and operation of membrane distillation with feed recirculation for high recovery brine concentration, *Desalination*, **445** (2018) 51–62. doi:10.1016/j.desal.2018.07.018.
36. D. Winter, J. Koschikowski, M. Wieghaus, Desalination using membrane distillation: Experimental studies on full scale spiral wound modules, *J. Memb. Sci.*, **375** (2011) 104–112.
37. H. B. Parmar *et al.*, Nanofluids improve energy efficiency of membrane distillation, *Nano Energy* (2021).
38. J. Swaminathan, J. H. Lienhard, Energy efficiency of sweeping gas membrane distillation desalination cycle. in (Proceedings of the 22nd National and 11th International ISHMT-ASME Heat and Mass Transfer Conferencer, 2013).
39. J. Swaminathan *et al.* Energy efficiency of permeate gap and novel conductive gap membrane distillation, *J. Memb. Sci.*, **502** (2016) 171–178.
40. A. Deshmukh, M. Elimelech, Understanding the impact of membrane properties and transport phenomena on the energetic performance of membrane distillation desalination, *J. Memb. Sci.*, **539** (2017) 458–474.
41. K. H. Mistry, J. H. Lienhard, Generalized least energy of separation for desalination and other chemical separation processes, *Entropy*, **15** (2013) 2046–2080.
42. G. P. Thiel, E. W. Tow, L. D. Banchik, H. W. Chung, J. H. Lienhard, Energy consumption in desalinating produced water from shale oil and gas extraction, *Desalination*, **366** (2015) 94–112.

CHAPTER 9

Capacitive Deionization

S. A. Hawks[a], D. I. Oyarzun[a], A. Ramachandran[b], P. G. Campbell[a], J. G. Santiago[c], and M. Stadermann[a,*]

[a] *Physical and Life Sciences Directorate, Lawrence Livermore National Laboratory, Livermore, CA 94550, United States*
[b] *Department of Aeronautics & Astronautics, Stanford University, Stanford, CA, 94305, United States*
[c] *Department of Mechanical Engineering, Stanford University, Stanford, CA, 94305, United States*

9.1. Introduction to Capacitive Deionization

Capacitive deionization (CDI) is a desalination method that removes charged dissolved species from water by reversibly adsorbing them to a porous, conducting material, a process known as electrosorption.[1–5] CDI shows promise for treating near potable water (<2000 ppm) and for the selective removal of certain ions.[6–10] The scalability of the electrical methods suggests it should work well in small- and mid-scale devices.

While capacitive deionization was reported in 1960,[11] it only saw sporadic development until the mid-2000's. Since then, the technology has advanced more rapidly as the research community continues to grow, and there are multiple companies that offer CDI devices commercially today.

CDI is still a nascent technology: new device configurations[12–14] and materials are continuously emerging as the performance-limiting

*Corresponding author. stadermann2@llnl.gov

factors are characterized. Models of operation, both on the device scale and on the system scale, are still being developed and refined.[8,15–24] Relatively little operational data has been published, and not much is known yet about the lifetime and aging mechanism of CDI devices.

In this chapter, we will describe how CDI works and give an overview of the different device geometries that are currently of interest, give a detailed review of device performance parameters, discuss selectivity of CDI devices for the removal of specific ions, and conclude with an outlook on potential impacts of future developments on the performance of CDI.

9.2. How CDI Works

The distinguishing feature of CDI is that ions are removed from water by electrosorption.[25] Electrosorption may either be the primary removal mechanism, or it may be used as a means to drive an ionic current across ion selective membranes, similar to electrodeionization and electrodialysis. Electrosorption usually occurs in the electric double layer of the porous carbon electrodes that are used.[26] However, recently other materials have been used that are akin to the intercalation electrodes[27] used in lithium ion batteries, and these materials may be storing ions in the bulk of the electrode as well. While these materials hold some promise for future applications due to their high sorption capacity, they will not be further discussed in this chapter.

On a liquid-solid interface, there can be an accumulation of ions near the surface within a region called the electric double layer. If the surface is uncharged, the number of positive and negative ions in the double layer is equal. If the solid is charged, e.g. by adding electrons, the double layer net charge will become more positive to compensate for the charge on the surface. This can occur either by adsorbing an additional positive ion, or by ejecting a negative ion from the double layer. The former is the desired effect, and can lead ultimately to desalination of the liquid that the surface is immersed in. For a device, the two surfaces are called electrodes and one will be charge positively and the other one negatively. At each of those electrodes, the respective ions of opposite charge will accumulate until there are few left in the liquid. Co-ion ejection, however, increases the number of ions in the liquid. For practical purposes, both processes occur while the applied voltage to the surface is close to the uncharged state, and there is no desalination. Once all co-ions are ejected from the double layer, desalination begins. The efficiency of ion adsorption compared to charge added to the electrode will be discussed in greater detail in Section 9.3.4.

Electrodes are typically charged until they reach a specified voltage. At this point, ions can no longer be adsorbed, and the electrodes have to be discharged or regenerated so that more ions can be adsorbed. In typical operation, water that flows through the device during charging is desalinated, and water that flows during discharge will increase salt concentration and become the brine. This type of cyclical operation is one of the hallmarks of CDI.[25]

Ions may be adsorbed to the surface not only by electrical charge in the solid, but also by chemical charge on the surface (Fig. 9.1).

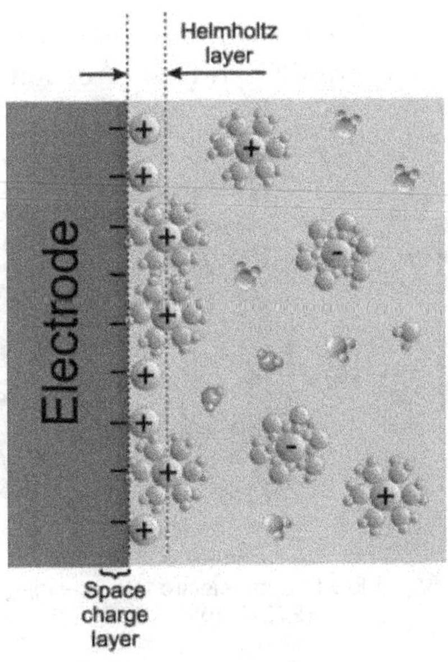

Fig. 9.1. Schematic of the electric double layer close to the electrode surface. Reproduced with permission from Ref. [139] with permission from The Royal Society of Chemistry.

In this case, charge must be added to the solid before the double layer composition is electrostatically neutral. All carbon electrode materials have native charged groups on the surface, typically in the form of carboxy and hydroxy groups,[29] and their presence usually reduces the performance of the device unless membranes are used (see Section 9.2.1). Device aging usually occurs by oxidation of the electrode,[30] which adds more charged groups and further reduces the performance. One type of device architecture makes use of this effect by adding more functional groups to each electrode and adsorbing ions while the electrode is electrically discharged, and the desorbing the ions by charging the electrodes[14,31–33] (see Section 9.2.3).

9.2.1. *Basic CDI architectures*

Based on the electrode and flow configurations, CDI systems can be broadly classified into flow by (FB-CDI), flow through electrode (FTE-CDI), and flow electrode (FCDI), see Fig. 9.2. We here describe salt transport mechanisms using these CDI architectures, and highlight some benefits offered by each of these systems. There are three major electrical and charge transport mechanisms for CDI devices.[12,20,24,34] These are drift/diffusion of ions across the spacer

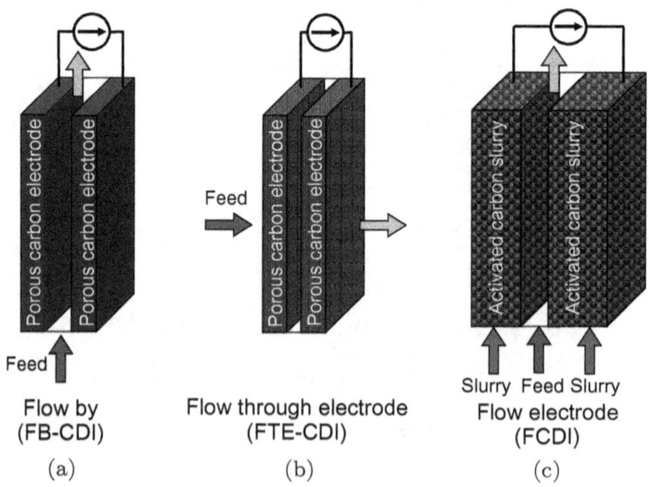

Fig. 9.2. Schematic of basic CDI architectures: (a) flow by (FB-CDI), (b) flow through electrode (FTE-CDI), and (c) flow electrode (FCDI).

and electrode, the resistive-capacitive charging of the electrodes, and the advection of feedwater in the spacer. In most practical desalination applications with sufficiently high flowrates, diffusion effects within CDI cells do not have to be spatially accounted for, and salt transport dynamics can be predicted well using a well-mixed reactor (i.e., continuously stirred tank reactor (CSTR)) approximation. Such an approximation couples net advective salt transport with electrical forcing source/sink.[26,34]

In a flow by CDI (FB-CDI) architecture, the feed stream to be desalinated flows in a spacer between two fixed porous charging electrodes. The applied electric field is in a direction perpendicular to feedwater flow. The first demonstration of FB-CDI was in the work of Blair and Murphy,[11] where a stack of porous carbon electrodes was used to demineralize saline water. This architecture has become the most widely adopted design for CDI cells in literature, primarily due to the ease of device fabrication and assembly. FB-CDI also allows for easy incorporation of ion-exchange membranes (cf. Section 9.2.2) between the electrodes, typical of membrane-based CDI applications. Salt transport across the spacer and electrode in FB-CDI can be limited by diffusion, especially when advection time scales are high (i.e., when feedwater flowrates are low) compared to the diffusion time scale.

In FTE-CDI, the feed stream flows directly through the fixed electrodes and spacer, so the applied electric field is parallel to the flow direction. This architecture was first demonstrated by Johnson *et al.*,[34] and was later revived in the work of Avraham *et al.*[35] However, the early works on FTE-CDI used unstable electrode materials,[34] and materials more suited for FB-CDI.[35] Suss *et al.*[12] demonstrated the use of hierarchical carbon aerogel monoliths, which are ideally suited for FTE-CDI. The design used specialized porous carbon materials with bimodal (micro- and nano-scale) pore structures to lower hydraulic resistance and provide high capacitance. An advantage of FTE-CDI architecture is that it eliminates the need for a spacer between the electrodes (except for preventing electrode contact), thus minimizing spacer thickness. The reduced spacer thickness can minimize diffusion time across the spacer to allow for better salt transport, lower ionic resistance, and allow for compact cells. Transport in FTE-CDI is typically limited by advection at high flowrates.[20] A drawback of FTE-CDI is a difficulty in scaling up in a parallel fashion.

More recently, a new class of architecture called flow electrode CDI (FCDI) was developed by Jeon *et al.*[36] In FCDI systems, a suspended (slurry) carbon electrode material is used in place of fixed electrodes (such as in FB-CDI and FTE-CDI). The electrode slurry is pumped through the electrode compartments while the salt water is pumped through a spacer. Note that some versions of FCDI cells use ion-selective membranes (c.f. Section 9.2.2). The major advantage of the FCDI architecture is that it allows for continuous desalination of feedwater (unlike charging-discharging cycling operation). The salt trapped in the flowing carbon particles can be recovered in a separate downstream process. Furthermore, a continuous flow of electrode slurry in FCDI replenishes the electrode, allowing for a high capacitance which is sufficient to even desalinate sea water.[36]

So far, all of the discussed CDI architectures involve a combination of native activated carbon electrodes (static and flow) and a spacer channel only. In the next section, we review CDI systems that combine these basic architectures with ion-selective membranes and/or chemically functionalized electrodes.

9.2.2. *CDI with membranes*

A drastic improvement in overall CDI performance was achieved with the addition of ion-selective membranes to the basic CDI architecture, where such desalination units are called as membrane capacitive deionization (MCDI) systems.[37–40] The most commonly used MCDI architecture consists of a standalone cation-exchange membrane placed in between the cathode and spacer, and a standalone anion-exchange membrane placed in between the anode and spacer. Alternately, membranes can be directly deposited on the electrodes to make a more compact cell.[41] As in FB-CDI cells, feed water in MCDI flows in a spacer channel in between the electrode-membrane stack, and this type of system is sometimes referred to as flow between MCDI (FB-MCDI). Membranes in MCDI play an important role in excluding co-ions from electromigrating towards each electrode on application of electric field — a feature absent in the basic CDI architectures. Thus, MCDI cells have high charge efficiency (>90%) when compared to basic CDI cells (with typical charge efficiencies of 50–80%). This improvement in charge efficiency directly improves overall CDI performance, which includes maximum achievable desalination depth, volumetric energy consumption, and productivity. For more details on these metrics, see Section 9.3. A comparison of the performance of basic CDI cells versus FB-MCDI cells in a recent work[42] showed that an FB-MCDI device has an approximately three times higher productivity than an FTE-CDI cell at equivalent energy consumption, salt removal, and water recovery conditions. However, membranes add to the total cost of the device, and thus these devices must have a higher throughput to be competitive.

Ion-selective membranes can also been combined with flow electrode CDI units, as was demonstrated in Jeon *et al.*,[36] to improve overall performance. No practical way has been found yet to combine membranes with a flow through CDI architecture. A recent interesting application[43] of MCDI is that of selective ion removal, which is achieved by designing ion-exchange membranes that allow selective transport of particular ions of interest (even with the same sign of electric charge); cf. Section 9.4 for further discussion on selective salt removal using CDI.

9.2.3. *Inverted CDI*

Conventional CDI electrodes suffers from long-term stability as a result of oxidation of the carbon electrodes in an aqueous solution on application of voltage. To mitigate this problem, an "inverted CDI" (i-CDI) scheme was recently demonstrated.[14,31] In i-CDI, the anode (positive electrode) is modified with net negative surface charges and the cathode (negative electrode) is modified with net positive surface charges in solution (see Fig. 9.4). Thus, salt removal and regeneration in i-CDI happens in a manner opposite to conventional CDI systems. Specifically, salt removal occurs when the cell is short-circuited (electrical discharging), while ions are desorbed on

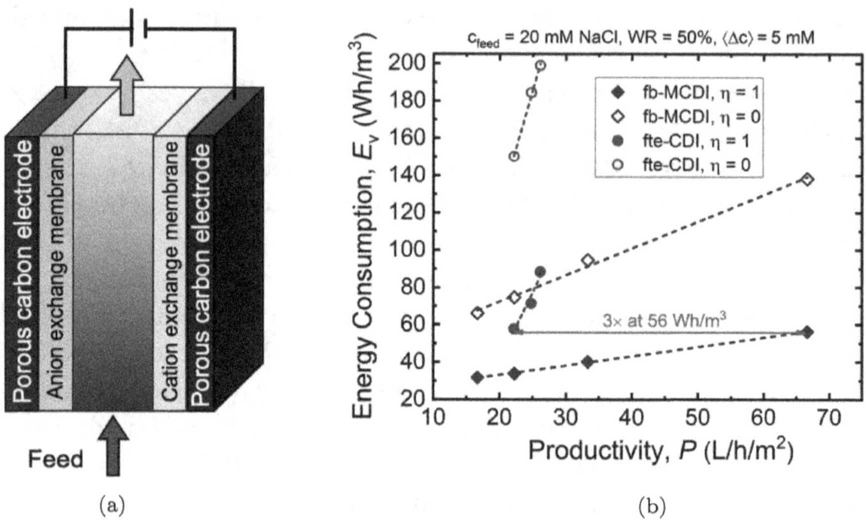

Fig. 9.3. (a) Schematic of a flow by MCDI cell (FB-MCDI). (b) Performance comparison between FB-MCDI and FTE-CDI with perfect energy recovery and reuse ($\eta = 1$) and no energy recovery ($\eta = 0$). The separation conditions are identical for each data point, average concentration reduction of 5 mM, feed concentration of 20 mM, and 50% water recovery. The plot in (b) is reproduced with permission from Elsevier from Hawks *et al.*[5] Copyright 2019, with permission from Elsevier.

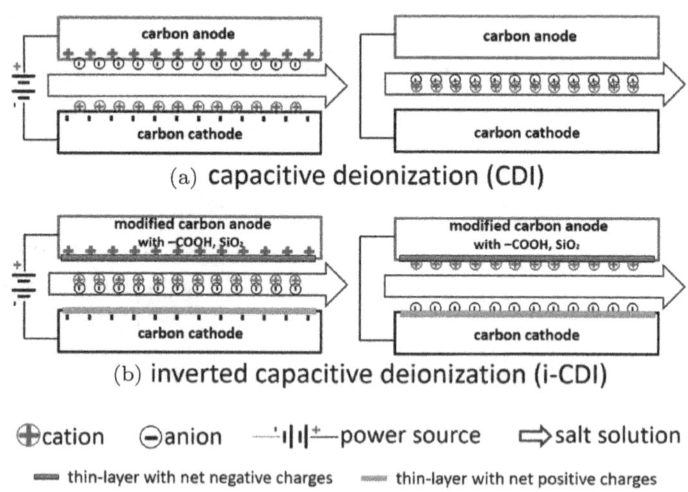

Fig. 9.4. Schematic of inverted CDI (i-CDI) scheme, and comparison with conventional CDI scheme. Reproduced with permission from The Royal Society of Chemistry from Gao *et al.*[14]

application of voltage (electrical charging). Further, cations and anions are adsorbed/desorbed at the positive and negative electrodes, respectively, exactly in an opposite manner to conventional CDI. Gao *et al.*,[14] demonstrated that an i-CDI system was stable for about 600 hours, roughly ten times longer than a conventional CDI system operated under similar conditions.

9.2.4. *CDI operating methods*

For a given source of feedwater, there are mainly two kinds of independent control for a CDI unit, viz. the electrical input (current or voltage), and the flowrate of feed solution. Most of the existing research in CDI has been focused on the electrical operational schemes, and attention on flowrate control has gained interest only recently. We here discuss some of the commonly used electrical forcing and flowrate control methods for CDI.

9.2.4.1. *Electrical control methods*

The conventional electrical forcing schemes for CDI are constant current (CC) and constant voltage (CV). In CC, during the desalination phase, a constant positive current is applied to the CDI cell until the cell voltage increases to an upper threshold value (typically limited by Faradaic losses, to around 1.2 V). This is followed by a regeneration phase in which the direction of the constant current is reversed (with either the same or different magnitude compared to charging) to discharge the cell to a lower voltage threshold (typically, 0 V). In CV operation, salt removal occurs when the CDI cell is held at a constant, high value of voltage (again, around 1.2 V), followed by salt regeneration at a constant, low voltage value (typically, 0 V). A mixed scheme that has also been explored for CDI includes constant current charging to a maximum voltage value followed by constant voltage (typically, 0 V) discharging. There is currently no consensus on which operational method (CC or CV) is superior in performance to the other because the advantages of one operational scheme over the other depend on several factors which include achievable desalination depth, productivity, and energy recovery, among many others. Refer to these contributions[44–51] for detailed discussion on this topic.

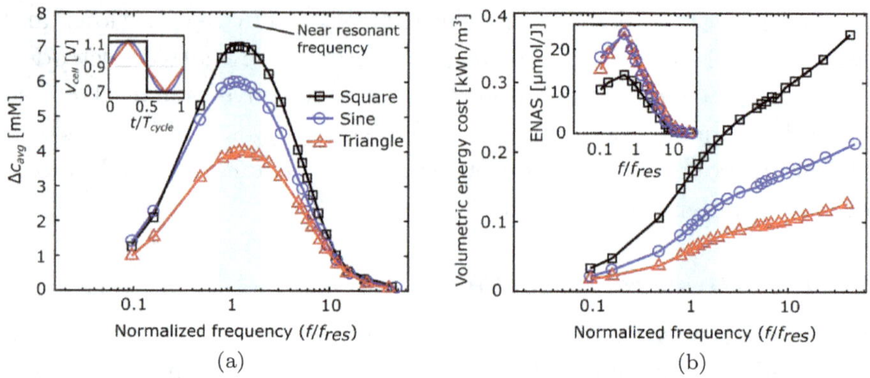

Fig. 9.5. Measured (a) effluent concentration reduction Δc_{avg}, and (b) energy consumption, as a versus forcing frequency normalized by the resonant frequency f_{res}. Data are shown for three different forcing waveforms: square wave (CV), triangular (CC), and sinusoidal (SV) voltages. Productivity is fixed at $10\,\mathrm{L/h/m^2}$. Forcing waveforms are shown in the inset, corresponding to 0.7–1.1 V voltage window. Figure (b) inset shows energy normalized adsorbed salt (ENAS) versus forcing frequency. All forcing waveforms show maximum Δc_{avg} near resonant frequency. At resonant frequency cycling, CV results in maximum Δc_{avg}, followed by SV and CC; however, CC consumes the least energy, followed by SV and CV. Figure (b) inset shows ENAS for SV and CC are nearly equal and around two times higher than square wave operation. Thus, SV balances energy and salt removal performance achieved by CC and CV, respectively. Reproduced with permission from Elsevier (2018) from Ramachandran *et al.*[52]

However, more recently, a DC-offset sinusoidal voltage (SV) operation was demonstrated for CDI, and this scheme was compared with CC and CV operation.[52] This experimental and theoretical study compared CC, CV, and SV operations for the same voltage window and productivity (see Fig. 9.5). The study found that CC consumes least energy, but also results in the least desalination depth, while CV consumes maximum energy and achieves maximum desalination depth. Interestingly, SV was found to achieve good energy performance (similar to CC), and simultaneously achieving significant desalination depth (similar to CV). Thus, SV has the potential to be used as an operation scheme which balances energy and desalination performance for CDI.

Finally, regardless of the electrical forcing waveform (CC, CV, or SV), the aforementioned study[52] also experimentally demonstrated and mathematically derived a unique (resonant) frequency of cycling for CDI which results in maximum desalination. For a DC-offset sine wave applied voltage, the resonant frequency *fres* is given by $1/f_{res} = 2\pi\sqrt{\tau RC}$, where τ is the fluid residence time within the CDI cell and RC is the resistive-capacitive charging time scale. Cycling at near-resonant frequency for CDI implies an optimal balance between absolute amount of salt removed (in moles) and dilution (depending on the feed volume processed). Thus, resonant cycling frequency is an important consideration in the design of operational strategies to achieve good overall performance for CDI.

9.2.4.2. *Flow rate control methods*

Most CDI cells are operated with a constant flow rate of feedwater. Some recent work[42,53] demonstrated the use of fluid pulsation-type operations wherein short periods of high flow rate (relative to a nominal, constant flow rate) were introduced during a CDI cycle. Fluid-pulsations were performed to primarily improve CDI cycle charge efficiency (specifically, the flow efficiency component see charge efficiency section below). However, most studies are currently limited to cases where the volume of desalted water produced was approximately equal to the volume of brine produced, i.e. water recovery of nearly 50%.

A recent study[49] showed that water recovery for CDI can be increased to about 90% by the use of different flow rate during CDI charging and discharging. Specifically, the aforementioned study[49] used a high (nominal) value of flow rate during electrical charging, followed by a low flow rate during electrical discharging, to maximize the fraction of desalted water relative to the total processed feedwater. High water recovery using this operation was achieved with minimal trade-offs in overall performance (see Fig. 9.6). Also, importantly, this study[49] highlighted that the electrical response of the CDI cell is nearly independent of flow rate changes during a cycle. This finding highlights that further optimization of CDI performance can be performed using flow rate control schemes, independent of electrical forcing schemes.

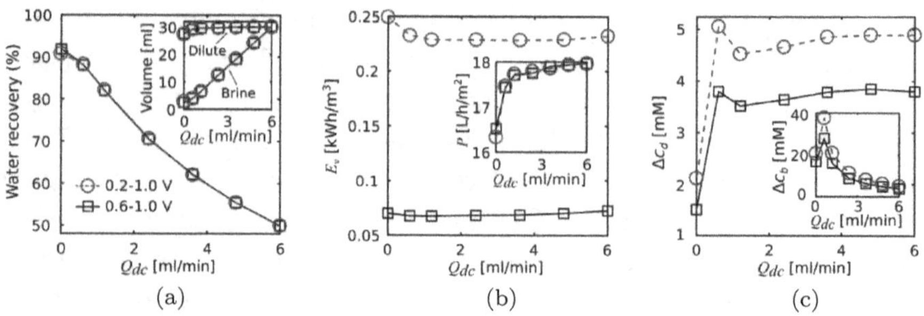

Fig. 9.6. Measured (a) water recovery, (b) volumetric energy consumption, and (c) average concentration reduction in the desalted water, for constant voltage operation between 0.2–1.0 V and 0.6–1.0 V, and different flow rate during charging and discharging. The flow rate during charging was fixed at 6 ml/min and the flow rate during discharge (Q_{dc}) was varied between 0 and 6 ml/min. The insets in (a), (b), and (c) respectively show the volume of desalted water (and brine) produced per cycle, freshwater productivity, and average concentration increase of brine for Q_{dc} between 0 and 6 ml/min. Figure reproduced with permission from Elsevier (2019) from Ramachandran *et al.*[49]

9.3. CDI Performance and Water Cost

Characterizing desalination performance in CDI is perhaps more difficult than with other water treatment technologies because the amount of salinity reduction and water recovery is highly variable and strongly dependent upon device characteristics and operation conditions. Thus, of the many performance measures that have emerged in CDI research,[25,50] many results are difficult to compare across systems due to a lack of attention toward separation conditions. This has inspired a recent work by the present authors to standardize performance metrics in CDI so to make them more comparable across systems as well as to other technologies.[5] Figure 9.7 schematically presents how such a performance system can be imagined. The left side of Fig. 9.7 shows the three critical separation conditions that must be carefully taken into account: the water recovery (WR), feed salinity (c_{feed}), and salinity reduction (Δc).[5] Below we discuss the limitations associated with CDI in terms of attainable WR and Δc, and others have estimated that CDI can only compete with RO when treating a feed below a concentration of $<\sim$50 mM NaCl salinity equivalent.[54]

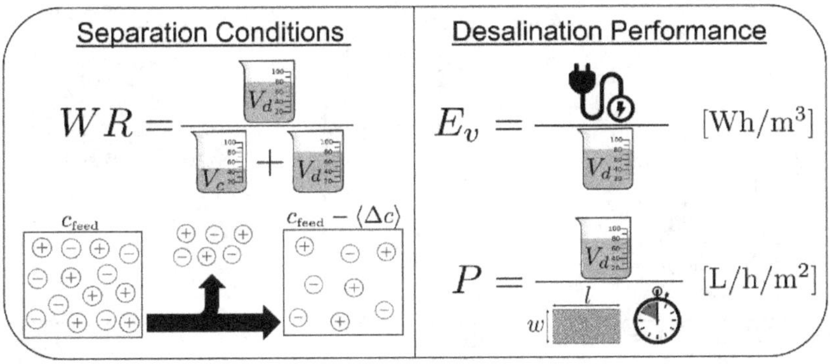

Fig. 9.7. Schematic of parameters to consider when comparing CDI desalination performance. Reproduced with permission from Elsevier (2019) from Hawks *et al.*[5]

Once the desired separation conditions are achieved, then the right-hand side of Fig. 9.7 can be considered. Here, CDI desalination performance is characterized in terms of specific energy consumption E_v (Wh/m^3) and throughput productivity P ($L/h/m^2$). A detailed argument for why such performance measures should be used over other potential metrics is given in Ref. [5]. Briefly though, E_v and P were chosen because they center around the volume of treated water produced, which is the primary product of any desalination operation. Furthermore, such metrics are more comparable to other desalination technologies, which facilitate understanding as to where CDI is competitive in the water treatment space.

9.3.1. *Energy consumption and thermodynamic efficiency*

Much research has gone into understanding energy efficiency in CDI systems, including in terms of operation mode, device architecture, and electrode materials. Of particular recent interest is the thermodynamic energy efficiency (TEE, symbol η) in Figs. 9.8 and 9.9 below.

9.3.2. *Salinity reduction and water recovery*

In order to better understand the relationship between Δc and WR, here we detail a highly simplified model that illustrates the fundamental tradeoff between the maximally achievable Δc and

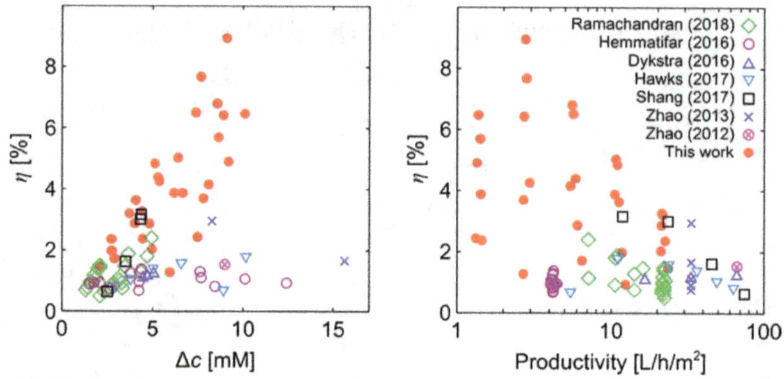

(a) (b)

Fig. 9.8. (a) TEE vs. concentration reduction and (b) TEE vs. throughput productivity. Reproduced with permission from Hemmatifar *et al.*[21] Copyright 2018 American Chemical Society.

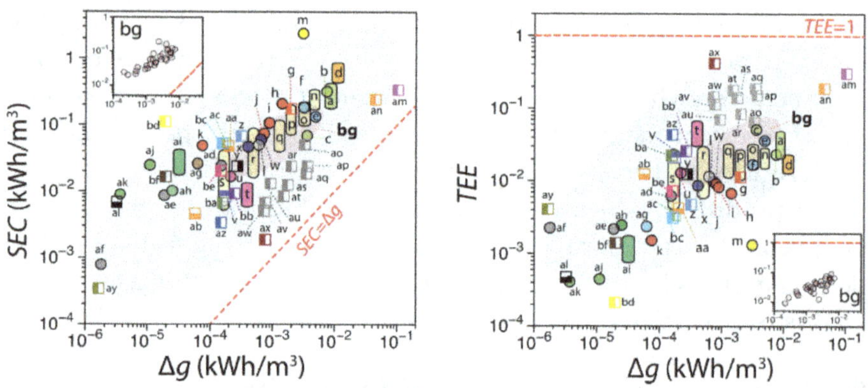

(a) (b)

Fig. 9.9. (a) Experimental specific energy cost (i.e. E_v) vs thermodynamic energy cost (Δg). (b) Thermodynamic energy cost (TEE) vs thermodynamic energy cost (Δg). Reproduced with permission from Wang *et al.*[55] Copyright 2019 American Chemical Society.

WR. We note that experimental results, like those discussed below, are significantly more complex and depend on considerably more variables (e.g., operation mode flow efficiency, etc.).[42] Thus, the relations derived here are primarily meant to act as a conceptual guide that allow one to form general expectations for achievable Δc and *WR* with a CDI cell.

As a starting point, we normalize the equation for water recovery by the CDI reactor active fluid volume ($\forall$),[22] giving

$$WR = \frac{\widetilde{V}_d}{\widetilde{V}_d + \widetilde{V}_c} \qquad (1)$$

where $\widetilde{V}_d = V_d/\forall$ is the volume of diluate produced (V_d) divided by the CDI reactor active fluid volume, and $\widetilde{V}_c = V_c/\forall$ is the volume of concentrate produced (V_c) divided by the CDI reactor active fluid volume. Next, we assume that the minimum volume of brine generated for a CDI cycle is held constant at one reactor volume ($\widetilde{V}_c = 1$). Further, we assume that the volume of treated water produced can be equal to or greater than one reactor volume, yielding $\widetilde{V}_d \geq 1$. The rationale behind these minimum-volume assumptions is that cell must be fully filled with water to operate; thus, the smallest-possible treatable volume is $\forall$. Given such an assumption, the salinity reduction achieved by the cell can be written as a fraction of the maximum possible salinity reduction ($\Delta c_{\max}$), which is equivalent to fully desalinating one cell volume of feed water,

$$\Delta c = \frac{\Delta c_{\max}}{\widetilde{V}_d} \qquad (2)$$

We further assume that there is a mechanism in place by which no mixing occurs between the diluate and the brine, which is a highly simplifying and idealized assumption and not representative of real CDI devices[52] Nevertheless, such an assumption allows one to establish an upper limit curve for salinity reduction and water recovery,

$$WR_{\max} = \frac{1}{1 + \frac{\Delta c}{\Delta c_{\max}}} \qquad (3)$$

where $WR_{\max}$ is the maximum attainable water recovery, and the maximum salinity reduction can be estimated as

$$\Delta c_{\max} = \frac{CV\Lambda}{F\forall} \qquad (4)$$

where C is the cell capacitance (F), V is the applied cell voltage (V), Λ is the charge efficiency, and F is Faraday's constant (C/mol).

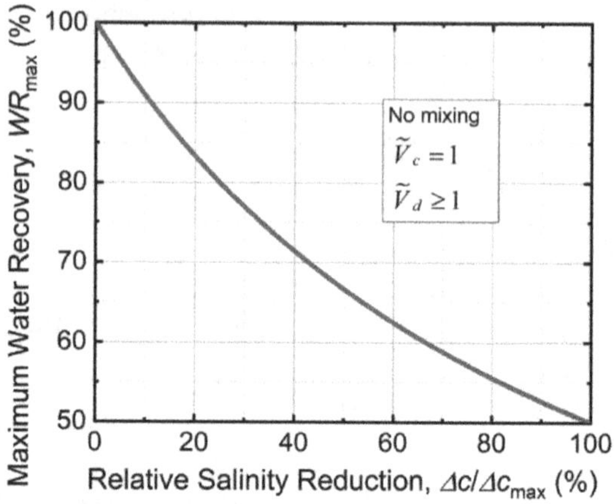

Fig. 9.10. Maximum achievable water recovery as a function of salinity reduction for the simplified model presented above.

Figure 9.10 plots the maximum attainable water recovery for a given salinity reduction against relative salinity reduction. Figure 9.10 and Eq. (3) show that, for example, to achieve 90% *WR*, one can only produce a diluate with Δc that is ~10% of the maximum possible Δc of the cell. Of course, such a tradeoff will generally not be achievable by real cells, which are subject to further mixing losses (i.e., non-unity flow efficiency). As an example, a 22.4 cm^2 flow-through electrode CDI device might have numbers like $C = 10\,F$, $\forall = 1.2\,ml$, $V = 1\,V$, and $\Lambda = 0.86$ would give $\Delta c_{max} = 74\,mM$ (~4320 ppm assuming NaCl as the salt). Thus, such a cell would not be expected to achieve a salinity reduction any greater than 7.4 mM (~432 ppm) at 90% *WR*. Ultimately, all the parameters in Eq. (4) are important to optimize for a given cell in order to achieve the largest possible Δc at a given *WR*.

As discussed above, relatively few studies in the CDI literature have examined *WR* explicitly.[39,49,56,57]

9.3.3. *Water cost*

While desalination characteristics and material properties are important to study, ultimately the most important comparator is produced water cost. To assess this in CDI, one must define a feed salinity of interest and desired final effluent salinity, then derive device performance characteristics for such a separation, and finally compute the required capital and operating cost to arrive at a cost of produced water in $\$/m^3$. Accurately performing such a technoeconomic analysis is challenging because it involves estimating the system lifetime, the annual materials and labor costs, and the waste disposal cost amongst many other variable economic factors. A recent attempt at such an analysis was performed by Hand *et al.*[58] who concluded that the primary economic driver for CDI is capital cost, and that consequently regular CDI had an economic advantage over the more expensive but higher performing MCDI. Such results are dependent on a number of assumptions, but highlight the general expectation that the desired volume or throughput of water treatement is typically large enough that system cost outweighs operating expense. Furthermore, Hand *et al.*[58] found that treating water at the lower end of brackish salinity ($\sim$10 meq/L) is likely where CDI has the largest economic advantage because smaller salinity reductions can be performed at higher productivity (Fig. 9.11). Finally, since capital cost is so significant, Hand *et al.*[58] also found that achieving a device service lifetime of >2–3 yr was crucial in dermining economic viability (Fig. 9.11).

While studies like Hand *et al.*[58] present a more thorough technoeconomic assessment of CDI, here we present a highly simplified water cost analysis that supports their conclusions and further highlights the importance of capital cost over, for example, energy cost. For a more detailed example of technoeconomic feasibility and models, though, we refer the reader to the literature.[58–60,140]

To begin our simple analysis, we define a total water cost $K(\$/m^3)$ that is the sum of the capital cost and energy,

$$K = K_c + K_e \tag{5}$$

where K_c is the capital cost per unit volume of water produced ($\$/m^3$) and K_o is the operating cost per unit volume of water produced ($\$/m^3$). The capital cost K_c can be amortized over the lifetime of the device τ (years) and normalized by the volume of

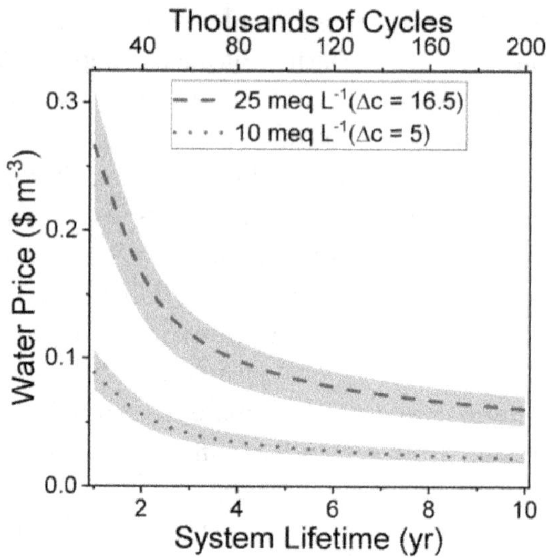

Fig. 9.11. The calculated median equivalent annual water price of CDI-treated brackish water by Hand *et al.*,[58] showing that the cost drops significantly for treating lower salinity waters with devices that last >3 years. Reproduced with permission from Hand *et al.*[58] Copyright 2019 American Chemical Society.

water produced as

$$K_c = \frac{k_c A}{\dot{V}\tau} \tag{6}$$

where k_c is the total capital cost per projected face area of the CDI device ($\$/m^2$), A is the projected face area of the CDI device (m^2), and $\dot{V}$ is the volume of treated water produced by the system per unit time (L/h). This calculation of capital cost is highly simplified and does account, for example, for the added cost of paying interest on a loan.[60] In total, k_c includes materials, fabrication, and deployment costs for the CDI device. Recognizing that $\dot{V}/A$ is, by definition, the throughput productivity P (L/h/m^2), we can rewrite Eq. (6) as

$$K_c = \frac{k_c}{P\tau}. \tag{7}$$

In terms of the operating cost K_o, we assume for simplicity that energy consumption is the primary driver of recurring expenses and therefore can be written as

$$K_e = k_e E_v. \tag{8}$$

where E_v is the energy consumed per volume of treated water (Wh/m^3) as discussed/defined above (Fig. 9.7), and k_e is the energy unit cost ($/Wh). Notably, this equation ignores many other operating costs, including brine disposal costs, which can be substantial and are expected to scale with 1 – *WR*.

Combining Eqs. (7) and (8), the total cost per unit volume of water treated can be written as

$$K = \frac{k_c}{P\tau} + k_e E_v. \tag{9}$$

Note that E_v is implicitly a function of the throughput productivity P,[5] where the relationship between E_v and P can be directly measured or computed by a CDI device model for a given operation mode and set of device parameters. Together with assumptions for the economic parameters k_c, k_e, and the assumed device operational lifetime τ, the estimated cost of produced water can be computed.

As an example of the estimations that one can do with Eq. (9), Fig. 9.12 plots the percentage of total CDI water cost due to capital cost (K_c/K) as a function of device cost (k_c) for different specific energy consumptions (E_v). Figure 9.12 shows that, because energy is relatively cheap and energy consumption is relatively low for treating brackish water, most of the total cost is expected to come from capital expenses. For example, the difference in total water cost between the green dot and light-blue dot in Fig. 9.12 is 782 $/acre-ft, which is entirely due to reducing the device cost of the system from ~425 $/m^2 to ~50 $/m^2. The capital cost and energy operating cost, as modeled here, only begin to become comparable for an extremely low device cost (~50 $/m^2) and reasonably large energy consumption (800 Wh/m^3). As evidence to support these conclusions, Voltea, Inc. reports that ~50% of the annual cost to treat 2,000 ppm water with membrane CDI is due to maintenance, cleaning, and module replacement (i.e. low lifetime in Eq. (7)), whereas energy

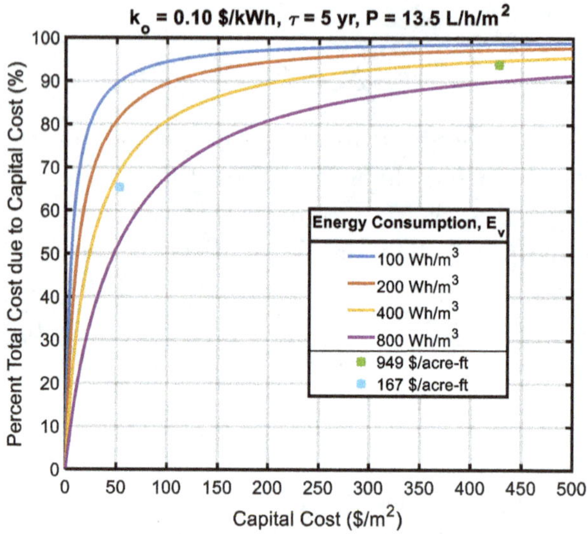

Fig. 9.12. Percentage of total CDI water cost due to capital cost ($100K_c/K$) as a function of capital cost (k_c) computed from Eq. (9) assuming an energy cost of \$0.10/kWh, a productivity of $13.5\,\mathrm{L/h/m^2}$, and a device lifetime of 5 yr.

cost only accounts for ~25% of the annual expense[61] Notably, it is unclear if the initial capital cost is factored into their calculations. It is also important to note that Voltea estimates that ~25% of the total annual operational cost is due to brine disposal, which is not accounted for in these calculations.[61]

Overall, these simple estimations are in full agreement with more detailed studies like Hand *et al.*,[58] and suggest that increasing operational lifetime, increasing throughput productivity, and reducing device capital expense will be the most significant drivers of lowering the cost of water treated by CDI.

9.3.4. *Performance indicators: Charge efficiency*

An important consideration in CDI is the relationship among applied electrical charge versus time in a cycle, the amount of salt trapped inside and then released from the cell in the cycle, and the time-varying amount of salt which is deficit and in excess in the effluent stream. The first and third of these are of most practical

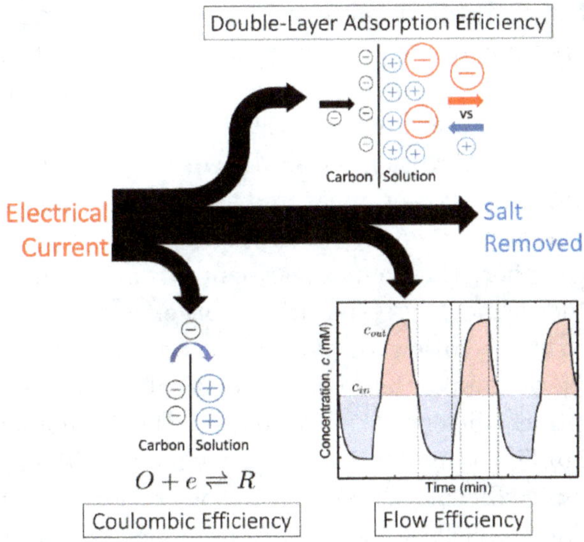

Fig. 9.13. Schematic of typical charge efficiencies of interest in CDI.

importance, as they are primary inputs and outputs of a CDI cell applied to desalination.

Broadly speaking, there are three main reasons for a disparity between applied (and withdrawn) electrical charge and the deficit/excess of salt in the effluent (Fig. 9.13). First, undesirable electrochemical reactions can divert electric charge via Faraday reactions at one or both electrodes (first bottom loss arrow in Fig. 9.13), which is occurs to a less extent for low potentials across the EDLs.[62,63] Second, if salt is trapped via a distrubted electric double layer (EDL), then at low values of potential across the EDLs, the balancing of electric surface charge causes by both ion repulsion and attraction. This can lead to a situation where ions repulsed from both electrodes cause shielding of electric surface charge but no trapping of net salt (top loss arrow in Fig. 9.13). Note that we define net salt as the sum of anion and cation pairs. This effect is reduced at higher potentials across EDLs. Finally, water from which salt is removed and trapped in the cell is not necessarily recoverable downstream. This is because there is a time delay and mixing with feed associated with water flowing in the cell and its recovery at a downstream location. Hence, desalted water can mix with the first volumes of water to be become brine at the beginning of regeneration. Similarly, incomplete removal of brine during regeneration results in some high salt concentration water mixing with the first volumes of input water at the start of the adsorption phase. The result is additional disparity between electric charge and measured salt downstream, and is called the flow efficieny (final loss arrow in Fig. 9.13).[42,64]

Taken together, these efficencies can strongly impact CDI perforamnce and achievable separation space.[25,28,35,65–70] For example, long term operation of typical carbon electrodes results in charged surface goups that can dramatically reduce the double-layer adsorption efficiency.[71–75] Moreover, the Coulombic efficiency can be a significant limiter to voltage window and achieving high thermodynamic efficiency,[21] while the flow efficiency ultimately puts a physical limit on achievable salinity reduction.[42] Charge efficiency limitions are ultimately what inspired the use of membranes in the technology.[18,37,39,40,51] Overall, the charge efficiency in CDI is a parameter that must be carefully accounted for when evaluating an operation mode or cell design.

9.3.5. *Performance indicators: Capacitance and salt adsorption capacity*

One of the most active areas in CDI research is improving the (maximum) salt adsorption capacity, (m)SAC, of the electrode materials.[25,39,47,65,76–87] As shown above in Fig. 9.10, though, the most important factor in determining the achievable separation space (i.e., Δc and WR) is Δc_{max}, not mSAC. Here we will show why this is an important distinction due to electrode mass density and cell construction considerations.

To understand the relation between Δc_{max} and mSAC, one can show that Δc_{max} in units of g/L is related to mSAC by

$$\Delta c_{\text{max}} = mSAC\frac{m}{\forall}. \tag{10}$$

where m is the total electrode mass in the cell (both electrodes). By assuming the device is composed of two identical solid electrodes and a porous separator, we can rewrite the factor $m/\forall$ by breaking down the terms into several common cell characteristics

$$\forall = 2p_{ma}V_{el} + p_{sp}V_{sp} \tag{11}$$

$$p_{ma} = 1 - \frac{\rho}{\rho_0} - p_{mi} \tag{12}$$

$$m = 2V_{el}\rho \tag{13}$$

where ρ is the mass density of the porous electrodes (g/cc), ρ_0 is the mass density of the pure solid phase of the electrodes (g/cc), p_{sp} is the separator porosity, p_{ma} is the electrode macroporosity, p_{mi} is the electrode microporosity, V_{el} is the geometric volume of the electrode (face area multiplied by thickness), and V_{sp} is the separator geometric volume (face area multiplied by thickness). Finally, combining Eqs. (11–13) we arrive at an alternative version of Eq. (10)

$$\Delta c_{\max} = mSAC \frac{\rho}{1 - \frac{\rho}{\rho_0} - p_{mi} + \frac{p_{sp}}{2}\frac{V_{sp}}{V_{el}}}. \tag{14}$$

Figure 9.14 plots $\Delta c_{\max}$ vs. porous electrode density ρ for several different mSAC values using reasonable numbers for the rest of the parameters in Eq. (14). Figure 9.14 shows that a wide range of mSAC values ranging from excellent to poor can give rise to the exact same $\Delta c_{\max}$ depending on the electrode density. While the geometric and fluid-flow details that determine $\forall$ are different for different cell configurations (e.g., MCDI vs. CDI; fteCDI vs. fbCDI),[42,52] this derivation still illustrates how mSAC is broadly related to actual desalination capability ($\Delta c_{\max}$). In particular, Fig 9.14 implies that mSAC may not be the most reliable indicator of desalination capability. If an electrode material has low density but reasonably high capacitance, then it is possible to have a large mSAC but low desalination capability. Similar to the conclusions from the metrics section,[5] we find that it is essential to start with the metrics of most importance (e.g., salinity reduction, water recovery, productivity, etc.) and work backwards towards essential indicators/characteristics that can be studied and optimized through research and development.

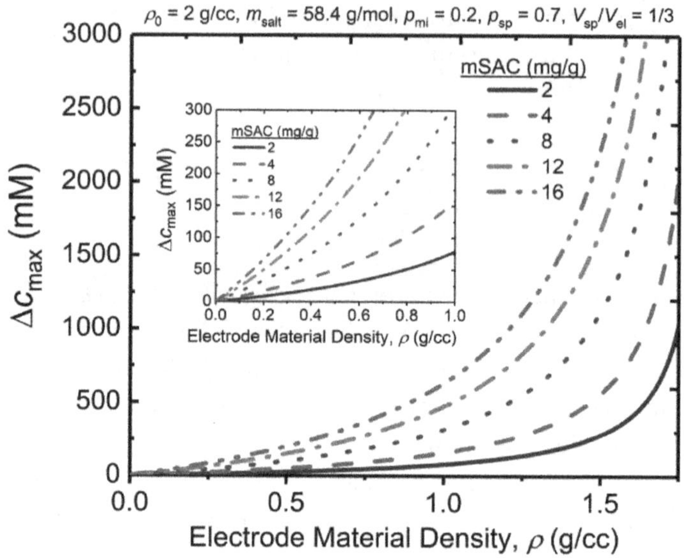

Fig. 9.14. Relationship between maximum salinity reduction and mSAC, showing that many mSAC values are possible for a given Δc_{max} due to different electrode densities.

In terms of the importance of capacitance in CDI, Section 9.3.2 Eq. (4) already highlights its role in widening the achievable separation space in the form of capacitance per reactor volume ($C/\forall$).[52] Eqs. (3) and (4) show that if a CDI cell or electrode system can achieve higher $C/\forall$, then the achievable salinity reduction is larger for a given water recovery. Conversely, if the desired salinity reduction and water recovery are already met, then increasing $C/\forall$ would allow one to decrease the applied voltage V in Eq. (4) to use less energy on the separation. Finally, Hawks *et al.*[42] showed that it is critical to treat many cell volumes of feed ($V_d >\sim 5\forall$) to achieve a high flow efficiency with a typical CDI reactor. Since capacitance allows one to treat more fluid volume for a given separation, it implicitly factors into achieving a high flow efficiency operation.[42]

9.3.6. *Performance indicators: Resistance*

Resistance is a critical factor that underlies achievable separation space and desalination performance in CDI. In terms of separation space, voltage-controlled operations on high-resistance cells will flow less charge over a given time and therefore typically remove less salt when compared to a lower resistance device. For example, Ramachandran *et al.*[52] showed theoretically that for sinusoidal voltage charging, the maximum possible salinity removal with infinite cell capacitance was finite and scaled with $1/R$ due to how resistance limits current flow to the device. On the other hand, for current-controlled operations, a higher resistance reduces the charging time for a given voltage window, which can result in a charge/discharge frequency that is too fast relative to the fluid residence time and therefore a low flow efficiency.[42]

In terms of energy, there are primarily two irreversible energy loss pathways in CDI: resistive energy dissipation of the form I^2R, and Faradaic reactions leading to electrochemistry as opposed to salt adsorption. For large voltages ($>\sim 1\,V$) and long charging times ($\gg\sim 10RC$), Faradaic currents can have a significant impact on performance; otherwise, energy dissipation through the resistive elements of the system dominate energy consumption.[23] To illustrate the relevance of I^2R-type losses, Fig. 9.15(a) plots the average total electrical power consumed by a $22.4\,\text{cm}^2$ face area fte-CDI device vs. the square of the applied constant current for different voltage windows and feed concentrations. The linear dependence of average power consumption (total energy irreversibly consumed/total cycle time) on the square of the charge/discharge current suggest that, for multiple feed concentrations and voltage windows, the energy dissipation is well described by a single effective resistance and the equation $P = I^2R + P_0$, where P_0 is a small offset likely caused by Faradaic losses. Figure 9.15(b) further plots the measured R vs. the inverse of the NaCl feed concentration, which yields the expected linear relationship due to the effect of electrolyte conductivity on R. This simple analysis suggests that measured power dissipation scales well with the I^2R heating law.

While the data in Fig. 9.15 suggest that a single value for the resistance accurately describes how power consumption scales with applied current over a range of constant current CDI operation, this is likely not the case for all operation modes and CDI devices.

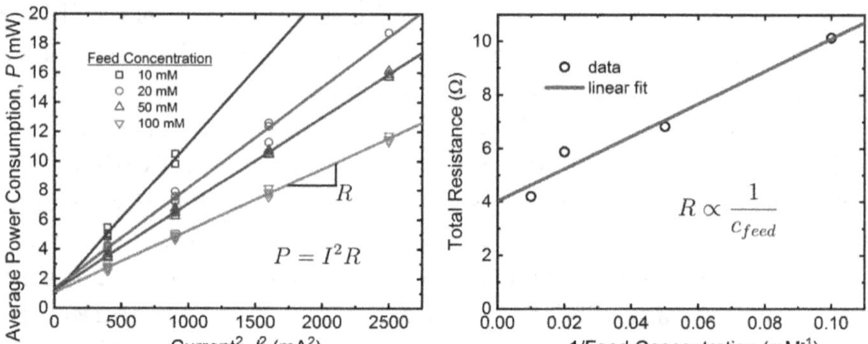

(a) (b)

Fig. 9.15. (a) Average power consumption during a DSS CDI cycle versus the square of the constant charging/discharging current. Multiple data points at each I^2 represent multiple voltage windows with either 0.2 V or 0.3 V discharge voltage and 0.9 V or 1.0 V charge voltage. The average concentration reduction was between 0–4.4 mM and the Coulombic efficiency was always >90%. (b) From the slopes of the linear fit from (a), the resistance can be plotted against the inverse of the feed concentration, giving an expected linear relationship.

For particularly high charging currents, like those found in constant voltage operation, the resistance dynamically changes throughout the cycle due to transmission line and electrolyte depletion effects.[23,88–90] Additionally, measured resistance values, like those in Fig. 9.15, can be further understood by breaking down the total resistance into its constituents with electrochemical impedance spectroscopy (EIS) and/or continuum model fitting.[46,91,92,141]

EIS in particular has been used extensively to characterize and understand the resistive elements of porous electrode systems,[93–106] including for CDI applications.[5,91] The work of Qu *et al.*[91] specifically used EIS to separate contact, ionic, and external (setup) resistances in a fte-CDI cell. Therein, the authors found that careful attention must be paid to the current collector contact, which can have significant parallel R/C characteristics if titanium is used with insufficient pressure applied to the contact. With knowledge from the EIS data in hand, Qu *et al.*[91] were able to solve their contact issue and thus dramatically reduce the resistance of their cell. Since resistance directly impacts energy performance and the achievable separation space, it is important to systematically understand all the resistive elements of a CDI cell so that each component can be ranked and the major contributors minimized as much as possible.[21]

9.4. Selective Removal of Contaminant Ions

9.4.1. *Introduction*

Contamination of groundwater with trace amounts of contaminants and/or rare metal ions makes the selective removal of ions an attractive alternative to decrease the cost of water treatments. A method capable of removing specific ions at high selectivity is thus highly relevant for point-of-use water treatment and industrial applications. Predominant methods to selectively remove organic and inorganic contaminants include reverse osmosis (RO),[107,108] forward osmosis (FO),[109] electrodialysis (ED), and ionic exchange resins.[110,111] Although these membrane technologies (such as RO or FO) show high rejection (98% or higher[112–115]), they are not yet optimized to desalinate low concentration streams such as brackish water or selectively remove trace amounts of contaminants.

An alternative technique is ionic exchange resins, where surface groups on the resin are initially electrostatically shielded by low-affinity ions (e.g., Cl^-) that can adsorb higher affinity ions (e.g., NO_3^-) via ion exchange. This affinity mechanism is presently not fully understood for first-principles predictions of quantitative affinity; however, ion size, electronic structure, and the associated interaction with ionic surface groups[116] are known to influence adsorption kinetics. Even though the selective removal of ionic species with exchange resins is promising, this technology suffers from important drawbacks, particularly the need for high concentration chemicals for regeneration and the disposal of the generated chemical waste.

Recently, there has been more focus on the selective removal of ionic species from water with CDI, inverted capacitive deionization (i-CDI), or membrane capacitive deionization (MCDI). An important advantage of i-CDI over ion exchange resins is that i-CDI requires no high concentration chemicals for regeneration. In i-CDI, an electrical bias is applied to regenerate the cell, which also shows the possibility to recover energy during adsorption. Multiple mechanisms have been attributed to selectivity, including the ion valence, steric effects, and interaction between nanostructure of the material and the hydrated ion. Experimental evidence of selective removal of ions with CDI has been demonstrated for nitrate, fluoride, and lithium. In these contributions, selectivity is achieved with either a selective membrane or due to the selectivity of the untreated activated carbon electrode.

In this section, we show an overview of the current status of selectivity in CDI and i-CDI, the key figures of merit to quantify selectivity in CDI, and experimental evidence of the competition between nitrate and chloride during adsorption into quaternary amine moieties with functionalized porous carbon electrode.

9.4.2. *Overview of selectivity in CDI and i-CDI*

Selective ion removal has been an active area of research in CDI community in recent years. The selectivity of adsorption has been attributed mainly to ion valence and steric effects,[9,10,117–121] and the interaction between nanoporous size and hydrated radius also seems to play a significant role.[117–120,122] Selectivity has been demonstrated experimentally for lithium,[123–126] nitrate,[127,128] and fluoride removal.[129–131] However, these studies relied on the intrinsic selectivity of pristine activated carbon surfaces.[117,132,133] Other studies[134,135] coated the surface of the activated carbon electrodes with an anion-exchange resin, which then served as an ion-selective barrier. In this section, we focus on selectivity of functionalized porous electrodes due to the high values of the cycle selectivity coefficient reported.

9.4.3. *Theory: Intrinsic and cycle ion selectivity*

The selectivity coefficient is defined in ionic exchange literature as the ratio of the enrichment factor between the high and the low-affinity ion, or equivalently as the ratio of equilibrium constants. The enrichment factor refers to the concentration ratio of the stationary phase (concentration in the exchanger resin) over the mobile phase (concentration in the external solution). We refer to this selectivity coefficient as intrinsic selectivity coefficient, since the coefficient is only valid near-equilibrium and in the absence of an externally applied bias. For instance, the CTA-AC electrode shows an intrinsic selectivity defined as

$$k_{int} = \frac{K_{\mathrm{NO_3}}}{K_{\mathrm{Cl}}} = \frac{c_{\mathrm{CTANO_3}}/c_{\mathrm{NO_3^-}}}{c_{\mathrm{CTACl}}/c_{\mathrm{Cl^-}}}, \tag{15}$$

where $c_{\mathrm{Cl^-}}$ and $c_{\mathrm{NO_3^-}}$ are concentrations in external solution when no applied voltage. Note that the Eq. (15) is useful to quantify the intrinsic selectivity coefficient of any material undergoing an adsorption competition among ions. We recognize that an external bias can either increase or decrease the value of the intrinsic selectivity coefficient. That is why in the presence of an externally applied bias, we propose a definition to determine the modified selectivity coefficient as:

$$\tilde{k}_{int} = \frac{(c_{\mathrm{CTANO_3}} + \Delta c_{\mathrm{NO_3^-}})/c_{\mathrm{NO_3^-}}}{(c_{\mathrm{CTACl}} + \Delta c_{\mathrm{Cl^-}})/c_{\mathrm{Cl^-}}}, \tag{16}$$

or, equivalently, using mass conservation equations, as

$$\tilde{k}_{int} = \frac{(c_{\mathrm{NO_3^-},0} - c_{\mathrm{NO_3^-}})/c_{\mathrm{NO_3^-}}}{(c_{\mathrm{Cl^-},0} - c_{\mathrm{Cl^-}})/c_{\mathrm{Cl^-}}}, \tag{17}$$

where $\Delta c_{\mathrm{NO_3^-}}$ and Δc_{Cl^-} correspond to the increase in concentration in the micropore due to an applied electric field and can be written as $\Delta \mathrm{c_i} = \mathrm{c_{m,i,A}} + \mathrm{c_{m,i,C}} - 2\mathrm{c_i}$ (for $\mathrm{i} = \mathrm{Cl^-}, \mathrm{NO_3^-}$). Note that for no applied external bias, the external solution and micropore concentrations are identical (i.e. $\mathrm{c_{m,i,j}} = \mathrm{c_i}$ for $\mathrm{i} = \mathrm{Cl^-}, \mathrm{NO_3^-}$ and $\mathrm{j} = \mathrm{A, C}$).

In CDI applications, however, the selective competition of the ionic species is determined from the dynamics of the interaction among chemical affinity, electrosorption, and mass transport, while the intrinsic selectivity coefficients ($\mathrm{k_{int}}$ and $\tilde{\mathrm{k}}_{\mathrm{int}}$) only show the effect of chemical affinity in selectivity. For example, for a cyclic operation, CDI requires a regeneration (and desorption) step, so that the possible selectivity coefficient is influenced by the operational parameters of the CDI cell. Here, we show an operation scheme of selective passive adsorption at V_{ads} (adsorption voltage) of nitrate, followed by a non-specific electrical regeneration step at V_{reg} (regeneration voltage). We introduce the cycle selectivity coefficient k_{cycle} as a new metric to quantify the selectivity in a full CDI cycle. In the case of a closed-loop system, cycle selectivity coefficient is defined as:

$$k_{cylce} = \frac{\left[c_{\mathrm{NO}_3^-}(V_{reg}) - c_{\mathrm{NO}_3^-}(V_{ads})\right] / c_{\mathrm{NO}_3^-,0}}{\left[c_{\mathrm{Cl}^-}(V_{reg}) - c_{\mathrm{Cl}^-}(V_{ads})\right] / c_{\mathrm{Cl}^-,0}}, \tag{18}$$

where the numerator corresponds to the relative change in concentration measured at the end of adsorption and regeneration steps of NO_3^- in the external solution. Similarly, the denominator is the relative change for Cl^-. For the case of open-loop systems, the cycle selectivity coefficient can be calculated as

$$k_{cylce} = \frac{c_{\mathrm{Cl}^-,0}}{c_{\mathrm{NO}_3^-,0}} \frac{\int_0^{t_{ads}} \left(c_{\mathrm{NO}_3^-,0} - c_{\mathrm{NO}_3^-}(t)\right) dt}{\int_0^{t_{ads}} \left(c_{\mathrm{Cl}^-,0} - c_{\mathrm{Cl}^-}(t)\right) dt}, \tag{19}$$

where t_{ads} is duration of the adsorption step and $c_{NO_3^-}(t)$ and $c_{Cl^-}(t)$ correspond to the measured effluent concentration of NO_3^- and Cl^-, respectively. Again, the expressions for k_{cycle} in a closed loop and cycle are applicable for any dynamic system under a competition between ions.

9.4.4. *Quantification of intrinsic selectivity*

In this section, we focus on a commercially available activated carbon electrodes (Material Methods LLC., PACMM^{TM} 203, Irvine, CA) functionalized with anionic (sodium dodecylbenzene sulfonate, SDBS) and cationic (cetrimonium bromide, CTAB) surfactants. We refer to the functionalized electrodes as CTA-AC and DBS-AC for the cation and anion PACMM electrode functionalization, respectively. To quantify the intrinsic selectivity coefficient, 50 mg of treated CTA-AC were placed into 8 containers filled with 24 mL $NaNO_3$ at several concentrations (0.3 to 3 mM). Figure 9.16 shows NO_3^- and

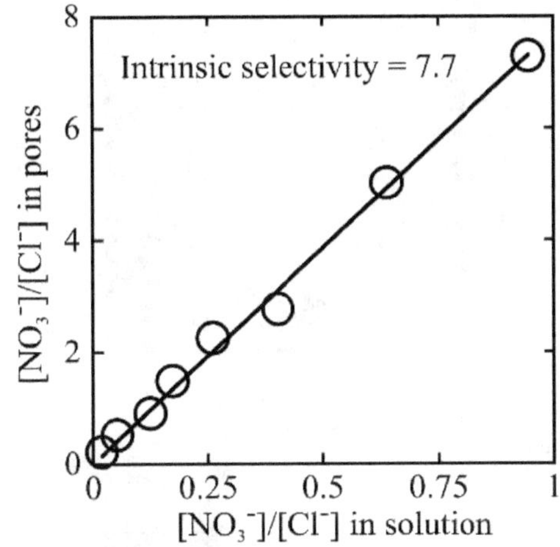

Fig. 9.16. Independent, controlled, ex-situ experiment to determine intrinsic selectivity of nitrate over chloride with CTA-AC electrode. The ordinate is the ratio of nitrate to chloride ions inside the micropores. The abscissa is the ratio of nitrate to chloride ions in solution measured with IC. Reproduced with permission from Elsevier from Ref. [8].

Cl^- in solution ($c_{NO_3^-}/c_{Cl^-}$) on the abscissa versus the adsorption equilibrium of NO_3^- and Cl^- in the micropores (c_{CTANO_3}/c_{CTACl}) on the ordinate. Ion chromatography (IC) was used to measure nitrate and chloride concentrations in solution after 24 h equilibration time. The micropore concentration of nitrate and chloride was calculated indirectly using the following relation:

$$\frac{c_{CTANO_3}}{c_{CTACl}} = \frac{n_{NO_3^-} - v_{sol}c_{NO_3^-}}{n_{Br^-} - v_{sol}c_{Cl^-}}, \tag{20}$$

where n_{Br^-} is the moles of bromide (initially bound to CTA-AC) released after the initial exchange with Cl^- and $n_{NO_3^-}$ is moles of NO_3^- added to the solution. The latter is roughly equal to the amount of exchanged Cl^-. Markers from Fig. 9.16 show the experimental results and the solid line correspond to a linear regression to minimize the mean squared error ($R^2 = 0.9952$). Intrinsic selectivity is obtained from the slope of the regression line and for the nitrate versus chloride case is approximately 7.7. This measured intrinsic selectivity coefficient related to the ratio of dissociation constants as $k_{int} = K_{Cl}/K_{NO_3} \approx 7.7$.

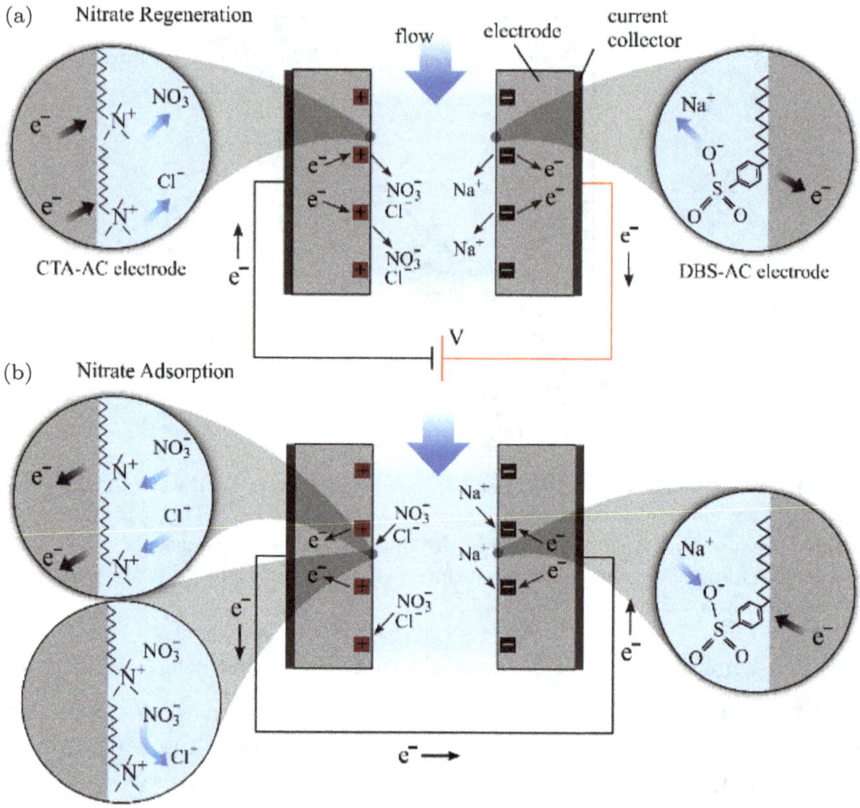

Fig. 9.17. Schematic of an inverted CDI with functionalized electrodes. The CTA-AC and DBS-AC electrodes are functionalized with CTAB and SDBS, respectively. (a) Schematic of the electrical desorption of ions at a fixed potential, V. The electric field generate a migration of electrons from the DBS AC to CTA-AC. Electrons that arrive the CTA-AC repel Cl^- and NO_3^- ions, while the holes generated in the DBS-AC repel Na^+ ions. (b) Schematic of a short-circuit adsorption mechanism. Ions are adsorbed by forcing migration of electrons from CTA AC to DBS AC. During adsorption exchange of NO_3^- and Cl^- occurs in the CTA-AC. Reproduced with permission from Elsevier from Ref. [8].

9.4.5. *Quantification of cycle selectivity in i-CDI*

Figure 9.17 shows a schematic of the adsorption and regeneration of an i-CDI cell. To perform the experiments, we assemble a flow-between i-CDI cell that consists of a radial inflow geometry. The cell contains a single pair of surfactant-treated electrodes (CTA-AC and DBS-AC). Each electrode was 3.5 cm diameter, ~300 μm thickness, and a total dry mass of 0.28 g. Titanium sheets, 50 μm thick, were used as current collectors, and a woven plastic mesh, 300 μm thick, (McMaster-Carr, Los Angeles, CA) were used as a spacer between the electrodes. A fresh functionalized CTA-AC electrode contains bromide, which is exchanged with chloride in a preconditioning experiment at 0 V, fixed flow rate (0.23 ml min^{-1}) and a feed solution of 2 mM $NaNO_3$ and 2 mM NaCl for 12 h. We set, in all experiments, the duration of the regeneration and adsorption to 2 h. A fraction collector (Model 2128, Bio-Rad, CA) was used to collect the effluent in aliquots accumulated over 10 min intervals and saved into separate containers. The concentration of each species in the container was measured using ion chromatography (DIONEX DX 500, DIONEX, CA)

Figure 9.18 shows experimental evidence (markers) of the adsorption dynamics between nitrate and chloride and the dynamic model predictions (solid lines) for nitrate and chloride. Details of the model

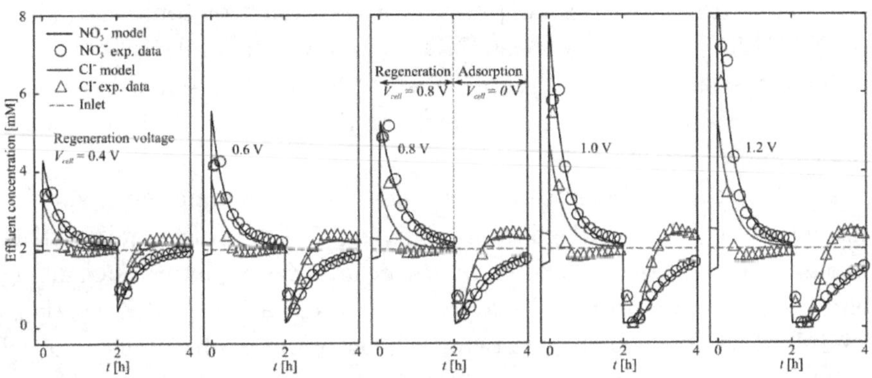

Fig. 9.18. Dynamic steady state of adsorption and electrical regeneration of NO_3^- and Cl^-. The independently measured values of NO_3^- and Cl^- concentration are shown versus time. The feed concentration ratio is 1:1 (2 mM$NaNO_3$ and 2 mMNaCl), and constant flow rate Q = 0.43 mL min^{-1}. Symbols correspond to experimental data by IC and solid lines correspond to predictions from the multi-species dynamic model. The label on each cycle corresponds to the applied bias in the regeneration step. To ensure dynamic steady state the experimental data correspond to the fourth cycle (first, second, and third cycle were discarded). Reproduced with permission from Elsevier from Ref. [8].

are available in Oyarzun *et al.*[8] The measurements were performed to effluent aliquots of our i-CDI cell for the case of a 1:1 feed solution (of 2 mM) ratio of $NaNO_3$ to NaCl. The applied regeneration bias was 0.4, 0.6, 0.8, 1.0, and 1.2 V, while the adsorption voltage was 0 V. In the regeneration step, a fraction of the electrically- and chemically-adsorbed ions are expelled from the CTA-AC electrode and flushed from the cell. Simultaneously, sodium ions are expelled from the counter electrode DBS-AC.

During the 0 V adsorption step, NO_3^- and Cl^- are adsorbed into the CTA-AC at different rates, which is consistent with the coupling between mass transport dynamics with the different intrinsic anion selectivity of CTA-AC. Nitrate and chloride effluent concentrations drop rapidly at the beginning of the adsorption step. However, at longer times chloride concentration approaches the inlet value faster than nitrate. Furthermore, soon after the start of the adsorption step, the concentration of chloride increases above the feed solution, while, in contrast, nitrate concentration stays below the feed value during the adsorption step. We conclude that this dynamic behavior corresponds to experimental evidence of ionic exchange on CTA-AC surface between $NO_3{}^-$ and Cl^-. Similarly, selective behavior is observed in adsorption of cations of different valence and hydrated radii (such as Na^+, Mg^{2+}, and Ca^{2+}) in traditional CDI operation.[133,136–138]

In Fig. 9.18, solid lines correspond to our dynamic multi-species model using $K_{Na} = 10\,\frac{mol}{m^3}$ and $K_{NO_3} = 77\,\frac{mol}{m^3}$ as the main fitting parameters as described in Oyarzun *et al.*[8] K_{Cl} is determined using the measured intrinsic selectivity $k_{int} \approx 7.7$ as $K_{Cl} = K_{NO_3}/k_{int}$. Mainly during the adsorption step, the dynamic multi-species model predictions show good agreement with measurements of effluent concentrations by IC. At longer adsorption times, the model captures the exchange between NO_3^- and Cl^-. During regeneration, at higher regeneration voltages, the model predictions start to deviate from the experimental measurements. We hypothesize that this discrepancy in the regeneration step at higher voltages is attributed to a generation of hydroxide which may be generated by Faradaic reactions.

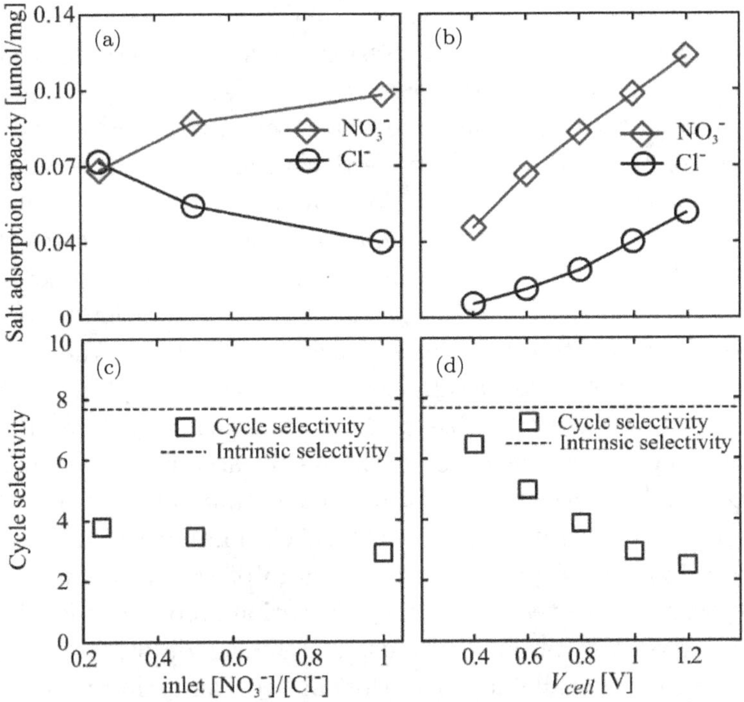

Fig. 9.19. Effect of several operational parameters in the cycle selectivity coefficient. (a) and (c) show salt adsorption capacity for nitrate and chloride, and cycle selectivity at 1:4, 1:2 and 1:1 ratios of initial concentrations. Chloride was 2, 4 and 8 mM while nitrate was fixed at 2 mM. The applied Regeneration voltage was fixed at 1 V. (b) and (d) show specific salt adsorption for nitrate and chloride as well as cycle selectivity at different regeneration voltages for an inlet concentration ratio 1:1 (2 mM nitrate and 2 mM chloride). A constant flow rate of 0.43 mL min 1 was supplied. Reproduced with permission from Elsevier from Ref. [8].

9.4.6. *Effect of regeneration voltage and the ratio of concentrations in cycle selectivity in i-CDI*

We summarize the adsorption capacity in μmol of anions (nitrate and chloride) for a fixed regeneration voltage of 1V (Fig. 9.19(a)) and three ratios of the feed solution, and regeneration voltages from 0.4 to 1.2 V at 1:1 ratio of feed concentration (Fig. 9.19(b)). Markers are measurements for each species in the full adsorption step. Cycle selectivity shows a slight increment (Fig. 9.19(c)) for feed chloride concentration from 2 to 8 mM at fixed (2 mM) feed nitrate concentration and regeneration voltage of 1 V.

Figure 9.19(d) presents the cycle selectivity coefficient for a 1:1 nitrate to chloride ratio of feed concentration at regeneration voltages from 0.4 to 1.2 V. Lower values of regeneration voltage lead to cycle selectivity values approaching the intrinsic selectivity value for CTA-AC. The experimental evidence shows a decrease in cycle selectivity with the magnitude of the regeneration voltage. In the regeneration step, the applied electronic charge in the CTA-AC carbon surface electrostatically repels both anionic species from the electrode with no noticeable preference, which might imply that, during regeneration, the quantity of ions released into the stream is a strong function of the applied bias and a weak function of the current micropore concentration in the CTA-AC. The available sites after regeneration are proportional to the ratio of species on the porous of the electrode at the end of the adsorption step. That is why this non-selective desorption mechanism, at higher desorption voltages, produces more available sites (for nitrate and chloride) on the CTA-AC. But nitrate is simultaneously re-adsorbed to the CTA-AC due to the quaternary ammonium moieties on the surface of the CTA-AC electrode. On the other hand, at high regeneration voltages, chloride shows an incomplete exchange with nitrate (Fig. 9.18). We believe that longer adsorption times for the cases at higher regeneration voltage will lead to a complete exchange, which in turn, increases the selectivity coefficient.

9.5. Outlook

While CDI is showing promise for a number of applications even in its current state, there is much room for improvement through further material development and device design. The cost of CDI devices compared to their productivity is still fairly low, and device lifetime must be extended. Enhancement of the selectivity would also be desirable.

Decreasing device resistance would increase the rate at which the device can be charged, and thereby increase productivity. Resistance is typically dominated by the ionic resistance of the liquid, which can be reduced by either reducing electrode thickness or by adding a conductive binder to the electrode.[76] Reducing electrode thickness would also reduce capacity, however, so concurrent development of the electrode material to increase capacity would be necessary to obtain an overall improvement using this method.

Functionalizing the electrode with groups that either increase selectivity of the process or inhibit undesirable reaction would also increase performance. The former would aim at applications that target specific ions and reduce treatment cost for those ions, while the latter could prevent electrode corrosion and extend device lifetime. Stable functional groups might also be used in iCDI configuration, where improvements are achieved by reducing the necessary voltage for the operation, or could be used in an extended CDI configuration to increase the effective voltage window that is used and thereby increase the capacity. Functionalization can be achieved both by directly modifying the carbon electrode or by coating the electrode.

Finally, performance could be improved by increasing electrode capacity. MCDI has sufficient capacity for most applications that CDI will reasonably be used in the near future, but CDI has insufficient capacity to achieve both a high concentration reduction and a high water recovery even at feed concentrations of 1,000 ppm. Electrode materials with enhance capacity would eliminate this problem, but these new materials would have to have cost and lifetimes comparable to or better than the currently used carbon materials.

In summary, CDI is a water treatment technique with a removal mechanism and limitations that are very different from reverse osmosis, and is likely to provide a complementary tool for treating low salinity water.

Acknowledgments

Work at LLNL was performed under the auspices of the US DOE by LLNL under Contract DE-AC52-07NA27344.

References

1. J. Landon, X. Gao, A. Omosebi, K. Liu, Progress and outlook for capacitive deionization technology, *Curr. Opin. Chem. Eng.*, **25** (2019) 1–8.

2. A. M. Ahmed, S. Tewari, Capacitive deionization: Processes, materials and state of the technology, *J. Electroanal. Chem.*, **813** (2018) 178–192.
3. C. Huyskens, J. Helsen, A. B. de Haan, Capacitive deionization for water treatment: Screening of key performance parameters and comparison of performance for different ions, *Desalination*, **328** (2013) 8–16.
4. Y. Oren, Capacitive deionization (Cdi) for desalination water treat — Past, present and future (a Review), *Desalination*, **228** (2008) 10–29.
5. A. S. Hawks, A. Ramachandran, S. Porada, G. P. Campbell, E. M. Suss, M. P. Biesheuvel, G. J. Santiago, M. Stadermann, Performance metrics for the objective assessment of capacitive deionization systems, *Water Res.*, **152** (2019) 126–137.
6. M. Valiskó, D. Boda, D. Gillespie, Selective adsorption of ions with different diameter and valence at highly charged interfaces, *J. Phys. Chem. C*, **111** (2007) 15575–15585.
7. Z. Huang, L. Lu, Z. Cai, Z. J. Ren, Individual and competitive removal of heavy metals using capacitive deionization, *J. Hazard Mater.*, **302** (2016) 323–331.
8. I. D. Oyarzun, A. Hemmatifar, W. J. Palko, M. Stadermann, J. G. Santiago, Ion selectivity in capacitive deionization with functionalized electrode: Theory and experimental validation, *Water Res. X*, **1** (2018) 100008.
9. A. S. Hawks, M. R. Cerón, I. D. Oyarzun, A. T. Pham, C. Zhan, K. C. Loeb, D. Mew, A. Deinhart, C. B. Wood, J. G. Santiago, Using ultramicroporous carbon for the selective removal of nitrate with capacitive deionization, *Environ. Sci. Technol.*, **53** (2019) 10863–10870.
10. M. T. Mubita, E. J. Dykstra, M. P. Biesheuvel, A. van der Wal, S. Porada, Selective adsorption of nitrate over chloride in microporous carbons, *Water Res*, **164** (2019) 114885.
11. W. J. Blair, G. W. Murphy, Electrochemical Demineralization of Water with Porous Electrodes of Large Surface Area. In *Saline Water Conversion*, American Chemical Society: 1960; Vol. 27, pp 206–223.
12. E. M. Suss, F. T. Baumann, L. W. Bourcier, M. C. Spadaccini, A. K. Rose, G. J. Santiago, M. Stadermann, Capacitive desalination with flow-through electrodes, *Energy Environ. Sci.*, **5** (2012) 9511.
13. C. Zhang, L. Wu, J. Ma, N. A. Pham, M. Wang, T. D. Waite, Integrated flow-electrode capacitive deionization and microfiltration system for continuous and energy-efficient brackish water desalination. *Environ. Sci. Technol.*, **53** (2019) 13364–13373.

14. X. Gao, A. Omosebi, J. Landon, K. Liu, Surface charge enhanced carbon electrodes for stable and efficient capacitive deionization using inverted adsorption–desorption behavior, *Energy Environ. Sci.*, **8** (2015) 897.
15. L. Zhang, M. P. Biesheuvel, I. I. Ryzhkov, Theory of ion and water transport in electron-conducting membrane pores with ph-dependent chemical charge, *Phys. Rev. Appl.*, **12** (2019) 014039.
16. E. J. Dykstra, J. K. Keesman, M. P. Biesheuvel, A. van der Wal, Theory of Ph changes in water desalination by capacitive deionization, *Water Res.*, **119** (2017) 178–186.
17. N. E. Guyes, N. A. Shocron, A. Simanovski, M. P. Biesheuvel, M. E. Suss, A one-dimensional model for water desalination by flow-through electrode capacitive deionization, *Desalination*, **415** (2017) 8–13.
18. M. P. Biesheuvel, R. Zhao, S. Porada, A. van der Wal, Theory of membrane capacitive deionization including the effect of the electrode pore space, *J. Colloid Interface Sci.*, **360** (2011) 239–248.
19. P. M. Biesheuvel, Thermodynamic cycle analysis for capacitive deionization, *J. Colloid Interface Sci.*, **332** (2009) 258–264.
20. Y. Qu, G. P. Campbell, A. Hemmatifar, M. J. Knipe, K. C. Loeb, J. J. Reidy, A. M. Hubert, M. Stadermann, J. G. Santiago, Charging and transport dynamics of a flow-through electrode capacitive deionization system, *J. Phys. Chem. B*, **122** (2018) 240–249.
21. A. Hemmatifar, A. Ramachandran, K. Liu, I. D. Oyarzun, Z. M. Bazant, J. G. Santiago, Thermodynamics of ion separation by electrosorption, *Environ. Sci. Technol.*, **52** (2018) 10196–10204.
22. A. Ramachandran, A. Hemmatifar, A. S. Hawks, M. Stadermann, J. G. Santiago, Self similarities in desalination dynamics and performance using capacitive deionization, *Water Res.*, **140** (2018) 323–334.
23. A. Hemmatifar, W. J. Palko, M. Stadermann, J. G. Santiago, Energy breakdown in capacitive deionization, *Water Res.*, **104** (2016) 303–311.
24. A. Hemmatifar, M. Stadermann, J. G. Santiago, Two-dimensional porous electrode model for capacitive deionization, *J. Phys. Chem. C*, **119** (2015) 24681–24694.
25. E. M. Suss, S. Porada, X. Sun, M. P. Biesheuvel, J. Yoon, V. Presser, Water desalination via capacitive deionization: What is it and what can we expect from it? *Energy Environ. Sci.*, **8** (2015) 2296–2319.
26. M. P. Biesheuvel, B. van Limpt, A. van der Wal, Dynamic adsorption/desorption process model for capacitive deionization, *J. Phys. Chem. C*, **113** (2009) 5636–5640.
27. B. M. Dixit, D. Moreno, X. Xiao, C. M. Hatzell, K. B. Hatzell, Mapping charge percolation in flowable electrodes used in capacitive deionization, *ACS Mater. Lett.*, **1** (2019) 71–76.

28. R. Zhao, M. P. Biesheuvel, H. Miedema, H. Bruning, A. van der Wal, Charge efficiency: A functional tool to probe the double-layer structure inside of porous electrodes and application in the modeling of capacitive deionization, *J. Phys. Chem. Lett.*, **1** (2010) 205–210.
29. Z. Ding, V. Trouillet, S. Dsoke, Are functional groups beneficial or harmful on the electrochemical performance of activated carbon electrodes? *J. Electrochem. Soc.*, **166** (2019) A1004–A1014.
30. Y. Bouhadana, E. Avraham, M. Noked, M. Ben-Tzion, A. Soffer, D. Aurbach, Capacitive deionization of Nacl solutions at non-steady-state conditions: Inversion functionality of the carbon electrodes, *J. Phys. Chem. C*, **115** (2011) 16567–16573.
31. X. Gao, A. Omosebi, J. Landon, K. Liu, Enhanced salt removal in an inverted capacitive deionization cell using amine modified microporous carbon cathodes, *Environ. Sci. Technol.*, **49** (2015) 10920–10926.
32. I. D. Oyarzun, A. Hemmatifar, W. J. Palko, M. Stadermann, J. G. Santiago, Adsorption and capacitive regeneration of nitrate using inverted capacitive deionization with surfactant functionalized carbon electrodes, *Sep. Purif. Technol.*, **194** (2018) 410–415.
33. W. J. Palko, I. D. Oyarzun, B. Ha, M. Stadermann, J. G. Santiago, Nitrate removal from water using electrostatic regeneration of functionalized adsorbent, *Chem. Eng. J.*, **334** (2018) 1289–1296.
34. M. A. Johnson, W. A. Venolia, G. R. Wilbourne, J. Newman, *The electrosorb process for desalting water*; U.S. Dept. of the Interior, 1970.
35. E. Avraham, M. Noked, Y. Bouhadana, A. Soffer, D. Aurbach, Limitations of charge efficiency in capacitive deionization Ii. On the behavior of Cdi cells comprising two activated carbon electrodes, *J. Electrochem. Soc.*, **156** (2009) P157–P162.
36. S.-I. Jeon, H.-R. Park, J.-G. Yeo, S. Yang, H. C. Cho, H. M. Han, D. K. Kim, Desalination via a new membrane capacitive deionization process utilizing flow-electrodes, *Energy Environ. Sci.*, **6** (2013) 1471–1475.
37. M. P. Biesheuvel, A. van der Wal, Membrane capacitive deionization, *J. Membr. Sci.*, **346** (2010) 256–262.
38. J.-B. Lee, K.-K. Park, H.-M. Eum, C.-W. Lee, Desalination of a thermal power plant wastewater by membrane capacitive deionization, *Desalination*, **196** (2006) 125–134.
39. R. Zhao, O. Satpradit, M. H. H. Rijnaarts, M. P. Biesheuvel, A. van der Wal, Optimization of salt adsorption rate in membrane capacitive deionization, *Water Res.*, **47** (2013) 1941–1952.

40. R. Zhao, S. Porada, M. P. Biesheuvel, A. van der Wal, Energy consumption in membrane capacitive deionization for different water recoveries and flow rates, and comparison with reverse osmosis, *Desalination*, **330** (2013) 35–41.
41. Y.-J. Kim, J.-H. Choi, Enhanced desalination efficiency in capacitive deionization with an ion-selective membrane, *Sep. Purif. Technol.*, **71** (2010) 70–75.
42. A. S. Hawks, M. J. Knipe, G. P. Campbell, K. C. Loeb, A. M. Hubert, G. J. Santiago, M. Stadermann, Quantifying the flow efficiency in constant-current capacitive deionization, *Water Res.*, **129** (2018) 327–336.
43. Y.-J. Kim, J.-H. Choi, Selective removal of nitrate ion using a novel composite carbon electrode in capacitive deionization, *Water Res.*, **46** (2012) 6033–6039.
44. J.-H. Choi, Comparison of Constant voltage (CV) and Constant current (CC) operation in the membrane capacitive deionisation process, *Desalination Water Treat.*, **56** (2015) 921–928.
45. P. Długołecki, A. Van Der Wal, Energy recovery in membrane capacitive deionization, *Environ. Sci. Technol.*, **47** (2013) 4904–4910.
46. E. J. Dykstra, S. Porada, A. van der Wal, P. M. Biesheuvel, Energy consumption in capacitive deionization — Constant current versus constant voltage operation, *Water Res.*, **143** (2018) 367–375.
47. J. Kang, T. Kim, K. Jo, J. Yoon, Comparison of salt adsorption capacity and energy consumption between constant current and constant voltage operation in capacitive deionization, *Desalination*, **352** (2014) 52–57.
48. Y. Qu, G. P. Campbell, L. Gu, M. J. Knipe, E. Dzenitis, G. J. Santiago, M. Stadermann, Energy consumption analysis of constant voltage and constant current operations in capacitive deionization, *Desalination*, **400** (2016) 18–24.
49. A. Ramachandran, I. D. Oyarzun, A. S. Hawks, M. Stadermann, J. G. Santiago, High water recovery and improved thermodynamic efficiency for capacitive deionization using variable flowrate operation, *Water Res.*, **155** (2019) 76–85.
50. L. Wang, S. Lin, Intrinsic tradeoff between kinetic and energetic efficiencies in membrane capacitive deionization, *Water Res.*, **129** (2018) 394–401.
51. R. Zhao, M. P. Biesheuvel, A. van der Wal, Energy consumption and constant current operation in membrane capacitive deionization, *Energy Environ. Sci.*, **5** (2012) 9520.
52. A. Ramachandran, A. S. Hawks, M. Stadermann, J. G. Santiago, Frequency analysis and resonant operation for efficient capacitive deionization, *Water Res.*, **144** (2018) 581–591.

53. X. Shang, D. R. Cusick, K. C. Smith, A combined modeling and experimental study assessing the impact of fluid pulsation on charge and energy efficiency in capacitive deionization, *J. Electrochem. Soc.*, **164** (2017) E536–E547.
54. A. Hemmatifar, Energy consumption and salt adsorption in capacitive deionization. Stanford, 2018.
55. L. Wang, E. J. Dykstra, S. Lin, Energy efficiency of capacitive deionization, *Environ. Sci. Technol.*, **53** (2019) 3366–3378.
56. W. M. Saleem, C. A. Y. Jande, M. Asif, W.-S. Kim, Hybrid CV-CC operation of capacitive deionization in comparison with constant current and constant voltage, *Sep. Sci. Technol.*, **51** (2016) 1063–1069.
57. A. Ramachandran, I. D. Oyarzun, A. S. Hawks, G. P. Campbell, M. Stadermann, J. G. Santiago, Comments on "comparison of energy consumption in desalination by capacitive deionization and reverse osmosis", *Desalination*, **461** (2019) 30–36.
58. S. Hand, S. J. Guest, R. D. Cusick, Technoeconomic analysis of brackish water capacitive deionization: Navigating tradeoffs between performance, lifetime, and material costs, *Environ. Sci. Technol.*, **53** (2019) 13353–13363.
59. C. Bales, P. Kovalsky, J. Fletcher, T. D. Waite, Low cost desalination of brackish groundwaters by capacitive deionization (Cdi) — Implications for irrigated agriculture, *Desalination*, **453** (2019) 37–53.
60. J. Newman, G. P. Hoertz, A. C. Bonino, J. A. Trainham, Review: An economic perspective on liquid solar fuels, *J. Electrochem. Soc.*, **159** (2012) A1722–A1729.
61. Voltea Capdi Vs Ro Technical Comparison 2,000 Ppm. https://voltea.com/wp-content/uploads/2016/02/CapDI-vs-RO-2000-ppm.pdf.
62. S. Shanbhag, F. J. Whitacre, M. S. Mauter, The origins of low efficiency in electrochemical de-ionization systems, *J. Electrochem. Soc.*, **163** (2016) E363–E371.
63. B. Pillay, J. S. Newman, The influence of side reactions on the performance of electrochemical double-layer capacitors, *J. Electrochem. Soc.*, **143** (1996) 1806.
64. M. A. Johnson, J. Newman, Desalting by means of porous carbon electrodes, *J. Electrochem. Soc.*, **118** (1971) 510.
65. T. Kim, E. J. Dykstra, S. Porada, A. van der Wal, J. Yoon, P. M. Biesheuvel, Enhanced charge efficiency and reduced energy use in capacitive deionization by increasing the discharge voltage, *J. Colloid Interface Sci.*, **446** (2015) 317–326.

66. X. Gao, A. Omosebi, J. Landon, K. Liu, Enhancement of charge efficiency for a capacitive deionization cell using carbon xerogel with modified potential of zero charge, *Electrochem. Commun.*, **39** (2014) 22–25.
67. H. Li, C. Nie, L. Pan, Z. Sun, The study of membrane capacitive deionization from charge efficiency, *Desalination Water Treat.*, **42** (2012) 210–215.
68. I. Cohen, E. Avraham, M. Noked, A. Soffer, D. Aurbach, Enhanced charge efficiency in capacitive deionization achieved by surface-treated electrodes and by means of a third electrode, *J. Phys. Chem. C*, **115** (2011) 19856–19863.
69. E. Avraham, M. Noked, Y. Bouhadana, A. Soffer, D. Aurbach, Limitations of charge efficiency in capacitive deionization processes Iii: The behavior of surface oxidized activated carbon electrodes, *Electrochim. Acta*, **56** (2010) 441–447.
70. E. Avraham, Y. Bouhadana, A. Soffer, D. Aurbach, Limitation of charge efficiency in capacitive deionization, *J. Electrochem. Soc.*, **2009**, *156*, P95.
71. E. N. Holubowitch, A. Omosebi, X. Gao, J. Landon, K. L. Liu, Membrane-free electrochemical deoxygenation of aqueous solutions using symmetric activated carbon electrodes in flow-through cells, *Electrochim. Acta*, **297** (2019) 163–172.
72. X. Gao, A. Omosebi, Z. Ma, F. Zhu, J. Landon, M. Ghorbanian, N. Kern, K. Liu, Capacitive deionization using symmetric carbon electrode pairs, *Environmental Science: Water Res. & Technology*, **5** (2019) 660–671.
73. X. Gao, A. Omosebi, J. Landon, K. Liu, Voltage-based stabilization of microporous carbon electrodes for inverted capacitive deionization, *J. Phys. Chem. C*, **122** (2018) 1158–1168.
74. X. Gao, A. Omosebi, N. Holubowitch, J. Landon, K. Liu, Capacitive deionization using alternating polarization: Effect of surface charge on salt removal, *Electrochim. Acta*, **233** (2017) 249–255.
75. E. Guyes, T. Malka, M. E. Suss, Enhancing the ion size-based selectivity of capacitive deionization electrodes, *Environ. Sci. Technol.* (2019).
76. P. A. Bhat, R. E. Reale, M. del Cerro, C. K. Smith, R. D. Cusick, Reducing impedance to ionic flux in capacitive deionization with bi-tortuous activated carbon electrodes coated with asymmetrically charged polyelectrolytes, *Water Res. X*, **3** (2019) 100027.
77. M. Kim, D. M. Cerro, S. Hand, R. D. Cusick, Enhancing capacitive deionization performance with charged structural polysaccharide electrode binders, *Water Res.*, **148** (2019) 388–397.

78. X. Gao, S. Porada, A. Omosebi, K.-L. Liu, M. P. Biesheuvel, J. Landon, Complementary surface charge for enhanced capacitive deionization, *Water Res.*, **92** (2016) 275–282.
79. Y. Gendel, E. K. A. Rommerskirchen, O. David, M. Wessling, Batch mode and continuous desalination of water using flowing carbon deionization (Fcdi) technology, *Electrochem. Commun.*, **46** (2014) 152–156.
80. Z.-H. Huang, M. Wang, L. Wang, F. Kang, Relation between the charge efficiency of activated carbon fiber and its desalination performance, *Langmuir*, **28** (2012) 5079–5084.
81. J. Kang, T. Kim, H. Shin, J. Lee, I. J. Ha, J. Yoon, Direct energy recovery system for membrane capacitive deionization, *Desalination*, **398** (2016) 144–150.
82. C. Kim, P. Srimuk, J. Lee, S. Fleischmann, M. Aslan, V. Presser, Influence of pore structure and cell voltage of activated carbon cloth as a versatile electrode material for capacitive deionization, *Carbon*, **122** (2017) 329–335.
83. P. Liu, H. Wang, T. Yan, J. Zhang, L. Shi, D. Zhang, Grafting sulfonic and amine functional groups on 3D graphene for improved capacitive deionization, *J. Mater. Chem. A*, **4** (2016) 5303–5313.
84. S. Porada, L. Weinstein, R. Dash, A. van der Wal, M. Bryjak, Y. Gogotsi, P. M. Biesheuvel, Water desalination using capacitive deionization with microporous carbon electrodes, *ACS Appl. Mater. Inter.*, **4** (2012) 1194–1199.
85. S. Porada, L. Borchardt, M. Oschatz, M. Bryjak, S. J. Atchison, J. K. Keesman, S. Kaskel, M. P. Biesheuvel, V. Presser, Direct prediction of the desalination performance of porous carbon electrodes for capacitive deionization, *Energy Environ. Sci.*, **6** (2013) 3700.
86. T. Kim, J. Yoon, Cdi ragone plot as a functional tool to evaluate desalination performance in capacitive deionization, *RSC Adv.*, **5** (2015) 1456–1461.
87. D.-J. Yoon, J.-H. Choi, A new standard metric describing the adsorption capacity of carbon electrode used in membrane capacitive deionization, *Water Res.*, **148** (2019) 126–132.
88. A. F. Posey, T. Morozumi, Theory of potentiostatic and galvanostatic charging of the double layer in porous electrodes, *J. Electrochem. Soc.*, **113** (1966) 176–184.
89. R. de Levie, On porous electrodes in electrolyte solutions — Iv, *Electrochim. Acta*, **9** (1964) 1231–1245.
90. R. de Levie, On porous electrodes in electrolyte solutions: I. capacitance effects, *Electrochim. Acta*, **8** (1963) 751–780.

91. Y. Qu, F. T. Baumann, G. J. Santiago, M. Stadermann, Characterization of resistances of a capacitive deionization system, *Environ. Sci. Technol.*, **49** (2015) 9699–9706.
92. E. J. Dykstra, R. Zhao, M. P. Biesheuvel, A. van der Wal, Resistance identification and rational process design in capacitive deionization, *Water Res.*, **88** (2016) 358–370.
93. J. Bisquert, Infuence of the boundaries in the impedance of porous film electrodes, *Phys. Chem. Chem. Phys.*, **2** (2000) 4185–4192.
94. S. Devan, R. V. Subramanian, R. E. White, Analytical solution for the impedance of a porous electrode, *J. Electrochem. Soc.*, **151** (2004) A905.
95. M. González-Cuenca, W. Zipprich, B. A. Boukamp, G. Pudmich, F. Tietz, Impedance studies on chromite-titanate porous electrodes under reducing conditions, *Fuel Cells*, **1** (2001) 256–264.
96. A. Hasbach, U. Retter, K. Siegler, W. Kautek, On the impedance of porous electrodes — Double-layer charging and charge transfer on an inhomogeneous inside electrode surface, *J. Electroanal. Chem.*, **561** (2004) 29–35.
97. R. Kant, M. B. Singh, Theory of the electrochemical impedance of mesostructured electrodes embedded with heterogeneous micropores, *J. Phys. Chem. C*, **121** (2017) 7164–7174.
98. A. Lasia, Impedance of porous electrodes, *J. Electroanal. Chem.*, **397** (1995) 27–33.
99. P. J. Meyers, M. Doyle, M. R. Darling, J. Newman, The impedance response of a porous electrode composed of intercalation particles, *J. Electrochem. Soc.*, **147** (2000) 2930.
100. H. P. Nguyen, G. Paasch, Transfer matrix method for the electrochemical impedance of inhomogeneous porous electrodes and membranes, *J. Electroanal. Chem.*, **460** (1999) 63–79.
101. G. Paasch, K. Micka, P. Gersdore, Theory of the electrochemical impedance of macrohomogeneous porous electrodes, *Electrochim. Acta*, **38** (1993) 112–120.
102. J. Segalini, B. Daffos, L. P. Taberna, Y. Gogotsi, P. Simon, Qualitative electrochemical impedance spectroscopy study of ion transport into sub-nanometer carbon pores in electrochemical double layer capacitor electrodes, *Electrochim. Acta*, **55** (2010) 7489–7494.
103. E. M. Suss, F. T. Baumann, A. M. Worsley, A. K. Rose, F. T. Jaramillo, M. Stadermann, J. G. Santiago, Impedance-based study of capacitive porous carbon electrodes with hierarchical and bimodal porosity, *J. Power Sources*, **241** (2013) 266–273.

104. U. Tröltzsch, O. Kanoun, Generalization of transmission line models for deriving the impedance of diffusion and porous media, *Electrochim. Acta*, **75** (2012) 347–356.
105. C. Yang, C.-Y. V. Li, F. Li, K.-Y. Chan, Impedance analysis of ion transportation in hierarchical porous core-shell carbons with transmission line model, *ECS Transactions*, **45** (2013) 51–63.
106. C. Yang, C.-y. V. Li, F. Li, K.-y. Chan, Complex impedance with transmission line model and complex capacitance analysis of ion transport and accumulation in hierarchical core-shell porous carbons, *J. Electrochem. Soc.*, **160** (2013) H271–H278.
107. A. L. Richards, S. B. Richards, A. I. Schäfer, Renewable energy powered membrane technology: Salt and inorganic contaminant removal by nanofiltration/reverse osmosis, *J. Membr. Sci.*, **369** (2011) 188–195.
108. M. Xie, D. L. Nghiem, E. W. Price, M. Elimelech, Comparison of the removal of hydrophobic trace organic contaminants by forward osmosis and reverse osmosis, *Water Res.*, **46** (2012) 2683–2692.
109. R. Kumar, P. Pal, A novel forward osmosis-nano filtration integrated system for coke-oven wastewater reclamation, *Chem. Eng. Res. Des.*, **100** (2015) 542–553.
110. S. Samatya, N. Kabay, Ü. Yüksel, M. Arda, M. Yüksel, Removal of nitrate from aqueous solution by nitrate selective ion exchange resins, *React. Funct. Polym.*, **66** (2006) 1206–1214.
111. X. Xu, B. Gao, Y. Zhao, S. Chen, X. Tan, Q. Yue, J. Lin, Y. Wang, Nitrate removal from aqueous solution by Arundo Donax L. Reed based anion exchange resin, *J. Hazard. Mater.*, **203** (2012) 86–92.
112. S. Jamaly, N. Darwish, I. Ahmed, S. Hasan, A short review on reverse osmosis pretreatment technologies, *Desalination*, **354** (2014) 30–38.
113. A. Subramani, J. G. Jacangelo, Emerging desalination technologies for water treatment: A critical review, *Water Res.*, **75** (2015) 164–187.
114. L. Chekli, S. Phuntsho, E. J. Kim, J. Kim, Y. J. Choi, J.-S. Choi, S. Kim, H. J. Kim, S. Hong, J. Sohn, A comprehensive review of hybrid forward osmosis systems: Performance, applications and future prospects, *J. Membr. Sci.*, **497** (2016) 430–449.
115. Y. Zhao, Z. Zhang, L. Dai, H. Mao, S. Zhang, Enhanced both water flux and salt rejection of reverse osmosis membrane through combining isophthaloyl dichloride with biphenyl tetraacyl chloride as organic phase monomer for seawater desalination, *J. Membr. Sci.*, **522** (2017) 175–182.
116. C. E. Harland, *Ion Exchange: Theory and Practice*; Royal Society of Chemistry, 1994; Vol. 6.

117. L. Han, G. K. Karthikeyan, A. M. Anderson, K. B. Gregory, Exploring the impact of pore size distribution on the performance of carbon electrodes for capacitive deionization, *J. Colloid Interface Sci.*, **430** (2014) 93–9.
118. C.-H. Hou, P. Taboada-Serrano, S. Yiacoumi, C. Tsouris, Electrosorption selectivity of ions from mixtures of electrolytes inside nanopores, *J. Chem. Phys.*, **129** (2008) 224703.
119. S.-J. Seo, H. Jeon, K. J. Lee, G.-Y. Kim, D. Park, H. Nojima, J. Lee, S.-H. Moon, Investigation on removal of hardness ions by capacitive deionization (CDI) for water softening applications, *Water Res.*, **44** (2010) 2267–2275.
120. Y. Li, C. Zhang, Y. Jiang, T.-J. Wang, H. Wang, Effects of the hydration ratio on the electrosorption selectivity of ions during capacitive deionization, *Desalination*, **399** (2016) 171–177.
121. C. Zhan, M. R. Cerón, A. S. Hawks, M. Otani, C. B. Wood, A. T. Pham, M. Stadermann, P. G. Campbell, Specific ion effects at graphitic interfaces, *Nature Commun.*, **10** (2019) 1–8.
122. C. Nie, Y. Zhan, L. Pan, H. Li, Z. Sun, Electrosorption of different cations and anions with membrane capacitive deionization based on carbon nanotube/nanofiber electrodes and ion-exchange membranes, *Desalination Water Treat.*, **30** (2011) 266–271.
123. D.-H. Lee, T. Ryu, J. Shin, C. J. Ryu, K.-S. Chung, Y. H. Kim, Selective lithium recovery from aqueous solution using a modified membrane capacitive deionization system, *Hydrometallurgy*, **173** (2017) 283–288.
124. A. Siekierka, M. Bryjak, In *Hybrid Capacitive Deionization with Anion-Exchange Membranes for Lithium Extraction*, E3S Web of Conferences, EDP Sciences: 2017; p 00157.
125. A. Siekierka, M. Bryjak, J. Wolska, The use of activated carbon modified with polypyrrole as a supporting electrode for lithium ions adsorption in capacitive deionization, *Desalin. Water Treat.*, **64** (2017) 251–254.
126. A. Siekierka, J. Wolska, M. Bryjak, W. Kujawski, Anion exchange membranes in lithium extraction by means of capacitive deionization system, *Desalin. Water Treat.*, **75** (2017) 331–341.
127. W. Tang, P. Kovalsky, B. Cao, D. He, T. D. Waite, Fluoride removal from brackish groundwaters by constant current capacitive deionization (CDI), *Environ. Sci. Technol.*, **50** (2016) 10570–10579.
128. W. Tang, P. Kovalsky, B. Cao, T. D. Waite, Investigation of fluoride removal from low-salinity groundwater by single-pass constant-voltage capacitive deionization, *Water Res.*, **99** (2016) 112–121.

129. J. J. Lado, R. E. Pérez-Roa, J. J. Wouters, M. I. Tejedor-Tejedor, C. Federspill, M. J. Ortiz, M. A. Anderson, Removal of nitrate by asymmetric capacitive deionization, *Sep. Purif. Technol.*, **183** (2017) 145–152.
130. W. Tang, P. Kovalsky, D. He, T. D. Waite, Fluoride and nitrate removal from brackish groundwaters by batch-mode capacitive deionization, *Water Res.*, **84** (2015) 342–349.
131. J.-H. Yeo, J.-H. Choi, Enhancement of nitrate removal from a solution of mixed nitrate, chloride and sulfate ions using a nitrate-selective carbon electrode, *Desalination*, **320** (2013) 10–16.
132. H. C. Hou, C. Y. Huang, A comparative study of electrosorption selectivity of ions by activated carbon electrodes in capacitive deionization, *Desalination*, **314** (2013) 124–129.
133. J. S. Seo, H. Jeon, K. J. Lee, Y. G. Kim, D. Park, H. Nojima, J. Lee, S. H. Moon, Investigation on removal of hardness ions by capacitive deionization (CDI) for water softening applications, *Water Res.*, **44** (2010) 2267–2275.
134. Y.-J. Kim, J.-H. Choi, Selective removal of nitrate ion using a novel composite carbon electrode in capacitive deionization, *Water Res.*, **46** (2012) 6033–6039.
135. Y.-J. Kim, J.-H. Kim, J.-H. Choi, Selective removal of nitrate ions by controlling the applied current in membrane capacitive deionization (MCDI), *J. Membr. Sci.*, **429** (2013) 52–57.
136. E. J. Dykstra, J. Dijkstra, A. Van der Wal, M. V. H. Hamelers, S. Porada, On-line method to study dynamics of ion adsorption from mixtures of salts in capacitive deionization, *Desalination*, **390** (2016) 47–52.
137. M. E. Suss, Size-based ion selectivity of micropore electric double layers in capacitive deionization electrodes, arXiv, 2017, 1–13.
138. R. Zhao, van M. Soestbergen, M. H. H. Rijnaarts, A. van der Wal, Z. M. Bazant, P. M. Biesheuvel, Time-dependent ion selectivity in capacitive charging of porous electrodes, *J. Colloid Interface Sci.*, **384** (2012) 38–44.
139. J. Biener, M. Stadermann, M. Suss, M. A. Worsley, M. M. Biener, K. A. Rose, T. F. Baumann, J. S. King, A. Wittstock, Advanced carbon aerogels for energy applications. *Energy & Environmental Science*, **4** (2011) 656.
140. T. D. Hasseler *et al.*, Process design tools and techno-economic analysis for capacitive deionization. *Water Research*, **183** (2020) 116034.
141. H. A. Kuo *et al.*, Understanding resistances in capacitive deionization devices. *Environmental Science: Water Research & Technology*, **6.7** (2020) 1842–1854.

CHAPTER 10

Surface Modified Reverse Osmosis Membranes

Soomin Kim and Yoram Cohen*

University of California, Los Angeles; Los Angeles, CA 90095

10.1. Overview of Reverse Osmosis (RO)

10.1.1. *Water and salt transport in RO membranes*

RO desalination separates water from a saline feed solution by applying an external pressure that is greater than the osmotic pressure (which causes flow of water to the feed-side of a RO membrane) across a semi-permeable membrane. RO desalination is typically operated in a crossflow filtration mode in which a saline feed solution flows tangentially across the feed-side of the membrane surface. High pressure applied at the feed-side of the membrane surface forces the solution through the membrane, which selectively permeates water (Fig. 10.1). Buildup of rejected salt ions near and at the feed-side of the membrane surface leads to concentration polarization, which increases the feed-side osmotic pressure, and thus decreases the net driving pressure. The osmotic pressure increases in the feed flow direction feed flow (x-direction) due to water permeation through the membrane which then results in an axial increase of salt concentration in the bulk feed. Therefore, permeate flux correspondingly decreases along the same direction. As the

*Corresponding author. Yoram Cohen (profyc@gmail.com)

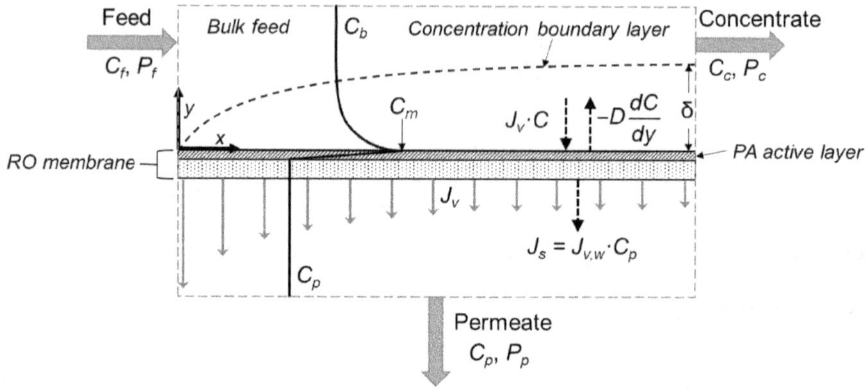

Fig. 10.1. Schematic representation of concentration polarization in a crossflow RO membrane channel. The black solid line shows salt concentration profile, black dotted arrows indicate directions of salt fluxes, and gray arrows show direction of permeate fluxes. C_f, C_b, C_c, and C_p are the concentrations of salt in the feed water, bulk of the solution, concentrate, and permeate. C_m is the salt concentration at the membrane surface. J_v is the volumetric permeate flux, J_s is the salt flux, D is the salt diffusivity in water, dC/dy is the salt concentration driving force in the y-direction. P_f, P_c, and P_p are the pressures of the feed, concentrate, and permeate streams.

feed flows axially along the membrane surface, the increased solute concentration can be quantified, for example, based on the classical Film Theory, which assumes one-dimensional fluid flow (x-direction in Fig. 10.1) and a fully-developed concentration boundary layer.[1] Accordingly, the concentration polarization modulus (CP), which can be related to salt concentration at the feed-side of the membrane surface (C_m) relative to the bulk solution (C_b) is given as:

$$CP = \frac{C_m - C_p}{C_b - C_p} = \exp\left(\frac{J_v}{k}\right) \tag{1}$$

in which C_p, C_b and C_m are the permeate, bulk and membrane salt concentration, respectively, J_v is the volumetric permeate flux, and k is the salt mass transfer coefficient. The membrane level of salt rejection is generally quantified in terms of the observed (R_o) or intrinsic (R_i) salt rejection defined as $R_o = 1 - C_p/C_b$ and $R_i = 1 - C_p/C_m$, respectively. These parameters are highly dependent on RO operating conditions (e.g., feed pressure, feed solution salinity,

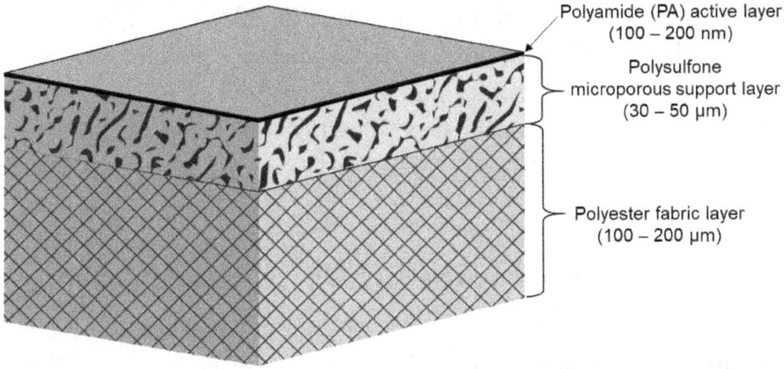

Fig. 10.2. Illustration of a polyamide (PA) thin film composite (TFC) membrane.

and permeate flux); thus, meaningful comparisons of the performance of different RO membranes must be done at the same operating conditions.

Transport of water and salt ions through highly selective non-porous membranes (such as PA TFC SWRO membranes where salt transport essentially occurs primarily via diffusion) is generally described by the solution-diffusion model,[2] whereby the volumetric water flux (J_v) and salt flux (J_s) across a RO membrane are expressed as[3] $J_v = L_p(\Delta P - \Delta\pi_m)$ and $J_s = J_vC_p = B(C_m - C_p)$ in which L_p and B are the water and salt permeability coefficients, respectively, ΔP is the transmembrane pressure, and $\Delta\pi_m$ is the osmotic pressure difference between the membrane surface at the feed and permeate sides, respectively. L_p and B are influenced by the properties of the PA active layer such as layer properties thickness and free volume.

10.1.2. *Polyamide (PA) thin film composite (TFC) membrane*

RO technology utilizes semi-permeable membranes that allow for water permeation while rejecting dissolved salts in the feed water. Currently, RO processes predominantly utilize polyamide (PA) based thin film composite (TFC) membranes, which consist of three layers (Fig. 10.2). The top layer is an ultrathin PA active layer (100–200 nm)[4] and is formed via a polycondensation reaction of diamine and acyl chloride[5] on top of a microporous membrane support layer that is typically made of polysulfone (30–50 μm). The PA active layer that dictates the membrane water and salt permeabilities, while the underlying microporous membrane support provides mechanical stability as well as high water flux. The bottom layer is a nonwoven or woven polyester fabric layer (100–200 μm), which provides additional mechanical strength.[6]

10.1.3. *Water permeability and water/salt selectivity tradeoff*

According to the solution-diffusion model, water permeability (P_w) and salt permeability (P_s) of a PA TFC membrane can be expressed as follows,[3]

$$P_w = D_w K_w \tag{2}$$

$$P_s = D_s K_s \tag{3}$$

where D_w and D_s are the diffusion coefficients of water and salt in the PA active layer, respectively, and K_w and K_s are the partition coefficients (solubility) of water and salt between the saline solution and the PA active layer. Accordingly, water/salt selectivity can be expressed by the ratio of P_w and P_s (i.e., P_w/P_s). The water permeability coefficient, L_p, and salt permeability coefficient, B, respectively, can be related to P_w and P_s by the following relations:

$$L_p = \frac{P_w}{l} \cdot \frac{V_m}{RT} \tag{4}$$

$$B = \frac{P_s}{l} \tag{5}$$

where l is the active layer thickness, V_m is the partial molar volume of water, R is the universal gas constant, and T is the absolute temperature. It is noted that P_w and P_s are intrinsic material properties of the PA active layer, unlike L_p and B, which depend on the active layer thickness as shown in Eq. 4 and 5. However, accurate determination of the PA active layer thickness is a challenging task. Thus, L_p and B are used more frequently instead of P_w and P_s to quantify the PA TFC membrane desalination performance.

Consequently, water/salt selectivity of a PA TFC membrane is often quantified by the ratio of L_p and B (i.e., L_p/B).

For RO desalination applications, it is desirable to have membranes with high water permeability and high water/salt selectivity. However, PA TFC membranes display a trade-off between water permeability and water/salt selectivity.[7–9] It has been shown that this trade-off behavior is attributed to the structural properties of PA active layers (e.g., layer thickness, free volume) which lead to a high level of salt permeation through the membrane (lower salt retention) that increases with increasing water permeability.[10] For example, a greater water permeation rate can be achieved by reducing the PA active layer thickness, which is obvious from Eq. (4). Analysis of commercial PA TFC membranes via positron annihilation lifetime spectroscopy (PALS) have shown that PA active layers contain subnanometer scale free volume cavities (~0.2–0.3 nm in radius) that are postulated to strongly influence both water and salt transport in the membranes.[4,7,11] In general, reducing free volume in the PA active layer is expected to decrease water and salt permeabilities; however, membrane permeability can also be impacted by size and spatial distributions, as well as the total number of such cavity sites that constitute the free volume within the PA active layer.

Surface properties of PA active layers (e.g., surface hydrophilicity, charge density, roughness) have also been shown to have a measurable impact on RO membrane desalination performance. For example, increasing surface hydrophilicity has been linked to increased membrane water permeability via increased water sorption in the PA active layers.[12] However, increased water sorption in the PA layer has been associated with increased free volume in the layer, which can in turn also increase salt permeability.[13] It has been also shown that increasing surface roughness of PA active layers may be a contributing factor to increased water permeability by providing a higher effective membrane surface area for water permeation or via a thinner active layer that provides a shorter diffusion path in the active layer.[14]

In order to overcome the RO membrane performance tradeoff, various means have been proposed to tailor the PA selective layer characteristics (e.g., thickness, extent of crosslinking) as well as the resulting membrane surface properties (e.g., roughness, hydrophilicity). It was demonstrated that PA TFC membrane desalination performance could be tuned by: optimizing the conventional PA selective layer synthesis conditions (e.g., solvent type, reaction temperature, use of additives) the post-synthesis curing conditions,[15,16] tuning the porosity/structure of the porous support layer,[17] or employing alternative PA selective layer synthesis methods (e.g., molecular layer-by-layer assembly[18]). In recent years, it has been suggested that membrane surface modification, which is a popular approach, primarily employed for membrane fouling reduction, can also be used to tune the performance of polyamide desalination membranes.[19–21]

10.1.4. *Polyamide interfacial polymerization*

The conventional PA TFC membrane fabrication method, first developed by Cadotte,[22] involves synthesis of an ultra-thin polyamide selective layer via interfacial polymerization of diamine and acyl chloride monomers *in situ* at the surface of a microporous polymer support layer (usually a polysulfone or polyethersulfone ultrafiltration membrane).[23,24] The surface of a microporous support layer is first coated with a layer of an aqueous solution of diamine (typically *m*-phenylenediamine or MPD), which penetrates into the underlying porous support layer to some degree. The MPD layer is subsequently contacted with an acyl chloride (typically trimesoyl chloride, TMC) solution in an organic phase (usually hexane) to initiate interfacial polymerization[23] (Fig. 10.3). The low solubility of acyl chlorides in water and good solubility of MPD in organic solvents lead to diffusion of the MPD monomers from the aqueous phase to the organic phase where polymerization is initiated within the water/organic interfacial region.[25] It has been shown that the interfacial polymerization conditions (e.g., monomer concentration, organic solvent type, solvent temperature, polymerization time) and post membrane curing conditions can significantly impact the resulting PA active layer structure (e.g., thickness and degree of cross-linking) and membrane desalination performance.[23,26,27]

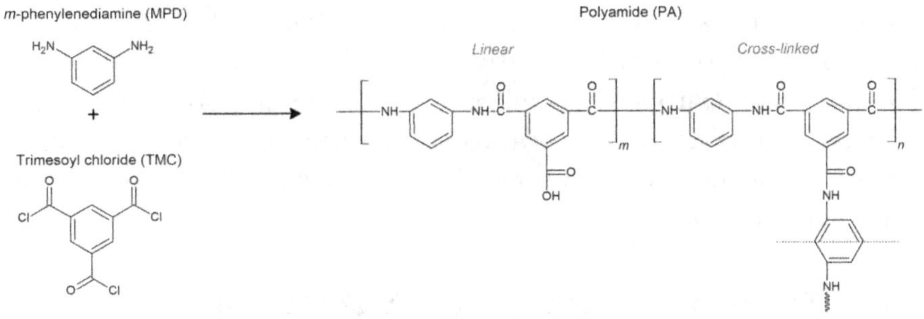

Fig. 10.3. Synthesis of an aromatic polyamide (PA) selective layer via interfacial polymerization of m-phenylenediamine (MPD) and trimesoyl chloride (TMC).

10.1.5. *Seawater reverse osmosis (SWRO) membrane fouling*

In seawater RO desalination, membrane biofouling mitigation is a major challenge.[28] Biofouling is caused by surface adhesion of microorganisms (e.g., bacteria and algae) that leads to the formation of a biofilm.[29,30] Biofilms are highly undesirable because they can be difficult to remove once they are formed even via disinfection or chemical cleaning.[31] The process of biofilm development on a membrane surface is initiated by the surface attachment of organic polymers and colloids present in natural water that forms a conditioning film (composed of polysaccharides, proteins, lipids, nucleic acids, and humic and fulvic acids) that promotes surface attachment of microbial cells.[32] The surface attached microbial cells grow and produce extracellular polymeric substances (EPS), which strongly bind the cells to the membrane surface.[29] Subsequent microbial proliferation and EPS secretion leads to the formation of a mature biofilm on the membrane surface. In recent years, it has been suggested that the transparent exopolymer particles (TEP) are the major constituents of the initial conditioning film on the membrane surface and play a significant role in both initiation and development of a biofilm. TEP are clear, gel-like polysaccharides (0.4–300 μm) that exist as free-floating particles that are ubiquitous in marine and freshwater environments.[32] It was suggested that more effective removal of TEP can be achieved by utilizing UF for SWRO pretreatment (compared to the conventional pretreatment methods such as double media filtration (DMF).[32] However, UF does not provide 100% retention of TEP, therefore, TEP in the UF permeate can cause membrane fouling in the downstream RO process. Therefore, it is critical to also reduce RO membrane fouling propensity as has been proposed in numerous studies on membrane surface modification.

10.2. The Impact of PA Surface Modification on RO Desalination Performance

10.2.1. *Overview*

In recent years, it has been suggested that membrane surface modification, a popular approach primarily employed for membrane fouling reduction, can also be used to tune PA-TFC membrane desalination performance.[19–21] For example, affixing a hydrophilic polymer layer to a PA-TFC membrane surface, which can be accomplished via various means including coating (via physical adsorption)[33] and polymer grafting ("grafting to"), can be effective in reducing membrane fouling. In some cases, the hydrophilic polymer layers can also increase salt rejection but at the cost of reduced water permeability. However, as reviewed in Section 10.2, the impact of PA TFC membrane surface modification on membrane desalination performance (typically evaluated in terms of water permeability coefficient and nominal salt rejection) can vary significantly depending on the surface modification method and its specific application conditions. Accordingly, Sections 10.2.2–10.2.6 provide a review of various membrane modification approaches reported for PA-TFC membranes:

(1) coating
(2) layer-by-layer assembly
(3) initiated chemical vapor deposition
(4) polymer grating, and
(5) grafting polymerization

Also, details are provided in the various summary Tables (Tables 10.1–10.5) regarding the specific membrane synthesis protocols and performance determination conditions.

10.2.2. *Coating (physical adsorption)*

Surface attachment of a layer of hydrophilic material (usually polymer) via non-covalent interactions (e.g., van der Waals, hydrogen bonding, electrostatic interactions) is the most commonly reported membrane surface modification approach to reduce the fouling propensity of PA TFC membranes.[33–43] Synthesis of such surface coating layers can be accomplished by either immersing a membrane in a polymer solution for a prescribed period (e.g., dip-coating)[40] or by drawdown coating the membrane surface with a layer of polymer solution using a device such as a wet film applicator.[35] Subsequently, the wet polymer layer on the membrane surface is typically cured to evaporate the solvent, and, in some cases, to also thermally induce cross-linking of polymer chains and form a solid continuous polymer film on the membrane surface. For example, surfaces of commercial PA-TFC BWRO (Hydranautics ESPA1 and ESPA3) and SWRO (Hydranautics SWC4) membranes were coated with a 0.3 μm-thick layer of PEBAX® 1657 (a hydrophilic block copolymer of nylon-6 and poly(ethylene glycol)) via dip-coating and subsequently cured at 60°C.[34] It was reported that the resulting PEBAX surface coating layers reduced fouling of SWC4 and ESPA1 membranes in filtration of oil/surfactant/water emulsions. For example, the cumulative permeate volume over 100 days of oil/surfactant/water emulsion was 47% greater for the PEBAX coated SWC4 membrane than for the native SWC4 membrane. However, the fouling test of the native (base) membrane was conducted at an initial permeate flux that was greater than for the PEBAX coated membranes. Therefore, the initial fouling rate of the base membrane was greater than for the PEBAX coated membranes; hence, the claim that the PEBAX coated membranes had a lower fouling propensity based on the above tests may be misleading. It was also reported that the block copolymer layers significantly reduced L_p (evaluated with DI water) of ESPA1 (from $6.8\,\mathrm{L \cdot m^{-2} \cdot h^{-1} \cdot bar^{-1}}$ to $1.8\,\mathrm{L \cdot m^{-2} \cdot h^{-1} \cdot bar^{-1}}$) and ESPA3 (from $5.8\,\mathrm{L \cdot m^{-2} \cdot h^{-1} \cdot bar^{-1}}$ to $1.1\,\mathrm{L \cdot m^{-2} \cdot h^{-1} \cdot bar^{-1}}$) membranes by ~74% and ~81%, respectively. The authors attributed the significant reductions in L_p of the above membranes to the PEBAX polymer chains plugging "porous" regions of the PA selective layers of the base membranes. It was also reported that the PEBAX coating layer increased salt rejection (evaluated with a feed water containing 1500 mg/L NaCl at $\Delta P = 11.3$ bar) of the ESPA1 membrane from 97.8% to 99%, while the same coating layer resulted in a slight salt rejection reduction for the ESPA3 membrane (from 97.4% to 97%). However, no explanation was given for the difference in the observed salt rejection changes for the above two membranes, and the stability of the PEBAX coating was not reported.

In order to increase the stability of physically coated polymer layers on the membrane surface, glutaraldehyde (GA) was utilized in a number of studies to cross-link surface adsorbed polymer layers.[40,41,43] For example, sericin, a water-soluble protein that has polar side groups (i.e., hydroxyl, carboxyl, and amino), was adsorbed (primarily via hydrogen bonding) onto the surface of a PA-TFC membrane via dip-coating, followed by *in situ* cross-linking with GA.[40] Membrane fouling tests, conducted at the same initial permeate flux and crossflow velocity, showed that the sericin coated membrane had a lower permeate flux reduction (~23%) relative to the base PA membrane (~42%) after 100 h of BSA solution filtration (100 mg/L BSA aqueous solution at pH 6.8). Moreover, the sericin surface coating was reported to be stable over 300 h filtration of saline water (500 mg/L NaCl feed water at $\Delta P = 5$ bar) while maintaining membrane salt rejection in the range of 98.5–98.9% and permeate flux in the range of 25.5–26.5 $\mathrm{L \cdot m^{-2} \cdot h^{-1}}$. The sericin coating layers resulted in a 7–22% reduction of the base PA membrane L_p ($L_p^{Base} = 6.5\,\mathrm{L \cdot m^{-2} \cdot h^{-1} \cdot bar^{-1}}$) while salt rejection increased from 97.6% to 98.2–98.5% (tested with 500 mg/L NaCl feed water at $\Delta P = 5$ bar). It was also reported that the sericin coating layers reduced the membrane salt permeability coefficient (B) by ~31–49% relative to the base membrane ($B^{Base} = 0.75\,\mathrm{L \cdot m^{-2} \cdot h^{-1}}$). However, the membrane B values were evaluated based on the feed water salt concentration (C_f) instead of salt concentration at the membrane surface (C_m). Therefore, the reported membrane B values, which were evaluated at different permeate fluxes (different CP levels), may deviate significantly from the actual membrane salt permeability coefficient.

In another study, a (low energy) BWRO membrane (Dow Filmtec LC LE) surface was modified by dip-coating to form a layer of a random terpolymer, poly(methylacryloxyethyldimethyl benzyl ammonium chloride-*r*-acrylamide-*r*-2-hydroxyethyl methacrylate (P(MDBAC-*r*-Am-*r*-HEMA)) and GA, which was subsequently cross-linked by curing the terpolymer and GA coated membrane at 50°C for 2 h.[41] The terpolymer modified membranes had ~48–60% reduced L_p (reduced from 5.0 to 2.0–2.6 $\mathrm{L \cdot m^{-2} \cdot h^{-1} \cdot bar^{-1}}$) and increased salt rejection (98.6–99.1%) relative to the base membrane (98.1%) when tested with low salinity feed water (2000 mg/L NaCl) at $\Delta P = 15$ bar. When compared to a standard commercial BWRO membrane (Dow Filmtec BW30), the modified membranes had similar or ~4–26% lower L_p but achieved greater salt rejection (98.6–99.1%) than the BW30 membrane ($L_p = 2.7\,\mathrm{L \cdot m^{-2} \cdot h^{-1} \cdot bar^{-1}}$, salt rejection = 98.5%). In the above study, chlorine resistance of the terpolymer coated membranes was tested with an aqueous feed solution of 500 ppm sodium hypochlorite (NaOCl). The chlorine tolerance limit was defined as the NaOCl exposure time at which a significant (≥10%) salt rejection drop was observed due to membrane oxidation/degradation. It was reported that the chlorine tolerance limit for the terpolymer coated membrane was much greater (23,000 ppm · h) relative to the base (6,000 ppm · h) and BW30 (10, 000 ppm · h) membranes. It was also reported that the terpolymer coated membrane maintained the same initial permeate flux after alternate filtration cycles of a BSA solution (i.e., 100 mg/L BSA in 2000 mg/L NaCl aqueous solution, pH 4.7) and saline water (2000 mg/L NaCl aqueous solution) over 30 hours. However, the base and BW30 membranes lost 20% of their initial permeate fluxes. It is noted that the above tests were conducted at the same transmembrane pressure of 15 bar, and thus performance comparison at different initial fluxes (given that the tested membranes were reported to have a significantly lower L_p relative to the base membrane).

Crosslinked polymer coating layers were also prepared on a commercial SWRO membrane (CSM SHN) surface by immersing the membrane in aqueous solutions of homopolymers of hydroxyl poly(oxyethylene) methacrylate (HPOEM) or carboxylated polyethyleneimine (PEI), respectively, followed by exposing the polymer coated membrane surface to GA to cross-link the polymer layers on the membrane surface.[43] Membrane performance, evaluated with feed water containing 32,000 mg/L NaCl and 70 mg/L $CaCl_2$ at 55.2 bar, indicated that the HPOEM homopolymer and carboxylated PEI coating resulted in membranes that had L_p lower by 8% ($0.54\,L \cdot m^{-2} \cdot h^{-1} \cdot bar^{-1}$) or higher by 2% ($0.60\,L \cdot m^{-2} \cdot h^{-1} \cdot bar^{-1}$), respectively, relative to the base membrane ($0.59\,L \cdot m^{-2} \cdot h^{-1} \cdot bar^{-1}$). Nominal salt rejection either slightly increased to 99.3% or decreased to 99.1% after coating with a HPOEM polymer or carboxylated PEI layer, respectively, relative to the base membrane salt rejection of 99.2%. Both HPOEM homopolymer and carboxylated PEI coating layers reduced fouling, evaluated with alginic acid and bovine serum albumin relative to the base membrane due to their amphiphilic and zwitterionic properties. For example, fouling tests with an alginic acid solution (30 mg/L alginic acid in 32,000 mg/L NaCl and 70 mg/L $CaCl_2$ aqueous solution at pH 8) were conducted for the above polymer coated and base membranes at initial permeate fluxes in the range of 29.8–32 $L \cdot m^{-2} \cdot h^{-1}$. After 20 h of alginic acid solution filtration, permeate flux reduction (relative to the initial flux) for the carboxylated PEI (PEI-COOH) coated membrane was ~9%; this was slightly lower relative to flux reduction observed for the HPOEM coated and base membranes (both reduced by ~15%). However, long term stability of the cross-linked polymer coating layers on the PA membrane surfaces was not reported in the above study.

One of the most popular coating materials used for membrane surface modification is poly(vinyl alcohol) (PVA). PVA coating layers were shown to be effective for increasing hydrophilicity of PA TFC membranes and mitigating membrane fouling. It has been reported that membrane manufacturers also utilized PVA as surface coating layers for commercial PA TFC RO membranes in some cases.[44] It was shown that the structure of surface coated PVA films may be affected by the humidity to which the film is exposed during the post-modification membrane curing step, which can also have an impact on PA membrane performance. For example, a solution containing PVA cross-linked with GA (PVA-GA) was allowed to contact the active side of a laboratory synthesized base PA membrane, followed by curing the solution coated membrane at 60°C for 10 min at two different relative humidity levels (30% or 90%) to form a PVA-GA layer on the membrane surface.[33] Membrane performance tests (performed with 32.8 g/L aqueous NaCl solution at 55 bar) revealed that the PVA-GA layer (prepared from 0.3% PVA-GA solution), cured at a 90% relative humidity level, resulted in ~18% L_p reduction (from $0.93\,\mathrm{L \cdot m^{-2} \cdot h^{-1} \cdot bar^{-1}}$ to $0.76\,\mathrm{L \cdot m^{-2} \cdot h^{-1} \cdot bar^{-1}}$) and 0.2% salt rejection increase (99.4% to 99.6%), relative to the base PA membrane. In contrast, curing of the same coating layer (prepared from the same PVA-GA concentrated solution) at a relative humidity of 30% resulted in a significantly greater (~76%) L_p reduction (from $0.93\,\mathrm{L \cdot m^{-2} \cdot h^{-1} \cdot bar^{-1}}$ to $0.22\,\mathrm{L \cdot m^{-2} \cdot h^{-1} \cdot bar^{-1}}$), while salt rejection decreased by 1.2% (i.e., from the base membrane level of 99.4% to 98.2%) relative to the base PA membrane. The above was attributed to the low relative humidity leading to the crystalline structure in the PVA-GA film relative to the amorphous structure for the film cured in a high humidity environment (inferred from wide angle X-ray diffraction analysis). The study also reported reduced B values for the PVA coated membranes (by up to ~67%) relative to the base PA membrane ($B = 0.31\,\mathrm{L \cdot m^{-2} \cdot h^{-1}}$). However, it is noted that salt permeability coefficients for different membranes were evaluated at a fixed ΔP of 55 bar; thus, the above comparisons were not at the same permeate flux (or the CP level) for all membranes; in fact, some of the PVA coated membranes had significantly lower L_p (by as much as 76%) relative to the base membrane. Membrane fouling tests were performed with an aqueous solution of 32,800 mg/L NaCl containing 150 mg/L of dodecyl trimethyl ammonium bromide (DTAB) as a model foulant. It was reported that after performing two cycles of the DTAB solution filtration (8 hours), followed by flushing the membrane with a DTAB-free NaCl (32,800 mg/L) solution (4 hours), the PVA coated membranes maintained a higher permeate flux (~81–83%) relative to the base membrane (72%). However, the above comparisons were not conducted at the same initial permeate flux. It is noted that the selected PVA coated membranes for the fouling studies had ~22–34% reduced L_p relative to the base membranes; thus, the initial permeate fluxes in the fouling study, conducted at a fixed $\Delta P = 55$ bar, were lower relative to the base membrane.

In another study, a low-energy BWRO membrane (Dow Filmtec LE) surface was coated utilizing three types of aqueous solutions containing either PVA, polyhexamethylene guanidine hydrochloride (PHMG), or a mixture of the two polymers, using a wet film applicator.[39] The polymer coated membranes were cured at 110°C for 2 min to evaporate the water and form cross-linked polymer layers on the membrane surface. Performance evaluation of the polymer coated membranes (with an aqueous solution containing 584 mg/L NaCl at $\Delta P = 27.6$ bar) indicated that the PHMG layer reduced the base membrane L_p ($L_p^{Base} \sim 4.4\,\mathrm{L \cdot m^{-2} \cdot h^{-1} \cdot bar^{-1}}$) most significantly (~83%) and reduced salt rejection to the greatest degree (i.e., reduced from 97.2% for the base membrane to 91.6%). Membranes with PVA-PHMG surface coated layers, prepared from solutions containing a mixture of PVA and PHMG in ratios of 99:1 and 95:5, had 45% and 62% reduced L_p and increased or decreased salt rejections of 98.2% and 95.8%, respectively, relative to the base membrane (97.2%). Among the four types of polymer layers, the PVA coating layer reduced L_p by the least amount (~15%), while increasing salt rejection to 98.5% (from 97.2% for the base membrane). The study also reported salt permeability coefficients for the above polymer coated membranes that were ~34–56% reduced relative to the base membrane ($3.47\,\mathrm{L \cdot m^{-2} \cdot h^{-1}}$). However, salt permeability coefficients were not evaluated at the same permeate flux. The study also reported that all three types of polymer coatings on the LE membranes reduced surface attachment of bacteria (*Pseudomonas aeruginosa*); however, PHMG, which is an antimicrobial polymer, was shown to be the most effective of the above surface modifiers for reducing bacterial surface attachment. For example, PHMG coating layer reduced surface attachment of *Pseudomonas aeruginosa* (evaluated after immersing membranes in a bacterial suspension at room temperature for a day) by ∼53% relative to the base.

Hydrogel coating layers were also shown to be effective at reducing membrane fouling due to oil/surfactants. For example, various types of poly(ethylene glycol) based hydrogels were synthesized onto a BWRO membrane (GE Osmonics AG) surface with either acrylic acid (AA), 2-hydroxyethyl acrylate (HEA), or poly(ethylene glycol) acrylate (PEGA) with a cross-linker (i.e., poly(ethylene glycol) diacrylate (PEGDA)).[35] The base AG membrane surfaces were coated with an aqueous monomer solution (containing either AA, HEA, or PEGA with PEGDA and a photoinitiator (1-hydroxycyclohexyl phenyl ketone)) and subsequently underwent UV-initiated polymerization to form a hydrogel surface layer. Salt rejection (evaluated with 2000 mg/L NaCl feed water at 15.5 bar) of the above hydrogel modified membranes was greater (99.1–99.3%) relative to the base membrane (99.0%), while DI water permeance was reduced by ∼36–42%. Fouling tests conducted with a cationic surfactant, DTAB, showed that the hydrogel layer coated membranes maintained a higher percentage of the initial permeate flux (36–43%) after 24 h filtration of a DTAB solution (200 ppm DTAB in 2000 mg/L NaCl feed water) relative to the base membrane (30%). However, in these tests the initial permeate flux of the base membrane was greater than for the hydrogel coated membranes by up to a factor of 1.5, and thus such comparison was deficient.

It is noted that polyamide membrane surfaces are inherently negatively charged due to the carboxylic acid groups produced via hydrolysis of the acyl chlorides during the polymerization step forming the active polymer layer.[12] Thus, cationic polymers may be adsorbed onto a PA membrane surface via electrostatic attraction, forming a more robust surface coating layer than coating layers affixed onto a membrane surface via van der Waals interactions. For example, a layer of a cationic phosphorylcholine polymer, poly[2-methacryloyloxyethyl phosphorylcholine (MPC)-*co*-2-aminoethylmethacrylate (AEMA)] (p(MPC-*co*-AEMA)), was adsorbed onto an ultralow pressure RO membrane (Nitto Denko

ES20) surface via electrostatic interactions (i.e., between cationic amino groups of p(MPC-*co*-AEMA) and the anionic carboxyl groups present on polyamide membrane surface).[42] The phosphorylcholine polymer layer was reported to reduce bacteria surface attachment, quantified after immersion of the membranes in a bacteria suspension (*Sphingomonas paucimobilis* as a model strain of Gram-negative bacteria) at 30°C for 24 h. The maximum thickness of the surface attached bacteria film (quantified by bacteria staining and observed via confocal laser scanning microscopy) was ∼48% lower for the p(MPC-*co*-AEMA) coated membrane relative to the base membrane. In addition, nominal salt rejection increased from 94.7% to 96.9% at the cost of a ∼24% L_p reduction (tested with 500 mg/L NaCl feed water at 7.5 bar).

In order to increase electrostatic attraction between a cationic biopolymer (i.e., chitosan) and the RO membrane surface, a laboratory synthesized PA TFC membrane (that had performance comparable to that of a commercial NF/ULPRO membrane) was first exposed to chlorine to increase the negative membrane surface charge.[37] It is noted that extensive exposure to chlorination may result in oxidation of polyamide and cause membrane degradation that increases water permeability and salt passage.[45] However, it was shown that surface treatment with chlorine at mild conditions (e.g., a low chlorine concentration and short exposure period) may be utilized to increase membrane permeability (while minimizing loss of membrane selectivity) to partly compensate for membrane permeability loss due to the subsequently coated surface polymer layers. It was reported that chlorination of the base PA membrane at the optimized conditions (exposure to a 200 mg/L sodium hypochlorite (NaClO) solution for 2 min) increased L_p by ∼16% and decreased salt rejection from 91.0% to 88.7% (evaluated with a feed water containing 1500 mg/L NaCl at 8 bar). The chlorine treated base PA membranes were then exposed to a chitosan solution (1000 mg/L) for 30 min to deposit a chitosan layer on the membrane surface. L_p for the chitosan layer modified membranes was reduced by ∼7% relative to the chlorine treated base membrane; but L_p was ∼8% greater relative to the native base PA membrane (i.e., prior to chlorination).

Moreover, the chitosan coated PA membrane achieved a higher level of salt rejection (94.5%) relative to the base PA membrane before (91.0%) and after chlorination (88.7%).

Changes in PA TFC membrane performance due to surface adsorbed coating layers reported in the literature are summarized in Table 10.1. Surface adsorbed coating layers typically resulted in L_p reduction relative to the base PA membrane, which can be quite significant in some cases (up to 83%), while being was accompanied by a rise in salt rejection. Simultaneous increase in L_p and salt rejection were also reported for chitosan coated PA membranes;[37] however, as discussed in Section 10.3, salt rejection of membranes evaluated at different permeate flux values can lead to under or overestimation of membrane salt rejection. Membrane salt permeability coefficient was also reported in a number of studies[33,40] where, it was evaluated at different permeate flux levels; thus, under such test conditions the performance comparisons among membranes are questionable. Moreover, due to the noncovalent nature of interactions between the polymer modifiers and the membrane surface (e.g., van der Waals, electrostatic, hydrogen bonding interactions), such polymer films are generally considered to be less stable relative to covalently attached counterparts.[46]

10.2.3. *Layer-by-layer (LbL) assembly*

In layer-by-layer (LbL) assembly, oppositely charged materials (e.g., polyelectrolyte, nanoparticle, etc.) are alternatively deposited onto a substrate surface to form a multilayered film. Since the surface of the polyamide active layer of TFC NF/RO membranes is anionic (due to carboxylic acid groups[47]), a cationic film is typically first deposited onto the native PA membrane surface. For example, an ultralow pressure RO membrane (Nitto Denko ES20) surface was modified by first depositing a cationic poly(allylamine hydrochloride) (PAH) layer, followed by surface deposition of an anionic poly(sodium 4-styrenesulfonate) (PSS) layer. It was shown that the surface charge of the LbL film is determined by the charge of the last deposited layer and thus can be controlled by the number of surface deposited layers.[48] Moreover, it was reported that increasing the number of

Table 10.1. Change in PA TFC membrane performance due to membrane surface modification by coating (physical adsorption).

Ref	Membrane surface modification method	Base PA TFC membrane[a]	Feed water; filtration mode[b]	ΔP (bar)[c]	L_p ($L \cdot m^{-2} \cdot h^{-1} \cdot bar^{-1}$)[d]		% Change in L_p^e	R_o(%)[f]		Change in R_o(%)[g]	Comment
					Base	Modified		Base	Modified		
[34]	Surface coating with a PEBAX® 1657 layer via dip-coating, followed by curing at 60°C	Hydranautics ESPA1 (BWRO)	1500 mg/L NaCl (21.5°C); crossflow	11.3	6.8[h]	1.8[h]	−74%	97.8	99.0	+1.2	All three base membranes were coated with a layer of PEBAX® 1657 at the same coating condition(1)
		Hydranautics ESPA3 (BWRO)			5.8[h]	1.1[h]	−81%	97.4	97.0	−0.4	
		Hydranautics SWC4 (SWRO)			0.9[h]	0.6[h]	−29%				
[40]	Surface coating of a sericin layer via dip-coating, followed by cross-linking with GA(2)	PA TFC membrane supplied by Development Center of Water Treatment Technology, Hangzhou, China	500 mg/L NaCl (25°C, pH 6.8); crossflow	5	6.5	6.0	−7%	97.6	98.2	+0.6	50 mg/L sericin(3)
						5.5	−16%		98.5	+0.9	100 mg/L sericin(3)
						5.0	−22%		98.4	+0.8	200 mg/L sericin(3)

Table 10.1. (*Continued*)

Ref	Membrane surface modification method	Base PA TFC membrane[a]	Feed water; filtration mode[b]	ΔP (bar)[c]	L_p ($L \cdot m^{-2} \cdot h^{-1} \cdot bar^{-1}$)[d]		% Change in L_p^e	$R_o(\%)^f$		Change in $R_o(\%)^g$	Comment
					Base	Modified		Base	Modified		
[41]	Surface coating of a hydrophilic terpolymer P(MDBAC-r-Am-r-HEMA)[(4)] layer via dip-coating, followed by cross-linking with GA[(2)]	Dow Filmtec LC LE (BWRO)	2000 mg/L NaCl (25°C, pH 7); crossflow	15	5.0	2.0	−60%	98.1	98.6	+0.5	200 mg/L terpolymer[(5)]
						2.3	−53%		98.7	+0.6	500 mg/L terpolymer[(5)]
						2.6	−48%		98.8	+0.7	1000 mg/L terpolymer[(5)]
						2.4	−52%		98.9	+0.8	3,000 mg/L terpolymer[(5)]
						2.1	−59%		99.1	+1.0	10,000 mg/L terpolymer[(5)]
		Dow Filmtec BW30 (BWRO)			2.7			98.5			Membrane used for comparison
[43]	Surface coating of HPOEM[(6)] homopolymer or carboxylated PEI[(7)] layer via dip-coating, followed by cross-linking with GA[(2)]	Toray SHN (SWRO)	32,000 mg/L NaCl + 70 mg/L $CaCl_2$ (25°C, pH 8); crossflow	55.2	0.6	0.5	−8%	99.2	99.3	+0.1	HPOEM[(6)]
						0.6	+2%		99.1	−0.1	Carboxylated PEI[(7)]

(*Continued*)

Table 10.1. *(Continued)*

Ref	Membrane surface modification method	Base PA TFC membrane(a)	Feed water; filtration mode(b)	ΔP (bar)(c)	L_p (L · m^{-2} · h^{-1} · bar^{-1})(d) Base	L_p Modified	% Change in L_p^{e}	R_o(%)(f) Base	R_o(%) Modified	Change in R_o(%)(g)	Comment
[33]	Membrane active side coating with solution containing pre-crosslinked PVA(8) with GA(2), followed by curing (60°C) to solidify the PVA(8) film on the membrane surface	Laboratory synthesized PA based TFC membrane	32,800 mg/L NaCl (25°C); crossflow			0.3	−66%		98.5	−0.9	[PVA]=0.1%, RH=30%(9)
						0.2	−76%		98.2	−1.2	[PVA]=0.3%, RH=30%(9)
						0.8	−12%		99.5	+0.1	[PVA]=0.1%, RH=90%(9)
					0.9	0.8	−18%	99.4	99.6	+0.2	[PVA]=0.3%, RH=90%(9)
				55		0.7	−24%		99.7	+0.3	[PVA]=0.5%, RH=90%(9)
						0.6	−33%		99.7	+0.3	PVA]=1.0%, RH=90%(9)
						1.0	+6%		99.5	+0.1	[PVA]=0.1%, RH=90%(9),(10)
		Dow Filmtec SW30HRLE (SWRO)			0.8			99.4			Membrane used for comparison
[39]	Surface coating of PVA(8), PHMG(11) or PVA:PHMG layer via dip-coating, followed by curing at 110°C	Dow Filmtec LE-400 (BWRO)	584 mg/L NaCl (23°C); crossflow			3.7	−15%		98.5	+1.3	PVA(12)
						2.4	−45%		98.2	+1.0	PVA:PHMG (99:1)(12)
				27.6	4.4	1.7	−62%	97.2	95.8	−1.4	PVA:PHMG (95:5)(12)
						0.8	−83%		91.6	−5.6	PHMG(12)

Table 10.1. (*Continued*)

Ref	Membrane surface modification method	Base PA TFC membrane[a]	Feed water; filtration mode[b]	ΔP (bar)[c]	L_p (L · m^{-2} · h^{-1} · bar^{-1})[d] Base	Modified	% Change in L_p^e	R_o(%)[f] Base	Modified	Change in R_o(%)[g]	Comment
[35]	Drawdown coating of aqueous monomer solution containing PEGDA[13], HPK[14] and either AA[15], HEA[16], or PEGA[17] followed by surface exposure to UV to initiate polymerization to form a hydrogel layer	GE Osmonics AG (BWRO)	2000 mg/L NaCl (25°C, pH 6.5–7.5); crossflow	15.5	4.5[h]	2.6[h]	−43%	99.0	99.3	+0.3	PEGDA[18]
						2.7[h]	−40%		99.1	+0.1	AA:PEGDA (50:50)[18]
						2.8[h]	−37%		99.2	+0.2	HEA: PEGDA (50:50)[18]
						2.9[h]	−36%		99.2	+0.2	PEGA:PEGDA (50:50)[18]
[42]	Surface coating of p(MPC-co-AEMA)[19] layer via dip-coating	Nitto Denko ES20 (ULPRO)	500 mg/L NaCl; crossflow	7.5	5.5	4.2	−24%	94.7	96.9	+2.2	

(*Continued*)

Table 10.1. (*Continued*)

Ref	Membrane surface modification method	Base PA TFC membrane[a]	Feed water; filtration mode[b]	ΔP (bar)[c]	L_p (L · m^{-2} · h^{-1} · bar^{-1})[d] Base	Modified	% Change in L_p[e]	R_o(%)[f] Base	Modified	Change in R_o(%)[g]	Comment
[37]	Membrane exposure to chlorine, followed by coating a positively charged chitosan layer	PA TFC membrane supplied by Hangzhou Tianchuang Environmental Technology Co., LTD.	1500 mg/L NaCl; crossflow	8	6.3	6.8	+8%	91.0	94.5	+3.5	2 min chlorination, 1000 mg/L chitosan[(20)]
						7.2	+15%		95.4	+4.4	5 min chlorination, 1000 mg/L chitosan[(20)]
						7.1	+12%		95.7	+4.7	5 min chlorination, 2000 mg/L chitosan[(20)]

[a]Commercial or laboratory synthesized PA TFC membranes used as base membranes for surface modification

[b]Feed water condition and filtration mode used to evaluate the membrane L_p and R_o

[c]Transmembrane pressure used to evaluate the membrane L_p and R_o

[d]Water permeability coefficient, L_p, defined as $L_p = J_v/(\Delta P - \Delta\pi_m)$ where J_v is volumetric permeate flux and $\Delta\pi_m$ is the osmotic pressure difference between the membrane surface at the feed and permeate sides.

[e]Percentage change in L_p relative to base membrane defined as $\Delta L_p = \left[\left(L_p^{Modified} - L_p^{Base}\right)/L_p^{Base}\right] \cdot 100\%$

[f]Percent observed (nominal) salt rejection defined as $R_o = \left(1 - C_p/C_f\right) \cdot 100\%$

[g]Change in the percent salt rejection R_o relative to base membrane defined as $\Delta R_o = R_o^{Modified} - R_o^{Base}$

[h]$L_p$was evaluated with deionized water.

(1)PEBAX®1657 is a hydrophilic block copolymer of nylon-6 and polyethylene glycol.

(2)GA = glutaraldehyde

(3)Sericin is a water-soluble globular protein. Base membranes were immersed in an aqueous solution (25°C) of sericin at the indicated concentrations (50–200 mg/L) for 5 min, followed by removing the excess solution from the membrane surface. Subsequently, the membrane surface was exposed to an aqueous solution (25°C) containing 0.2% (w/v) glutaraldehyde and 0.1% (w/v) sulfuric acid for 5 min; the excess solution was removed from the membrane surface. Lastly, the membranes were cured at 40°C for 5 min and rinsed with deionized water.

Table 10.1. (*Continued*)

[4] P(MDBAC-*r*-Am-*r*-HEMA) is a random terpolymer synthesized via free radical copolymerization of methylacryloxyethyldimethyl benzyl ammonium chloride (MDBAC), acrylamide (Am), and 2-hydroxyethyl methacrylate (HEMA) carried out in dimethyl sulfone (DMSO).

[5] Membranes were modified by immersing in a solution of P(MDBAC-r-Am-r-HEMA) at the indicated concentrations (200–10,000 mg/L) and GA (0.3 wt%) for 2 min, then the P(MDBAC-r-Am-r-HEMA) and GA adsorbed on the membrane surface was cross-linked at 50°C for 2 h.

[6] HPOEM = hydroxy poly(oxyethylene) methacrylate; The base SWRO membrane was immersed in a solution of laboratory synthesized HPOEM homopolymer (M_n 15,200 and M_w 38,600) for 1 min, and the excess solution was removed from the surface using a rubber roller. Subsequently, the HPOEM coated SWRO membranes were exposed to an aqueous GA solution (0.01 wt%) for 30 s to affix the coating layer onto the membrane surface.

[7] PEI = poly(ethyleneimine); The base SWRO membranes were immersed in an aqueous solution of PEI (0.1 wt%) for 1 min, followed by removal of excess solution from the membrane. The membranes were then exposed to an aqueous GA solution (0.01 wt%) at pH 9 for 30 s to form cross-linked polymer layers on the membrane surface. Finally, the membranes were immersed in an aqueous bromoacetic acid (0.1 wt%) in the presence of potassium hydroxide (0.08 wt%) at 30°C for 10 h to form carboxyl end groups (from the reaction between the PEI amine groups and bromoacetic acid).

[8] PVA = poly(vinyl alcohol)

[9] PVA layers were casted onto base PA membrane surfaces by exposing the PA membrane active side to an aqueous PVA solution (PVA concentration, [PVA], of 0.1 wt% or 0.3 wt%) containing GA for 3 min, followed by draining of the excess PVA solution from the membrane surface and curing the PVA coated membranes at 60°C for 10 min at a fixed relative humidity (RH) or 30% or 90%.

[10] Following the membrane active surface exposure to the PVA solution (0.1 wt%) containing GA for 3 min, the membrane sample was rinsed with deionized water prior to curing at 60°C and RH = 90%.

[11] PHMG = polyhexamethylene guanidine hydrochloride

[12] Membrane sample was drawdown coated with a 4 μm-thick film of an aqueous solution of PVA (2 wt%), PVA:PHMG (2 wt%, ratios of PVA and PHMG are 99:1 or 95:5), or PHMG (2 wt%) via a wet film applicator. Subsequently, the wet film coated membranes were cured at 110°C for 2 min to evaporate the water and induce thermal cross-linking of the polymers to increase the film stability.

[13] PEGDA = poly(ethylene glycol) diacrylate, M_w: 698 g/mol

[14] HPK = 1-hydroxycyclohexyl phenyl ketone, a photoinitiator

[15] AA = Acrylic acid

[16] HEA = 2-hydroxyethyl acrylate

Table 10.1. (*Continued*)

(17)PEGA = poly(ethylene glycol) acrylate (M_w: 380 g/mol)

(18)Pre-polymerization mixtures containing 50 mol% cross-linker (PEGDA) and 50 mol% monomer (AA, HEA, or PEGA) were prepared with 1 wt% (based on total cross-linker and monomer content) photoinitiator (HPK). The base membranes were drawdown coated with the above pre-polymerization mixture using a wet film applicator and exposed to UV light for 90 s at an intensity of 3,000 $\mu W/cm^2$ to initiate polymerization of the coating layer.

(19)p(MPC-co-AEMA) = poly[2-methacryloyloxyethyl phosphorylcholine (MPC)-co-2-aminoethylmethacrylate (AEMA)], a cationic phosphorylcholine polymer; The base membrane was immersed in an aqueous solution of 0.1 wt% p(MPC-co-AEMA) polymer for 3 h (in a refrigerator) and rinsed with an aqueous solution of 3.5 wt% NaCl for 1 h.

(20)The base membranes were immersed in dilute aqueous solutions of sodium hypochlorite (NaClO concentration of 200 mg/L) for up to 20 minutes at 19°C and pH 5.5 (in order to oxidize the PA membrane surface and generate negatively charged surface groups on the membrane surface), followed by rinsing with deionized water. Subsequently, the active sides of the chlorine-treated base membranes were exposed to a chitosan solution (1000 mg/L or 2000 mg/L chitosan in a dilute acetic acid solution) at 19°C for 30 min to allow for the chitosan (exists as positively charged chitosanium in acidic solutions) to adsorb onto the membrane surface interaction), followed by rinsing with deionized water.

bilayers (i.e., a pair of cationic and anionic layers) monotonously decreased L_p while increasing salt rejection. Membrane performance testing (with 500 mg/L feed concentration at $\Delta P = 7.5$ bar) showed that increasing the number of PAH/PSS bilayers up to 6 led to increased salt rejection from 98.3% up to 99.3%, while L_p was reduced by up to ~48% (reduced from 4.7 to 2.5 $\text{L} \cdot \text{m}^{-2} \cdot \text{h}^{-1} \cdot \text{bar}^{-1}$).

Change in the RO membrane performance was also reported for a BWRO membrane (CSM RE4021-TE), which was coated with anionic polyacrylic acid (PAA) and cationic tobramycin (TOB) layers via LbL assembly.[49] It was reported that three bilayers of PAA/TOB resulted in increased salt rejection from 99.2% to 99.5% and increased L_p by ~17% (3.0 to 3.5 $\text{L} \cdot \text{m}^{-2} \cdot \text{h}^{-1} \cdot \text{bar}^{-1}$) relative to the base BWRO membrane (2000 mg/L NaCl feed concentration at $\Delta P = 15.5$ bar). Increasing the number of bilayers from three to ten resulted in a ~21% reduction in L_p (3.5 to 2.8 $\text{L} \cdot \text{m}^{-2} \cdot \text{h}^{-1} \cdot \text{bar}^{-1}$) but did not increase salt rejection above 99.5%. It was also reported that TOB, which is an aminoglycoside antibiotic, increased the mortality of *E. Coli* and *B. Subtilis* significantly to 99.6–100% relative to the base BWRO membrane (10–15%). Moreover, the surface deposited PAA-TOB LbL films increased membrane fouling resistance, evaluated with to BSA and alginic acid as model foulants. It was reported that after 3 cycles of foulant solution filtration (each cycle consisted of filtration of a foulant solution (100 ppm BSA or alginic acid + 2000 mg/L NaCl) for 4 hours, followed by 1 hour of DI water flushing) the LbL film (with 3 PAA-TOB bilayers) modified membrane retained 75% of its initial permeate flux, relative to 38–49% flux recovery for the native BWRO membrane.

The LbL assembly technique was also utilized to deposit alternating layers of anionic and cationic nanoparticles (LUDOX® HS-30 and CL-30 colloidal silica; nominal diameters of 12 nm and 22 nm, respectively) onto a commercial NF membrane (Dow Filmtec NF270).[50] The NF270 membrane had a water permeability coefficient of ~10.8 $\text{L} \cdot \text{m}^{-2} \cdot \text{h}^{-1} \cdot \text{bar}^{-1}$ and salt rejection of ~31% when evaluated with 150 mg/L NaCl concentration feed water at $\Delta P = 17.2$ bar. It was reported that even a single layer of either anionic or cationic silica nanoparticles deposited onto the base NF membrane greatly reduced the membrane water permeability coefficient (~57%) and increased salt rejection to ~52%. Deposition of one or two nanoparticle bilayers resulted in greater L_p reduction by ~61% and ~68%, respectively, relative to the base NF membrane; however, salt rejection remained the same as for the single nanoparticle layer coated membrane (~51–52%). Increasing the number of bilayers to 4 further reduced L_p (~76%) relative to the base NF membrane; however, salt rejection of 47% was achieved with these membranes, which was slightly lower than that for the membranes coated with ≤2 bilayers. It is noted that stability of the nanoparticle bilayers was not investigated in the above study.

In another study, layers of polyethyleneimine (PEI) coated copper nanoparticles (PEI-CuNPs) and polyacrylic acid (PAA) were deposited onto a commercial SWRO membrane (Toray UTC-82V) via spray- and spin-assisted LbL assembly.[51] It was reported that surface deposition of 10 bilayers of PEI-CuNPs/PAA onto the base SWRO membrane resulted in L_p (evaluated with DI water) reduction by ~13% (reduced from 1.59 to $1.38\,\mathrm{L \cdot m^{-2} \cdot h^{-1} \cdot bar^{-1}}$), while salt rejection (evaluated with 2,922 mg/L NaCl concentration feed water and $\Delta P = 27.6$ bar) was unaltered. The LbL film was shown to also reduce biofouling propensity of the base membrane due to the antimicrobial property of the copper nanoparticles deposited onto the membrane surface. Permeate flux reduction (relative to the initial permeate flux), following a 24 h period of bacterial solution filtration, for the LbL film (10 bilayers of PEI-CuNPs/PAA) coated and the native SWRO base membranes was 43% and 66%, respectively. However, it was reported that the surface deposited copper nanoparticles, embedded in the LbL films, were nearly depleted from the membrane surface after just 7 days of immersion in a 32 g/L NaCl solution. Therefore, re-deposition of LbL films or copper nanoparticles would be required on a frequent basis which is not a practical approach to mitigate membrane fouling.

NF/RO membrane performance changes due to various types of LbL films reported in the literature are summarized in Table 10.2. In general, increasing the number of bilayers deposited onto a membrane surface led to increased surface film thickness and hydraulic

Table 10.2. Change in PA TFC membrane performance due to membrane surface modification by layer-by-layer (LbL) assembly.

Ref	Surface modification method	Base PA TFC membrane[a]	Feed water; filtration mode[b]	ΔP (bar)[c]	L_p ($L \cdot m^{-2} \cdot h^{-1} \cdot bar^{-1}$)[d] Base	Modified	% Change in L_p^e	R_o(%)[f] Base	Modified	Change in R_o (%)[g]	Comment
[48]	Surface deposition of PAH(1) and PSS(2) layers via LbL	Nitto Denko ES20 (ULPRO)	500 mg/L NaCl; crossflow	7.5	4.7	2.5	−48%	98.3	99.3	+1.1	6 bilayers of PAH/PSS(3)
[49]	Surface deposition of PAA(4) and TOB(5) layers via LbL	CSM RE4021-TE (BWRO)	2000 mg/L NaCl (25°C); crossflow	15.5	3.0	3.5	+17%	99.2	99.5	+0.3	3 bilayers of PAA/TOB(6)
[50]	Surface deposition of anionic (Ludox HS-30) and cationic (Ludox CL-30) colloidal silica nanoparticles(7) via LbL	Dow Filmtec NF270 (NF)	150 mg/L NaCl; dead-end	17.2	10.8	∼4.7	−57%	∼31	∼52	+21	a single layer of either PAA or TOB(8),(9)
						∼4.2	−61%		∼51	+20	1 bilayer of PAA/TOB(8)
						∼3.5	−68%		∼52	+21	2 bilayers of PAA/TOB(8)

(*Continued*)

Table 10.2. (*Continued*)

Ref	Surface modification method	Base PA TFC membrane[a]	Feed water; filtration mode[b]	ΔP (bar)[c]	L_p ($L \cdot m^{-2} \cdot h^{-1} \cdot bar^{-1}$)[d] Base	Modified	% Change in L_p[e]	R_o(%)[f] Base	Modified	Change in R_o (%)[g]	Comment
						~2.6	−76%		~47	+16	4 bilayers of anionic and cationic nanoparticles(8)
[51]	Surface deposition of PEI-CuNPs(10) and PAA(4) layers via LbL	Toray UTC-82C (SWRO)	2,922 mg/L NaCl (20°C); crossflow	27.6	1.6	1.4	−13%			No change	10 bilayers of PEI-CuNPs/PAA(11)

[a]Commercial or laboratory synthesized PA TFC membranes used as base membranes for surface modification
[b]Feed water condition and filtration mode used to evaluate the membrane L_p and R_o
[c]Transmembrane pressure at which the membrane L_p and R_o were evaluated
[d]Water permeability coefficient, L_p, defined as $L_p = J_v/(\Delta P - \Delta\pi_m)$ where J_v is volumetric permeate flux and $\Delta\pi_m$ is the osmotic pressure difference between the membrane surface at the feed and permeate sides.
[e]Percentage change in L_p relative to base membrane defined as $\Delta L_p = \left[\left(L_p^{Modified} - L_p^{Base}\right)/L_p^{Base}\right] \cdot 100\%$
[f]Percent observed (nominal) salt rejection defined as $R_o = \left(1 - C_p/C_f\right) \cdot 100\%$
[g]Change in the percent salt rejection R_o relative to base membrane defined as $\Delta R_o = R_o^{Modified} - R_o^{Base}$

Table 10.2. (*Continued*)

(1)PAH = poly(allylamine hydrochloride), a cationic polyelectrolyte
(2)PSS = poly(sodium 4-styrenesulfonate), an anionic polyelectrolyte
(3)Surface deposition of PAH and PSS layers performed by utilizing a cross-flow membrane cell. The base membrane surface was exposed to PAH and PSS solutions (10 g/L PAH or PSS in a 10 mM Tris buffer solution containing 1 M NaCl at pH 7) alternatively (for 30 min each). After deposition of each layer, the membrane surface was rinsed with deionized water for 5 minutes. A maximum of six PAH/PSS bilayers were deposited.
(4)PAA = poly(acrylic acid), an anionic polyelectrolyte
(5)TOB = tobramycin, an aminoglycoside antibiotic
(6)The active side of the base membrane was first exposed to a 0.5 mg/L TOB solution for 15 min, followed by rinsing with deionized water. The membrane active side was then exposed to 1 mg/L PAA solution for 15 min, followed by rinsing with deionized water. The above two steps were repeated up to 5 times to acquire 10 bilayers of TOB/PAA. After surface deposition of the bilayers, the bilayers were cross-linked by immersing the membrane in a solution containing 5 mg/L EDC and 4.5 mg/L NHS for 12 h. The membranes were then subsequently rinsed with deionized water.
(7)Anionic Ludox HS-30 and cationic Ludox CL-30 colloidal silica nanoparticles had nominal diameters of 12 nm and 22 nm, respectively.
(8)The base membranes were immersed in aqueous nanoparticle solutions (0.03 wt%, pH 3) for 10 min, followed by two deionized water rinses, each lasting for 2 min. The step was repeated to acquire 1, 2, and 4 bilayers (each bilayer consists of one cationic and one anionic nanoparticle layer).
(9)A single layer of either anionic Ludox HS-30 or cationic Ludox CL-30 nanoparticle
(10)Positively charged poly(ethyleneimine) (PEI) coated copper nanoparticles (PEI-CuNPs) were synthesized.
(11)The spray- and spin-assisted PEI-CuNPs/PAA LbL coating process was accomplished by spinning the base membrane (using a spin-coater) samples at 2000 rpm while alternatively spraying solutions of positively charged PEI-CuNPs (pH 8.3) and negatively charged PAA (1 g/L, pH 3) onto the membrane surface at 2.1 bar. Between each layer deposition, the base membrane was rinsed with deionized water for 3 s and air-dried for 10 s (via spinning the membrane sample). The above steps were repeated to obtain a desired number of PEI-CuNPs/PAA bilayers.

resistance, hence reduced water permeability. Salt rejection, however, did not monotonously increase with increasing number of bilayers. In most cases, salt rejection of LbL modified membranes increased relative to the base membrane, and it seems that there was an optimal number of bilayers at which membrane salt rejection was maximized.

10.2.4. *Initiated chemical vapor deposition (iCVD)*

A hydrophilic polymer film may be deposited onto a membrane surface via initiated chemical vapor deposition (iCVD) technique (Fig. 10.4).[52] In iCVD, vinyl monomers and initiators in a vapor phase are deposited and polymerized (via free radical polymerization) simultaneously onto a substrate surface.[52] Polymerization

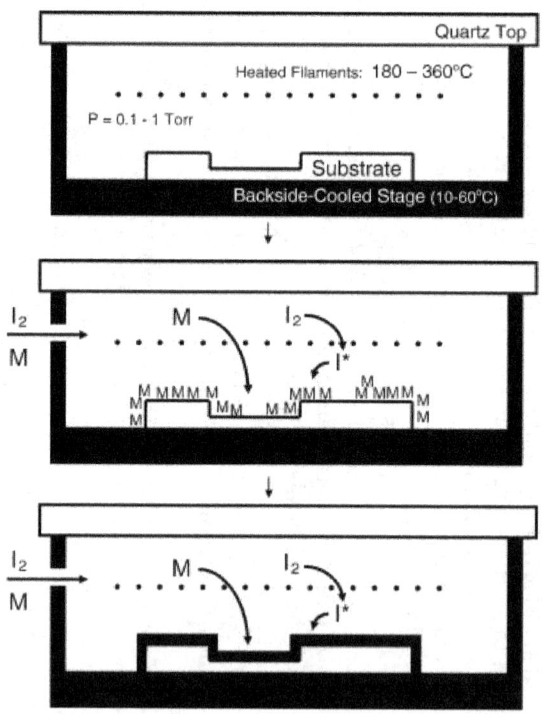

Fig. 10.4. Illustration of the iCVD process.[52]

is initiated by using chemical initiators (e.g., such as tert-butyl peroxide) which thermally dissociate into radicals (upon passing through a heated filament at 200–360°C in the reactor) that can initiate polymerization from the surface adsorbed monomers.[52] In the iCVD process, polymerization can proceed, solvent free, at a relatively low substrate surface temperature (i.e., 20–30°C)[53–57] and thus should maintain the integrity of the polyamide membrane.[58] Moreover, the technique allows for precise control of the deposited polymer film thickness in the order of a tenth to a hundredth nanometer.[58] However, it was reported that deposited polymer films of thickness below 20 nm do not form robust coating layers and could not withstand high applied pressure ($\Delta P \geq 48.3$ bar), as evidenced by the gradually increased permeate flux during filtration tests of the polymer film coated polyamide RO membranes.[54]

The iCVD technique can be applied to any type of substrate; however, in some cases, an adhesion layer is required to prevent delamination of surface deposited polymer films from the polyamide membrane surface. For example, prior to iCVD deposition of a copolymer (2-(dimethylamino)ethyl methacrylate (DMAEMA) and ethylene glycol diacrylate (EGDA)) layer, maleic anhydride (MA) was covalently anchored to a commercial BWRO membrane (Koch TFC-HR) via reaction between the anhydride and amide functional groups present in the RO membrane surface.[53] The subsequently deposited p(DMAEMA-*co*-EGDA) copolymer layer was covalently bonded to the MA groups on the membrane surface, thus forming a robust copolymer layer that resisted delamination. The copolymer layer was then treated with 1,3-propanesulfone to obtain zwitterionic sulfobetaine surface functional groups. The zwitterionic functionalized p(DMAEMA-*co*-EGDA) film effectively reduced surface attachment of *Escherichia coli* cells. It was also reported that surface deposition of the zwitterionic copolymer film (30 nm in thickness) reduced L_p of the TFC-HR membrane by ~17% (reduced from 5.3 to 4.4 $\text{L} \cdot \text{m}^{-2} \cdot \text{h}^{-1} \cdot \text{bar}^{-1}$) and also reduced salt rejection from 97% to 95% (tested with 2000 mg/L NaCl feed water at $\Delta P = 20.7$ bar).

Other types of zwitterionic and amphiphilic copolymer films, synthesized onto commercial PA TFC membranes via the iCVD technique, were also reported to reduce membrane biofouling.[54,56–58] In the above studies, the thickness of the surface deposited polymer films was in the range of 20–40 nm, which was deemed to be optimal given the low impact on water permeability (~10–20% reduction relative to the base membrane; Table 10.3).[54–56,58] Salt rejection, on the other hand, increased or remained the same after surface deposition of the polymer films via iCVD. For example, an amphiphilic polymer film, synthesized by copolymerization of a hydrophilic monomer, hydroxyethyl methacrylate (HEMA), and a hydrophobic monomer, perfluorodecyl acrylate (PFA), via the iCVD technique, was shown to be effective for reducing surface adhesion of *Escherichia coli*.[54] Surface deposition of the amphiphilic copolymer p(HEMA-*co*-PFA) film (~20 nm thickness) onto a BWRO membrane (Koch TFC-HR) reduced L_p (evaluated with DI water) by ~10% (reduced from 3.5 to 3.2 $L \cdot m^{-2} \cdot h^{-1} \cdot bar^{-1}$), while salt rejection remained at the same level of 98% (evaluated with 12,000 mg/L NaCl feed water at $\Delta P = 48.3$ bar). Surface deposition of the same film (~20 nm thickness), via iCVD, onto SWRO membranes (Dow Filmtec SW30HR, GE AD, and Toray UTC80B) also resulted in a similar degree of L_p reduction (~8–11%), while salt rejection (evaluated with 2000 mg/L NaCl feed water at $\Delta P = 50$ bar) was either decreased (99.5% to 99.3% for SW30HR) or increased (99.2% to 99.4% for AD and 99.3% to 99.7% for UTC80B), depending on the base membrane.[55]

In other studies, polymerization of 4-vinylpyridine (4-VP) was carried out via iCVD with a cross-linker, ethylene glycol diacrylate (EGDA) or divinylbenzene (DVB), in order to increase the mechanical stability of the resulting copolymer (i.e., p(4-VP-*co*-EGDA) or p(4-VP-*co*-DVB), respectively) film formed onto the BWRO membrane (Koch TFC-HR) surface.[56–58] The resulting copolymers were reacted with 3-bromopropionic acid or 1,3-propanesultone at the solid-gas interface to form zwitterionic surface functional groups (carboxybetaine or sulfobetaine, respectively).[56–58] It was reported

Table 10.3. Change in PA TFC membrane performance due to membrane surface modification by iCVD.

Ref	Surface modification technique	Base PA TFC membrane[a]	Feed water; filtration mode[b]	ΔP (bar)[c]	L_p(L· $m^{-2} \cdot h^{-1} \cdot bar^{-1}$)[d] Base	L_p(L· $m^{-2} \cdot h^{-1} \cdot bar^{-1}$)[d] Modified	% Change in L_p^e	R_o(%)[f] Base	R_o(%)[f] Modified	Change in R_o (%)[g]	Comment
[53]	Surface deposition of p(DMAEMA-*co*-EGDMA) film via iCVD, followed by surface treatment with 1,3-propanesultone to form zwitterionic sulfobetaine surface functional groups(1)	Koch TFC-HR (BWRO)	2000 mg/L NaCl; crossflow	20.7	5.3	4.4	−17%	97.0	95.0	−2.0	30 nm thick film(2)
[54]	Surface deposition of a p(HEMA-*co*-PFA) film via iCVD(3)	Koch TFC-HR (BWRO)	12,000 mg/L NaCl (25°C); crossflow	48.3	3.5^h	3.2^h	−10%	98.0	98.0	No change	20 nm thick film(4)

(*Continued*)

Table 10.3. *(Continued)*

Ref	Surface modification technique	Base PA TFC membrane[a]	Feed water; filtration mode[b]	ΔP (bar)[c]	L_p(L· m^{-2} · h^{-1} · bar^{-1})[d] Base	Modified	% Change in L_p^e	R_o(%)[f] Base	Modified	Change in R_o (%)[g]	Comment
[55]	Surface deposition of p(HEMA-co-PFA) film by via iCVD(5)	Dow Filmtec SW30HR (SWRO)	2000 mg/L NaCl (25°C; pH 6); crossflow	50	0.74	0.68	−8%	99.5	99.3	−0.2%	
		GE Osmonics AD (SWRO)			0.50	0.45	−10%	99.2	99.4	+0.2%	20 nm thick film(4)
		Toray UTC80B (SWRO)			0.94	0.84	−11%	99.3	99.7	+0.4%	
[56]	Surface deposition of p(4VP-*co*-EGDA) film via iCVD, followed by surface treatment with 3-bromopropionic acid to form zwitterionic carboxybetaine surface functional groups(6)	Koch TFC-HR (BWRO)	2000 mg/L NaCl (21°C); crossflow	20.7	2.20[h]	1.82[h]	−17%	96.0	97.9	+1.9	38 nm thick film(7)

Table 10.3. (*Continued*)

Ref	Surface modification technique	Base PA TFC membrane[a]	Feed water; filtration mode[b]	ΔP (bar)[c]	L_p(L· $m^{-2} \cdot h^{-1} \cdot bar^{-1}$)[d] Base	L_p Modified	% Change in L_p^e	R_o(%)[f] Base	R_o(%)[f] Modified	Change in R_o (%)[g]	Comment
[57]	Surface deposition of a p(4VP-*co*-EGDA) film via iCVD, followed by surface treatment with 1,3-propanesultone to form zwitterionic sulfobetaine surface functional groups(8)	Koch TFC-HR (BWRO)	2000 mg/L NaCl; crossflow	20.7	2.16^h	1.76^h	−19%	96.1	97.6	+1.5	35 nm thick film(9)
[58]	Surface deposition of a p(4VP-*co*-DVB) film via iCVD, followed by surface treatment with 1,3-propanesultone to form zwitterionic sulfobetaine surface functional groups(10)	Koch TFC-HR (BWRO)	35,000 mg/L NaCl; dead-end	48.3	0.48	0.39	−19%	92.0	92.0	No change	30 nm thick film(11)

(*Continued*)

Table 10.3. (*Continued*)

[a]Commercial or laboratory synthesized PA TFC base membranes for surface modification

[b]Feed water condition and filtration mode for determination of membrane permeability and observed rejection (i.e., L_p and R_o, respectively)

[c]Transmembrane pressure at which membrane L_p and R_o were determined

[d]Water permeability coefficient, L_p, defined as $L_p = J_v/(\Delta P - \Delta\pi_m)$ where J_v is the volumetric permeate flux and $\Delta\pi_m$ is the osmotic pressure difference between the membrane surface at the feed and permeate sides.

[e]Percentage change in L_p relative to the base membrane defined as $\Delta L_p = \left[\left(L_p^{Modified} - L_p^{Base}\right)/L_p^{Base}\right] \cdot 100\%$

[f]Percent observed (nominal) salt rejection defined as $R_o = \left(1 - C_p/C_f\right) \cdot 100\%$

[g]Change in the percent salt rejection R_o relative to base membrane defined as $\Delta R_o = R_o^{Modified} - R_o^{Base}$

[h]$L_p$was evaluated with deionized water.

(1)The base membrane was first treated with maleic anhydride, which covalently bonded to the base membrane surface. The surface grafted MA was critical for preventing delamination of the subsequently deposited zwitterionic films when placed in water. The MA grafted base membrane surface was deposited with a random copolymer p(DMAEMA-*co*-EGDMA) film via iCVD during which tert-butyl peroxide (initiator) and monomers, 2-(dimethylamino)ethyl methacrylate (DMAEMA) and ethylene glycol dimethacrylate (EGDMA), were delivered into the reactor resulting in deposited copolymer film at a filament temperature of 250°C and stage temperature of 20°C. The pressure in the vacuum chamber was maintained at ~29 Pa. Following the film deposition, the p(DMAEMA-*co*-EGDMA) film surface was exposed to 1,3-propane sultone vapor to form sulfobetaine zwitterionic groups on the membrane surface.

(2)The reported performance was for the base membrane modified with a p(DMAEMA-*co*-EGDMA) film of 30 nm thickness (after treatment with 1,3-propane sultone).

(3)The base membrane was modified by depositing a random amphiphilic copolymer p(HEMA-*co*-PFA) film via iCVD. The p(HEMA-*co*-PFA) film was deposited by delivering tert-buryl peroxide (initiator), a hydrophilic monomer (2-hydroxyethyl methacrylate (HEMA)), and a hydrophobic monomer (perfluorodecyl acrylate (PFA)) to the iCVD reactor at a pressure of 27 Pa, while maintaining the filament and stage temperatures at 210°C and 30°C, respectively.

Table 10.3. (*Continued*)

[4]The reported performance is for a base membrane modified with a 20-nm thick p(HEMA-*co*-PFA) film.
[5]A random amphiphilic copolymer p(HEMA-*co*-PFA) film was deposited onto the membrane surface via iCVD. The p(HEMA-*co*-PFA) film was deposited by delivering tert-buryl peroxide (initiator), a hydrophilic monomer (2-hydroxyethyl methacrylate (HEMA)), and a hydrophobic monomer (perfluorodecyl acrylate (PFA)) to the iCVD reactor at a pressure of 27 Pa, while maintaining the filament and stage temperatures at 220°C and 30°C, respectively.
[6]A random copolymer p(4-VP-*co*-EGDA) film deposition onto the base membrane surface, via iCVD, was accomplished by delivering tert-buryl peroxide (initiator), 4-vinylpyridine (4-VP), and ethylene glycol diacrylate (EGDA) to the iCVD reactor where temperatures of the filament and substrate were maintained at 200°C and 20°C, respectively, and pressure of 60 Pa. Subsequently, the deposited copolymer was exposed to 3-bromopropionic acid vapor to form zwitterionic carboxybetaine surface groups.
[7]Reported performance for a base membrane modified with a p(4-VP-*co*-EGDA) film of a final film thickness of 38 nm (after treatment with 3-bromopropionic acid).
[8]A random copolymer p(4-VP-*co*-EGDA) film deposition onto the base membrane surface, via iCVD, was accomplished by delivering tert-buryl peroxide (initiator), 4-vinylpyridine (4-VP), and ethylene glycol diacrylate (EGDA) to the iCVD reactor with a setting of filament and substrate temperatures of 220°C and 20°C, respectively, and total pressure of 60 Pa. Subsequently, the deposited copolymer was exposed to 1,3-propane sultone vapor to form zwitterionic sulfobetaine surface groups.
[9]Reported performance for a base membrane modified with a p(4-VP-*co*-EGDA) film of a 35 nm final film thickness (after treatment with 1,3-propane sultone).
[10]A random copolymer p(4VP-*co*-DVB) film deposition onto the base membrane surface, via iCVD, was performed by delivering tert-buryl peroxide (initiator), 4-vinylpyridine (4-VP), and divinylbenzene (DVB) to the iCVD reactor with filament and substrate temperature settings of 250°C and 20°C, respectively, and total pressure of 107 Pa. Subsequently, the deposited copolymer was exposed to 1,3-propane sultone vapor to form zwitterionic sulfobetaine surface groups.
[11]Reported performance is for a base membrane modified with a p(4-VP-*co*-DVB) film of 30 nm final film thickness (after treatment with 1,3-propane sultone).

that both carboxybetaine and sulfobetaine functionalized p(4-VP-*co*-EGDA) copolymer films (38 nm and 35 nm in thickness, respectively) were effective for lowering surface attachment of bacteria cells (*Escherichia coli* and *Pseudomonas aeruginosa*) and increasing salt rejection of the base membrane (Koch TFC-HR) from 96.0–96.1% to 97.6–97.9% (evaluated with feed water containing 2000 mg/L NaCl at $\Delta P = 20.7$ bar) while reducing L_p (evaluated with DI water) by ~17–19% (reduced from ~2.2 to ~1.8 $\text{L} \cdot \text{m}^{-2} \cdot \text{h}^{-1} \cdot \text{bar}^{-1}$).[56,57] However, for the case of desalting a high salinity feed water (i.e., 35,000 mg/L NaCl at $\Delta P = 48.3$ bar), surface modification of the same base membrane with a sulfobetaine functionalized p(4-VP-*co*-DVB) copolymer film (30 nm in thickness) did not improve salt rejection (remained at ~92%) and reduced L_p by ~19% (i.e., from 0.48 to 0.39 $\text{L} \cdot \text{m}^{-2} \cdot \text{h}^{-1} \cdot \text{bar}^{-1}$).[58]

Studies on PA TFC membrane surface modification via iCVD indicate that the approach is suitable for deposition of nano-scale polymer films capable of mitigating bacterial surface attachment. Such films have been reported to have a relatively moderate to low impact on the base membrane water permeability (Table 10.3) compared to other membrane surface modification approaches. However, scale up of the iCVD approach for commercial membrane applications is likely to be hindered given the requirement of a vacuum chamber (≤107 Pa)[54,58] which would limit the treatable membrane surface area.

10.2.5. *Polymer grafting ("Grafting to")*

In polymer grafting, preformed polymer chains are covalently attached to a membrane surface via a reaction between the polymer chain ends and the membrane surface. For example, polyamide active layer surfaces of TFC NF/RO membranes contain free carboxylic acid and primary amine groups that may be used to react with epoxy end groups of poly(ethylene glycol) diglycidyl ether (PEGDE).[59] Commercial BWRO (Dow XLE) and NF (Dow NF90) membranes were grafted with PEGDE by exposing active sides of the membranes to 1% w/w or 15% w/w aqueous PEGDE (MW 200 or 1000) solutions at 40°C for 10 minutes. Membrane performance evaluation (with 2000 mg/L NaCl feed water at $\Delta P = 10.3$ bar) revealed that PEGDE (MW 200) grafted XLE membranes had 15–31% lower L_p and intrinsic salt rejection (99.1–99.3%) that was similar or higher relative to the base XLE membrane ($L_p^{XLE} \sim 6.9\,\mathrm{L \cdot m^{-2} \cdot h^{-1} \cdot bar^{-1}}$, intrinsic salt rejection = 99.1%). Surface modification with a higher molecular weight PEGDE (MW 1000) resulted in increased reductions of L_p (37–44%) relative to the base XLE membrane and correspondingly higher intrinsic salt rejection (99.4–99.5%), which was attributed to the longer PEGDE chains sealing surface defects present on the native base membrane surface more effectively than the shorter chains. When the same PEGDE (MW 1000) molecules were grafted onto a NF base membrane (NF90), reduction in L_p (45–48%) and the intrinsic salt rejection increased from 97.8% to 98.8% relative to the base membrane ($L_p^{NF} \sim 9.0\,\mathrm{L \cdot m^{-2} \cdot h^{-1} \cdot bar^{-1}}$, intrinsic salt rejection = 97.8%) was more substantial compared to the PEGDE modified XLE membranes. It was conjectured that NF90, being a nanofiltration membrane, has more surface defects than the XLE BWRO membrane. Therefore, it was suggested that water permeability reduction due to plugging of base membrane surface defects by the surface grafted PEGDE chains was more significant for the NF90 than for the XLE membrane.

It is noted that the polyamide active layer surface also contains acyl chlorides (which is the precursor to carboxylic acids), which can be used to covalently anchor hydrophilic molecules that contain either a hydroxyl or amine end group.[60,61] It was shown that 3-Monomethylol-5,5-dimethylhydantoin (MDMH), a hydantoin derivative, can be anchored to a PA membrane surface via the reaction between a hydroxyl end group in MDMH and either the acyl chloride or carboxylic acid groups present on the PA active layer surface.[61] In the study, laboratory synthesized PA TFC membrane surfaces were exposed to an aqueous solution containing MDMH of varying concentrations (2–20 wt%) at 40°C for 5 min, followed by curing at a high temperature (103°C) for 10 min.[61] Membrane performance evaluation (with 2000 mg/L NaCl feed water at $\Delta P =$ 15 bar) revealed that surface grafting of MDMH onto the base PA membrane resulted in ~16–43% increased L_p (increased from 5.5 to 6.3–7.5 $L \cdot m^{-2} \cdot h^{-1} \cdot bar^{-1}$) but at reduced nominal salt rejection (reduced from 95.6% to 91.8–94.3%). The MDMH grafted membranes exhibited a higher resistance to chlorine oxidation relative to the base membrane. Exposing the MDMH grafted membranes to 2000 ppm · h of chlorine (at pH 4) led to 9–26% L_p increase and 2.8–4.7% lower salt rejection. The same chlorine exposure resulted in a greater increase in L_p (~39%) and correspondingly greater reduction of salt rejection (by ~7%) due to membrane degradation. The antimicrobial character of the surface grafted MDMH layers was tested by immersing the chlorine treated membranes (exposed to 500 ppm · h at pH 4) in an *Escherichia coli* suspension and incubating the solution at 37°C for 48 hours. After 48 h exposure to the microbial cells, permeate flux of the base membrane was reduced by ~25%, whereas the MDMH grafted membrane flux was reduced slightly by ~13–22%. However, the bacteria fouling tests for the above membranes were conducted at different initial permeate flux values.

Covalent tethering of hydrophilic macromolecules on a PA membrane surface is also possible by utilizing a mediating chemical that allows for a coupling reaction between a functional group present on the polyamide surface and a chemical modifier. For example, primary amines ($R–NH_2$) such as imidazolidinyl urea, poly(ethyleneimine), poly(vinylamine), and polyether diamines (JEFFAMINE® ED series) have been grafted onto PA membrane surfaces via carbodiimide-mediated coupling reaction.[62–65] A water soluble carbodiimide compound such as 1-ethyl-3-(3-dimethylaminopropyl) carbodiimide hydrochloride (EDC) can be employed to convert carboxylic acid groups present on PA membrane surfaces to *O*-acylisourea ester intermediates (Fig. 10.5). The *O*-acylisourea ester intermediate is then displaced via a nucleophile attack by a primary amine to form a surface amide group.[66,67] However, *O*-acrylisourea is highly unstable in aqueous solutions and rapidly undergoes hydrolysis or rearranges to form the more stable *N*-acylurea, thus resulting in a low yield coupling reaction.[66,67] The reaction yield can be significantly increased by use of *N*-hydroxysuccinimide (NHS) (or its derivative

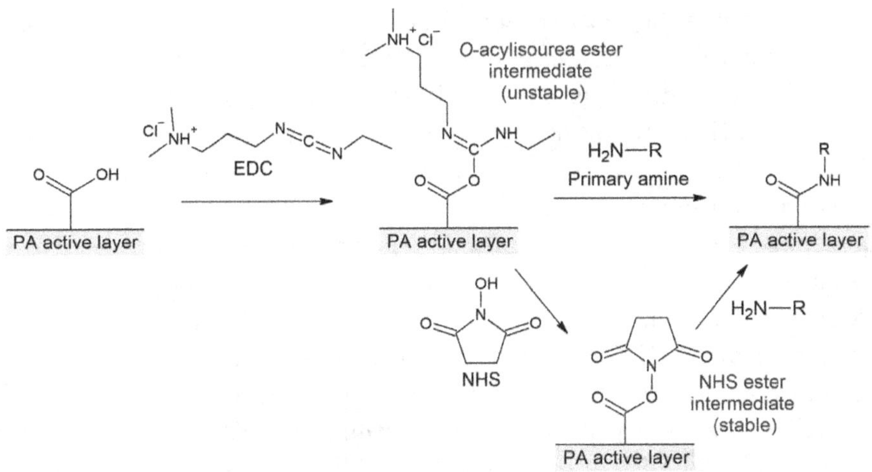

Fig. 10.5. 1-Ethyl-3-(3-dimethylaminopropyl) carbodiimide hydrochloride (EDC) mediated coupling reaction between carboxylic acid of a polyamide active layer surface and a primary amine modifier facilitated by use of *N*-hydroxysuccinimide (NHS) (side reactions and byproducts are not shown).

N-hydroxysulfosuccinimide) in conjunction with EDC to convert the *O*-acylisourea intermediate to a more stable NHS ester intermediate, which hydrolyzes slowly relative to the rate at which it reacts with a primary amine (Fig. 10.5).[66,67]

The above approach (i.e., carbodiimide-mediated coupling reaction) was utilized to graft imidazolidinyl urea (IU) and polyethyleneimine (PEI) onto an ultralow pressure RO membrane (CSM RE4021-TL).[62,63] Surface grafting of IU and PEI (MW 70,000) both resulted in a similar degree of L_p reduction (~25%) relative to the base membrane (~6.3–6.8 $\mathrm{L \cdot m^{-2} \cdot h^{-1} \cdot bar^{-1}}$). However, the PEI grafted membrane had a higher salt rejection (~98.1%) than the IU grafted membrane (96.9%), which had a lower salt rejection than the base membrane (~97.0–97.1%) (both L_p and salt rejection were evaluated with 2000 mg/L NaCl feed water at $\Delta P = 15.5$ bar). The surface grafted IU layer significantly reduced membrane biofouling propensity and also reduced the PA active layer degradation by chlorine.[62] The surface grafted PEI layers reduced membrane fouling by positively charged model foulants (i.e., lysozyme, dodecyl trimethyl ammonium chloride (DTAC), and cetyl trimethyl ammonium chloride (CTAC)).[63] For example, filtration of feed solutions containing 100 mg/L lysozyme and 2000 mg/L NaCl at pH 4–10 for 500 min resulted in 20–39% flux reduction for the PEI grafted membranes, whereas the permeate flux of the base membrane was reduced by 41–60%. The above was partly attributed to a positively charged PEI surface layer, which effectively reduced attachment of the above foulants by electrostatic repulsion over a wider range of pH compared to the base PA membrane. The surface grafted PEI layer increased the isoelectric point of the base membrane from 3.1 to ~8. However, in the above fouling tests, the initial permeate flux of the base membrane was ~28% higher relative to that of the PEI grafted membrane. Therefore, the assertion of inferiority of the base membrane is questionable.

Glutaraldehyde was utilized as a mediating compound for surface grafting of polyvinyl alcohol (PVA) onto a commercial PA TFC ULPRO membrane (supplied by Hangzhou Tianchuang Environmental Technology Co., LTD).[68] Glutaraldehyde, which possesses two aldehyde end groups, was grafted onto the PA membrane surface via reaction between one of the aldehyde groups in the GA and the amide or free primary amines present on the PA membrane surface.[68] Subsequently, PVA chains ($M_W \sim 22{,}000$) were covalently tethered to the PA membrane surface via a reaction between the hydroxyl PVA groups and free ends of surface tethered GA molecules with aldehyde groups. The resulting PVA modified membranes L_p was in the range of 5.4–5.9 $L \cdot m^{-2} \cdot h^{-1} \cdot bar^{-1}$ which was 4% lower or 4–11% greater than that of the base PA membrane (5.6 $L \cdot m^{-2} \cdot h^{-1} \cdot bar^{-1}$), while salt rejection increased up to 98.5% from 98.1% for the base PA membrane (tested with feed water containing 500 mg/L NaCl at $\Delta P = 5$ bar). The PVA grafted membrane had a lower fouling propensity with respect to various types of model foulants including bovine serum albumin (BSA), sodium dodecyl sulfate (SDS), and dodecyltrimethyl ammonium bromide (DTAB). Following the filtration of the foulant solutions (containing either 100 mg/L BSA, 200 mg/L SDS, or 10 mg/L DTAB) for up to 40 hours, permeate flux for the PVA grafted membranes was reduced by ~19–33%, whereas ~27–54% permeate flux decline was observed for the native ULPRO membrane. It was also reported that membrane cleaning of the fouled membranes via DI water flushing for 60 minutes (at a crossflow velocity of 20 cm/s) achieved permeability recovery for the PVA grafted membranes (79–93%), which was greater than for the native membrane (59–82%).

In another study, the ULPRO base membrane (supplied by Hangzhou Tianchuang Environmental Technology Co., Ltd.) was modified by surface grafting PVA chains using potassium persulfate ($K_2S_2O_8$) as a thermal initiator.[69] The base membrane surface was coated with a layer of an aqueous solution containing PVA ($M_w \sim 22{,}000$) and $K_2S_2O_8$ and then subsequently heated to 60°C for 2 minutes. Upon heating to 60°C the potassium persulfate is expected to dissociate to generate sulfate radicals ($\cdot SO_4^-$), which can abstract hydrogen atoms from the PA surface and PVA to generate free radicals. Subsequent combination of the free radicals generated on the PA membrane surface and PVA macroradicals were reasoned to lead to grafting of PVA chains onto the PA membrane surface.[69] The PVA grafted membranes (via potassium persulfate initiation) were also tested with 500 mg/L NaCl solution at $\Delta P = 5$ bar demonstrating ~3–12% reduced L_p (5.3–$5.8\ \mathrm{L \cdot m^{-2} \cdot h^{-1} \cdot bar^{-1}}$) and increased salt rejection (98.6–98.9%) relative to the base membrane ($L_p^{Base} \sim 6\ \mathrm{L \cdot m^{-2} \cdot h^{-1} \cdot bar^{-1}}$, $R_o^{Base} = 98.3\%$).

Gamma radiation has been utilized to graft preformed polyvinylpyrrolidone (PVP) chains onto a laboratory synthesized PA TFC membrane surface.[20] In this approach, the PA membrane surface was first covered with a layer of aqueous solution of PVP (M_w of 8000, 24,000, or 58,000), followed by surface irradiation with gamma rays under nitrogen. Gamma irradiation of the aqueous polymer solution causes radiolysis of water to produce OH radicals that can abstract hydrogen atoms from PVP in the solution.[70] The resulting PVP macroradicals react and can covalently bond with the underlying PA membrane surface. It was shown that the PVP grafted membrane performance may be tuned by adjusting the polymer grafting conditions including the concentration of PVP in the aqueous solution, PVP molecular weight, and gamma ray dose. Membrane performance evaluation (with feed water containing 2000 mg/L NaCl at $\Delta P = 15.5$ bar) indicated that PVP surface grafting could achieve a significant increase in salt rejection (from 97.2% to 99.4%), while reducing L_p by only 2% (from 1.9 to $1.8\,\text{L}\cdot\text{m}^{-2}\cdot\text{h}^{-1}\cdot\text{bar}^{-1}$).

Change in Alteration of PA TFC membrane performance due to polymer grafting reported in the current literature is summarized in Table 10.4. The impact of grafted polymer chains on the membrane surface can vary significantly depending on the approach utilized for polymer grafting. Polymer grafting typically yields a low surface chain density due to steric hindrance. Thus, when a high surface density is important, which is the case for generating fouling-resistant membrane surfaces, graft polymerization is preferred over polymer grafting.

10.2.6. *Graft polymerization ("Grafting from")*

In graft polymerization, end-tethered polymer chains are grown from surface initiation sites via sequential monomer addition, thus enabling a high surface chain density. Surface initiation sites (e.g., free radical, peroxide, etc.) on a polyamide membrane can be generated by surface activation via a chemical initiator (e.g., redox,[71] thermal[72] or atom transfer radical polymerization (ATRP)[73]), UV irradiation[74] or low pressure plasma (LPP).[75] Among the above, redox-initiated free radical graft polymerization has been most frequently utilized to graft tethered hydrophilic polymer layers on a polyamide membrane surface.[71,76–80] Typical redox systems utilized to generate surface initiation sites (radicals) on PA membrane surfaces include, for example, potassium persulfate ($K_2S_2O_8$) paired with either sodium metabisulfite ($Na_2S_2O_5$),[76,77,79] sodium bisulfate ($NaHSO_3$),[80] or potassium metabisulfite ($K_2S_2O_5$).[19,71] In earlier work, methacrylic acid (MA) and poly(ethylene glycol) methacrylate (PEG-MA) were grafted onto two types of commercial

Table 10.4. Change in PA TFC membrane performance due to membrane surface modification by polymer grafting ("grafting to").

Ref	Surface modification technique	Base PA TFC membrane[a]	Feed water; filtration mode[b]	ΔP (bar)[c]	L_p(L· $m^{-2} \cdot h^{-1} \cdot bar^{-1}$)[d] Base	Modified	% Change in L_p^e	$R_o(\%)^f$ Base	Modified	Change in R_o (%)[g]	Comment
[59]	Surface grafting of poly (ethylene glycol) diglycidyl ether PEGDE[(1)]	Dow Filmtec XLE (BWRO)	2000 mg/L NaCl (25°C, pH 8); crossflow	10.3	6.9	5.8	−15%	99.1	99.1	+0.0	1% w/w PEGDE, MW 200[(2)]
						4.8	−31%		99.3	+0.2	15% w/w PEGDE, MW 200[(2)]
						4.4	−37%		99.4	+0.3	1% w/w PEGDE, MW 1000[(2)]
						3.9	−44%		99.5	+0.4	15% w/w PEGDE, MW 1000[(2)]
		Dow Filmtec NF90 (NF)			9.0	5.0	−45%	97.8	98.8	+1.0	1% w/w PEGDE, MW 1000[(2)]
						4.7	−48%		98.8	+1.0	15% w/w PEGDE, MW 1000[(2)]
[61]	Surface grafting of 3-monome thylol-5,5-dimethylh-ydantoin (MDMH)[(3)]	Laboratory-made PA TFC membrane	2000 mg/L NaCl (25°C); crossflow	15	5.5	6.3	+16%	95.6	94.3	−1.3	modified w/2 wt% MDMH[(4)]
						7.5	+43%		91.8	−3.8	modified w/ 20 wt% MDMH[(4)]

(*Continued*)

Table 10.4. (*Continued*)

Ref	Surface modification technique	Base PA TFC membrane[a]	Feed water; filtration mode[b]	ΔP (bar)[c]	L_p(L· m^{-2} · h^{-1} · bar^{-1})[d] Base	Modified	% Change in L_p^e	R_o(%)[f] Base	Modified	Change in R_o (%)[g]	Comment
[62]	Carbodiimide-mediated grafting of imidazolidinyl urea[(5)]	CSM RE4021-TL (ULPRO)	2000 mg/L NaCl (25°C); crossflow	15.5	6.8	5.1	−25%	97.1	96.9	−0.2	
[63]	Carbodiimide-mediated grafting of poly(ethyleneimine) (PEI)[(6)]	CSM RE4021-TL (ULPRO)	2000 mg/L NaCl (25°C)	15.5	6.3	4.7	−25%	97.0	98.1	+1.1	PEI, MW 70,000[(7)]
[64]	Carbodiimide-mediated grafting of poly (vinylamine) (PVAm)[(8)]	CSM RE4021-TE (ULPRO)	2000 mg/L NaCl (25°C, pH 7); crossflow	15.5	4.7	4.5	−4%	97.9	97.9	+0.1	0.05 w/v% PVAm[(8)]
						4.2	−10%		97.8	−0.1	0.10 w/v% PVAm[(8)]
						4.1	−12%		98.4	+0.5	0.20 w/v% PVAm[(8)]
						3.7	−22%		98.2	+0.4	0.25 w/v% PVAm[(8)]
[65]	Carbodiimide-mediated grafting of PEG derivatives[(9)]	Hangzhou Beidouxing Membrane Co., Ltd.	1500 mg/L NaCl (25°C); crossflow	10.5	4.8^h	3.0^h	−36%	96.7	96.4	−0.3	Jeffamine ED600[(9)]
						3.4^h	−28%		96.5	−0.2	Jeffamine ED2001[(9)]

Table 10.4. (*Continued*)

Ref	Surface modification technique	Base PA TFC membrane[a]	Feed water; filtration mode[b]	ΔP (bar)[c]	L_p(L· m^{-2} · h^{-1} · bar^{-1})[d] Base	L_p Modified	% Change in L_p^e	R_o(%)[f] Base	R_o Modified	Change in R_o (%)[g]	Comment
[68]	Surface grafting of poly(vinyl alcohol) (PVA) via glutaraldehyde(10)	Hangzhou Tianchuang Environmental Technology Co., LTD. (ULPRO)	500 mg/L NaCl (25°C, pH 7); crossflow	5	5.6	5.9	+5%	98.1	98.3	+0.2	[PVA]= 50 mg/L(10)
						6.2	+11%		98.5	+0.4	[PVA]= 100 mg/L(10)
						5.7	+4%		98.5	+0.4	[PVA]= 150 mg/L(10)
						5.4	−4%		98.4	+0.4	[PVA]= 200 mg/L(10)
[69]	Surface grafting of poly(vinyl alcohol) (PVA) via using a thermal initiator ($K_2S_2O_8$)(11)	Hangzhou Tianchuang Environmental Technology Co., LTD. (ULPRO)	500 mg/L NaCl (25°C, pH 7); crossflow	5	6.0	5.8	−3%	98.3	98.9	+0.6	[PVA]= 100 mg/L(11)
						5.6	−7%		98.8	+0.5	[PVA]= 300 mg/L(11)
						5.4	−10%		98.7	+0.4	[PVA]= 500 mg/L(11)
						5.3	−12%		98.6	+0.3	[PVA]= 1000 mg/L(11)
[20]	Surface grafting of poly(vinyl alcohol) (PVA) via gamma irradiation(12)	Laboratory synthesized PA TFC membrane	2000 mg/L NaCl (25°C); crossflow	15.5	1.9	1.8	−2%	97.2	99.4	+2.2%	2 w/v% PVP (MW 58,000) and 20 kGy radiation dose(13)

(*Continued*)

Table 10.4. *(Continued)*

[a]Commercial or laboratory synthesized base PA TFC membranes for surface modification

[b]Feed water condition and filtration mode used to evaluate the membrane L_p and R_o

[c]Transmembrane pressure at which membrane L_p and R_o were determined.

[d]Water permeability coefficient, L_p, defined as $L_p = J_v/(\Delta P - \Delta\pi_m)$, where J_v is the volumetric permeate flux and $\Delta\pi_m$ is the osmotic pressure difference between the membrane surface at the feed and permeate sides.

[e]Percentage change in L_p relative to base membrane defined as $\Delta L_p = \left[\left(L_p^{Modified} - L_p^{Base}\right)/L_p^{Base}\right] \cdot 100\%$

[f]Percent observed (nominal) salt rejection defined as $R_o = \left(1 - C_p/C_f\right) \cdot 100\%$

[g]Change in the observed percent salt rejection, relative to the base membrane, defined as $\Delta R_o = R_o^{Modified} - R_o^{Base}$

[h]L_p evaluated with deionized water.

(1)The base membranes were modified by exposing the active side of the membranes to aqueous solutions (40°C) of poly(ethylene glycol) diglycidyl ether (PEGDE) for 10 min, followed by rinsing with deionized water and a 25 vol% isopropyl solution.

(2)Reported performance is for a membrane modified with a 1% w/w or 15% w/w PEGDE solution prepared with PEGDE that has PEG MW of 200 or 1000.

(3)3-monomethylol-5,5-dimethylhydantoin (MDMH) solutions were prepared with 2-20 wt% MDMH, 0.1 wt% sodium dodecyl sulfate, and 3 wt% triethyl amine in deionized water where solution pH was adjusted to 9.5 using (+)-10-camphorsulfonic acid. The active sides of the base membranes were exposed to MDMH solutions (40°C) for 5 min, followed by draining the excess solution. The membranes were then cured at 103°C for 10 min.

(4)Reported performance for the base membrane modified with 2 or 20 wt% MDMH solution.

(5)The active side of the base membrane was first exposed to a *N*-(3-dimethylaminopropyl)-*N*-ethylcarbodiimide hydrochloride (EDC · HCl) solution for 10 min, and *N*-hydroxysuccinimide (NHS) was then added to the EDC · HCl solution. The membrane surface was exposed to a solution containing EDC · HCl and NHS in a dark place at room temperature for 16 h, prior to addition of a 5 wt% ethylenediamine (EDA) solution to a solution containing EDC · HCl and NHS. Subsequently, the membrane surface was exposed to an aqueous solution containing 5 wt% imidazolidinyl urea (IU), EDC · HCl, and NHS for 12 h and then rinsed with deionized water.

Table 10.4. (*Continued*)

(6)The base membrane active site was first exposed to an EDC · HCl solution for 10 min, and NHS was added to the EDC · HCl solution. After 15 minutes of membrane surface exposure to a solution containing EDC · HCl and NHS, an aqueous solution containing 5 wt% poly(ethyleneimine) (PEI) of MW in the range of 600–70,000 was added to the solution containing EDC · HCl and NHS. The membrane was then placed in a dark room at room temperature for 12 h, followed by rinsing with deionized water.

(7)The reported performance is for a base membrane modified with PEI of MW 70,000.

(8)The base membranes were modified following a previously reported method.(5),(6) Poly(vinylamine) (PVAm) was grafted onto the base membrane surface in the presence of EDC · HCl and NHS for 4 hours, followed by membrane rinsing with deionized water. Aqueous PVAm solutions (PVAm concentration in the range of 0.05–0.25 w/v%) were utilized for the polymer grafting reactions.

(9)Poly(ethylene glycol) (PEG) derivatives with amine end groups, Jeffamine ED600 and Jeffamine ED2001, were grafted onto the base membrane surface via the carbodiimide-induced method. The base membranes were immersed in 0.1 wt% EDC aqueous solution at 4°C for 3 h, followed by rinsing with deionized water. The membranes were then placed in an aqueous solution containing ED600 or ED2001 at 4°C for 24 h and rinsed with deionized water.

(10)The base membranes were sequentially treated with an aqueous solution of glutaraldehyde (GA) and an aqueous poly(vinyl alcohol) (PVA) solution. First, the base membrane surface was exposed to a 0.04 wt% GA aqueous solution (25°C, pH 3) for 5 min, and the excess solution was drained from the membrane surface afterwards. Subsequently, the membranes were rinsed with deionized water and dried in air to remove surface moisture. The GA-coated membrane surface was subsequently exposed to an aqueous PVA solution (PVA concentration in the range of 50–200 mg/L, 25°C, pH 3) for 2 min, followed by draining the excess solution and membrane surface drying under hot air flow (50°C).

(11)The base membrane active side was exposed to an aqueous solution (25°C, pH 7.6) of 1 wt% potassium persulfate ($K_2S_2O_8$) and PVA (MW 22,000; concentration in the range of 100–1000 mg/L) for 2 min, followed by draining the excess solution and air drying the membrane at the room temperature. Subsequently, the membrane surface was dried under hot air flow (60°C) for 2 min.

(12)The base membrane active side was exposed to an aqueous poly(vinyl alcohol) (PVA) solution (MW 8,000, 24,000, or 58,000; concentration up to 5 w/v%) in a glass radiation tube. The membrane and grafting solutions were degassed and irradiated by γ-ray at a radiation dose (up to 100 kGy).

(13)Reported membrane performance is for a base membrane modified with a 2 w/v% PVP (MW 58,000) solution and 20 kGy irradiation dose.

BWRO membranes (Hydranautics CPA2 and Dow BW30) using a $K_2S_2O_8/Na_2S_2O_5$ redox pair.[77] Surface-initiated graft polymerization reactions were carried out by immersing the base membrane samples in aqueous MA (10%) and PEG-MA (16%) solutions containing the redox initiators for periods of 10–60 minutes and 15–30 minutes, respectively. Membrane performance evaluation (with feed water containing 1500 mg/L NaCl at $\Delta P = 15.5$ bar) showed that MA surface graft polymerization reduced the base membrane water permeability coefficient ($L_p^{CPA2} = 2.3\ \mathrm{L{\cdot}m^{-2}{\cdot}h^{-1}{\cdot}bar^{-1}}$) by ~9–36%, and that increasing the reaction time of MA graft polymerization led to a greater L_p reduction. Graft polymerization of PEG-MA onto the BW30 base membrane ($L_p^{BW30} = 2.7\,\mathrm{L \cdot m^{-2} \cdot h^{-1} \cdot bar^{-1}}$) led to a relatively small reduction in L_p(~5%). Graft polymerization of MA and PEG-MA also impacted salt rejection of the base membranes, which either increased (from 95.5% to 96.9–98% for CPA2) or decreased (from 94.8% to 94.2–94.6% for BW30) to various degrees (Table 10.5).

In recent years, there has been growing interest in utilizing surface tethered zwitterionic polymer layers due to their effectiveness in reducing membrane fouling.[76,80] For example, a low pressure PA TFC RO membrane (CSM RE4021-TE) was modified by redox-initiated graft polymerization of N,N'-dimethylaminoethyl methacrylate (DMAEMA) using a $K_2S_2O_8/Na_2S_2O_5$ redox pair, followed by a surface quarternization reaction with 3-bromopropionic acid to obtain zwitterionic poly(carboxybetaine methacrylate) (PCBMA) chains grafted onto the membrane surface.[76] It was reported that permeate flux decline after 4 hour filtration of BSA and LYS solutions (1000 mg/L BSA or LYS + 2000 mg/L NaCl at pH 6.8) was reduced by only ≤10% due to the surface grafted PCBMA layer (evaluated at the same initial flux as the base membrane). However, permeate flux recovery after membrane cleaning with DI water flushing (for 30 minutes) for the PCBMA grafted membranes was 92–95%, whereas only 71–74% of initial permeate flux recovery was achieved for the base membrane. It was also reported that

Table 10.5. Change in PA TFC membrane performance due to membrane surface modification by graft polymerization.

Ref	Surface modification technique	Base PA-TFC membrane[a]	Feed water; filtration mode[b]	ΔP (bar)[c]	L_p(L· m^{-2} · h^{-1} · bar^{-1})[d] Base	Modified	% Change in L_p^e	R_o(%)[f] Base	Modified	Change in R_o (%)[g]	Comment
[76]	Redox-initiated graft polymerization of DMAEMA (via $K_2S_2O_8/Na_2S_2O_5$ redox initiators), followed by surface quaternization reaction with 3-BPA to obtain PCBMA[(1)]	CSM RE4021-TE (BWRO)	2000 mg/L NaCl (25°C); crossflow	15	4.7	5.1	+9%	97.9	98.0	+0.1	TE-PDMAEMA5 ([DMAEMA] = 0.05 M)[(2)]
						5.7	+22%		98.0	+0.1	TE-PDMAEMA10 ([DMAEMA] = 0.1 M)[(2)]
						5.3	+13%		97.8	−0.1	TE-PDMAEMA20 ([DMAEMA]=0.2 M)[(2)]
						5.6	+19%		97.1	−0.8	TE-PDMAEMA30 ([DMAEMA]=0.3 M)[(2)]
						5.8	+23%		97.9	No change	TE-PCBMA obtained after surface quaternization of TE-PDMAEMA10 membrane[(3)]

(*Continued*)

Table 10.5. *(Continued)*

Ref	Surface modification technique	Base PA-TFC membrane[a]	Feed water; filtration mode[b]	ΔP (bar)[c]	L_p(L· m^{-2} · h^{-1} · bar^{-1})[d] Base	Modified	% Change in L_p^e	R_o(%)[f] Base	Modified	Change in R_o (%)[g]	Comment
[77]	Redox-initiated graft polymerization of MA and PEG-MA (via $K_2S_2O_8$/ $Na_2S_2O_5$ redox initiators)[(4)]	Hydranautics CPA2 (BWRO)	1500 mg/L NaCl	15.5	2.3	2.1	−9%	95.5	97.5	+2.0	10% MA (10 min)[(4)]
						2.1	−9%		98.0	+2.5	10% MA (20 min)[(4)]
						1.6	−30%		97.8	+2.3	10% MA (40 min)[(4)]
						1.5	−36%		96.9	+1.4	10% MA (60 min)[(4)]
		Dow Filmtec BW30 (BWRO)			2.7	2.6	−5%	94.8	94.6	−0.2	16% PEGMA (15 min)[(4)]
						2.6	−5%		94.2	−0.6	16% PEGMA (30 min)[(4)]
[80]	Redox-initiated graft polymerization of SVBP (via $K_2S_2O_8$/$NaHSO_3$ redox initiators)[(5)]	Dow Filmtec XLE (BWRO)	2000 mg/L NaCl (25°C, pH 6.8); crossflow	15	6.1	4.7	−23%	98.0	99.6	+1.6	The graft polymerization reaction was carried out for 18 h[(5)]

Table 10.5. (*Continued*)

Ref	Surface modification technique	Base PA-TFC membrane[a]	Feed water; filtration mode[b]	ΔP (bar)[c]	L_p(L· m^{-2} · h^{-1} · bar^{-1})[d] Base	Modified	% Change in L_p^e	R_o(%)[f] Base	Modified	Change in R_o (%)[g]	Comment
[78]	Redox-initiated graft co-polymerization of SPMA and MBA (via Ce(IV)/PVA redox initiators)[(6)]	Laboratory synthesized PA-TFC membrane	2000 mg/L NaCl; crossflow	17.2	3.4	3.2	−4%	97.0	97.5	+0.5	[SPMA]=20 mM, [MBA]=9.5 mM, [PVA]=10 mM[(7)]
						2.3	−32%		95.0	−2.0	[SPMA]=4 mM, [MBA]=1.9 mM, [PVA]=20 mM[(7)]
						2.6	−24%		96.0	−1.0	[SPMA] = 8 mM, [MBA] = 1.9 mM, [PVA] = 20 mM[(7)]
						3.7	+9%		94.0	−3.0	[SPMA] = 28 mM, [MBA] = 13 mM, [PVA] = 10 mM[(7)]
						1.2	−64%		94.0	−3.0	[SPMA] = 20 mM, [MBA] = 9.5 mM, [PVA] = 0 mM[(7)]
[79]	Redox-initiated graft polymerization of NIPAm and AA (via $K_2S_2O_8/Na_2S_2O_5$ redox initiators)[(8)]	Development Center of Water Treatment Technology (Hangzhou, China)	500 mg/L NaCl (25°C, pH7); crossflow	5	5.6	6.6	+16%	98.2	98.0	−0.2	NIPAm GP period=30 min[(9)]
						7.1	+26%		97.6	−0.6	NIPAm GP period=60 min[(9)]
						6.4	+13%		97.2	−1.0	NIPAm GP period=90 min[(9)]
						5.5	−2%		96.8	−1.4	NIPAm GP period=120 min[(9)]

(*Continued*)

Table 10.5. (*Continued*)

Ref	Surface modification technique	Base PA-TFC membrane[a]	Feed water; filtration mode[b]	ΔP (bar)[c]	L_p(L· m^{-2} · h^{-1} · bar^{-1})[d] Base	Modified	% Change in L_p^{e}	R_o(%)[f] Base	Modified	Change in R_o (%)[g]	Comment
						6.6	+16%		98.0	−0.2	NIPAm GP period=60 min, AA GP period=30 min[(10)]
						6.1	+8%		98.3	+0.1	NIPAm GP period=60 min, AA GP period=60 min[(10)]
						5.8	+4%		98.6	+0.4	NIPAm GP period=60 min, AA GP period=90 min[(10)]
						5.4	−4%		98.8	+0.6	NIPAm GP period=60 min, AA GP period=120 min[(10)]
[71]	Redox-initiated concentration polarization enhanced graft polymerization of METMAC, SPE, and GMS (via $K_2S_2O_8/K_2S_2O_5$ redox initiators)[(11)]	Hydranautics ESPA1 (LPRO)	1500 m/L NaCl; crossflow				−20% to −40%	95.0	95.0	No change	METMAC (positively charged monomer)[(11)]
									96.5	+1.5	SPE (zwitterionic monomer)[(11)]
									96.5	+1.5	GMS (negatively charged monomer)[(11)]

Table 10.5. (*Continued*)

Ref	Surface modification technique	Base PA-TFC membrane[a]	Feed water; filtration mode[b]	ΔP (bar)[c]	L_p(L· m^{-2} · h^{-1} · bar^{-1})[d] Base	Modified	% Change in L_p^e	R_o(%)[f] Base	Modified	Change in R_o (%)[g]	Comment
[72]	Thermal-induced graft polymerization of ADMH (via AIBA thermal initiator)(12)	CSM RE4021-TL (ULPRO)	2000 mg/L NaCl (25°C); crossflow	15	10.1	11.1	+10%	96.6	95.8	−0.8	GP period=10 min(12)
						12.3	+22%		91.8	−4.8	GP period=100 min(12)
[84]	SI-ATRP of MEDSAH (via BIBB initiator)(13)	Nitto Denko ES20 (ULPRO)	500 mg/L NaCl; crossflow	7.5	6.7	4.7	−31%	97.3	~96.2	−1.1	GP period=10 min(13)
						4.3	−37%		~96.6	−0.7	GP period =20 min(13)
						4.1	−39%		~95.3	−2.0	GP period=30 min(13)
						3.1	−54%		~96.0	−1.3	GP period=60 min(13)
						2.2	−67%		~93.5	−3.8	GP period=120 min(13)
[73]	SI-ATRP of MPC (via BIBB initiator)(14)	Laboratory synthesized PA-TFC membrane	500 mg/L NaCl	7.5	1.5	0.6	−63%	~96	~90	−6.0	GP period=120 min(14)

(*Continued*)

Table 10.5. (*Continued*)

[a]Commercial or laboratory synthesized PA-TFC base membranes for surface modification

[b]Feed water condition and filtration mode at which membrane L_p and R_o were determined

[c]Transmembrane pressure at which membrane L_p and R_o were determined

[d]Water permeability coefficient, L_p, defined as $L_p = J_v/(\Delta P - \Delta\pi_m)$ where J_v is volumetric permeate flux and $\Delta\pi_m$ is the osmotic pressure difference between the membrane surface at the feed and permeate sides.

[e]Percentage change in L_p relative to the base membrane defined as $\Delta L_p = \left[\left(L_p^{Modified} - L_p^{Base}\right)/L_p^{Base}\right] \cdot 100\%$

[f]Percent observed (nominal) salt rejection defined as $R_o = (1 - C_p/C_f) \cdot 100\%$

[g]Change in percent salt rejection, R_o, relative to the base membrane defined as $\Delta R_o = R_o^{Modified} - R_o^{Base}$

(1)The active sides of the base membranes were exposed to a solution (30°C) containing 0.05–0.3 mol/L of N,N′-dimethylaminoethyl methacrylate (DMAEMA) and the redox initiators (i.e.,$K_2S_2O_8$ and $Na_2S_2O_5$) for 1 h to graft poly(DMAEMA) (PDAMEMA) chains onto the membrane surface. The modified membranes were dried, and the membrane active surfaces were subsequently exposed to a solution (30°C) containing 3-bromopropionic acid (3-BPA) for 2 days to obtain surface grafted zwitterionic carboxybetaine methacrylate (CBMA) polymer chains.

(2)The PDMAEMA grafted base TE membranes synthesized using DMAEMA solutions of concentrations 0.05 M, 0.1 M, 0.2 M, and 0.3 M were labelled as TE-PDMAEMA5, TE-PDMAEMA10, TE-PDMAEMA20, and TE-PDMAEMA30, respectively.

(3)The TE-PDMAEMA10 membrane was further modified with 3-BPA to obtain the TE-PCBMA membrane.

(4)Graft polymerization of methacrylic acid (MA) and poly(ethylene glycol) methacrylate (PEGMA) were carried out by immersing the base membranes in aqueous solutions of 10% MA and 16% PEGMA (10–20% monomer concentration) containing the redox initiators ($K_2S_2O_8$ and $Na_2S_2O_5$) for 10–60 min and 15–30 min, respectively.

(5)The base membranes were modified by immersion in an aqueous solution of 4-(2-sulfoethyl)-1-(4-vinylbenzyl) (pyridinium betaine) (SVBP) (2 mmolL^{-1}), $K_2S_2O_8$ (1 mmolL^{-1}) and $NaHSO_3$ (1 mmolL^{-1}) at 25°C for up to 18 h. The reported performance is for the base membrane that was graft polymerized with SVBP for 18 h.

(6)The base membrane surfaces were modified via Ce(IV)-PVA redox system initiated free radical graft co-polymerization of 3-sulphopropyl methacrylate (SPMA) and methylene-bis-acrylamide (MBA). The base membrane active side was exposed to aqueous solutions containing SPMA, MBA, poly(vinyl alcohol) (PVA), and ceric ammonium nitrate in HNO_3 for 15 min, followed by drying at 50°C for 10 min and rinsing with deionized water.

(7)Reported performances for base membranes modified with co-polymers of SPMA and MBA, synthesized at various concentrations of SPMA, MBA, and PVA as indicated in the table.

(8)Graft polymerization reactions of N-isopropylacrylamide (NIPAm) and acrylic acid (AA) were performed sequentially as follows. The base membranes were immersed in aqueous solutions (30°C) containing 2 wt% NIPAm, 0.5 wt% AA and redox initiators ($K_2S_2O_8$ and $Na_2S_2O_5$) for 30–120 min. The poly(NIPAm) (PNIPAm) grafted membranes were rinsed with deionized water for at least 12 hours and subsequently immersed in aqueous AA monomer solutions (30°C) for 30–120 min. The modified membranes were thoroughly rinsed with deionized water.

(9)Reported performances are for base membranes modified via surface graft polymerization (GP) of NIPAm at the indicated NIPAm GP reaction periods (30–120 min).

Table 10.5. (*Continued*)

(10) Reported performances for base membranes modified via surface graft polymerization of NIPAm (reaction period of 60 min), followed by surface graft polymerization (GP) of AA at the indicated AA GP reaction periods (30–120 min).

(11) Surface modification of the base membranes were carried out inside a laboratory scale crossflow membrane cell where the base membrane surfaces were contacted with solutions of three monomers, [2-(methacryloyloxy)ethyl]trimethylammonium chloride (METMAC), [2-(methacryloyloxy)ethyl]dimethyl-(3-sulfopropyl)ammonium (SPE), and glycidyl methacrylate (GMA), containing $K_2S_2O_8/K_2S_2O_5$ redox initiators, which were delivered to the membrane cell at pressure of 20 bar while keeping the retentate line closed. The concentrations of the monomers and redox initiators and graft polymerization period employed for graft polymerization (GP) reactions of METMAC, SPE, and GMA were as follows:

GP of METMAC: [METMAC] = 30 mM, $[K_2S_2O_8]$ = 2 mM, $[K_2S_sO_5]$ = 2 mM, GPperiod = 35 min

GP of SPE: [SPE] = 65 mM, $[K_2S_2O_8]$ = 2mM, $[K_2S_sO_5]$ = 2 mM, GPperiod = 60 min

GP of GMA: [GMA] = 2.5 mM, $[K_2S_2O_8]$ = 2 mM, $[K_2S_sO_5]$ = 4 mM, GPperiod = 30 min

It is noted that the GP of GMA was followed by conversion of the epoxy groups of GMA to negatively charged sulfonate groups via reaction with 10% Na_2SO_3 in 15% aqueous isopropyl solution at 35–40°C for 12–16 h. The resulting membrane surfaces were grafted with polymers of sulfonated glycidyl methacrylate (GMS).

(12) The base membrane active side was exposed to an aqueous solution containing 0.02 wt% 2,2-azobis(isobutyramidine) dihydrochloride (AIBA) for 15 min, followed by surface coating with aqueous solutions of 3-allyl-5,5-dimethylhydantoin (ADMH) monomer (5 wt%). Graft polymerization was carried out at 70°C for the indicated periods (10–100 min).

(13) The base membranes were modified by immersion in aqueous solutions of 3-aminopropyltrimethoxysilane (APTES), followed by drying in a vacuum chamber at room temperature for 10 min. Subsequently, the membranes were immersed in a hexane solution of 2-bromoisobutyryl bromide (BIBB) at room temperature for 1 min to immobilize BIBB initiator molecules on the membrane surface. The BIBB-immobilized membranes were then immersed in a mixture of 14 mL of DI water and methanol (1:1, v/v) containing 10 mmol of [(2-methacryloyloxy)ethyl]dimethyl[3-sulfopropyl]ammonium hydroxide (MEDSAH) and 0.8 mmol of ascorbic acid. After bubbling nitrogen through the mixture for 10 min, 4 mL of DI water/methanol (1:1, v/v) containing 0.02 mmol of copper (II) bromide ($CuBr_2$) and 0.04 mmol of tris(2-pyridylmethyl)amine (TPMA) was added to initiate the ATRP reaction. The solution mixture was then stirred for the indicated polymerization time.

(14) The base membrane surface was first modified with diethanolamine (DEA) by immersing the membrane in an aqueous solution containing 2 wt% DEA, 2 wt% triethylamine, and 4 wt% (±)-10-camphorsulfonic acid for 5 min, and then drying at 80°C for 10 min. The DEA-modified membrane was then immersed in 20 mL of a hexane solution containing 4.2 mmol of dehydrated pyridine and 4.2 mmol of 2-bromoisobutyryl bromide (BIBB) for 1h to immobilize BIBB initiators on the membrane surface. Subsequently, the BIBB-immobilized base membrane was immersed in a solution containing 18 mL of deionized water/methanol (1:4 v/v) containing 2-methacryloyloxyethyl phosphorylcholine (MPC) and bubbled with nitrogen for 15 min, followed by addition of 2 mL of deionized water/methanol (1:4, v/v) containing 0.02 mmol of 2,2′-bipyridine, 0.002 mmol of $CuBr_2$, and 0.02 mmol of ascorbic acid was added to the solution to initiate the ATRP reaction under constant stirring for a period of 120 min.

the surface grafted PCBMA layer increased the base membrane L_p ($4.7\,\text{L} \cdot \text{m}^{-2} \cdot \text{h}^{-1} \cdot \text{bar}^{-1}$) by 23% while maintaining salt rejection at 97.9% (evaluated with feed water containing 2000 mg/L NaCl at $\Delta P = 15$ bar).

It has been shown that surface tethered zwitterionic polymers, such as poly (4-(2-sulfoethyl)-1-(4-vinylbenzyl) pyridinium betaine) (PSVBP), which exhibit increased swelling with increased salt concentration, could allow for improve membrane cleaning by flushing with a solution of salinity higher than that of the raw feed water being treated.[80] For example, surface tethered PSVBP chains synthesized onto a PA-TFC brackish water RO (BWRO) membrane (Dow Filmtec XLE) surface, via redox-initiated graft polymerization increased salt rejection from 98.0% up to 99.6% but at the cost of ~23% permeate flux reduction (reduced from 6.1 to $4.7\,\text{L} \cdot \text{m}^{-2} \cdot \text{h}^{-1} \cdot \text{bar}^{-1}$), evaluated with 2000 mg/L NaCl feed water at $\Delta P = 15$ bar. Membrane fouling tests with BSA showed that the surface tethered PSVBP layer did not reduce BSA fouling appreciably after filtration of a BSA solution over 100 hours (tested with a solution containing 100 mg/L BSA and 2000 mg/L NaCl at pH 6.8). BSA solution filtration was accompanied by ~24% and ~28% permeate flux reduction for the PSVBP grafted membrane and the native XLE membrane, respectively. Subsequent DI water flushing (for 2 hours) of the BSA fouled membranes restored the PSVBP membrane to ~83% of its initial permeability, while the fouled XLE membrane permeability was not recovered. Additional membrane cleaning (for 2 hours) with alternating cycles of DI water and saline water (1.5 M NaCl solution) flushing restored up to ~90% of the initial permeability for the PSVBP grafted membrane, while the native XLE membrane permeability declined further from 72% to 57% of the initial membrane permeability. The latter result was attributed to dehydration of BSA after exposure to a high salinity water that led to the aggregation and precipitation of BSA molecules and subsequent formation of a dense cake layer on the membrane surface. The above results demonstrated that the tethered

PSVBP layer could be beneficial for increasing membrane cleaning efficiency. However, it is noted that with the tethered PSVBP layer the initial membrane permeability (which remained at ~90%) was not restored even after a very long membrane cleaning period (4 hours).

A temperature-responsive polymer has also been utilized to improve RO membrane cleaning efficiency.[79] Redox-initiated graft polymerization reactions of N-isopropylacrylamide (NIPAm) and acrylic acid (AA) were performed sequentially onto a commercial PA-TFC RO membrane (Development Center of Water Treatment Technology in Hangzhou, China) by exposing the membrane surface first to an aqueous NIPAM monomer solution (30°C) containing redox initiators ($K_2S_2O_8/Na_2S_2O_5$), followed by exposure to an aqueous AA monomer solution (30°C) also containing the same redox initiators. Membranes that were surface grafted with poly(N-isopropylacrylamide) (PNIPAm) led to salt rejection reduction from 98.2% to 96.8–98.0% and up to 26% increased L_p from $5.6\,\mathrm{L \cdot m^{-2} \cdot h^{-1} \cdot bar^{-1}}$ up to $7.1\,\mathrm{L \cdot m^{-2} \cdot h^{-1} \cdot bar^{-1}}$ (membrane performance was tested with 500 mg/L NaCl feed water at $\Delta P =$ 5 bar). Subsequent grafting of poly(acrylic acid) (PAA), however, increased salt rejection up to 98.6% (which is 0.4% greater than the base membrane) while maintaining 4% increased L_p relative to the base membrane. Membrane fouling tests conducted with BSA (at the same initial permeate flux), showed that after 40 h filtration of a BSA solution (200 mg/L BSA + 500 mg/L NaCl at pH 6.8) permeate fluxes for the PNIPAm and PAA grafted membranes (PA-g-NIPAm60-AA90) and the base PA membrane were reduced by ~45% and ~21%, respectively. Membrane cleaning effectiveness for the PA-g-NIPAm60-AA90 and base PA membranes were assessed to undergo essentially the same level of permeate flux reduction (40% of initial permeate flux) due to BSA fouling. Membrane cleaning, performed by membrane soaking in DI water (45°C) for 3 h, followed by DI water (25°C) flushing for 30 min showed that membrane cleaning of the NIPAm60-AA90, recovered 93% of initial membrane permeability;

this was greater than the 82% permeability recovery achieved for the base PA membrane. The higher permeability recovery achieved for the NIPAm60-AA90 membrane was attributed to the PNIPAm chain segments that collapse upon heating above the critical solution temperature of 32°C. It was also reported that the surface grafted PNIPAm and PAA chains improved membrane resistance to chlorine. Exposure to a sodium hypochlorite solution (1000 mg/L NaClO at pH 7 and 25°C), for up to 5 hours, resulted in salt rejection of the base membrane being significantly reduced from 98.2% to 75.8%. In contrast, salt rejection of the NIPAm60-AA90 membrane decreased from 98.5% to 96.5%. However, the scalability of the above approach for commercial membrane fabrication may present a formidable challenge given the requirement for two different polymerization reactions, and a lengthy reaction period.

Improved chlorine resistance was also reported for a PA membrane surface modified with a copolymer of 3-sulphopropyl methacrylate (SPMA) and methylene-bis-acrylamide (MBA) synthesized via redox-initiated graft polymerization using a Ce(IV)/polyvinyl alcohol (PVA) redox system.[78] Ce(IV) rapidly generates radical sites on PVA, which then initiates copolymerization of SPMA and MBA. Polymerization of SPMA in the presence of MBA (which served as a cross-linker) yielded a cross-linked copolymer layer of PVA-g-poly(SPMA-co-MBA) on the PA membrane surface. The surface grafted PVA-g-poly(SPMA-co-MBA) layer, synthesized at various concentrations of PVA, SPMA, and MBA, however, resulted in salt rejection decrease from 97.0% to 94.0–96.0%, except for a case in which salt rejection was slightly increased to 97.5% (membrane performance was evaluated with and aqueous 2000 mg/L NaCl solution at $\Delta P = 17.2$ bar). After graft polymerization the membrane L_p declined by up to 64% (base membrane L_p was $3.4\,\mathrm{L \cdot m^{-2} \cdot h^{-1} \cdot bar^{-1}}$). At the optimal graft polymerization reaction conditions, salt rejection of the synthesized membrane (TFC-M1) increased slightly to 97.5% and L_p was reduced (by 4%) to $3.2\,\mathrm{L \cdot m^{-2} \cdot h^{-1} \cdot bar^{-1}}$ relative to the base PA membrane. Also, it was shown that the surface grafted PVA-g-poly(SPMA-co-MBA) layer retarded PA membrane oxidation due to chlorine attack and the associated PA membrane degradation. Membrane exposure to 1000 mg/L NaOCl aqueous solution led to permeate flux increase by up to a factor of 4.5 base membrane of and reduced salt rejection from 97% to 30% after 6 h exposure period. The same chlorine exposure period resulted in 20% increase in permeate flux and salt rejection reduction from 96% to 65% for the TFC-M1 membrane.

It is typically difficult to achieve a high degree of surface grafting with redox initiators due to the slow redox-initiation step which does not preferentially take place at the membrane surface (i.e., homopolymerization also occurs in the bulk solution).[81] It was proposed that a higher rate of surface graft polymerization can be achieved by carrying out the polymerization reaction at a relatively moderate monomer concentration, while the monomer solution is being filtered through the membrane.[81] Typical RO/NF membranes would typically reject monomers and initiators, which would then accumulate at the membrane surface resulting in concentration polarization. The increased monomer and initiator concentrations at the membrane surface would thus lead to the polymerization reaction occurring preferentially near the membrane surface. Such a concentration polarization enhanced graft polymerization approach was reported for surface initiated graft polymerization of [2-(methacryloyloxy)ethyl]dimethyl-(3-sulfopropyl)ammonium (SPE), sulfonated glycidyl methacrylate (GMS), and [2-(methacryloyloxy)ethyl]trimethylammonium chloride (METMAC) onto a commercial LPRO membrane (Hydranautics ESPA1) using a redox initiator pair ($K_2S_2O_8/K_2S_2O_5$).[71] Membrane modification with the above polymers resulted in ∼20–40% reduced L_p, while salt rejection (assessed for 1500 mg/L NaCl feed water) either remained the same (95%) or increased to 96.5%. The above was attributed to the surface grafted polymer plugging of the polyamide membrane defects.

A thermal initiator, 2,2′-azobis(isobutyramidine) dihydrochloride (AIBA), was also utilized to carry out free radical graft polymerization of 3-allyl-5,5-dimethylhydantoin (ADMH) onto a commercial ULPRO membrane (CSM RE4021-TL).[72] It was reported that upon exposure to chlorine, surface grafted polymers of ADMH formed surface *N*-halamine groups that exhibit antimicrobial property. Thus, it was suggested that the above polymer would also produce a surface coating layer that would protect the underlying polyamide active layer from chlorine attack. For example, chlorine exposure at 1500 ppm · h led to a 29% increase in L_p and 9% lower salt rejection for the base membrane. At the same chlorine exposure, the ADMH graft polymerized membrane had a lower L_p rise of 14% and about 4% reduced salt rejection. Following chlorine exposure, membrane biofouling propensity was tested by exposing the chlorine treated membranes to a bacterial suspension of *Escherichia coli*, followed by incubation at 37°C for 60 h. After 60 h exposure to microbial cells, the base membrane permeate flux decreased by ~31%, whereas only ~9% permeate flux reduction was observed for the ADMH polymer grafted membrane. It was reported that the resulting surface grafted polymer layer led to increased base membrane L_p ($10.1\,\mathrm{L \cdot m^{-2} \cdot h^{-1} \cdot bar^{-1}}$), which was up to a 22% increase for graft polymerization carried out over a period of 100 minutes (membrane performance was evaluated with an aqueous 2000 mg/L NaCl solution at $\Delta P = 15$ bar). However, post-grafting, the membrane L_p increase was accompanied by 4.8% reduction of salt rejection (from 96.6% to 91.8%).

As revealed by the above examples, free radical graft polymerization (FRGP) is a versatile approach that is applicable to various types of monomers and surface initiators. However, the main drawback of FRGP is a broad chain size or molecular weight distribution reflecting a highly polydisperse surface grafted polymer chains.[82] Controlled radical polymerization such as atom transfer radical polymerization (ATRP), on the other hand, enables better control over the surface grafted polymer chains length and enabling a narrower size distribution of surface grafted polymer chains relative to FRGP.[83] In order to carry out a surface-initiated ATRP (SI-ATRP) reaction from a polyamide membrane surface, an ATRP initiator such as 2-bromoisobutyryl bromide (BIBB) must be first immobilized onto the membrane surface.[73,84] BIBB is typically employed because its acyl halide groups can react readily with hydroxyl and amino groups that are present on the polyamide membrane surface.[84] However, the number of hydroxyl and amino groups present on a commercial PA-TFC membrane surface is typically too small to achieve a high density of surface immobilized ATRP initiators.[84] Thus, an additional step is required to increase the density of hydroxyl and/or amino groups on the PA membrane surface prior to surface immobilization of ATRP initiators. For example, 3-aminopropyltrimethoxysilane (APTES) was used to generate amine groups on the surface of a commercial PA-TFC ULPRO membrane (Nitto Denko ES20) to increase the density of immobilized BIBB initiators on the membrane surface.[84] Following initiator immobilization, poly[(2-methacryloyloxy)ethyl]dimethyl[3-sulfopropyl]ammonium hydroxide (pMEDSAH) was grafted onto the membrane surface via SI-ATRP. It was reported that the pMEDSAH grafted membrane (graft polymerization period $\geq$ 1 h) maintained its initial permeate flux over 20 h filtration of a saline solution (0.85 wt% NaCl) containing bacteria (*Sphingomonas paucimobilis*), thus demonstrating resistance to biofouling. In contrast, permeate flux for the base membrane dropped to below 40% of the initial permeate flux after bacteria solution filtration for a period of 20 hours. However, surface grafting of the pMEDSAH layer led to a significant reduction of membrane L_p by ~31–67% relative to the base membrane ($L_p^{base} \sim 6.7\,\mathrm{L \cdot m^{-2} \cdot h^{-1} \cdot bar^{-1}}$), while salt rejection decreased by from 97.3% to 93.5–96.6% (evaluated with aqueous 500 mg/L NaCl solution at $\Delta P = 7.5$ bar).

In order to increase the number density of ATRP initiators grafted onto the PA-TFC membrane surface (laboratory synthesized), it was reported that the membrane surface could be first treated with diethanolamine (DEA) to generate surface hydroxyl groups to immobilize BIBB initiator molecules onto the membrane surface.[73] BIBB initiators then served to graft poly(2-methacryloyloxyethyl phosphorylcholine) (pMPC) onto the base membrane surface via SI-ATRP.[73] The surface grafted pMPC layer was reported to be effective in reducing bacterial (*Sphingomonas* paucimobilis) fouling by ≤5% (by mass) relative to the base membrane (after filtration of the bacterial suspension for 14 h). However, the surface grafted pMPC layer led to deterioration of membrane performance. Surface grafting of pMPC chains resulted in membrane L_p reduction by ~63%($L_p^{base} \sim 1.5\,\mathrm{L \cdot m^{-2} \cdot h^{-1} \cdot bar^{-1}}$) relative to the base membrane, and also significantly reduced salt rejection (from ~96% to ~90% as assessed with aqueous 500 mg/L NaCl feed solution at $\Delta P = 7.5$ bar).

Examples of graft polymerization techniques for modifying PA-TFC RO membrane surfaces are summarized in Table 10.5. In general, it can be concluded that redox-initiated graft polymerization (performed with the persulfate initiator systems) was most effective for maintaining or improving RO membrane performance. However, when the redox initiators are present in the solution phase during graft polymerization, initiation of homopolymerization in solution can dominate over membrane surface-initiated graft polymerization. Thus, the surface density of grafted polymer chains is expected to be lower compared to approaches that generate surface initiation sites directly on the membrane surface prior to the monomer addition reaction.

Surface initiation sites on a PA membrane surface can also be generated by membrane surface exposure to an impinging stream atmospheric pressure plasma (APP). Surface activation via APP has been shown to be is effective in producing a high density of surface active sites, in a relatively short plasma exposure time (≪1 min[85,86]); this enables the formation of a high-density layer of tethered chains upon graft polymerization in a solution of the target monomer.[87,88] Previous studies on APP induced graft polymerization (APPIGP) have focused primarily on reducing membrane fouling propensity; but it has also been documented that membrane permeability and salt rejection are impacted by the base membrane choice.[87,89,90] Recently, it was demonstrated that surface tethered poly(acrylic acid) (PAA) layers synthesized onto PA-TFC membranes via the APPIGP (Fig. 10.6) can be used to tune seawater RO desalination performance.[90]

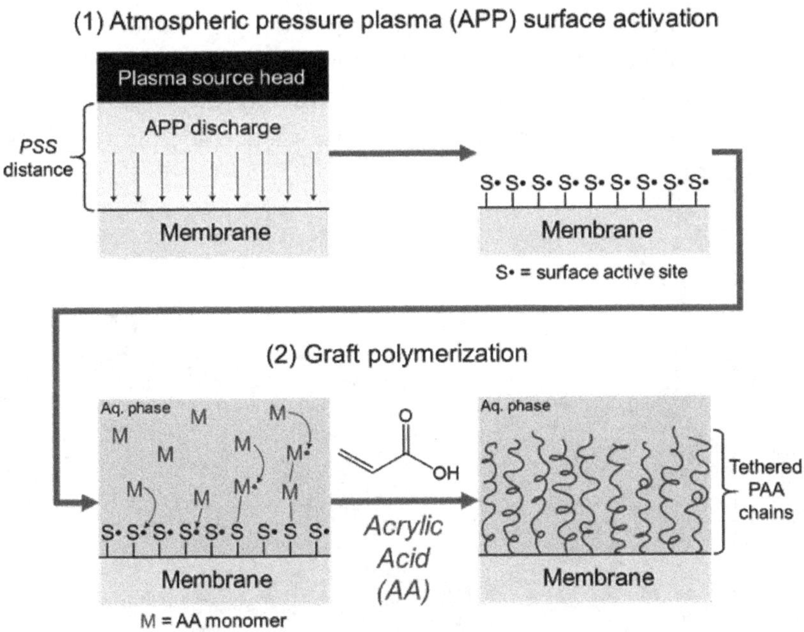

Fig. 10.6. Two-step process for membrane structuring with tethered polymer chains of (1) surface activation by plasma, followed by (2) membrane graft polymerization.

10.3. Salt Rejection-Permeability Tradeoff

The impact of membrane surface modification on PA-TFC membrane performance is generally reported in terms of changes in water permeability coefficient (or permeate flux) and nominal (observed) salt rejection of the PA-TFC membranes before and after PA surface modification. However, nominal salt rejection is not a definitive metric for membrane performance comparison given that salt rejection depends on permeate flux (which is a function of water permeability, transmembrane pressure, and crossflow velocity). The water permeability coefficient and salt rejection are typically evaluated at a fixed transmembrane pressure. However, since water permeability is generally reduced as a consequence of membrane surface modification, evaluation of membrane performance, pre- and post-modification, at the same transmembrane pressure would lead to the modified-membrane salt rejection being determined at a lower permeate flux and thus lower salt rejection[91] relative to the base membrane. When the salt rejection-permeability tradeoff is utilized to compare surface-modified PA-TFC membranes the results are diverse without a discernable indication of a trade-off relationship as illustrated in Fig. 10.7.

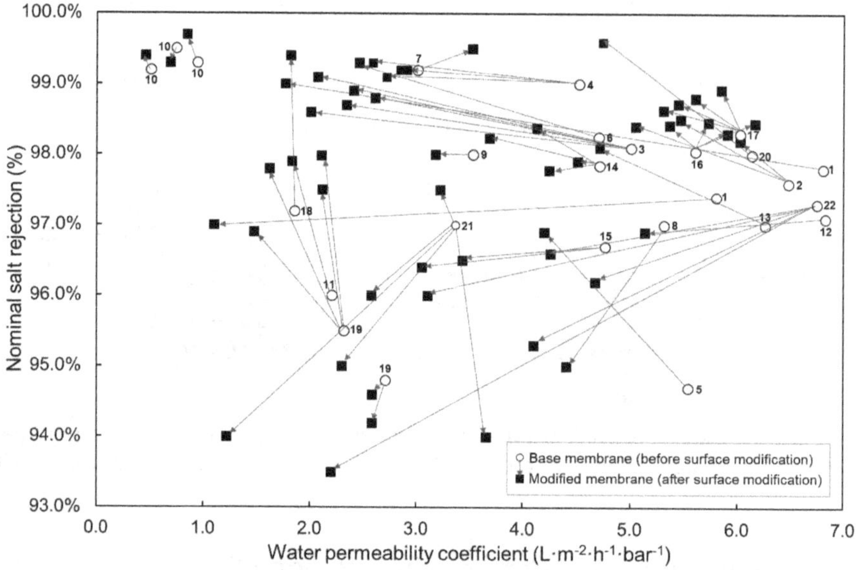

1. Louie et al., J. Membr. Sci. 280 (2006) 762-770. (coating)
2. Yu et al., Sep. Pur. Tech. 118 (2013) 285-293. (coating)
3. Ni et al., J. Membr. Sci. 451 (2014) 205-215. (coating)
4. Sagle et al., J. Membr. Sci. 240 (2009) 92-108. (coating)
5. Saeki et al., Colloids Surf., A 443 (2014) 171-176. (coating)
6. Ishigami et al., Sep. Pur. Tech. 99 (2012) 1-7. (LbL assembly)
7. Wang et al., J. Membr. Sci. 539 (2017) 403-411. (LbL assembly)
8. Yang et al., Chem. Mater. 23 (2011) 1263-1272. (iCVD)
9. Ozaydin-Ince et al., Thin Solid Films 539 (2013) 181-187. (iCVD)
10. Matin et al., Desalination 343 (2014) 128-139. (iCVD)
11. Shafi et al., Desalination 362 (2015) 93-103. (iCVD)
12. Xu et al., J. Membr. Sci. 435 (2013) 80-91. (polymer grafting)
13. Xu et al., Desalination 365 (2015) 398-406. (polymer grafting)
14. Wu et al., J. Membr. Sci. 495 (2015) 1-13. (polymer grafting)
15. Kang et al., Desalination 275 (2011) 252-259. (polymer grafting)
16. Hu et al., J. Membr. Sci. 501 (2016) 209-219. (polymer grafting)
17. Liu et al., Desalination 367 (2015) 11-20. (polymer grafting)
18. Liu et al., RSC Adv. 9 (2019) 17766. (polymer grafting)
19. Belfer et al., J. Membr. Sci. 139 (1998) 175-181. (graft polymerization)
20. Meng et al., J. Membr. Sci. 461 (2014) 123-129. (graft polymerization)
21. Rana et al., Desalination 375 (2015) 93-103. (graft polymerization)
22. Yang et al., J. Membr. Sci. 550 (2018) 332-339. (graft polymerization)

Fig. 10.7. Water permeability coefficient and nominal salt rejection values for base membranes and modified membranes reported by the studies that performed various types of membrane surface modification (the arrows show the direction at which water permeability coefficient and salt rejection changed from the base membrane to the modified membrane).

Comparisons of RO membranes in terms of water permeability and water/salt selectivity have been proposed in various studies for assessing improvements in membrane performance.[3,93] However, for PA-TFC membranes, accurate PA active layer thickness measurements are required (for determination of water permeability) which is a challenging task.[13] Thus, RO membrane performance comparison is often accomplished based on correlating the water permeability coefficient (note that L_p is the membrane water permeability which is dependent on the active layer thickness) with water/salt selectivity (evaluated as the L_p/B ratio).[92–94] However, membrane water and salt permeability coefficients can vary with operating and environmental conditions[95] (e.g., permeate flux, cross flow velocity, pressure, salinity, pH, temperature, and membrane conditioning), and thus the challenge for comparison of membrane performance data from different studies.

The assessment of water permeability and water/salt selectivity tradeoff for PA-TFC membranes nano-structured with a hydrophilic polymer (poly(acrylic) acid) was reported recently[96] for three different base PA-TFC RO membranes. Synthesis of these nanos-structured membranes followed the approach indicated in Fig. 10.6. Once created, the L_p and B parameter values for the membranes pre- and post-surface nano-structuring were determined using high salinity feed water (32,000 mg/L) at the same crossflow velocity and permeate flux, thereby comparing the performance of the membranes at the same level of concentration polarization. The L_p and B for the Base-PA1 were within the typical ranges for commercial BWRO membranes, whereas Base-PA2 and Base-PA3 membranes were seawater membranes having 29% and 33% lower L_p and 61% and 72% lower B, respectively, relative to Base-PA1 membrane. The water/salt selectivity (i.e., L_p/B ratio) was evaluated for the SNS-PAA-PA membranes as a function of L_p as shown in Fig. 10.8. Selectivity increased only for the SNS-PAA-PA1 membranes for which L_p reduction was lower than the reduction in B relative to the Base-PA1 membrane. Upon surface-nano-structuring, the Base-PA1 membrane selectivity improved by up to 56%, with a relatively low water permeability reduction ($\leq$5%). In contrast, about 41% water/salt selectivity decrease was observed for the SNS-PAA-PA2 membranes. A higher L_p reduction, in the range of 16–40%, was reported relative to the Base-PA2 base membrane. It is noted that selectivity of the SNS-PAA-PA2 membranes was maintained at about the same level as its base PA membrane. On the other

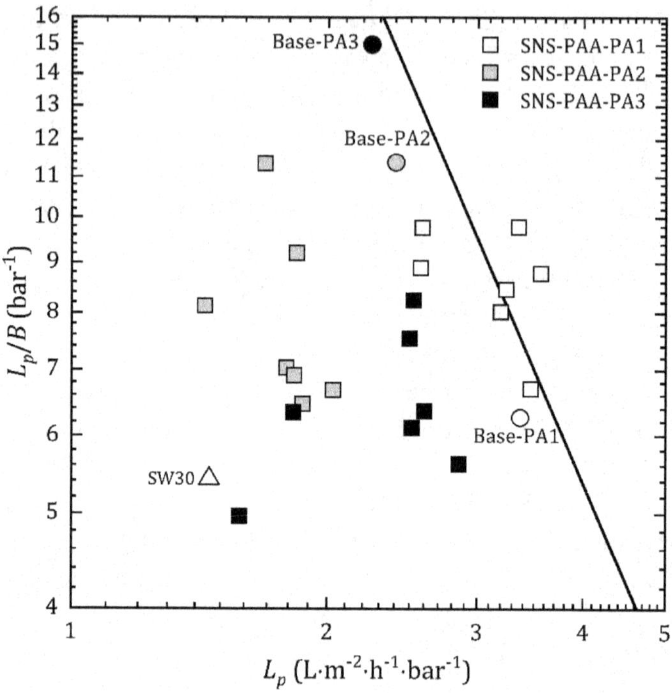

Fig. 10.8. Water permeability coefficient (L_p) and water/salt selectivity (i.e., L_p/B ratio) for polyamide base membranes (Base-PA1: white circle, Base-PA2: gray circle, and Base-PA3: black circle) and the respective SNS-PAA-PA membranes (SNS-PAA-PA1: white squares, SNS-PAA-PA2: gray squares and SNS-PAA-PA3: black squares), and the Dow Filmtec SW30 membrane (white triangle). The black solid line is the selectivity-permeability tradeoff for PA-TFC membranes proposed in Ref. 92. (Adapted from 96).

hand, a significant selectivity loss was noticed for the SNS-PAA-PA3 membranes associated with increased B value (~249%) relative to much lower L_p change (~19%–27%). The above study demonstrated that one can derive membranes of a given selectivity but with a range of water permeability, or membranes of a given permeability with a range of selectivity. This performance objective was achieved by fine-tuning RO membrane performance owing to the tethered PAA layers that were synthesized at different plasma treatment and graft polymerization conditions.

References

1. I. Sutzkover, D. Hasson, R. Semiat, Simple technique for measuring the concentration polarization level in a reverse osmosis system, *Desalination*, **131**(1–3) (2000) 117–127.
2. R. W. Baker, *Reverse Osmosis*, in *Membrane Technology and Applications*. 2004, John Wiley & Sons, Ltd. p. 191–235.
3. G. M. Geise *et al.*, Water permeability and water/salt selectivity tradeoff in polymers for desalination, *J. Membr. Sci.*, **369**(1–2) (2011) 130–138.
4. J. Lee *et al.*, Water vapor sorption and free volume in the aromatic polyamide layer of reverse osmosis membranes, *J. Membr. Sci.*, **425** (2013) 217–226.
5. V. Freger, Nanoscale heterogeneity of polyamide membranes formed by interfacial polymerization, *Langmuir*, **19**(11) (2003) 4791–4797.
6. T. Kamada *et al.*, Controlled surface morphology of polyamide membranes via the addition of co-solvent for improved permeate flux, *J. Membr. Sci.*, **467** (2014) 303–312.
7. T. Fujioka *et al.*, Rejection of small and uncharged chemicals of emerging concern by reverse osmosis membranes: The role of free volume space within the active skin layer, *Sep. Purif. Technol.*, **116** (2013) 426–432.
8. L. Lin *et al.*, Investigating the void structure of the polyamide active layers of thin-film composite membranes, *J. Membr. Sci.*, **497** (2016) 365–376.
9. K. Kosutic, L. Kastelan-Kunst, B. Kunst, Porosity of some commercial reverse osmosis and nanofiltration polyamide thin-film composite membranes, *J. Membr. Sci.*, **168**(1–2) (2000) 101–108.
10. H. Zhang, G. M. Geise, Modeling the water permeability and water/salt selectivity tradeoff in polymer membranes, *J. Membr. Sci.*, **520** (2016) 790–800.
11. S. H. Kim, S. Y. Kwak, T. Suzuki, Positron annihilation spectroscopic evidence to demonstrate the flux-enhancement mechanism in morphology-controlled thin-film-composite (TFC) membrane, *Environ. Sci. Technol.*, **39**(6) (2005) 1764–1770.
12. J. R. Werber, S. K. Bull, M. Elimelech, Acyl-chloride quenching following interfacial polymerization to modulate the water permeability, selectivity, and surface charge of desalination membranes, *J. Membr. Sci.*, **535** (2017) 357–364.
13. G. M. Geise, D. R. Paul, B. D. Freeman, Fundamental water and salt transport properties of polymeric materials, *Progress in Polymer Science*, **39**(1) (2014) 1–42.

14. M. C. Y. Wong *et al.*, Impact of liquid-filled voids within the active layer on transport through thin-film composite membranes, *J. Membr. Sci.*, **500** (2016) 124–135.
15. A. K. Ghosh *et al.*, Impacts of reaction and curing conditions on polyamide composite reverse osmosis membrane properties, *J. Membr. Sci.*, **311**(1–2) (2008) 34–45.
16. M. L. Lind *et al.*, Tailoring the structure of thin film nanocomposite membranes to achieve seawater RO membrane performance, *Environ. Sci. Technol.*, **44**(21) (2010) 8230–8235.
17. T. H. Lee *et al.*, Highly porous carbon nanotube/polysulfone nanocomposite supports for high-flux polyamide reverse osmosis membranes, *J. Membr. Sci.*, **539** (2017) 441–450.
18. W. Choi *et al.*, Tailor-Made Polyamide Membranes for Water Desalination, *ACS Nano*, **9**(1) (2015) 345–355.
19. R. Bernstein, S. Belfer, V. Freger, Improving performance of spiral wound RO elements by in situ concentration polarization-enhanced radical graft polymerization, *J. Membr. Sci.*, **405** (2012) 79–84.
20. C. Liu *et al.*, Tailor-made high-performance reverse osmosis membranes by surface fixation of hydrophilic macromolecules for wastewater treatment, *RSC Advances*, **9**(31) (2019) 17766–17777.
21. Y. L. Zhao *et al.*, Surface modification of polyamide reverses osmosis membrane by phosphonic acid group with improved performance, *J. Appl. Polym. Sci.*, **136**(2) (2019) 46931.
22. J. M. Gohil, P. Ray, A review on semi-aromatic polyamide TFC membranes prepared by interfacial polymerization: Potential for water treatment and desalination, *Sep. Purif. Technol.*, **181** (2017) 159–182.
23. W. Xie *et al.*, Polyamide interfacial composite membranes prepared from m-phenylene diamine, trimesoyl chloride and a new disulfonated diamine, *J. Membr. Sci.*, **403** (2012) 152–161.
24. B. Khorshidi *et al.*, A novel approach toward fabrication of high performance thin film composite polyamide membranes, *Sci. Rep.*, **6** (2016).
25. R. Nadler, S. Srebnik, Molecular simulation of polyamide synthesis by interfacial polymerization, *J. Membr. Sci.*, **315**(1–2) (2008) 100–105.
26. J. Jegal, S. G. Min, K. H. Lee, Factors affecting the interfacial polymerization of polyamide active layers for the formation of polyamide composite membranes, *J. Appl. Polym. Sci.*, **86**(11) (2002) 2781–2787.
27. Y. Jin, Z. H. Su, Effects of polymerization conditions on hydrophilic groups in aromatic polyamide thin films, *J. Membr. Sci.*, **330**(1–2) (2009) 175–179.
28. M. T. Khan *et al.*, Kinetic study of seawater reverse osmosis membrane fouling, *Environ. Sci. Technol.*, **47** (2013) 10884–10894.

29. H. C. Flemming *et al.*, Biofouling — the Achilles heel of membrane processes, *Desalination*, **113**(2–3) (1997) 215–225.
30. American Water Works Association and Knovel (Firm), *Microfiltration and ultrafiltration membranes for drinking water*, in *AWWA Manual M53*. 2005, American Water Works Association,: Denver, CO. p. xxii, 257 p.
31. M. F. A. Goosen *et al.*, Fouling of reverse osmosis and ultrafiltration membranes: A critical review, *Separation Science and Technology*, **39**(10) (2004) 2261–2297.
32. E. Bar-Zeev *et al.*, Transparent Exopolymer Particles: From Aquatic Environments and Engineered Systems to Membrane Biofouling, *Environ. Sci. Technol.*, **49**(2) (2015) 691–707.
33. Q. F. Zhang *et al.*, Effect of poly(vinyl alcohol) coating process conditions on the properties and performance of polyamide reverse osmosis membranes, *Desalination*, **379** (2016) 42–52.
34. J. S. Louie *et al.*, Effects of polyether-polyamide block copolymer coating on performance and fouling of reverse osmosis membranes, *J. Membr. Sci.*, **280**(1–2) (2006) 762–770.
35. A. C. Sagle *et al.*, PEG-coated reverse osmosis membranes: Desalination properties and fouling resistance, *J. Membr. Sci.*, **340**(1–2) (2009) 92–108.
36. Y. Zhou *et al.*, Surface modification of thin film composite polyamide membranes by electrostatic self deposition of polycations for improved fouling resistance, *Sep. Purif. Technol.*, **66**(2) (2009) 287–294.
37. J. Xu, X. S. Feng, C. J. Gao, Surface modification of thin-film-composite polyamide membranes for improved reverse osmosis performance, *J. Membr. Sci.*, **370**(1–2) (2011) 116–123.
38. Y. N. Kwon *et al.*, Surface modification of a polyamide reverse osmosis membrane for chlorine resistance improvement, *J. Membr. Sci.*, **415** (2012) 192–198.
39. J. Nikkola *et al.*, Surface modification of thin film composite RO membrane for enhanced anti-biofouling performance, *J. Membr. Sci.*, **444** (2013) 192–200.
40. S. C. Yu *et al.*, Improving fouling resistance of thin-film composite polyamide reverse osmosis membrane by coating natural hydrophilic polymer sericin, *Sep. Purif. Technol.*, **118** (2013) 285–293.
41. L. Ni *et al.*, Surface coating on the polyamide TFC RO membrane for chlorine resistance and antifouling performance improvement, *J. Membr. Sci.*, **451** (2014) 205–215.
42. D. Saeki, T. Tanimoto, H. Matsuyama, Prevention of bacterial adhesion on polyamide reverse osmosis membranes via electrostatic interactions using a cationic phosphorylcholine polymer coating, *Colloids*

and Surfaces a-Physicochemical and Engineering Aspects, **443** (2014) 171–176.
43. H. Choi *et al.*, Surface modification of SWRO membranes using hydroxyl poly (oxyethylene) methacrylate and zwitterionic carboxylated polyethyleneimine, *J. Membr. Sci.*, **486** (2015) 97–105.
44. C. Y. Y. Tang, Y. N. Kwon, J. O. Leckie, Effect of membrane chemistry and coating layer on physiochemical properties of thin film composite polyamide RO and NF membranes I. FTIR and XPS characterization of polyamide and coating layer chemistry, *Desalination*, **242**(1–3) (2009) 149–167.
45. J. H. Lee *et al.*, Correlating chlorine-induced changes in mechanical properties to performance in polyamide-based thin film composite membranes, *J. Membr. Sci.*, **433** (2013) 72–79.
46. D. J. Miller *et al.*, Surface modification of water purification membranes, *Angew. Chem. Int.*, **56**(17) (2017) 4662–4711.
47. G. Hurwitz, G. R. Guillen, E. M. V. Hoek, Probing polyamide membrane surface charge, zeta potential, wettability, and hydrophilicity with contact angle measurements, *J. Membr. Sci.*, **349**(1–2) (2010) 349–357.
48. T. Ishigami *et al.*, Fouling reduction of reverse osmosis membrane by surface modification via layer-by-layer assembly, *Sep. Purif. Technol.*, **99** (2012) 1–7.
49. Y. Wang *et al.*, Improved flux and anti-biofouling performances of reverse osmosis membrane via surface layer-by-layer assembly, *J. Membr. Sci.*, **539** (2017) 403–411.
50. E. P. Chan *et al.*, Tailoring the Permselectivity of Water Desalination Membranes via Nanoparticle Assembly, *Langmuir*, **30**(2) (2014) 611–616.
51. W. Ma *et al.*, Spray- and spin-assisted layer-by-layer assembly of copper nanoparticles on thin-film composite reverse osmosis membrane for biofouling mitigation, *Water Res.*, **99** (2016) 188–199.
52. T. P. Martin *et al.*, Initiated chemical vapor deposition (iCVD) of polymeric nanocoatings, *Surface & Coatings Technology*, **201**(22–33) (2007) 9400–9405.
53. R. Yang *et al.*, Surface-tethered zwitterionic ultrathin antifouling coatings on reverse osmosis membranes by initiated chemical vapor deposition, *Chemistry of Materials*, **23**(5) (2011) 1263–1272.
54. G. Ozaydin-Ince *et al.*, Surface modification of reverse osmosis desalination membranes by thin-film coatings deposited by initiated chemical vapor deposition, *Thin Solid Films*, **539** (2013) 181–187.

55. A. Matin *et al.*, Surface modification of seawater desalination reverse osmosis membranes: Characterization studies & performance evaluation, *Desalination*, **343** (2014) 128–139.
56. H. Z. Shafi *et al.*, Surface modification of reverse osmosis membranes with zwitterionic coating for improved resistance to fouling, *Desalination*, **362** (2015) 93–103.
57. H. Z. Shafi *et al.*, Synthesis of surface-anchored stable zwitterionic films for inhibition of biofouling, *Materials Chemistry and Physics*, **239** (2020) 121971.
58. R. Yang *et al.*, Synergistic prevention of biofouling in seawater desalination by zwitterionic surfaces and low-level chlorination, *Advanced Materials*, **26**(11) (2014) 1711–1718.
59. E. M. Van Wagner *et al.*, Surface modification of commercial polyamide desalination membranes using poly(ethylene glycol) diglycidyl ether to enhance membrane fouling resistance, *J. Membr. Sci.*, **367**(1–2) (2011) 273–287.
60. G. D. Kang *et al.*, A novel method of surface modification on thin-film composite reverse osmosis membrane by grafting poly(ethylene glycol), *Polymer*, **48**(5) (2007) 1165–1170.
61. X. Y. Wei *et al.*, A novel method of surface modification on thin-film-composite reverse osmosis membrane by grafting hydantoin derivative, *J. Membr. Sci.*, **346**(1) (2010) 152–162.
62. J. Xu *et al.*, A novel reverse osmosis membrane with regenerable anti-biofouling and chlorine resistant properties, *J. Membr. Sci.*, **435** (2013) 80–91.
63. J. Xu *et al.*, Positively charged aromatic polyamide reverse osmosis membrane with high anti-fouling property prepared by polyethylenimine grafting, *Desalination*, **365** (2015) 398–406.
64. J. H. Wu *et al.*, Polyvinylamine-grafted polyamide reverse osmosis membrane with improved antifouling property, *J. Membr. Sci.*, **495** (2015) 1–13.
65. G. D. Kang *et al.*, Surface modification of a commercial thin film composite polyamide reverse osmosis membrane by carbodiimide-induced grafting with poly(ethylene glycol) derivatives, *Desalination*, **275**(1–3) (2011) 252–259.
66. J. V. Staros, R. W. Wright, D. M. Swingle, Enhancement by N-hydroxysulfosuccinimide of water-soluble carbodiimide-mediated coupling reactions, *Anal. Biochem.*, **156**(1) (1986) 220–222.
67. D. Sehgal, I. K. Vijay, A method for the high-efficiency of water-soluble carbodiimide-mediated amidation, *Anal. Biochem.*, **218**(1) (1994) 87–91.

68. Y. T. Hu *et al.*, Enhancing the performance of aromatic polyamide reverse osmosis membrane by surface modification via covalent attachment of polyvinyl alcohol (PVA), *J. Membr. Sci.*, **501** (2016) 209–219.
69. M. H. Liu *et al.*, Improving fouling resistance and chlorine stability of aromatic polyamide thin-film composite RO membrane by surface grafting of polyvinyl alcohol (PVA), *Desalination*, **367** (2015) 11–20.
70. J.-C. An *et al.*, Radiation-induced synthesis of poly(vinylpyrrolidone) nanogel, *Polymer*, **52**(25) (2011) 5746–5755.
71. R. Bernstein, S. Belfer, V. Freger, Bacterial attachment to RO membranes surface-modified by concentration-polarization-enhanced graft polymerization, *Environ. Sci. Technol.*, **45**(14) (2011) 5973–5980.
72. X. Y. Wei, *et al.*, Surface modification of commercial aromatic polyamide reverse osmosis membranes by graft polymerization of 3-allyl-5,5-dimethylhydantoin, *J. Membr. Sci.*, **351**(1–2) (2010) 222–233.
73. D. Saeki, T. Tanimoto, H. Matsuyama, Anti-biofouling of polyamide reverse osmosis membranes using phosphorylcholine polymer grafted by surface-initiated atom transfer radical polymerization, *Desalination*, **350** (2014) 21–27.
74. T. H. A. Ngo, D. T. Tran, C. H. Dinh, Surface photochemical graft polymerization of acrylic acid onto polyamide thin film composite membranes, *J. Appl. Polym. Sci.*, **134**(5) (2017).
75. M. Y. Teng *et al.*, Plasma deposition of acrylamide onto novel aromatic polyamide membrane for pervaporation, *Eur. Polym. J.*, **36**(4) (2000) 663–672.
76. J. Wang *et al.*, Improving the water flux and bio-fouling resistance of reverse osmosis (RO) membrane through surface modification by zwitterionic polymer, *J. Membr. Sci.*, **493** (2015) 188–199.
77. S. Belfer *et al.*, Surface modification of commercial composite polyamide reverse osmosis membranes, *J. Membr. Sci.*, **139**(2) (1998) 175–181.
78. H. H. Rana *et al.*, Low fouling and improved chlorine resistant thin film composite reverse osmosis membranes by cerium(IV)/polyvinyl alcohol mediated surface modification, *Desalination*, **357** (2015) 93–103.
79. Q. B. Cheng *et al.*, Surface modification of a commercial thin-film composite polyamide reverse osmosis membrane through graft polymerization of N-isopropylacrylamide followed by acrylic acid, *J. Membr. Sci.*, **447** (2013) 236–245.
80. J. Q. Meng *et al.*, A novel salt-responsive TFC RO membrane having superior antifouling and easy-cleaning properties, *J. Membr. Sci.*, **461** (2014) 123–129.

81. R. Bernstein, S. Beller, V. Freger, surface modification of dense membranes using radical graft polymerization enhanced by monomer filtration, *Langmuir*, **26**(14) (2010) 12358–12365.
82. G. T. Lewis, Y. Cohen, Controlled nitroxide-mediated styrene surface graft polymerization with atmospheric plasma surface activation, *Langmuir*, **24**(22) (2008) 13102–13112.
83. C. H. Liu *et al.*, Antifouling thin-film composite membranes by controlled architecture of zwitterionic polymer brush layer, *Environ. Sci. Technol.*, **51**(4) (2017) 2161–2169.
84. Z. Yang, D. Saeki, H. Matsuyama, Zwitterionic polymer modification of polyamide reverse-osmosis membranes via surface amination and atom transfer radical polymerization for anti-biofouling, *J. Membr. Sci.*, **550** (2018) 332–339.
85. G. T. Lewis *et al.*, Inorganic surface nanostructuring by atmospheric pressure plasma-induced graft polymerization, *Langmuir*, **23**(21) (2007) 10756–10764.
86. Y. Hara *et al.*, Relationship between peroxide radical species on plasma-treated PFA surface and adhesion strength of PFA/electroless copper-plating film, *Curr. Appl. Phys.*, **12** (2012) S38–S41.
87. N. H. Lin *et al.*, Polymer surface nano-structuring of reverse osmosis membranes for fouling resistance and improved flux performance, *Journal of Materials Chemistry*, **20**(22) (2010) 4642–4652.
88. K. J. Moses, Y. Cohen, Wettability of terminally anchored polymer brush layers on a polyamide surface, *J. Colloid Interface Sci.*, **436** (2014) 286–295.
89. M.-M. Kim *et al.*, Surface nano-structuring of reverse osmosis membranes via atmospheric pressure plasma-induced graft polymerization for reduction of mineral scaling propensity, *J. Membr. Sci.*, **354**(1–2) (2010) 142–149.
90. K. J. Varin, N. H. Lin, Y. Cohen, Biofouling and cleaning effectiveness of surface nanostructured reverse osmosis membranes, *J. Membr. Sci.*, **446** (2013) 472–481.
91. L.-Y. Hung, S. J. Lue, J.-H. You, Mass-transfer modeling of reverse-osmosis performance on 0.5–2% salty water, *Desalination*, **265**(1) (2011) 67–73.
92. N. Y. Yip, M. Elimelech, Performance limiting effects in power generation from salinity gradients by pressure retarded osmosis, *Environ. Sci. Technol.*, **45**(23) (2011) 10273–10282.
93. J. R. Werber, C. O. Osuji, M. Elimelech, Materials for next-generation desalination and water purification membranes, *Nat. Rev. Mater.*, **1**(5) (2016) 16018.

94. X. Lu *et al.*, Tuning the permselectivity of polymeric desalination membranes via control of polymer crystallite size, *Nature Communications*, **10**(1) (2019) 2347.
95. E. M. Van Wagner *et al.*, Effect of crossflow testing conditions, including feed pH and continuous feed filtration, on commercial reverse osmosis membrane performance, *J. Membr. Sci.*, **345**(1–2) (2009) 97–109.
96. S. Kim *et al.*, Upgrading polyamide TFC BWRO and SWRO membranes to higher SWRO membrane performance via surface nanostructuring with tethered poly (acrylic acid), *J. Membr. Sci.*, **597** (2020) 117736.
97. J. R. Werber, A. Deshmukh, M. Elimelech, The critical need for increased selectivity, not increased water permeability, for desalination membranes, *Environ. Sci. Technol. Letters*, **3**(4) (2016) 112–120.

CHAPTER 11

Scale-up of Nanocomposite Membranes Embedded with Silver Nanoparticles: From Laboratory Scale to Production Scale

Xiaobo Dong[a], Lillian Banks[a], Tequila A.L. Harris[b], and Isabel C. Escobar[a,*]

[a]*Department of Chemical and Materials Engineering, University of Kentucky, Lexington 40506, KY, USA*
[b]*George W. Woodruff School of Mechanical Engineering, Georgia Institute of Technology, Atlanta 30332, GA, USA*

11.1. Background and Discussion

Biofouling results from the growth of microorganisms present on membrane surface[1,2] that produce a biofilm. Many strains of bacteria, such as *Staphylococcus*, *Salmonella*, *Enterococcus* and *Pseudomonas*, produce extracellular polymeric substances (EPS), which create a tough matrix on the membrane surface and may lead to membrane biodegradation, clogging of pores, higher operating costs, increased pressure drop, and other performance issues.[2–4] Biofilms continue to grow at the expense of available nutrients in the feed water and create microbial aggregates that adhere to membrane surfaces.[3,5] It is extremely difficult to completely eradicate biofouling. Even if 99.9–99.99% of all bacteria are removed from the membranes, the remaining cells can become entangled into the system, remain protected, attach to the surface, start to colonize and multiply at the expense of available nutrients.[5–7]

*Corresponding author. Isabel.Escobar@uky.edu; Tel.: +1-859-257-7990

In order to control membrane biofouling, feed waters can be pretreated,[8] particles can be removed,[9] the membrane can be cleaned via backwash (or reverse flow) and/or chemical treatment, or membranes can be modified to create antifouling materials via graft polymerization, UV irradiation and other processes.[10] Pretreatment allows the efficiency and life expectancy of membranes to be maintained through removal of foulants before these reach the main process membranes.[11] Processes such as UF and MF are predominantly used as pretreatment options to remove larger particulates and microorganisms before more retentive processes (NF and RO) are used,[12,13] which helps reduce the need of chemicals.[2] For pretreatment, disinfectants (such as chlorine) may be used to control biofouling to boost membrane's life expectancy, but many membranes cannot withstand the corrosiveness of chlorine.[8] Thus, pretreatment can be performed with other disinfectants, such as ozone and potassium ferrate[7] to maintain membrane durability.[14] Polymers such as cellulose acetate and polyamide possess considerable resistant to chlorine, but are limited by operational pH ranges, salinities, molecular state, etc.[15] Functionalizing membranes with desirable properties, such as hydrophilicity and charge, can be performed using several different methods, such as surface coating and grafting.[16] Of focus here is the incorporation of nanoparticles on the membrane.[10]

Silver nanoparticles are chemically stable, possess catalytic and conductive properties, and are more antimicrobial than other metals.[17–19] Though the exact mechanisms of how AgNPs inhibit the bacteria growth are still unknown, there are some potential mechanisms that have been proposed to explain it and summarized in Fig. 11.1.[20]

However, nanoparticles tend to agglomerate, due to steric repulsion effect, attraction interaction, and the presence of ions and natural organic matter (NOM).[21] Agglomerating NPs leads to fewer surface area interactions available, which decreases their antimicrobial properties.[22–26] To minimize agglomeration, compounds, such as dextrin (made from maize starch), polyvinylpyrrolidone (PVP), or casein can be used to encapsulate nanoparticles.[21] Studies have

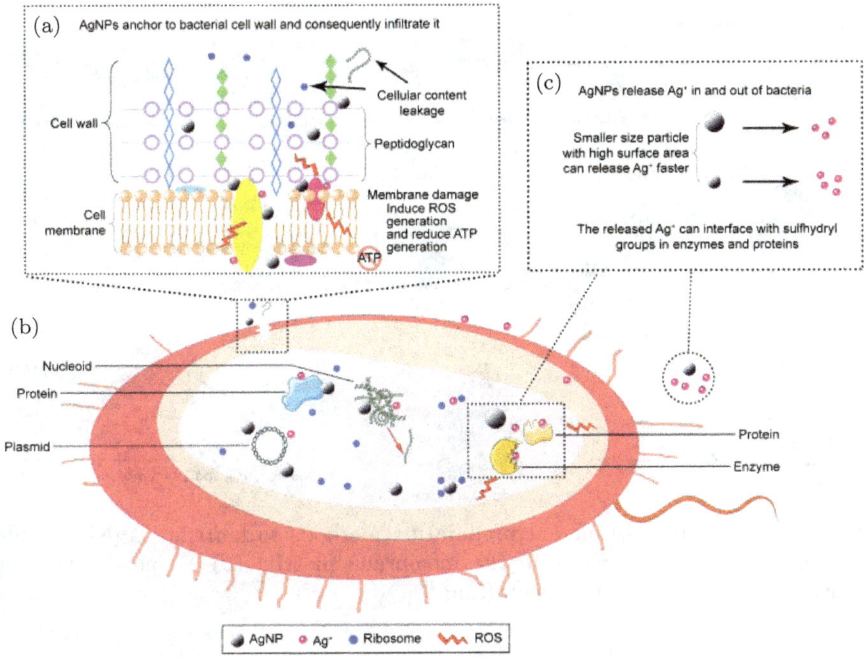

Fig. 11.1. Antibacterial mechanisms of AgNPs: (a) AgNPs anchor to the wall of bacterial cells, damage the cell wall and cause the leakage of cellular content. AgNPs can also bind to proteins on the cell membrane; (b) AgNPs penetrate into the bacterial cell, and then AgNPs release Ag^+ that can interact with cellular structure and macromolecules; (c) AgNPs can release Ag^+ in and out of bacteria, and these ions can deactivate bacterial proteins and enzymes.[20]

shown that these encapsulating agents have similar properties and performances, but their stabilities with Ag-NPs can be different based on the binding forces.[21]

11.1.1. *Incorporation of AgNPs into polymeric membranes*

Nanoparticles have been shown to not mix well with polymeric dope solutions leading to inconsistent mixing, increasing NP agglomeration, and decreasing membrane efficiency.[27,28] Even through consistent sonication, agglomeration is difficult to prevent. Solvent use for the preparation of dope solutions can also be a factor. Figure 11.2 shows different methods have been successfully developed for the

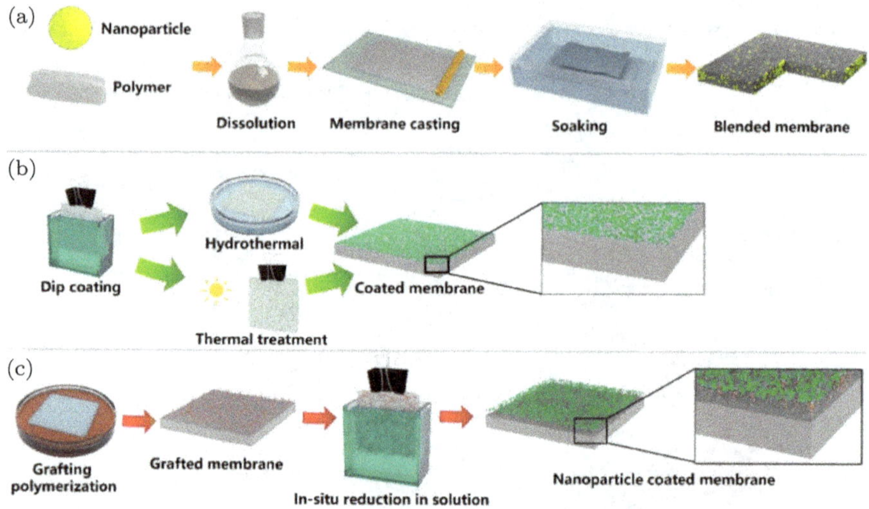

Fig. 11.2. Scheme of different preparation methods to fabricate AgNPs membranes. (a) blending AgNPs in the membrane matrix; (b) Membrane surface coating; (c) Membrane surface grafting.[29]

incorporation process, including blending AgNPs in the membrane matrix, surface modification, layer-by-layer (LBL) assembly, and other techniques.[29]

11.1.1.1. *Blending AgNPs into the membrane matrix*

There are two major methods involved in blending AgNPs into the membrane matrix, ex-situ and in-situ. The traditional blending method, the ex-situ method, has been used to physically blend AgNPs on membranes and proven to be effective. Polysulfone (PSf)[30] and polyethersulfone (PES)[31] have been investigated as the membrane materials to blend AgNPs into the dope solution. However, with this method it is difficult to control the NP distribution in the membranes, and thus, results in lowering the mechanical strength of the membranes and also decreases the water permeability.[32]

To address this issue, several materials, for instance, graphene oxide (GO), metal-organic framework (MOF), and TiO_2 have been investigated as carriers to incorporate AgNPs on membranes. Xu *et al.*[33] loaded $Mg(C_{10}H_{16}O_4)_2(H_2O_2)_2$ with AgNPs to form

Ag-MOF and subsequently blended with polyvinylidene fluoride (PVDF) dope solution to cast PVDF/PVP/Ag-MOF ultrafiltration membranes due to the highly porous and high inner surface areas of MOF. Via blending Ag-MOF, the hydrophilicity and water flux of membranes increased, along with the antibacterial activity. Wang *et al.*[34] employed GO as a carrier to incorporate AgNPs onto PVDF membranes. Specifically, GO-Ag/PVDF/F127 was prepared as the dope solution and then cast membranes, resulting in enhanced hydrophilicity, water flux and mechanical properties. Haghighat *et al.*[35] investigated physically blending AgNP and TiO_2, and then used 1wt% Ag/TiO2 NPs to fabricate polyvinyl chloride (PVC) membranes using nonsolvent induced phase separation (NIPS) method. The pure water flux of the nanocomposite membranes increased as compared to pristine PVC membranes, and the antimicrobial performance improved. Chen *et al.*[36] used dopamine as a reducing agent and a binder to immobilize AgNPs on TiO_2, and then fabricated the polysulfone/TiO2/AgNPs nanofiber membranes using electrospin. Results showed that the nanocomposite membranes exhibited enhanced performance in terms of water flux, salt rejection and antimicrobial activity. These methods using carriers all improved the water flux and mechanical properties and inhibited the agglomeration of AgNPs. However, a leaching study was not fully performed in the articles where a carrier was involved in fabricating AgNPs membranes.

11.1.1.2. *Surface modification*

Besides the blending method, surface modification has been investigated as a method to incorporate AgNPs on membrane surfaces for a more direct interaction between AgNPs and bacteria. Specifically, surface modification comprises of two methods, physical adsorption and covalent bonding.

For physical adsorption, Mosconi *et al.*[37] modified polypropylene (PP) films using chitosan-coated AgNPs. In this case, a polyacrylic acid grafted polypropylene (PAA-gPP) film was formed via photograft polymerization, and a CS/AgNP complex was adsorbed on the PAA-gPP films by electrostatic interactions to form the

CS/AgNP-coated PP films. The films exhibited excellent antimicrobial performance. Another case was developed by Ashraf *et al.*,[38] in which cellulose acetate polymer was blended with titanium dioxide (TiO_2) into a dope solution and then electrospun into nanofibers, followed by deacetylation to cellulose/TiO_2 nanofibers, AgNPs were then loaded on the surface of the nanofibers. Sawata *et al.*[32] grafted acrylamide onto a PES hollow fiber membrane to form a graft layer to improve the hydrophilicity, and then AgNPs were adsorbed in the graft layer to improve the antimicrobial performance. The main issue of physical adsorption is that the adsorption is unstable for long-term operation.

Several studies have focused on covalent bonding as a method to introduce AgNPs to polymeric membranes. Mauter *et al.*[39] encapsulated AgNPs in positively-charged polyethyleneimine (PEI). They functionalized PSf membranes using plasma and attached PEI to the PSf membrane surface by forming covalent bonds between the nucleophilic primary amine groups of PEI and the electrophilic carboxyl group of PSf. This method maximized the density of AgNPs on the membrane surface and presented improved antimicrobial performance. Wai *et al.*[40] blended PVP into PES to form a PVP/PES membrane, followed by coating polydopamine (PDA) onto the membrane surface to increase the hydrophilicity. Then, silver nitrate ($AgNO_3$) was reduced to AgNPs and then immobilized on the PDA layer. This method, in combination with physical rejection, showed 100% removal of *E. coli* during the filtration process. Furthermore, this method of incorporating AgNPs decreased membrane pore size and increased hydrophilicity to further improve the anti-biofouling performance.[41]

11.1.1.3. *Layer-by-layer (LBL) assembly*

The LBL technique has been applied to form membranes due to its ability to combine the functionality of multiple layers for a desired membrane function.[29] Liu *et al.*[42] fabricated nanofiltration (NF) and forward osmosis (FO) membranes using the LBL assembly method. Polyacrylonitrile (PAN) was used as the membrane substrate, and the top surface of the substrates was first soaked with positively charged poly(allylamine hydrochloride) (PAH) solution and then negatively charged poly(sodium 4-styrene-sulfonate) (PSS) solution. A controlled amount of AgNPs was added in both solutions, and then the LBL AgNP membranes was fabricated via doping in the PAH-AgNP or PSS-AgNP solutions. The membranes showed improved antibacterial properties against both gram-positive and gram-negative bacteria. Chitosan (CS) has also been used to fabricate LBL AgNP membranes,[43] and the membranes composed of $CS(Ag^0)/PMA(A^0)/CS(Ag^0)$ showed no metabolic activity after 48 hours of incubation, indicating that the membranes inhibited bacterial growth. However, the stability of different functional layers must be known in order to further investigate the method on a larger scale.

11.1.2. *Scalable fabrication of AgNPs membranes in a production scale*

Optimization and scale up are considered two of the most important components of process development.[44] Optimization focuses on the improvement of separation performance, which includes maximization of separation factor, mass transfer rates, and membrane lifetime, whereas scale up is fabricating a membrane of large enough area to process large volumes. Scale up includes designing of a membrane system to determine the area requirements and to integrate membranes into cost effective membrane modules.[44] Despite extensive research on the materials development for membranes, such as CA membranes, and nanocomposite membranes, such as AgNP membranes there is a dearth of studies or reports on scale-up. More research studies are required to develop the industrial production and application of nanocomposite membranes.[45] There are many research activities in laboratory and pilot scale on the application of nanocomposite membranes, however, many of these methods, such as spin coating, dip coating, etc., only work in a batch mode and cannot be transferred to large scale R2R methods, in contrast to roll coating, doctor blade casting, or slot die casting.[46] Only recently has there been studies on scale-up of plain membranes and nanocomposite membranes, using profile roller coating[46] and slot die casting embedded on roll-to-roll (R2R) systems.[47,48] Slot die casting is the most prominent method for scaling up thin film across a broad array of areas.

11.1.2.1. *Comparison of doctor blade casting and slot die casting of nanocomposite membranes*

Doctor blade casting has been a popular primary film casting method in the laboratory,[49] but it is not always the best method to fabricate membranes on an industrial scale because the membrane morphology is largely reliant on the viscosity of the dope solutions, and therefore, not suitable for continuous casting.[47] Slot die casting is a well-developed and commonly used method to manufacture polymer films. It is suitable for continuous casting of liquid films and, therefore, has been investigated to scale up polymeric membranes.[48] There is a constant demand to increase the processing speed of the thin film since the processing speed can increase the production output. At the same time, there is also a constant demand for maintaining low thicknesses of the film considering its applications in coating and polymer industry.[47] However, if these two key factors are not balanced, the polymer film quality cannot be maintained properly. Film quality is dependent upon several factors, such as the solution properties, fabrication process and processing parameters, and defects arise when these factors are altered.[47,48]

Scaling up from a small production scale, such as doctor blade, to a larger one, such as slot die, is not intuitively obvious, and several parameters must be identified in order to determine dope flow rate and substrate velocity. One such parameter is the surface tension of the polymer since coating speed depends on surface tension values, and solutions with lower surface tension can restrict the coating speeds to lower values.[47] Another key parameter is viscosity, which plays a significant role in determining the processing conditions because fabrication defects, such as air bubble entrapment, highly depends on the solution viscosity.[48] Air entrainment rate is lower at lower solution viscosities.[47] Further, contact angle quantifies the wettability of a surface.[50] For scale up, the sessile drop method was used to measure the contact angle of the membrane solution with respect to the tooling and substrate, and for membranes with respect to water.

11.1.2.2. *Advantage of slot die casting for scale up*

Often laboratory methods, such as spin coating and solution casting, are not scalable and/or do not introduce the same stress on the membrane as those formed using scalable approaches e.g., slot die casting. Furthermore, different manufacturing techniques may have significantly different membrane properties (e.g. mechanical, chemical, etc.), particle distribution and overall membrane functionality and its ability to treatment water. As alluded to above, slot die casting process can be tuned and tailored to the material of interest, e.g. AgNP membranes.

Integration of slot die casting into a R2R system allows for continuous casting of nanocomposite membranes. An illustration of a simple R2R system is shown in Fig. 11.3, which has been used to study scale up of AgNP membranes. The slot die coating process

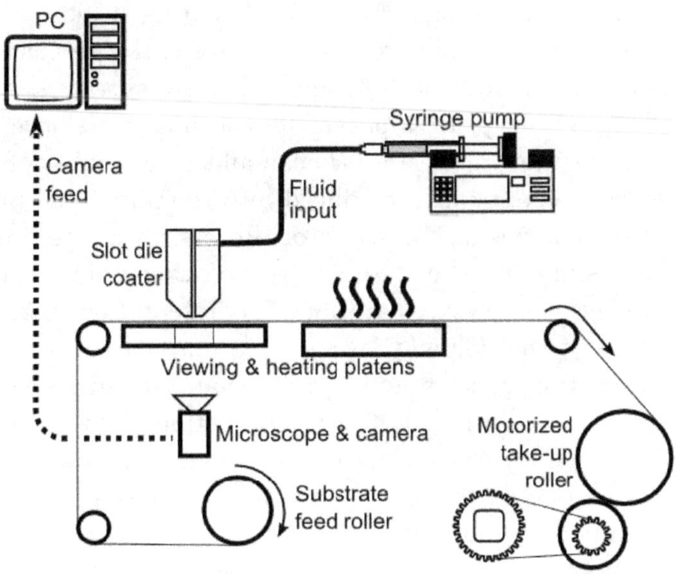

Fig. 11.3. Schematic of a simple R2R experimental setup with a slot die coater.[56]

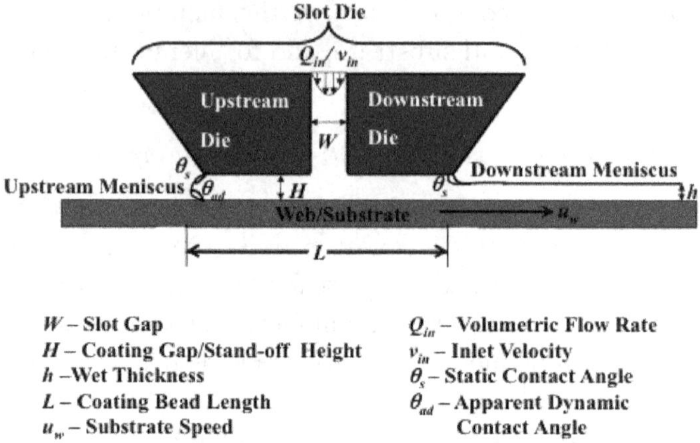

Fig. 11.4. Schematic of the lower section of a slot die coater with key parameters identified.[57]

consists of pumping a pre-metered flow rate of fluid between two die halves set apart by a small gap onto a substrate moving with a relative velocity to the dies, as depicted in Fig. 11.4. The slot die process offers the advantage of controlled coating thickness (h) as it is only a function of specific process parameters, the premetered flow rate of solution per unit die (Q') and substrate speed (u_w).[47] That is, $h = Q'/u_w$. Other process parameters such as the slot gap width (W) and the coating gap height (H) can affect the quality of the film as cast, with defects being introduced into the film if the process is not properly controlled. The space of process parameters in which defect-free casting can occur for a given coating fluid is called the casting/coating window.[51] A graphical representation of this space as a function of the volumetric flow rate and the substrate speed during the coating process and the associated defects seen at the boundaries is shown in Fig. 11.5. The competing viscous and surface tension forces acting in the coating bead under the dies determine when the onset of defects occurs due to increasing coating speed.[52,53] The most important non-dimensional parameter for predicting the onset of defects is the capillary number, which relates the relative strengths of viscous shear stress to the surface tension.[54]

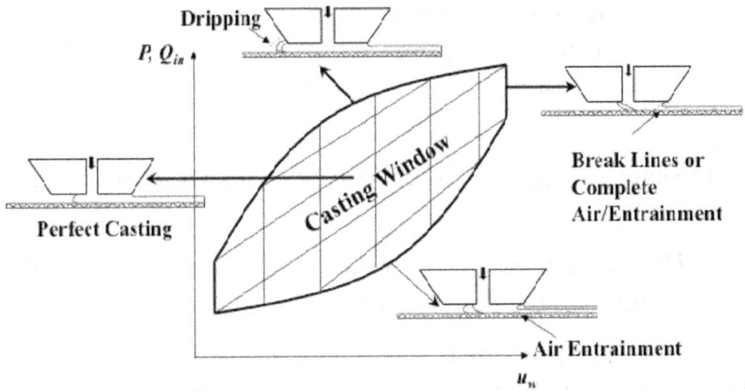

Fig. 11.5. Schematic of a generic casting window to illustrate the upper and lower boundaries of a slot die casting process before the onset of defects.[48,57]

For non-Newtonian fluids modeled using the power law relationship for the apparent viscosity (η) $\eta = k\dot{\gamma}^{n-1}$, the capillary number can be expressed for the slot die geometry by modeling the apparent viscosity of the fluid by the power law model as $Ca = \frac{k(\frac{u_w}{H})^{n-1}u_w}{\sigma}$. The capillary number can be related empirically to the non-dimensional film thickness parameter $h/H = f\ (Ca)$, at which defects such as air entrainment occur.[55]

Air entrainment and break line defects are of particular concern during scale up of AgNP membrane as such defects could compromise the integrity of the filtration membrane and are associated with higher coating speeds. Therefore, understanding the fluid characteristics to avoid such situations while increasing production speeds during manufacturing is critical, which can be done using slot die casting. A semi-empirical model, $u_{w,max} = [\frac{Q'H^{0.20-1.20}}{0.6}(\frac{\sigma}{k})^{0.20}]^{\frac{1}{(1+0.20n)}}$ developed by Bhamidipati *et al.*,[47,48] was used to approximate the maximum substrate speed ($u_{w,\max}$) to manufacture defect-free AgNP membranes, as a function of material properties and processing conditions.

During the slot die casting process experiments, the AgNP nanocomposite dope solution can be forced through slot die made of aluminum, steel or another compatible material. Although glass plates were used in the study, reels or flexible glass or modified flexible materials could be viable alternatives to make the process more robust and industrial scale.

11.2. Fabricate AgNPs Membranes from Laboratory to Production Scale

11.2.1. *Performance of AgNPs membranes fabricated at laboratory and production scales*

To study scaling up conditions for membranes, Sneha *et al.*[58] used cellulose acetate (CA) as the membrane material to compare the performance of membranes cast by doctor blade coating and slot die casting methods. CA was chosen because it is a widely studied material for membrane fabrication. The dope solution of 20 wt% CA was prepared and then cast using the doctor blade and slot die. For the doctor blade method, the thickness of membranes was set at 200 μm. For slot die casting, an aluminum slot die with a slot gap of 90 microns and a gap height of 178 microns above the substrate was used to deposit the CA membrane solution onto glass plates as it moved past the slot die at a given substrate velocity. The pre-metered volumetric flow rate into the die was controlled with a pump, to cast film of 150 to 200 μm thickness. A poly(ethylene terephthalate) (PET) film was used to convey glass plates used as substrates under the die from a feed roller and was collected by a take-up roller. Using the semi-empirical model, a 2-D flow rate of 0.13 mm^2/s with the speed of 1.3 mm/s[58] was chosen, to fabricate defect free film. The glass substrates were cleaned prior to casting with isopropyl alcohol to remove dust and foreign matter, and to ensure consistent wetting properties. After casting, the liquid film was converted into a membrane via phase inversion. All the casting conditions were set at 25°C and 48% humidity.

Results showed that the average pore size of production scale and laboratory scale membranes were 1.3 nm and 1.2 nm, respectively, showing no significant difference. Figure 11.6 shows that the DI water flux and the flux decline during the filtration of bovine serum albumin (BSA) and lipase solutions was not significantly different between the

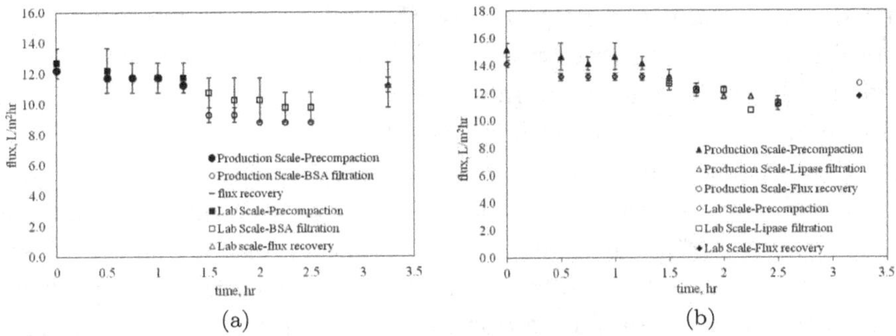

Fig. 11.6. Flux profiles of production and lab scale membranes during: (left) 1 g/L BSA protein filtration; (right) 1 g/L lipase filtration.[58]

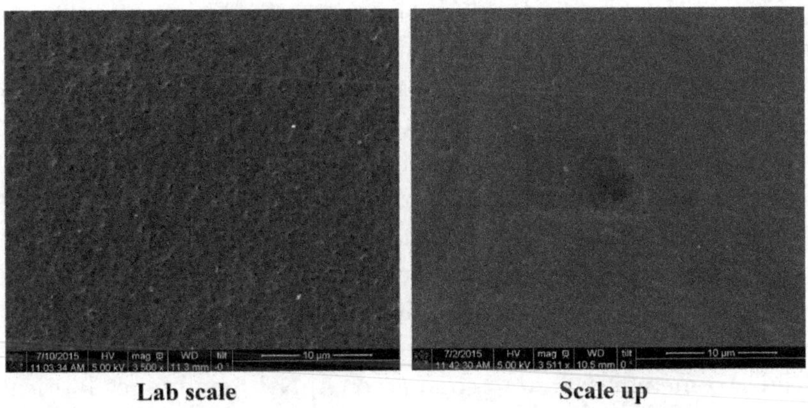

Fig. 11.7. SEM images of CA laboratory scale and scale up membranes before filtration.[58]

membranes cast at the different scales, doctor blade or laboratory scale and slot die or production scale.

Though flux profiles of the laboratory-scale and production-scale membranes were similar, the surface SEM images (Fig. 11.7) of these two membranes showed different textures. The production-scale membranes showed smoother surfaces compared to the lab scale membranes. It might be largely because the blade casting in the laboratory was performed with the dope solution poured on the substrate prior to casting, then a casting knife was manually controlled and thus velocity was not a fixed value. In slot die production scale, the flow rate of the polymer dope solution was set in the casting window while the substrate was moving at a fixed velocity.[58] The membranes fabricated under more controlled parameters showed a smoother top surface.

Does the casting method matter? From the experimental results reported in the literature,[58] though the surface morphologies of both membranes were significantly different, they presented similar average pore size and flux profiles. Overall, these two processes produced membranes with similar properties and performance. Therefore, since there was no significant difference in the performance of the membranes, the slot die method was then used to fabricate CA membranes incorporated with AgNPs.[59]

In a different study by the same group, Sneha *et al.*[59] physically blended AgNPs into the membrane matrix to incorporate AgNPs in a CA solution and then scaled the membranes up using slot-die casting. Physical and chemical attaching methods were both investigated. Specifically, in the physical attaching approach, casein-coated AgNPs were dispersed in the organic solvent, 1-methyl-2-pyrrolidinone (NMP). After sonication, CA was added and then stirred overnight to form the dope solution. The membranes were then cast on a glass substrate using two methods: the doctor blade casting and slot die casting, and then the BSA filtration performances of the two membranes were compared. AgNP membranes fabricated using doctor blade extrusion were within 2% accuracy of 115 μm thickness, with a length and width of 50 mm by 50 mm. For slot die casting, an aluminum slot die was used. Based on the material properties and desired thickness of 115 μm, the semi-empirical model given in Section 10.1.2.2 was used to determine a target velocity to process the membranes without defects. It was determined that a velocity of 1.20 mm/s was needed. Other processing conditions were as follows: coating gap 0.178 mm, slot gap 0.09 mm, and pressure 30 psi. The length and width of the samples were 305.4 mm by 100 mm, respectively, on each glass substrate. The slot die cast AgNP membranes had a thickness ranging from 110 to 125 μm. This error was due to various noise in the system and environmental conditions some of which could not be controlled, even so, the thickness was within 10% accuracy, using a pump would have allowed for better flow control and higher accuracy. This shows that the semi-empirical can be used to help set industrial scale fabrication parameters for defect free AgNP membranes. For the lab scale and production scale membranes, the final flux value at the end of 5 ml/L BSA filtration was $5.8 \pm 1.02\,L/m^2hr$ and $4.7 \pm 0.22\,L/m^2hr$, respectively; and the BSA rejection was $76.4 \pm 4.0\%$ and $77.1 \pm 1.5\%$, respectively. No significant difference was observed in the case of BSA filtration. However, leaching of AgNPs was an issue in this study, therefore, a method of covalently bonding AgNPs to membranes with minimal leaching and with the ability to be scaled via slot die was then investigated.

11.2.2. *Thiol-based covalent addition of AgNPs to membranes*

As discussed previously, chemically blending AgNPs into polymeric membranes is a potential approach to minimize leaching.[60] Glycidyl methacrylate (GMA) is a low-cost epoxy widely used in industrial products, such as coatings and adhesives[61] because its vinyl groups can be graft onto functional surfaces.[1,62] Epoxides are known to have highly reactive functional groups due to the strained three-membered ring of GMA, and are susceptible to a large range of nucleophiles.[63] With GMA's epoxy end, a large number of covalent attachments can be formed with ligands, such as amines, carboxylic acids, hydroxyls, pyridines, and peptides.[64] Wang *et al.* showed that poly-GMA can be attached to CA via the hydroxyl groups,[65] allowing additional possible functionalization to occur. The majority of primary and secondary amines were able to attach to the epoxide ring without the use of catalysts. However, solvent choice was a crucial factor to the S_N2 ring-opening reaction because the solvent polarity directly affected the conversion of the epoxide and vinyl terminal groups. DMSO has been found to act as a good solvent for many hydrophobic and hydrophilic primary amine reagents.[63] By using DMSO, a steady transition state of the S_N2 reaction progressed between the epoxy group of the GMA and amine group of the cysteamine (C_2H_7NS).[1] DMSO is an aprotic, polar solvent, where the solvent cannot donate hydrogens. Studies have shown that cysteamine is suitable for crosslinking reactions with GMA; however, cysteamine amine and thiol groups can unintentionally graft onto the epoxide groups of the poly-GMA.[66] Research shows that there are strong silver and thiol interactions through covalent bonding.[1,17,26,66,67] Therefore, inspired by the affinity of silver to the thiol groups of cysteine proteins in bacteria, Sprick *et al.*[60] developed a method to functionalize poly(glycidyl methacrylate) (polyGMA) via cysteamine, and utilize the thiol groups of cysteamine (CYS) to chemically attach casein-coated AgNPs to CA membranes. As shown in Fig. 11.8, CA/poly-GMA membranes were first fabricated using the NIPS method, and then the membrane surface was modified using a CYS bath. Lastly, AgNPs were attached to the thiol groups of CYS modified on the membrane surface.

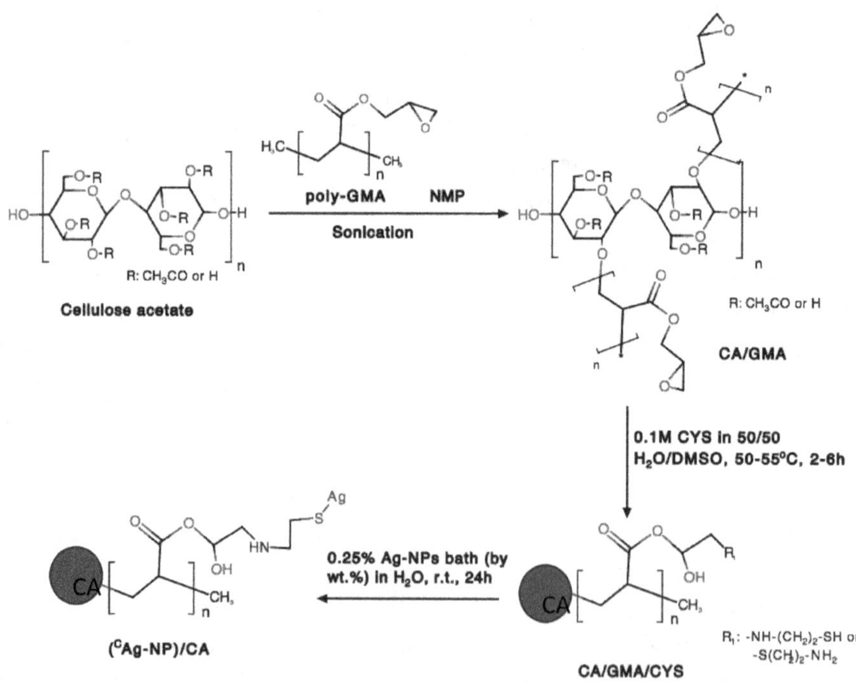

Fig. 11.8. Fabrication scheme of chemically attached AgNP/CA membranes.[60]

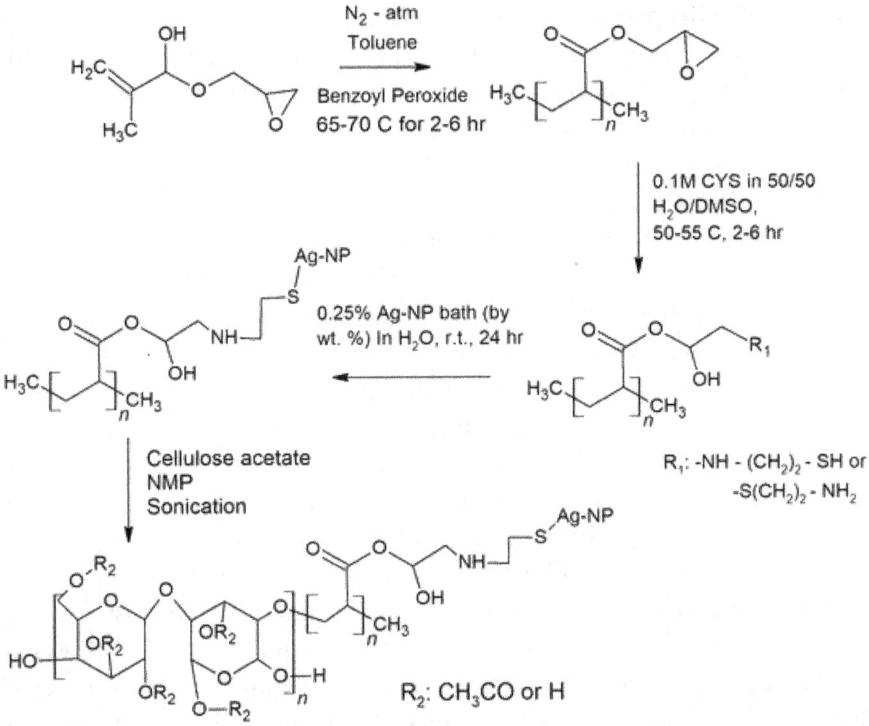

Fig. 11.9. Fabrication scheme of chemically attached CA/AgNP membranes that all chemical reactions were performed in prior to membrane casting.[68]

This study demonstrated that AgNPs could be chemically attached on CA membranes. Furthermore, minimal AgNP leaching was observed via this method; however, in order to scale up the membranes, surface modification techniques do not provide a suitable method since several steps are required. Therefore, a one-step method was developed to complete all chemical reactions prior to the membrane casting.[68] As shown in Fig. 11.9, instead of modifying membranes, the polyGMA was modified using CYS, and then, AgNPs were attached to polyGMA-CYS. The polyGMA-CYS-AgNPs complex was formed and blended with CA to develop the dope solution. Membranes cast using the blend of CA and the polymer complex showed improved antimicrobial performance as compared to baseline membranes, along with minimal AgNP leaching.[68] Given the conclusion that the slot die is not significantly different from lab casting, blending AgNPs in the dope solution prior to casting might be a potential approach to fabricate the CA/AgNP membranes on a production scale.

To characterize the polyGMA-CYS-AgNP membranes' low-biofouling properties, Dong *et al.*[68] developed an approach of using a visible bacterial lawn. *Serratia marcescens* was chosen because of its ability to produce red pigments under incubation conditions. A liquid agar solution mixed with a certain amount of bacterium was used to prepare agar plates, and a piece of membrane was gently laid on an agar plate for the experimental groups. Additionally, a liquid agar solution without bacterium was used to prepare agar plates for negative control. A liquid agar solution with the same amount of bacterium used in the experimental group was used to prepare agar plates for positive control. After incubation, the membranes were gently removed. The transparency of the bacterial lawn was visually compared between the experimental and control groups. A quantitative method was then used to calculate the color difference of the bacterial lawn and the color of the agar which had been covered by a piece of membrane.[68] Figure 11.10 shows that the positive control and negative control groups were at the opposite ends of the results, and CA membranes embedded with 2 wt% AgNPs presented a color close to the negative control, meaning that the bacterial growth was inhibited by the presence of the membranes.

11.2.3. *Scale-up work of thiol-based covalent addition of AgNPs to membranes*

Based on previous results discussed in Section 10.2.2, the affinity of AgNPs to thiol groups can be exploited to covalently bind the nanoparticles to a CA dope solution. This study then used dope solutions of CA and CA/1 wt% polyGMA-CYS-AgNPs to investigate factors that would control the scale-up fabrication of AgNP membranes using slot die casting. Shown in Fig. 11.11, the addition of polyGMA-CYS-AgNPs decreased the viscosity of dope solutions, and these dope solutions showed Newtonian fluid behavior;

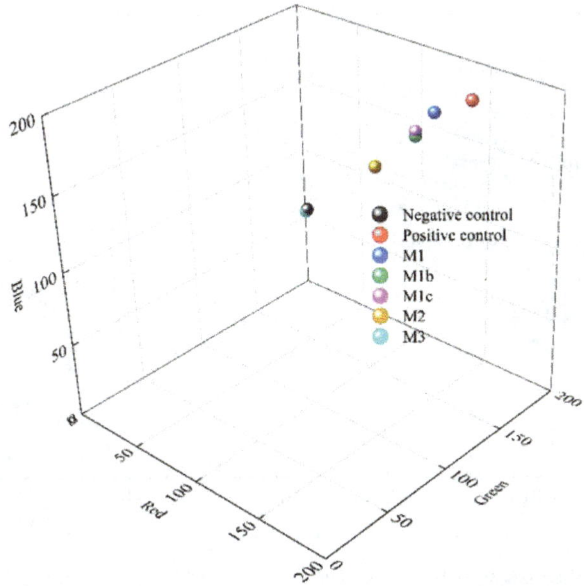

Fig. 11.10. Quantitative color differences after 24 h of bacterial growth study, where the x axis is red, y axis is green and z axis is blue: negative and positive control were at the two ends, the pristine membrane M1 was close to the positive control, and membrane embedded with AgNPs (M2 and M3) showed inhibition of bacterial growth.[68]

that is, the viscosity did not change significantly while the shear rate changed.[68] Therefore, fabrication using slot die casting was viable.

As previously established, to cast a defect free film, the dope solutions were cast through the slot die with a slot gap of 90 μm and a gap height of 178 μm between the slot die and the substrate. Within the casting window, a trial and error method was used to set the operating parameters at a 2-D flow rate of 0.13 mm^2/s with a substrate velocity of 1.3 mm/s.[58] Figure 11.12 shows that both the CA and CA/AgNPs membranes presented membrane surfaces without defects. However, the textures of surface morphology appeared to be different with CA/AgNPs membranes showing smoother surface. Given operating parameters of casting were the same, the difference may have been caused by the viscosity of dope solutions. Adding the polymer complex of polyGMA-CYS-AgNP decreased the viscosity of

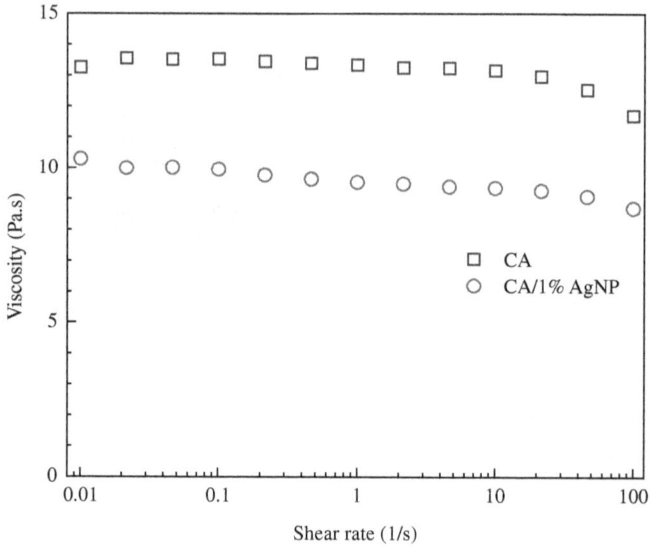

Fig. 11.11. Viscosity profiles of CA and CA/1 wt% AgNP dope solutions change as a function of the shear rate.

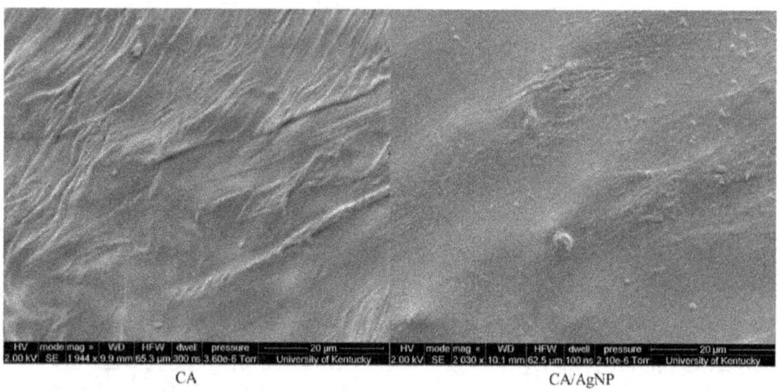

Fig. 11.12. SEM images of CA and CA/AgNP slot-die casting membranes before filtration.

dope solutions (Fig. 11.11) and thus produced a smoother surface texture under the same casting conditions.

Dead-end filtration was performed using a 50 mL Amicon stirred cell 8050 (EMD, Millipore, MA) at a constant pressure of 4.1 bar. After precompaction using DI water, the flux of the CA slot die

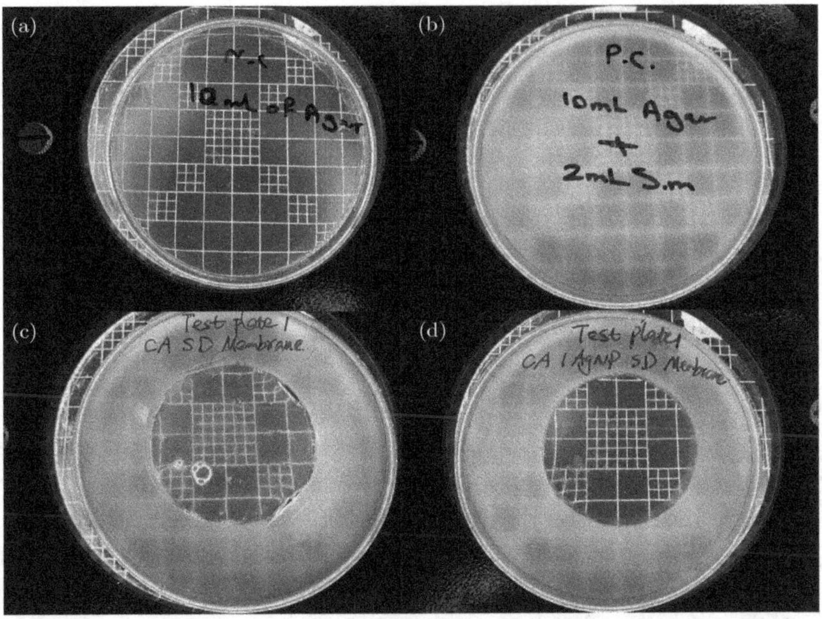

Fig. 11.13. The antimicrobial performance of membranes after 24 h of incubation: (a) negative control; (b) positive control; (c) CA slot-die membranes showing some suppression of bacterial growth; (d) CA/AgNPs slot-die membranes showing significant suppression of bacterial growth.

membrane was 38 ± 13 LMH while that of the CA/AgNPs slot die membrane was 20 ± 2 LMH. After filtration of 1 g/L BSA solution, the average flux of CA slot-die membranes was 18 ± 4 LMH, while that of CA/AgNPs slot-dic membranes was 12 ± 2 LMH. The lower flux values are related to presence of the polymer complex, which increased the polymer content of the membranes. The corresponding average BSA rejections were $73.1 \pm 0.7\%$ for CA slot-die membranes and $88.0 \pm 1.1\%$ for CA/AgNPs slot-die membranes. This again correlates with the addition of the polymer complex, which contained 1% polyGMA-CYS-AgNPs, since increasing the polymer content of the membrane matrix decreases the flux during filtration and increases rejection.

Figure 11.13 shows the results of the antimicrobial performed using *Serratia marcescens*, as previously described,[68] in which a bacterial lawn is developed on the positive control group and experimental groups. The CA slot-die membranes (Fig. 11.13(c)) showed some suppression of bacterial growth, likely due to the lack of air supply caused by the presence of the membrane on the bacterial lawn.[68] However, the CA/AgNP slot-die membranes (Fig. 11.13(d)) showed significant suppression as compared to all other samples. Therefore, filtration along with antimicrobial performance results show that the slot-die technique was effective to scale up CA membranes chemically embedded with AgNPs.

11.3. Conclusions

In conclusion, there are different procedures to embed AgNPs onto polymeric membranes, and different methods of manufacturing polymer films, such as membranes, on a production scale. In this Chapter, thiol-group affinity was exploited to covalently attach AgNPs to polymer dope solutions. Here, slot die was proven to be successful to scale up polymer membranes with and without AgNPs. Considering the practical factors, the combination of blending AgNPs in the dope solution and using slot die to fabricate the membranes might be a feasible approach.

As a bridge between the laboratory and production scale of membranes, this chapter can work as a guide for scaling up polymeric films embedded with nanoparticles. It not only provides an approach to scale up AgNP membranes for antimicrobial purposes, but also an approach to scale up polymeric films incorporated with other nanoparticles for other functions.

Acknowledgements

The authors acknowledge that this work was supported by the National Science Foundation under Cooperative Agreement (grant number 1355438) and by the NSF KY EPSCoR Program.

References

1. R. Hausman, I. C. Escobar, A comparison of silver- and copper-charged polypropylene feed spacers for biofouling control, *J. Appl. Polym. Sci.*, **128**(3) (2013) 1706–1714.

2. I. C. Escobar, E. H., C. Gabelich, F. DiGiano, Y. Le Gouellec, P. Berube, K. Howe, J. Allen, K. A. M. Benjamin, P. Brandhuber, J. Brant, Y. Chang, M. Chapman, A. Childress, W. C. T. Cooke, I. Crossley, G. Crozes, P. Huck, S. Kommineni, J. Jacangelo, A. K. J. Kim, D. Lawler, Q. Li, L. Schideman, S. Sethi, J. Tobiason, T. Tseng, S. V. A. Zander, Committee report: Recent advances and research needs in membrane fouling, *J. Am. WATER Work. Assoc.*, **97**(8) (2005) 79–89.
3. S. Asapu, S. Pant, C. L. Gruden, I. C. Escobar, An investigation of low biofouling copper-charged membranes for desalination, *Desalination*, **338** (2014) 17–25.
4. M. Ben-Sasson, X. Lu, E. Bar-Zeev, K. R. Zodrow, S. Nejati, G. Qi, E. P. Giannelis, M. Elimelech, In situ formation of silver nanoparticles on thin-film composite reverse osmosis membranes for biofouling mitigation, *Water Res.*, **62** (2014) 260–270.
5. R. Hausman, I. C. Escobar, A fourier transform infrared spectroscopic based biofilm characterization technique and its use to show the effect of copper-charged polypropylene feed spacers in biofouling control. In *Modern Applications in Membrane Science and Technology*, American Chemical Society: 2011; Vol. 1078, pp. 225–237.
6. H. C. Flemming, Biofouling in water systems — cases, causes and countermeasures, *Appl. Microbiol. Biotechnol.*, **59**(6) (2002) 629–40.
7. T. Nguyen, F. A. Roddick, L. Fan, Biofouling of water treatment membranes: A review of the underlying causes, monitoring techniques and control measures, *Membranes (Basel)*, **2**(4) (2012) 804–40.
8. K. Zodrow, L. Brunet, S. Mahendra, D. Li, A. Zhang, Q. Li, P. J. Alvarez, Polysulfone ultrafiltration membranes impregnated with silver nanoparticles show improved biofouling resistance and virus removal, *Water Res.*, **43**(3) (2009) 715–23.
9. S. Chede, I. C. Escobar, Fouling control using temperature responsive N-isopropylacrylamide (NIPAAm) membranes, *Environ. Prog. Sustain. Energy*, **35**(2) (2016) 416–427.
10. A. M. Ferreira, E. B. Roque, F. V. da Fonseca, C. P. Borges, High flux microfiltration membranes with silver nanoparticles for water disinfection, *Desalin. Water Treat.*, **56**(13) (2015) 3590–3598.
11. M. Daci-Ajvazi, B. Thaçi, N. Daci, S. Gashi, Membrane and adsorption processes for removing of organics and inorganics from urban wastewaters, *Orient. J. Chem.*, **32**(5) (2016) 2391–2400.
12. I. C. S. Escobar, Andrea, sustainability science and engineering. In *Sustainability Science and Engineering*, C. E. Isabel, I. S. Andrea, Eds. Elsevier: 2010; Vol. Volume 2, p. ii.

13. R. Fabris, E. K. Lee, C. W. K. Chow, V. Chen, M. Drikas, Pre-treatments to reduce fouling of low pressure micro-filtration (MF) membranes, *J. Membr. Sci.*, **289**(1–2) (2007) 231–240.
14. W.-L. Chou, D.-G. Yu, M.-C. Yang, The preparation and characterization of silver-loading cellulose acetate hollow fiber membrane for water treatment, *Polymers for Advanced Technologies*, **16**(8) (2005) 600–607.
15. J. Glater, S. Hong, M. Elimelech, The search for a chlorine-resistant reverse osmosis membrane, *Desalination*, **95**(3) (1994) 325–345.
16. G.-d. Kang, Y.-m. Cao, Development of antifouling reverse osmosis membranes for water treatment: A review, *Water Res.*, **46**(3) (2012) 584–600.
17. S. Y. Park, J. W. Chung, R. D. Priestley, S. Kwak, Covalent assembly of metal nanoparticles on cellulose fabric and its antimicrobial activity, *Cellulose*, **19** (2012) 2141–2151.
18. M. Sile-Yuksel, B. Tas, D. Y. Koseoglu-Imer, I. Koyuncu, Effect of silver nanoparticle (AgNP) location in nanocomposite membrane matrix fabricated with different polymer type on antibacterial mechanism, *Desalination*, **347** (2014) 120–130.
19. J. A. Prince, S. Bhuvana, K. V. K. Boodhoo, V. Anbharasi, G. Singh, Synthesis and characterization of PEG-Ag immobilized PES hollow fiber ultrafiltration membranes with long lasting antifouling properties, *J. Membr. Sci.*, **454** (2014) 538–548.
20. Y. a. Qing, L. Cheng, R. Li, G. Liu, Y. Zhang, X. Tang, J. Wang, H. Liu, Y. Qin, Potential antibacterial mechanism of silver nanoparticles and the optimization of orthopedic implants by advanced modification technologies, *Int J. Nanomedicine*, **13** (2018) 3311–3327.
21. H. Zhang, J. A. Smith, V. Oyanedel-Craver, The effect of natural water conditions on the anti-bacterial performance and stability of silver nanoparticles capped with different polymers, *Water Res.*, **46**(3) (2012) 691–9.
22. J. Pulit-Prociak, M. Banach, Silver nanoparticles — a material of the future...? In *Open Chem.*, 2016; Vol. 14, p. 76.
23. W. K. Son, J. H. Youk, W. H. Park, Antimicrobial cellulose acetate nanofibers containing silver nanoparticles, *Carbohydr. Polym.*, **65** (2006) 430–434.
24. H. Zhang, V. Oyanedel-Craver, Evaluation of the disinfectant performance of silver nanoparticles in different water chemistry conditions, *J. Environ. Eng.*, **138**(1) (2012) 58–66.
25. P. F. Andrade, A. F. de Faria, S. R. Oliveira, M. A. Zezzi Arruda, M. d. C. Goncalves, Improved antibacterial activity of nanofiltration polysulfone membranes modified with silver nanoparticles, *Water Res.*, **81** (2015) 333–342.

26. R. Behra, L. Sigg, M. J. D. Clift, F. Herzog, M. Minghetti, B. Johnston, A. Petri-Fink, B. Rothen-Rutishauser, Bioavailability of silver nanoparticles and ions: from a chemical and biochemical perspective, *J. R. Soc. Interface*, **10**(87) (2013) 20130396.
27. S. Zhao, Z. Wang, J. Wang, S. Wang, Poly(ether sulfone)/polyaniline nanocomposite membranes: Effect of nanofiber size on membrane morphology and properties, *Industrial & Engineering Chemistry Research*, **53**(28) (2014) 11468–11477.
28. A. Sotto, A. Boromand, R. Zhang, P. Luis, J. M. Arsuaga, J. Kim, B. Van der Bruggen, Effect of nanoparticle aggregation at low concentrations of TiO_2 on the hydrophilicity, morphology, and fouling resistance of PES–TiO_2 membranes, *Journal of Colloid and Interface Science*, **363**(2) (2011) 540–550.
29. J. Zhu, J. Hou, Y. Zhang, M. Tian, T. He, J. Liu, V. Chen, Polymeric antimicrobial membranes enabled by nanomaterials for water treatment, *J. Membr. Sci.*, **550** (2018) 173–197.
30. P. F. Andrade, A. F. de Faria, S. R. Oliveira, M. A. Z. Arruda, M. do Carmo Gonçalves, Improved antibacterial activity of nanofiltration polysulfone membranes modified with silver nanoparticles, *Water Res.*, **81** (2015) 333–342.
31. A. Ananth, G. Arthanareeswaran, A. Ismail, Y. S. Mok, T. Matsuura, Effect of bio-mediated route synthesized silver nanoparticles for modification of polyethersulfone membranes, *Colloids Surf. A Physicochem. Eng. Asp.*, **451** (2014) 151–160.
32. I. Sawada, R. Fachrul, T. Ito, Y. Ohmukai, T. Maruyama, H. Matsuyama, Development of a hydrophilic polymer membrane containing silver nanoparticles with both organic antifouling and antibacterial properties, *J. Membr. Sci.*, **387–388** (2012) 1–6.
33. W. Xu, H. Zhuang, Z. Xu, M. Huang, S. Gao, Q. Li, G. Zhang, Design and construction of Ag@ MOFs immobilized PVDF ultrafiltration membranes with anti-bacterial and antifouling properties, *Adv. Polym. Technol.*, **2020** (2020).
34. Y. Wang, G.-E. Chen, H.-L. Wu, Z.-L. Xu, J.-J. Wan, L.-J. Liu, S.-J. Xu, Y.-F. Kong, Q. Wu, J. Min, H.-F. Mao, Fabrication of GO-Ag/PVDF/F127 modified membrane IPA coagulation bath for catalytic reduction of 4-nitrophenol, *Sep. Purif. Technol.*, **235** (2020) 116143.
35. N. Haghighat, V. Vatanpour, M. Sheydaei, Z. Nikjavan, Preparation of a novel polyvinyl chloride (PVC) ultrafiltration membrane modified with Ag/TiO_2 nanoparticle with enhanced hydrophilicity and antibacterial activities, *Sep. Purif. Technol.*, **237** (2020) 116374.

36. H. Chen, M. Huang, Z. Wang, P. Gao, T. Cai, J. Song, Y. Zhang, L. Meng, Enhancing rejection performance of tetracycline resistance genes by a TiO_2/AgNPs-modified nanofiber forward osmosis membrane, *Chem. Eng. J.*, **382** (2020) 123052.
37. G. Mosconi, M. F. Stragliotto, W. Slenk, L. E. Valenti, C. E. Giacomelli, M. C. Strumia, C. G. Gomez, Original antifouling strategy: Polypropylene films modified with chitosan-coated silver nanoparticles, *J. Appl. Polym. Sci.*, **137**(10) (2020) 48448.
38. R. Ashraf, H. S. Sofi, T. Akram, H. A. Rather, A. Abdal-hay, N. Shabir, R. Vasita, S. H. Alrokayan, H. A. Khan, F. A. Sheikh, Fabrication of multifunctional cellulose/TiO_2/Ag composite nanofibers scaffold with antibacterial and bioactivity properties for future tissue engineering applications, *J. Biomed. Mater. Res. A n/a*, (n/a).
39. M. S. Mauter, Y. Wang, K. C. Okemgbo, C. O. Osuji, E. P. Giannelis, M. Elimelech, Antifouling ultrafiltration membranes via post-fabrication grafting of biocidal nanomaterials, *ACS Appl. Mater. Interfaces*, **3**(8) (2011) 2861–2868.
40. K. P. Wai, C. H. Koo, Y. L. Pang, W. C. Chong, W. J. Lau, In situ immobilization of silver on polydopamine-coated composite membrane for enhanced antibacterial properties, *J. Water Process. Eng.*, **33** (2020) 100989.
41. D. H. N. Perera, S. K. Nataraj, N. M. Thomson, A. Sepe, S. Hüttner, U. Steiner, H. Qiblawey, E. Sivaniah, Room-temperature development of thin film composite reverse osmosis membranes from cellulose acetate with antibacterial properties, *J. Membr. Sci.*, **453** (2014) 212–220.
42. X. Liu, S. Qi, Y. Li, L. Yang, B. Cao, C. Y. Tang, Synthesis and characterization of novel antibacterial silver nanocomposite nanofiltration and forward osmosis membranes based on layer-by-layer assembly, *Water Res.*, **47**(9) (2013) 3081–3092.
43. J. Kochan, M. Scheidle, J. van Erkel, M. Bikel, J. Büchs, J. E. Wong, T. Melin, M. Wessling, Characterization of antibacterial polyethersulfone membranes using the respiration activity monitoring system (RAMOS), *Water Res.*, **46**(16) (2012) 5401–5409.
44. J. Hildebrandt, Scale-up and optimization of membrane processes. In *Chromatographic and Membrane Processes in Biotechnology*, C. Costa, J. Cabral, Eds. Springer Netherlands: 1991; Vol. 204, pp. 415–428.
45. M. A. Alaei Shahmirzadi, A. Kargari, 9 — Nanocomposite membranes. In *Emerging Technologies for Sustainable Desalination Handbook*, V. G. Gude, Ed. Butterworth-Heinemann: 2018; pp. 285–330.
46. T. Bucher, V. Filiz, C. Abetz, V. Abetz, Formation of thin, isoporous block copolymer membranes by an upscalable profile roller coating

process-a promising way to save block copolymer, *Membranes (Basel)*, **8**(3) (2018) 57.
47. K. Bhamidipati, S. Didari, T. A. L. Harris, Experimental study on air entrainment in slot die coating of high-viscosity, shear-thinning fluids, *Chem. Eng. Sci.*, **80** (2012) 195–204.
48. K. L. Bhamidipati, S. Didari, T. A. L. Harris, Slot die coating of polybenzimiazole based membranes at the air engulfment limit, *J. Power Sources*, **239** (2013) 382–392.
49. C. F. Wang, Y. An, Q. H. Li, S. J. Wan, W. X. Chen, X. D. Liu, Nonsolvent effects on morphology of cellulose acetate films prepared by dry-cast process, *J. Macromol. Sci. Phys., Part B*, **51**(11) (2012) 2266–2275.
50. C. Jucker, M. M. Clark, Adsorption of aquatic humic substances on hydrophobic ultrafiltration membranes, *J. Membr. Sci.*, **97** (1994) 37–52.
51. K. J. Ruschak, Limiting flow in a pre-metered coating device, *Chem. Eng. Sci.*, **31**(11) (1976) 1057–1060.
52. O. J. Romero, L. E. Scriven, M. S. Carvalho, Slot coating of mildly viscoelastic liquids, *J. Non-Newton. Fluid Mech.*, **138**(2–3) (2006) 63–75.
53. M. S. Carvalho, H. S. Kheshgi, Low-flow limit in slot coating: Theory and experiments, *AIChE Journal*, **46**(10) (2000) 1907–1917.
54. Y.-R. Chang, H.-M. Chang, C.-F. Lin, T.-J. Liu, P.-Y. Wu, Three minimum wet thickness regions of slot die coating, *Journal of Colloid and Interface Science*, **308**(1) (2007) 222–230.
55. X. Ding, J. Liu, T. A. L. Harris, A review of the operating limits in slot die coating processes, *AIChE Journal*, **62**(7) (2016) 2508–2524.
56. A. Phillips, M. Ulsh, J. Mackay, T. Harris, N. Shrivastava, A. Chatterjee, J. Porter, G. Bender, The effect of membrane casting irregularities on initial fuel cell performance, *Fuel Cells n/a* (n/a).
57. K. L. Bhamidipati, Detection and elimination of defects during manufacture of high-temperature polymer electrolyte membranes. Georgia Institute of Technology, (2011).
58. S. Chede, P. Griffiths, I. C. Escobar, T. A. L. Harris, Does casting method matter in filtration membranes? A comparison in performance between doctor blade and slot-die extruded polymeric membranes, *J. Appl. Polym. Sci.*, **135**(24) (2018) 45563.
59. S. Chede, N. M. Anaya, V. Oyanedel-Craver, S. Gorgannejad, T. A. L. Harris, J. Al-Mallahi, M. Abu-Dalo, H. A. Qdais, I. C. Escobar, Desalination using low biofouling nanocomposite membranes: From batch-scale to continuous-scale membrane fabrication, *Desalination*, **451** (2019) 81–91.

60. C. Sprick, S. Chede, V. Oyanedel-Craver, I. C. Escobar, Bio-inspired immobilization of casein-coated silver nanoparticles on cellulose acetate membranes for biofouling control, *J. Environ. Chem. Eng.*, **6**(2) (2018) 2480–2491.
61. M. Benaglia, A. Alberti, L. Giorgini, F. Magnoni, S. Tozzi, Poly(glycidyl methacrylate): a highly versatile polymeric building block for post-polymerization modifications, *Polym. Chem.*, **4**(1) (2013) 124–132.
62. R. Hausman, T. Gullinkala, I. C. Escobar, Development of copper-charged polypropylene feedspacers for biofouling control, *J. Membr. Sci.*, **358**(1–2) (2010) 114–121.
63. K. A. McEwan, S. Slavin, E. Tunnah, D. M. Haddleton, dual-functional materials via CCTP and selective orthogonal thiol-Michael addition/epoxide ring opening reactions, *Polym. Chem.*, **4**(8) (2013) 2608.
64. Y. Zheng, H. Liu, P. V. Gurgel, R. G. Carbonell, Polypropylene non-woven fabrics with conformal grafting of poly(glycidyl methacrylate) for bioseparations, *J. Membr. Sci.*, **364**(1–2) (2010) 362–371.
65. J. Wang, P. Lei, H. Z. Liu, P. He, K. M. Wang, C. B. Liu, Y. K. Chen, M. M. Miao, C. C. Li, Preparation, characterization and properties of cellulose acetate-grafted-poly(glycidyl methacrylate) copolymers, *J. Macromol. Sci.: Pure & Applied Chemistry*, **52**(3) (2015) 226–233.
66. R. Groarke, D. Brabazon, Methacrylate polymer monoliths for separation applications, *Materials*, **9**(6) (2016) 446.
67. T. Sinclair, M. Zieba, S. Irusta, V. Sebastian, M. Arruebo, High-speed water sterilization using silver-containing cellulose membranes, *Nanotechnology*, **25**(30) (2014) 305101.
68. X. Dong, H. D. Shannon, A. Amirsoleimani, G. M. Brion, I. C. Escobar, Thiol-affinity immobilization of casein-coated silver nanoparticles on polymeric membranes for biofouling control, *Polymers*, **11**(12) (2019) 2057.

CHAPTER 12

Application of MOFs Nanomaterials and MOFs-Membranes in Water Treatment

Dipeshkumar D. Kachhadiya and Z. V. P. Murthy*

Department of Chemical Engineering, Sardar Vallabhbhai National Institute of Technology, Surat – 395007, Gujarat, India

12.1. Introduction

Water is a critical resource of life, social, cultural, and economic development. Water pollution is a major issue faced by the 21st century society due to increased population from the various industries.[1] Water pollution is caused when water bodies get contaminated by various activities like industrialization, urbanization, agricultural, etc (Fig. 12.1). There are two main sources of contamination are sewage and industrial wastes, and their treatment is becoming a must in this Industrial world. Nanotechnology provides a better solution to treat contaminated water. Nanoparticles have huge impact and potential to be used in the treatment of wastewater. Nanotechnology facilitated water and wastewater treatment reduces the difficulties faced by conventional methods and also provides new techniques that could give an economic solution.[2] Nanotechnology can provide a better solution by incorporating it in the various existing processes such as adsorption, membrane-based processes, photocatalysis, and detection and monitoring (Table 12.1).[3] The MOFs incorporated

*Corresponding author. E-mail: zvpm2000@yahoo.com, zvpm@ched.svnit.ac.in

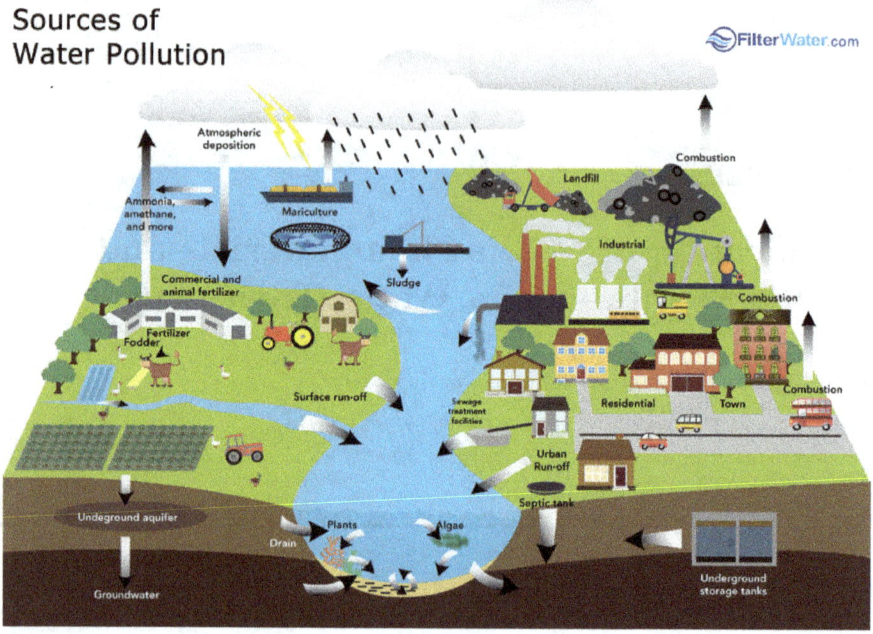

Fig. 12.1. Different sources of water pollution. (Reproduced with permission from: http://www.filterwater.com/images/upload/waterpollutionsources.jpg.)

membranes (say MOF membranes) exhibit outstanding performance in separating small molecules as they have uniform porous structures. The MOF membranes for separation can be classified as two main branches. First is higher surface polycrystalline membrane supported by porous substrates, and the second is a hybrid mixed matrix membrane (MMM).[4,5] The MOF membranes are normally used in pervaporation, gas permeation, nanofiltration, etc.[6] Particularly MOF membranes are used in gas permeation for H_2 removal, hydrocarbon separation from mixtures, and CO_2 separation. Likewise, in pervaporation, MOF membranes are used for solvent dehydration, organic-organic mixtures separation, and removal of organic dilution.[7] Furthermore, to simplify the process or operation, catalysis and separation were combined together, for which nanocatalysts are coated or deposited on the membrane to acquire catalytic features.

Table 12.1. Various nanomaterials, their properties and applications in wastewater treatment.

Application	Nanomaterials	Appropriate nanomaterials properties
Adsorption	CNTs	Higher effective surface area, higher adsorption sites, tuneable surface chemistry
	Nanoscale metal oxide	Higher surface area, higher active adsorption sites, compactable without reducing SA, easy recovery, super paramagnetic
Membranes processes	Nano-zeolites	Hydrophilicity, molecular sieve
	Nano-silver	Antimicrobial property, low toxicity, wide range of applicability
	Nano-TiO_2	Photocatalytic degradation, hydrophilicity, higher thermal and chemical stability
	Nano-magnetite	Tuneable surface structure, super paramagnetic
	CNTs	Small diameter, tuneable surface structure, atomic even inner surface, higher chemical and mechanical stability

12.2. Nanotechnology for Wastewater Treatment

12.2.1. *Nano-adsorbent*

Several groups of nanoparticles (NPs) have been utilized for adsorptive removal studies, namely carbon-based, iron-based, and aluminium based nanoparticles for the removal of various contaminants from water.[8]

12.2.1.1. *Carbon based nano-adsorbents*

Having advantageous features like high mechanical strength, higher active surface area, and electrical conductivity, carbon nanotubes

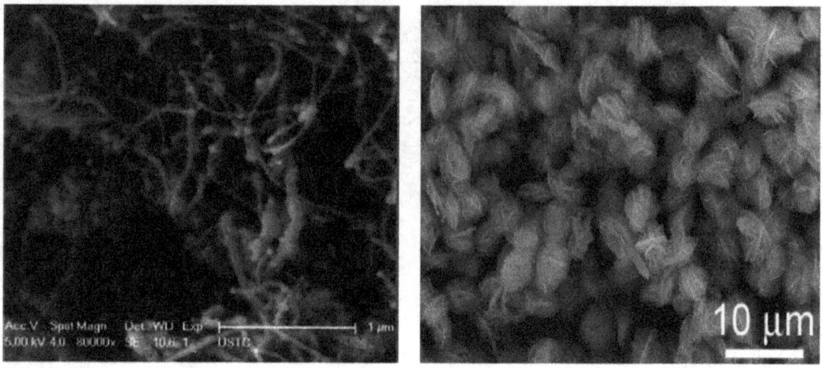

(a) (b)

Fig. 12.2. SEM images: (a) Fe NPs — functionalized MWCNTs, and (b) flowerlike 3D-structure of ceria precursor.[15,18] (Reproduced with permission from Royal Society of Chemistry; American Chemical Society.)

(CNTs) have been widely used to remove organic contents from wastewater. There are various articles on the application of surface-modified activated carbon materials, fullerenes, CNTs, and graphene/graphene-oxide for the removal of organic contaminants and various heavy metals and their ions, such as Cu, Cr, Fe, Ni, Zn, Pb, etc.[9] The CNTs have diverse contaminant interactions including, π–π interactions, hydrogen bond, covalent bond, electrostatic interactions, and hydrophobic effect. Due to these advantageous features, CNTs have shown higher efficiency towards the adsorption of various organic contaminants than activated carbon.[10,11] The CNTs have lager active adsorption sites for some bulky organic molecules on account of their bigger pores in groups and selective sorption sites.[12] Fe modified multi-walled CNTs (MWCNTs) have been widely used to remove a large number of heavy metals from water (Fig. 12.2(a)).[13–16] To remove or remediate Pb(II) ions from wastewater, graphene-based nanoparticles have been used.[17] Also, for metal ion removal and photocatalytic dye degradation graphene and its derivates with metals and metal oxides are listed as good agents. To remove heavy metals like Cr(VI), As(V), and azo dye orange II from wastewater, 3D flower type nanostructure of ceria and iron oxide have been developed for the simultaneous removal of metal ions and dyes (Fig. 12.2(b)).[18,19]

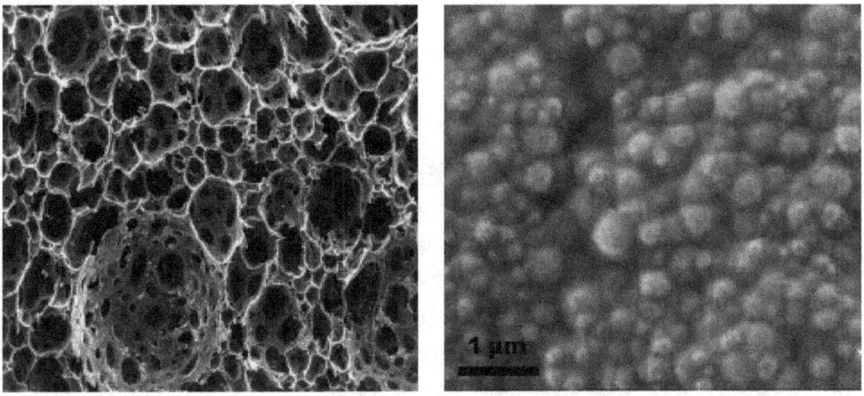

(a) (b)

Fig. 12.3. SEM images of (a) Fe_2O_3 functionalized carbon foam, (b) Cr^{+6} loaded magnetite nanospheres.[25,26] (Reproduced with permission from Elsevier Inc.)

12.2.1.2. *Metal-based nano-adsorbents*

In recent years, much attention has been drawn towards metal and its oxide-based nano-adsorbents for the removal of various contaminants from wastewaters.[20] For example, there are various metals and their oxides such as ceria, cobalt and its oxides, iron and its oxides, aluminum oxides, titanium dioxide, zinc oxide, and magnesium oxide have been widely used to purify the wastewater. These nanoparticles have irreplaceable advantageous features such as higher efficiency, tuneable surface chemistry, faster rate of adsorption, easy recovery after adsorption, and magnetic behaviour at nano-scale, and they have been widely used for separation of hazardous metal ions from wastewaters.[21] In the adsorption, for separation of various arsenic contaminants from wastewater, iron and its oxides like $Fe(OH)_2$ and $Fe(OH)_3$were cited as highly efficient and effective nanoparticles in the adsorption.[22–24] Furthermore, functionalized NPs have also been used to remove various metal ions such as copper, cobalt, nickel, chromium, and lead from wastewater (Fig. 12.3).[25,26] To remove highly toxic metal such as Hg(II), Pb ions from wastewater, hematite NPs coated graphene-oxide(GO) has been used.[27] Also, for the removal of other hazardous metal ions such as chromium, cadmium, and mercury, low-cost NPs such as alumina, zinc oxide and other nano-adsorbents have been used.[28,29]

12.2.1.3. *Polymer-based nano-adsorbents*

For adsorptive removal of various hazardous contaminants from water, different organic and inorganic or hybrid polymeric materials have been investigated and used in various techniques. Metal doped polymeric materials having various properties like appropriate functional group attachment, high active surface area, porous structure are described to be able to effectively bind with heavy metal ions such as Zn, Cd, Pb as well as organic dyes.[30] For effective adsorptive removal of various hazardous metal contaminants from wastewater, organic material functionalized mesoporous silica nano-adsorbents with higher surface area have been established using a sol-gel method.[31] For the removal of heavy metal ions like Zn^{2+}, Cu^{2+}, Cd^{2+} and Pb^{2+} from wastewater, Fe_3O_4 magnetic NPs modified with 3-aminopropyl)triethoxysilane (3-APTES) and copolymers of acrylic acid and crotonic acid were used (Fig. 12.4).[32] Furthermore, for the effective removal of metal ions from aqueous streams, a polymeric dendrimer-ultrafiltration nano-adsorbent system has been developed.[33]

12.2.2. *Membrane and membrane-based processes*

12.2.2.1. *MOFs*

ZIF-8: The zeolitic imidazolate framework (ZIF)-8 (i.e., ZIF-8) was successfully prepared by Fu *et al.* (2018).[34] Zinc nitrate hexahydrate

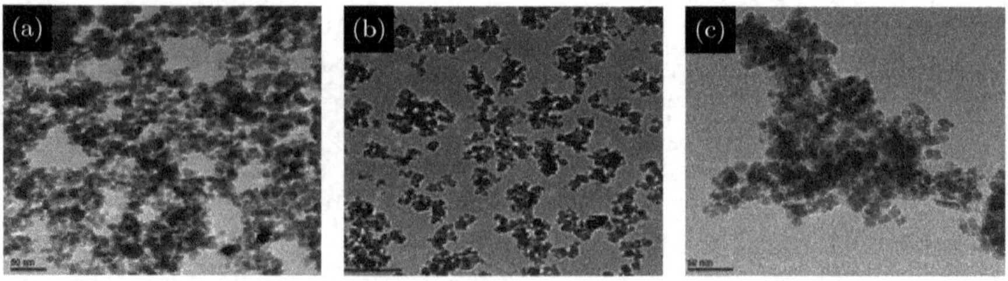

Fig. 12.4. TEM images: (a) Fe_3O_4, (b) APTES modified Fe_3O_4, and (c) APTES modified Fe_3O_4@AA-co-CA.[32] (Reproduced with permission from Elsevier Inc.)

($Zn(NO_3)_2 \cdot 6H_2O$)(1.8028 g, 1.5 mmol) and 2-methylimidazol (2-MeIM) (1.0052 g, 3 mmol) were dissolved in 80 mL of methanol separately. Then the $Zn(NO_3)_2 \cdot 6H_2O$ solution was added dropwise in 2-MeIM solution kept at continuous stirring. The mixture was stirred for 1 hour at 1000 rpm and kept it for crystallization at room temperature for 24 h. After that, the resulting ZIF-8 nanocrystals were collected by centrifugation at 10,000 rpm for 20 min and followed by methanol washing. Then final collected nanocrystals were dried at 75°C for overnight. The final product was collected and stored for further analysis.

CuBTC: The copper benzene-1,3,5-tricarboxylate (CuBTC) was successfully prepared by Azhar *et al.* (2016).[35] The 1,3,5 benzene tricarboxylic acid (2.1014 g) was dissolved in 40 mL of a mixture of DMF and methanol (1:1). Then the copper acetate monohydrate (2.995 g) was dissolved in 50 mL of mixture of distilled water/ethanol separately and stirred at room temperature for half-an-hour. After that solution I was added drop wise into solution II and sonicated for a half-hour. Then, the entire mixture was thermally treated in an autoclave at 100°C for 20 h. In the end, the obtained blue product was separated by centrifugation and washed with solvents. Finally, the final product was dried at 50°C for overnight in the oven.

UiO-66: Universitetet i Oslo (UiO-66) was successfully prepared by Zhu *et al.* (2014).[36] The appropriate amount of terephthalic acid (0.320 g), benzoic acid (2.44 g), HCl 37% (0.33 mL), and $ZrCl_4$ (0.466g) were dissolved in 36 mL of dimethylformamide (DMF). Then, the entire mixture was thermally treated in an autoclave at 120°C for 48 h. After completing the reaction, the mixture was cooled to room temperature and a white precipitate of UiO-66 was separated by centrifugation at speed of 5500 rpm for 20 min. Then the obtained product was washed with DMF, methanol and acetone to remove excess reactants. Finally, the white solid material obtained was dried in a vacuum oven at 80°C overnight.

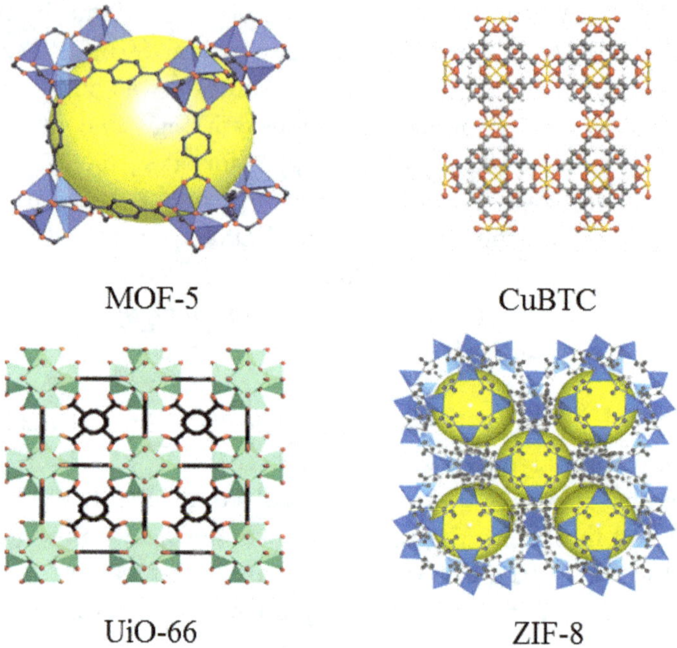

Fig. 12.5. Schematic representation of various MOF structures.[37–40]

12.2.2.2. *MOF and MOF based membrane application for wastewater treatment*

The MOFs composed of metal centres and organic binders have high surface areas and distinct pore structure.[41] The MOFs have recently gained attention as nano-adsorbents for various applications in wastewater treatment. By adding small-sized nanoparticles into the polymeric material, the properties like composition, surface structure, pore size and pore size distribution will be changed, resulting in the formation of good separation mixed-matrix membranes (MMMs).[42] Various methods such as direct crystallization, microfluidic processing, layer-by-layer assembly, interfacial/contra-diffusion synthesis, confinement conversion and chemical vapour deposition have been investigated to synthesis MOF membranes. Using MOF as a nano-adsorbent, various wastewater treatment approaches have been proposed. For example, for the removal of N,N-dimethylamine

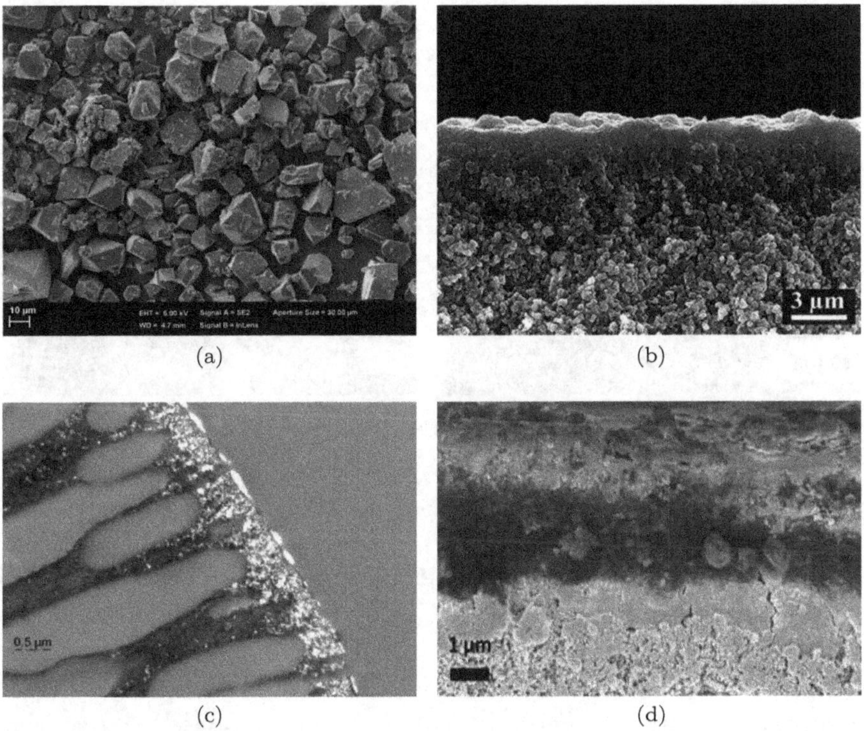

Fig. 12.6. SEM images: (a) HKUST-1 MOF, (b) Al_2O_3HF supported UiO-66, (c) TEM image of PES supported ZIF-8 film, and (d) C/S of CuBTC TFC membrane.[35,45,47,48] (Reproduced with permission from Elsevier Inc.; American Chemical Society; Royal Society of Chemistry.)

(DMA) or other carcinogenic by-products of pharmaceutical industries, water-resistant MOF-235 has exhibited excellent adsorptive nature towards DMA.[43] The excellent performance of Cu-MOFs has been reported for the treatment of toxic sulfonamide antibiotics from wastewater (Fig. 12.6(a)).[35] Lu *et al.* (2019)[44] reported a novel eco-friendly and sustainable cellulose supported Fe_3O_4 magnetic NPs and Cu-based MOF NPs catalyst CuBTC/Fe_3O_4/cellulose for adsorptive removal of water pollutants from wastewater. The MOF can be used in the removal of volatile components from aqueous mixtures and biofuel separation from fermentation broth. Liu *et al.* (2015)[45] prepared continuous 2.0 μm thick UiO-66 membrane on

Fig. 12.7. (a) TEM images of Cu/ZIF-8 nanocomposite, (b) SEM image of Fe/UiO-66.[50,51] (Reproduced with permission from Elsevier Inc.; Royal Society of Chemistry.)

ceramic hollow fibers by solvothermal method for the removal of metal ions like Ca^{2+}, Al^{3+}, and Mg^{2+} from water (Fig. 12.6(b)). Also, UiO-66 MOF was used in the adsorptive separation of perfluorooctanesulfonate from an aqueous mixture.[46] Li *et al.*[47] (2015) synthesized a ZIF-8 membrane used to remove Rose Bengal dye from water (Fig. 12.6(c)). Campbell *et al.*[48] (2015) prepared a CuBTC membrane used for the organic solvent nanofiltration (Fig. 12.6(d)). The polybenzimidazole (PBI)/zeolitic imidazolate frameworks (ZIF-8), that is PBI/ZIF-8, membrane was used for dehydration of various alcohols by adding ~50 nm size of ZIF-8 nanoparticles in PBI.[49] For the degradation of Rhodamine 6G dye, Chakraborty *et al.*[50] (2019) prepared composite photocatalyst of ~5 nm CuO NPs coated ZIF-8 using sol-gel method (Fig. 12.7(a)). Furthermore, UiO-66 supported Fe catalysts were used for the efficient catalytic removal of organic contaminants such as benzene derivatives and moderate for acetone removal from wastewater (Fig. 12.7(b)).[51] For the removal of organic pollutants such as phenolic contents, pharmaceuticals pollutants, dyes contaminants, and hazardous metal ions, it is reported that Fe-based MOFs demonstrated exceptional catalytic performance on the organic pollutants.[52] For catalytic adsorptive removal of 2,4,6-trinitrophenol from an aqueous mixture, using one-pot reflux

(a) (b) (c)

Fig. 12.8. SEM images: (a) MOF-199, (b) MIL-53(Fe)@CF MOF, and (c) MIL-101(Fe) coated with Fe_3O_4 @PDopa (Hamedi and Abdi, 2019; Le *et al.*, 2019; Mahmoodi and Abdi, 2019).[52–54] (Reproduced with permission from Elsevier Inc.; MDPI publishing.)

reaction, Guo *et al.*[53] (2019) prepared nanostructured composite hexagonal NH_2-MIL-88B(Fe) MOF. Using a modified solvothermal method, Mahmoodi and Abdi[54] (2019) prepared nanoporous MOF-199 as a nano-photocatalyst for removing Basic Blue-41 from wastewater (Fig. 12.8(a)). For the removal of total organic carbon and mineralization of Acid Orange 7 dye from wastewater, Le *et al.*[55] (2019) prepared catalytic electrodes on the porous carbon felt via carbonization of MIL-53(Fe) MOF (Fig. 12.8(b)). Hamedi *et al.*[56] (2019) prepared magnetic composite stable adsorbent MIL-101(Fe) coated with Fe_3O_4 NPs using eco-friendly and efficient binder 3,4-dihydroxy-L-phenylalanine (L-Dopa) for the adsorptive removal of malachite green and methyl red from wastewater (Fig. 12.8(c)). Rao *et al.*[57] (2017) prepared an IRMOF-3@GO composite material for the adsorptive removal of Cu^{+2}. For the removal of Au^{+3} from an aqueous mixture, Sun *et al.*[58] (2018) synthesized Fe coated MIL-100 on the para-phenylenediamine membrane.

12.3. Conclusions and Futuristic Aspects

It is well established that the traditional porous nanomaterials such as different types of nanoparticles (magnetic, polymeric, carbon based, etc) and MOFs (ZIF, copper-based, MIL-series, etc) are used for wastewater treatments. As the typical porous nanomaterials, it is well known that the materials such as metal-organic frameworks

(MOFs) and covalent organic frameworks (COFs) acquire fascinating surface properties like higher surface area, tuneable surface chemistry, a larger number of active sites accessible for various functional groups. These new classes of novel porous nanomaterials demonstrate an enormous potential in the removal of unwanted contaminants by adsorptive separation or by catalytic degradation from wastewater. The MOFs are not only used in the removal of organic contaminants but also used in the removal of dyes, microextractions, as a sensor material in detectors, and in the removal of hazardous and toxic metal ions from wastewaters. Further research is required to better understand the interactions between organic contaminants and MOFs. There is also a need to improve the adsorption capacities of available nanomaterials and MOFs. As a futuristic prospect, the future research orientations for NPs and newly developed MOF materials should be in terms of easy way of preparation; mechanical, thermal and chemical stability; effective and efficient adsorption; active adsorption or catalytic sites; eco-friendly and cost effective; recovery and reusability. Overall, great advances have been performed on the development of MOFs for the removal of emerging organic contaminants (EOCs) from water.[59]

References

1. S. Jamaly, A. Giwa, S. W. Hasan, Recent improvements in oily wastewater treatment: Progress, challenges, and future opportunities, *J. Environ. Sci.*, **37** (2015) 15–30. DOI: 10.1016/j.jes.2015.04.011
2. X. Qu, J. Brame, Q. Li, P. J. J. Alvarez, Nanotechnology for a safe and sustainable water supply: Enabling integrated water treatment and reuse, *Acc. Chem. Res.*, **46** (2013a) 834–843. DOI: 10.1021/ar300029v
3. X. Qu, P. J. J. Alvarez, Q. Li, Applications of nanotechnology in water and wastewater treatment, *Water Res.*, **47** (2013b) 3931–3946. DOI: 10.1016/j.watres.2012.09.058
4. I. Erucar, G. Yilmaz, S. Keskin, Recent advances in metal-organic framework-based mixed matrix membranes, *Chem. Asian J.*, **8** (2013) 1692–1704. DOI: 10.1002/asia.201300084
5. D. Zacher, O. Shekhah, C. Wöll, R. A. Fischer, Thin films of metal–organic frameworks, *Chem. Soc. Rev.*, **38** (2009) 1418–1429. DOI: 10.1039/b805038b

6. B. Shi, P. Marchetti, D. Peshev, S. Zhang, A. G. Livingston, Will ultra-high permeance membranes lead to ultra-efficient processes? Challenges for molecular separations in liquid systems, *J. Membr. Sci.*, **525** (2017) 35–47. DOI: 10.1016/j.memsci.2016.10.014
7. W. Li, Y. Zhang, Q. Li, G. Zhang, Metal—organic framework composite membranes: Synthesis and separation applications, *Chem. Eng. Sci.*, **135** (2015) 232–257. DOI: 10.1016/j.ces.2015.04.011
8. S. V. Jadhav, E. Bringas, G. D. Yadav, V. K. Rathod, I. Ortiz, K. V. Marathe, Arsenic and fluoride contaminated groundwaters: A review of current technologies for contaminants removal, *Journal of Environmental Management*, **162** (2015) 306–325. DOI: 10.1016/j.jenvman.2015.07.020
9. A. Stafiej, K. Pyrzynska, Adsorption of heavy metal ions with carbon nanotubes, *Sep. Purif. Technol.*, **58** (2007) 49–52. DOI: 10.1016/j.seppur.2007.07.008
10. B. Pan, B. Xing, Adsorption mechanisms of organic chemicals on carbon nanotubes, *Environ. Sci. Technol.*, **42** (2008) 9005–9013. DOI: 10.1021/es801777n
11. K. Yang, B. Xing, Adsorption of organic compounds by carbon nanomaterials in aqueous phase: Polanyi theory and its application, *Chem. Rev.*, **110** (2010) 5989–6008. DOI: 10.1021/cr100059s
12. L. Ji, W. Chen, L. Duan, D. Zhu, Mechanisms for strong adsorption of tetracycline to carbon nanotubes: A comparative study using activated carbon and graphite as adsorbents, *Environ. Sci. Technol.*, **43** (2009) 2322–2327. DOI: 10.1021/es803268b
13. W. Li, C. Y. Cao, L. Y. Wu, M. F. Ge, W. G. Song, Superb fluoride and arsenic removal performance of highly ordered mesoporous alumina, *J. Hazard. Mater.*, **198** (2011) 143–150. DOI: 10.1016/j.jhazmat.2011.10.025
14. H. J. Wang, A. L. Zhou, F. Peng, H. Yu, L. F. Chen, Adsorption characteristic of acidified carbon nanotubes for heavy metal Pb(II) in aqueous solution, *Mater. Sci. Eng. A*, **466** (2007) 201–206. DOI: 10.1016/j.msea.2007.02.097
15. H. Wang, N. Yan, Y. Li, X. Zhou, J. Chen, B. Yu, M. Gong, Q. Chen, Fe nanoparticle-functionalized multi-walled carbon nanotubes: One-pot synthesis and their applications in magnetic removal of heavy metal ions, *J. Mater. Chem.*, **22** (2012) 9230–9236. DOI: 10.1039/c2jm16584h
16. C. Zhang, J. Sui, J. Li, Y. Tang, W. Cai, Efficient removal of heavy metal ions by thiol-functionalized superparamagnetic carbon nanotubes, *Chem. Eng. J.*, **210** (2012) 45–52. DOI: 10.1016/j.cej.2012.08.062

17. G. Zhao, X. Ren, X. Gao, X. Tan, J. Li, C. Chen, Y. Huang, X. Wang, Removal of Pb(ii) ions from aqueous solutions on few-layered graphene oxide nanosheets, *Dalton Trans.*, **40** (2011) 10945. DOI: 10.1039/c1dt11005e
18. L.-S. Zhong, J.-S. Hu, H.-P. Liang, A.-M. Cao, W.-G. Song, L.-J. Wan, Self-assembled 3D flowerlike iron oxide nanostructures and their application in water treatment, *Adv. Mater.*, **18** (2006) 2426–2431. DOI: 10.1002/adma.200600504
19. L.-S. Zhong, J.-S. Hu, A.-M. Cao, Q. Liu, W.-G. Song, L.-J. Wan, 3D flowerlike ceria micro/nanocomposite structure and its application for water treatment and CO removal, *Chem. Mater.*, **19** (2007) 1648–1655. DOI: 10.1021/cm062471b
20. M. Hua, S. Zhang, B. Pan, W. Zhang, L. Lv, Q. Zhang, Heavy metal removal from water/wastewater by nanosized metal oxides: A review, *J. Hazard. Mater.*, **211–212** (2012) 317–331. DOI: 10.1016/j.jhazmat.2011.10.016
21. P. Xu, G. M. Zeng, D. L. Huang, C. L. Feng, S. Hu, M. H. Zhao, C. Lai, Z. Wei, C. Huang, G. X. Xie, Z. F. Liu, Use of iron oxide nanomaterials in wastewater treatment: A review, *Sci. Total Environ.*, **424** (2012) 1–10. DOI: 10.1016/j.scitotenv.2012.02.023
22. I. Andjelkovic, D. N. H. Tran, S. Kabiri, S. Azari, M. Markovic, D. Losic, Graphene aerogels decorated with α-FeOOH nanoparticles for efficient adsorption of arsenic from contaminated waters, *ACS Appl. Mater. Interfaces*, **7** (2015) 9758–9766. DOI: 10.1021/acsami.5b01624
23. S. Dixit, J. G. Hering, Comparison of arsenic(V) and arsenic(III) sorption onto iron oxide minerals: Implications for arsenic mobility, *Environ. Sci. Technol.*, **37** (2003) 4182–4189. DOI: 10.1021/es030309t
24. B. J. Lafferty, R. H. Loeppert, Methyl arsenic adsorption and desorption behavior on iron oxides, *Environ. Sci. Technol.*, **39** (2005) 2120–2127. DOI: 10.1021/es048701+
25. M. Kumari, C. U. Pittman, D. Mohan, Heavy metals [chromium (VI) and lead (II)] removal from water using mesoporous magnetite (Fe_3O_4) nanospheres, *J. Colloid Interface Sci.*, **442** (2015) 120–132. DOI: 10.1016/j.jcis.2014.09.012
26. C.-G. Lee, S. Lee, J.-A. Park, C. Park, S. J. Lee, S.-B. Kim, B. An, S.-T. Yun, S.-H. Lee, J.-W. Choi, Removal of copper, nickel and chromium mixtures from metal plating wastewater by adsorption with modified carbon foam, *Chemosphere*, **166** (2017) 203–211. DOI: 10.1016/j.chemosphere.2016.09.093
27. P. N. Diagboya, B. I. Olu-Owolabi, K. O. Adebowale, Synthesis of covalently bonded graphene oxide–iron magnetic nanoparticles and the kinetics of mercury removal, *RSC Adv.*, **5** (2015) 2536–2542. DOI: 10.1039/C4RA13126F

28. P. Z. Ray, H. J. Shipley, Inorganic nano-adsorbents for the removal of heavy metals and arsenic: a review, *RSC Adv.*, **5** (2015) 29885–29907. DOI: 10.1039/C5RA02714D
29. N. Sankararamakrishnan, M. Jaiswal, N. Verma, Composite nanofloral clusters of carbon nanotubes and activated alumina: An efficient sorbent for heavy metal removal, *Chem. Eng. J.*, **235** (2014) 1–9. DOI: 10.1016/j.cej.2013.08.070
30. G. Lofrano, M. Carotenuto, G. Libralato, R. F. Domingos, A. Markus, L. Dini, R. K. Gautam, D. Baldantoni, M. Rossi, S. K. Sharma, M. C. Chattopadhyaya, M. Giugni, S. Meric, Polymer functionalized nanocomposites for metals removal from water and wastewater: An overview, *Water Res.*, **92** (2016) 22–37. DOI: 10.1016/j.watres.2016.01.033
31. M. Li, M. Li, C. Feng, Q. Zeng, Preparation and characterization of multi-carboxyl-functionalized silica gel for removal of Cu (II), Cd (II), Ni (II) and Zn (II) from aqueous solution, *Appl. Surf. Sci.*, **314** (2014) 1063–1069. DOI: 10.1016/j.apsusc.2014.06.038
32. F. Ge, M.-M. Li, H. Ye, B.-X. Zhao, Effective removal of heavy metal ions Cd^{2+}, Zn^{2+}, Pb^{2+}, Cu^{2+} from aqueous solution by polymer-modified magnetic nanoparticles, *J. Hazard. Mater.*, **211–212** (2012) 366–372. DOI: 10.1016/j.jhazmat.2011.12.013
33. M. S. Diallo, L. Balogh, A. Shafagati, J. H. Johnson, W. A. Goddard, D. A. Tomalia, Poly(amidoamine) dendrimers: A new class of high capacity chelating agents for Cu(II) ions, *Environ. Sci. Technol.*, **33** (1999) 820–824. DOI: 10.1021/es980521a
34. F. Fu, B. Zheng, L. H. Xie, H. Du, S. Du, Z. Dong, Size-controllable synthesis of zeolitic imidazolate framework/carbon nanotube composites, *Crystals*, **8** (2018) 1–12. DOI: 10.3390/cryst8100367
35. M. R. Azhar, H. R. Abid, H. Sun, V. Periasamy, M. O. Tadé, S. Wang, Excellent performance of copper based metal organic framework in adsorptive removal of toxic sulfonamide antibiotics from wastewater, *J. Colloid Interface Sci.*, **478** (2016) 344–352. DOI: 10.1016/j.jcis.2016.06.032
36. X. Zhu, J. Gu, Y. Wang, B. Li, Y. Li, W. Zhao, J. Shi, Inherent anchorages in UiO-66 nanoparticles for efficient capture of alendronate and its mediated release, *Chem. Commun.*, **50** (2014) 8779–8782. DOI: 10.1039/C4CC02570A
37. S. S. Kaye, A. Dailly, O. M. Yaghi, J. R. Long, Impact of preparation and handling on the hydrogen storage properties of Zn_4O(1,4-benzenedicarboxylate)$_3$ (MOF-5), *J. Am. Chem. Soc.*, **129** (2007) 14176–14177. DOI: 10.1021/ja076877g

38. Y. Yang, P. Shukla, S. Wang, V. Rudolph, X.-M. Chen, Z. Zhu, Significant improvement of surface area and CO_2 adsorption of Cu–BTC via solvent exchange activation, *RSC Adv.*, **3** (2013) 17065. DOI: 10.1039/c3ra42519c
39. S. Yuan, L. Feng, K. Wang, J. Pang, M. Bosch, C. Lollar, Y. Sun, J. Qin, X. Yang, P. Zhang, Q. Wang, L. Zou, Y. Zhang, L. Zhang, Y. Fang, J. Li, H.-C. Zhou, Stable Metal-organic frameworks: Design, synthesis, and applications, *Adv. Mater.*, **30** (2018) 1704303. DOI: 10.1002/adma.201704303
40. K. Fujie, K. Otsubo, R. Ikeda, T. Yamada, H. Kitagawa, Low temperature ionic conductor: ionic liquid incorporated within a metal–organic framework, *Chem. Sci.*, **6** (2015) 4306–4310. DOI: 10.1039/C5SC01398D
41. G. Férey, Hybrid porous solids: past, present, future, *Chem. Soc. Rev.*, **37** (2008) 191–214. DOI: 10.1039/B618320B
42. M. Kalaj, K. C. Bentz, S. Ayala, J. M. Palomba, K. S. Barcus, Y. Katayama, S. M. Cohen, MOf-polymer hybrid materials: From simple composites to tailored architectures, *Chem. Rev.*, (2020) DOI: 10.1021/acs.chemrev.9b00575
43. R. Beltrán-Suito, A. Pinedo-Flores, F. Bravo-Hualpa, J. Ramos-Muñoz, M. Sun-Kou, R. del, Adsorption of N,N-dimethylamine from aqueous solutions by a metal organic framework, MOF — 235, *J. Dispers. Sci. Technol.*, **40** (2019) 901–908. DOI: 10.1080/01932691.2018.1489275
44. H. Lu, L. Zhang, B. Wang, Y. Long, M. Zhang, J. Ma, A. Khan, S. P. Chowdhury, X. Zhou, Y. Ni, Cellulose-supported magnetic Fe_3O_4–MOF composites for enhanced dye removal application, *Cellulose*, **26** (2019) 4909–4920. DOI:10.1007/s10570-019-02415-y
45. X. Liu, N. K. Demir, Z. Wu, K. Li, Highly water-stable zirconium metal–organic framework UiO-66 membranes supported on alumina hollow fibers for desalination, *J. Am. Chem. Soc.*, **137** (2015) 6999–7002. DOI: 10.1021/jacs.5b02276
46. C. A. Clark, K. N. Heck, C. D. Powell, M. S. Wong, Highly defective UiO-66 materials for the adsorptive removal of perfluorooctanesulfonate, *ACS Sustain. Chem. Eng.*, **7** (2019) 6619–6628. DOI: 10.1021/acssuschemeng.8b05572
47. Y. Li, L. H. Wee, A. Volodin, J. A. Martens, I. F. J. Vankelecom, Polymer supported ZIF-8 membranes prepared via an interfacial synthesis method, *Chem. Commun.*, **51** (2015) 918–920. DOI: 10.1039/C4CC06699E

48. J. Campbell, R. P. Davies, D. C. Braddock, A. G. Livingston, Improving the permeance of hybrid polymer/metal–organic framework (MOF) membranes for organic solvent nanofiltration (OSN) — development of MOF thin films via interfacial synthesis, *J. Mater. Chem. A*, **3** (2015) 9668–9674. DOI: 10.1039/C5TA01315A
49. J. G. Wijmans, R. W. Baker, The solution-diffusion model: A review, *J. Membr. Sci.*, **107** (1995) 1–21. DOI: 10.1016/0376-7388(95)00102-I
50. A. Chakraborty, D. A. Islam, H. Acharya, Facile synthesis of CuO nanoparticles deposited zeolitic imidazolate frameworks (ZIF-8) for efficient photocatalytic dye degradation, *J. Solid State Chem.*, **269** (2019) 566–574. DOI: 10.1016/j.jssc.2018.10.036
51. H. Zhuang, B. Chen, W. Cai, Y. Xi, T. Ye, C. Wang, X. Lin, UiO-66-supported Fe catalyst: a vapour deposition preparation method and its superior catalytic performance for removal of organic pollutants in water, *Royal Soc. Open Sci.*, **6** (2019) 182047. DOI: 10.1098/rsos.182047
52. V. K. Sharma, M. Feng, Water depollution using metal-organic frameworks-catalyzed advanced oxidation processes: A review, *J. Hazard. Mater.*, **372** (2019) 3–16. DOI: 10.1016/j.jhazmat.2017.09.043
53. H. Guo, B. Niu, X. Wu, Y. Zhang, S. Ying, Effective removal of 2,4,6-trinitrophenol over hexagonal metal-organic framework NH_2-MIL-88B(Fe), *Appl. Organomet. Chem.*, **33** (2019) e4580. DOI: 10.1002/aoc.4580
54. N. M. Mahmoodi, J. Abdi, Nanoporous metal-organic framework (MOF-199): Synthesis, characterization and photocatalytic degradation of Basic Blue 41, *Microchem. J.*, **144** (2019) 436–442. DOI: 10.1016/j.microc.2018.09.033
55. Le, Cowan, Drobek, Bechelany, Julbe, Cretin, Fe-Nanoporous carbon derived from MIL-53(Fe): A heterogeneous catalyst for mineralization of organic pollutants, *Nanomaterials*, **9** (2019) 641. DOI: 10.3390/nano9040641
56. A. Hamedi, M. B. Zarandi, M. R. Nateghi, Highly efficient removal of dye pollutants by MIL-101(Fe) metal-organic framework loaded magnetic particles mediated by Poly L-Dopa, *J. Environ. Chem. Eng.*, **7** (2019) 102882. DOI: 10.1016/j.jece.2019.102882
57. Z. Rao, K. Feng, B. Tang, P. Wu, Surface decoration of amino-functionalized metal–organic framework/graphene oxide composite onto polydopamine-coated membrane substrate for highly efficient heavy metal removal, *ACS Appl. Mater. Interfaces*, **9** (2017) 2594–2605. DOI: 10.1021/acsami.6b15873

58. D. T. Sun, L. Peng, W. S. Reeder, S. M. Moosavi, D. Tiana, D. K. Britt, E. Oveisi, W. L. Queen, Rapid, selective heavy metal removal from water by a metal–organic framework/polydopamine composite, *ACS Cent. Sci.*, **4** (2018) 349–356. DOI: 10.1021/acscentsci.7b00605
59. S. Rojas, P. Horcajada, Metal-organic frameworks for the removal of emerging organic contaminants in water, *Chem. Rev.*, (2020). Article ASAP. DOI: 10.1021/acs.chemrev.9b00797

CHAPTER 13

Light Transmitting Substrates for Convenient Solar Illumination of Nanophotocatalyst Coatings on Membranes for Low Pressure Water Filtration

Lavern T. Nyamutswa[a], Stephen F. Collins[b], Dimuth Navaratna[a,b] and Mikel C. Duke [a,*]

[a] *Institute for Sustainable Industries and Liveable Cities, Victoria University, Melbourne, Australia*
[b] *College of Engineering and Science, Victoria University, Melbourne, Australia*

13.1. Introduction

Membrane filtration technology has been widely adopted for water treatment with considerable success owing to several attributes such as ease of automation, compact design capability and effectiveness in water contaminant removal.[1,2] However, membrane processes suffer from contaminant deposition and build-up on the membrane surface, a phenomenon known as fouling. Since a membrane's functionality is based on being semi-permeable, anything that effectively blocks its pores renders it ineffective.

Membrane fouling leads to reduced process efficiency, such as reduced water flux, which in turn leads to reduced membrane

*Corresponding author: mikel.duke@.vu.edu.au

lifetime and higher costs associated with membrane cleaning and replacement.[3] It can also affect water quality and increase maintenance and operating costs.[4] In considering any membrane process, approaches to manage membrane fouling must be considered. The most common anti-fouling method of chemically cleaning the membrane brings challenges such as introducing secondary pollution, downtime when the cleaning is taking place, as well as increased costs through chemical procurement, transport, storage and disposal.

Meanwhile, heterogeneous semiconductor photocatalysis has been shown to be effective in degrading water contaminants such as organic compounds and deactivating microorganisms.[5,6] Despite being heavily studied, photocatalysis in water treatment has not seen meaningful commercial adoption due to several practical limitations. The most common heterogeneous semiconductor photocatalyst used in water treatment, nano-sized titania (TiO_2), is difficult to separate from water in slurry reactors after its role in treatment is complete. It also tends to agglomerate in water, reducing its activity. Photocatalysis is also widely understood to be too energy intensive for practical use in water treatment.

Researchers have shown that combining photocatalysis and membranes into one unit called a photocatalytic membrane can help to solve these challenges, hence there is a significant interest in this area of water treatment.[7,8] There is a symbiotic relationship between the membrane and the photocatalyst, whereby the membrane solves the separation challenge while the photocatalyst, through photocatalytic oxidation and induced super-hydrophilicity, helps to mitigate fouling. However, the practical application of photocatalytic membranes is still largely hindered by current configurations which make it difficult to integrate light sources into the setup. Despite publications on fundamental research on photocatalytic membranes continuing to rise each year, water treatment plants which apply this technology, or point of use water purification devices that are based on photocatalytic membranes, are still extremely hard to come across.

An alternative configuration, made possible by replacing conventional opaque membrane substrates with a novel light transmitting

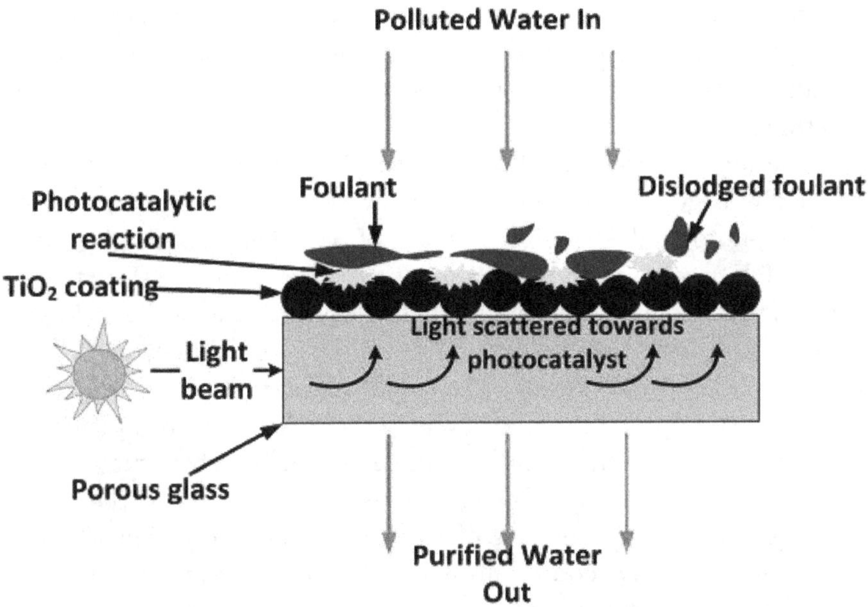

Fig. 13.1. The new concept for directing light through porous substrates to a functional photocatalytic membrane coating.[9]

substrate, is shown in Fig. 13.1.[9] This would allow light to be conveniently directed from the end of membrane modules, while also avoiding the inefficient traditional method of directing light through the light absorbing feed stream. The purpose of this chapter is to review and discuss progress in substrates that offer the desired transparency for directing light to a photocatalyst coating, while maintaining the required porosity to be a suitable support for a membrane.

13.2. Background

13.2.1. *Titania as a photocatalyst in water treatment*

Titania is one of the most studied semiconductor heterogeneous photocatalyst targeted for water treatment. The interest in titania has continued to rise since the discovery of the water splitting ability of a titania anode and platinum counter electrode nearly five decades ago.[10] Desirable properties such as chemical and biological inertness, photo-stability, non-toxicity, super-hydrophilicity, adsorption properties, and antimicrobial properties contribute to make titania the dominant heterogeneous photocatalyst studied for water treatment.[11]

The most photoactive phase of titanium dioxide, anatase, has a relatively large energy band gap of 3.2 eV, which means activation of the photocatalyst requires light of wavelengths in the ultraviolet (UV) region of the electromagnetic spectrum.[12] As such, solar activation of anatase titania is inefficient because sunlight largely consists of infrared radiation (52% to 55%), visible light (40%) and just 4–5% UV. This titania limitation has largely been addressed by impurity doping[13–18] and other methods such as sensitisation,[19] surface modification,[20] and blending it with other materials,[21] resulting in the band gap being successfully narrowed to the visible light region. Nitrogen doping of titania in a polyvinylidene fluoride (PVDF) dual layer hollow fibre membrane, for example, reduced the band gap energy to as low as 2.63 eV, which is well within the visible light region.[22] Simulated solar light from a 300 W Xenon lamp with emission wavelengths from 400–780 nm was also successfully applied on silver doped titania immobilised on a polyacrylonitrile (PAN) membrane to achieve almost 100% degradation of a 10 mg/L methylene blue solution in 1 hour.[23] The energy band gap in this case was lowered to 2.53 eV.

Even when the band gap has been successfully narrowed through modification, titania agglomeration in water and the requirement to separate it from water after the treatment process still pose challenges. Researchers have attempted to solve these problems by immobilizing nano-sized titania onto various substrates such as polymer films,[24] glass,[25] sand,[26] activated carbon[27] and clay.[28] Others have also attempted to reclaim suspended nano-sized titania through membrane filtration.[29] This has led to the development of a hybrid water treatment process which combines membrane filtration and photocatalytic oxidation, now widely known as a photocatalytic membrane reactor (PMR). The PMR can be configured in several ways. One of the most common configurations is having a slurry photocatalytic reactor that precedes a membrane filtration unit.

The membrane can either be placed in a separate unit that comes after the slurry reactor, or it can be submerged in the reactor itself. In the other common configuration, known as a photocatalytic membrane, a photocatalyst, such as titania, is coated onto a porous membrane substrate.[30] The photocatalytic membrane is interesting for future water treatment technologies because it combines photocatalysis and filtration into one unit.

13.2.2. *Photocatalytic membranes*

The first reported photocatalytic membranes were fabricated from alumina and titania through sol-gel techniques three decades ago.[31] However, the study was more concerned with determining the chemical and physical properties of the membranes rather than evaluating their photocatalytic activity. The now widely used term "photocatalytic membrane" was first mentioned in 1991,[32] even though the "membrane" was actually titania supported on glass. In this study, the photocatalytic degradation of an organic compound, 3-chlorosalicylic acid, facilitated by a supported photocatalyst, was investigated and validated. The dual function of photocatalytic membranes, namely permeability and photocatalytic activity, were first reported in the year 2006.[33] In the study, various titania sols were dip-coated onto alumina substrates to obtain a hierarchical mesoporous multilayer structure. The multilayer coating procedure improved water permeability while maintaining the retention of organics as well as the photocatalytic activity of the membranes. The key recommendations from this study was that the photocatalytic membranes had great potential in being applied in efficient water treatment and water reuse systems. Specifically, the study pointed out that the membranes could be applied for physical separation of water contaminants while exhibiting anti-fouling action, as well as decomposing organic contaminants and inactivating pathogenic microorganisms. Zhang *et al.* carried the same studies on titania-silica supported on alumina membranes in the same year, and came up with the same key recommendations.[34] They also demonstrated the benefits of coupling membrane technology with photocatalysis, whereby removal of Direct Black 168 from water by the photocatalytic membrane was 85% in 100 minutes, compared to 66% and 73% by photocatalysis and membrane separation as standalone methods, respectively. These works have inspired huge interest in this field, leading to the development of many synthesis routes for photocatalytic membranes followed by their evaluation for application in water treatment.

Substrates that have been used to fabricate photocatalytic membranes include metallic, organic, and inorganic materials, with organic and ceramic materials being the most common.[30] Those who propose alternatives to organic materials cite membrane structure deterioration and loss of the deposited titania layer over time. Oxidising conditions brought about by titania in the presence of light can also result in reactions which damage the organic polymeric host material, rendering the membrane non-functional through loss of rejection capability.[35] Moreover, if the titania is not chemically bonded to the polymer, which is common in blended membranes, it is more likely to detach and leave the host material over time. Non-chemically bonded titania also tends to aggregate into large clusters in polymer matrices, greatly reducing photocatalytic activity.[3]

The attraction to ceramics as membrane materials is due to desirable properties such as high chemical and thermal stability, high liquid flux, anti-fouling properties, low environmental impact as well as robustness.[36] Despite these excellent properties, ceramic materials are mostly opaque, which has meant light to activate the photocatalyst has to be directed through the water to be treated to reach the photocatalyst on the membrane, as shown in Fig. 13.2. Directing light this way can be a challenge in complex membrane element designs (e.g. ceramic monoliths). Integrating light sources into the photocatalytic membrane for practical purposes, especially in a large scale operation, can be a challenge. Organic water pollutants can also strongly absorb light before it reaches the photocatalyst,[37] rendering the process inefficient.

Because of these challenges, an interesting new concept which has been proposed is to use membrane materials which are transparent

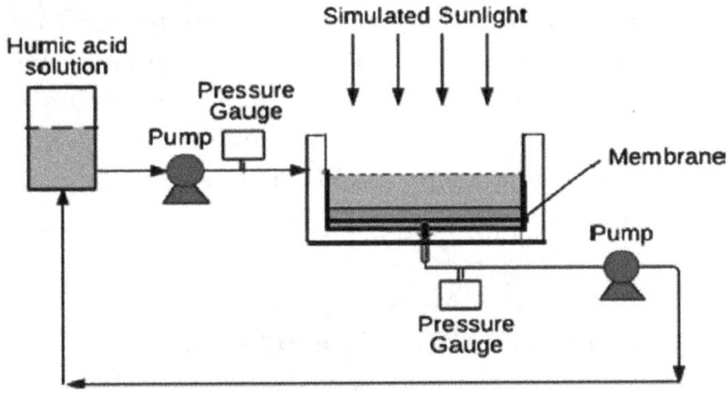

Fig. 13.2. Schematic showing the traditional means to direct light to photocatalytic membranes.[4]

to light in the UV and visible regions. This gives an opportunity to conduct light via the substrate instead of water, with the light being conveniently directed from the end of the element.[9] This could simplify module arrangement, whereby light sources, such as UV-LEDs attached to the end of a membrane monolith, direct light along its length and scatter it directly to the photocatalyst. The end of the monolith could simply be exposed to sunlight in the case of solar active photocatalysts. Light absorption by water contaminants is also avoided, as the light path does not need to pass through the untreated water.

13.3. Light Transmitting Photocatalytic Membrane

13.3.1. *Titania immobilisation methods*

Coupling photocatalysts with support substrates by immobilisation can be achieved through several well established methods. These include dip coating,[38] atomic layer deposition,[39] filtration[40] hydrothermal growth,[41,42] photo-deposition,[43] hot pressing,[44] drop casting,[45] electrospinning,[46] self-assembly,[47] electro-spraying,[44] chemical vapour deposition[48] and magnetron sputtering.[42] These and other methods have been critically reviewed and further explained in detail.[49] Wet chemical methods such as dip-coating and electrospinning are commonly used due to their cost effectiveness and scalability.[50] However, wet chemical methods are limited to thermally stable substrates because the final step in the preparation is high temperature calcination. On the other hand, even though micro-fabrication methods do not require high temperature treatment, they are carried out under stringent conditions in elaborate experimental facilities. As such, they may not be economically cost effective or scalable.[50]

13.3.2. *Choosing an appropriate substrate*

When choosing materials to develop a substrate to transmit light to a titania photocatalytic membrane coating, it follows that the most important thing to consider is to choose a light transmitting substrate. The next step is to select the attributes of a suitable substrate for immobilising a photocatalyst such as correct pore size and surface roughness. Generally, the substrate should be resistant to photocatalytic attack, and should offer strong and permanent attachment of the photocatalyst. Both the substrate and photocatalyst should also not undergo physical or chemical changes that alter their functionality during and after the fabrication process.[30] Other characteristics of a good membrane substrate include high porosity, high surface to volume ratio, adjustable morphology, as well as surface functionality that can be controlled or modified.[51] Functional groups such as carboxyl and carbonyl groups that exist on polymeric surfaces provide a means for titania attachment through oxygen bridges and hydrogen bonds.[52] When thermal stability is of importance, ceramic substrates are more ideal than polymers. Ceramic substrates are also more mechanically robust and can withstand harsh chemical environments.

13.3.3. *Light transparent substrates for photocatalytic membrane water treatment*

Several transparent materials that could be used as substrates for photocatalytic membranes are available in literature, even though the application for this purpose has not yet been evaluated. Among them are cellulose,[53–55] aramid nanofibers (ANFs) blended with bacterial cellulose (BC),[56] polylactic acid (PLA),[57] single-walled carbon nanotubes (SWCNTs),[58–60] silk fibroin,[61] poly(hydroxyethyl acrylamide) (PHEAAm) grafted onto polydimethylsiloxane (PDMS)[62] and free standing graphene oxide.[63] However, the common drawback that most of these substrates have is that they are biodegradable, hence stability in water could be a challenge.

Glass substrates, on the other hand, do not pose a biodegradation challenge. The application of glass substrates in membrane water treatment is not entirely new in the literature. Examples of glass types that have been used for this purpose include Vycor, Shirasu and sintered borosilicate.

13.3.3.1. *Sintered glass*

13.3.3.1.1. Preparation and properties

Particles of broken glass, known as glass cullet, have been applied in water and wastewater treatment as a substitute for filtration media such as sand, anthracite and garnet.[64] In membrane water treatment, glass ceramics are an attractive option due to the possibility of having high porosity membranes with high mechanical strength. Because of the high energy requirements for glass ceramics manufacturing, using glass cullet, which is otherwise regarded as waste, could offset some of these costs. The cullet can be pulverised into the desired particle size followed by a sintering process to form a membrane of the desired dimensions.

13.3.3.1.2. Potential advantages and disadvantages

The high porosity achievable by sintering glass particles into membranes is ideal for high flux filtration operations. However, a potential disadvantage of porous sintered glass membranes for light distribution is the low UV and visible light transmission that they exhibit due to the glass particles acting as light scattering centres. A previous study has found that wetting the membrane can improve the light transmission due to the refractive index of water being closer to that of the glass particles than that of air.[9] Filling the void space with water therefore reduces refraction and promotes transmission.

Sintered glass membranes with high porosity have more of these void spaces, which translates to better light transmission as well.

13.3.3.1.3. Application in membrane water treatment

Sintered glass filters in the form of flat discs are readily available on the market, albeit for small scale filtration applications only, such as laboratories. Sintered glass substrates made from borosilicate glass particles have been dip-coated with titania and applied in treatment of synthetic waste water.[9] It was found that the substrate offered enough UV light transmission (with a peak wavelength of 365 nm) to facilitate a photocatalytic effect on the membrane surface, as well as providing a means for membrane fouling control through photocatalytic degradation of the water contaminants and induced super-hydrophilicity. Although this work demonstrated a functional photocatalytic effect from light distributed via a porous glass substrate, custom designed materials with the correct transmission and scattering to the catalytic coating along a full length monolith in the order of hundreds of cm is required.

13.3.3.2. *Phase separated porous glass*

13.3.3.2.1. Preparation and properties

Porous glasses are prepared from leaching phase separated alkali borosilicate glasses. The phase separation in a borosilicate glass of suitable composition is initiated by heat treatment of the glass to between 500°C and 580°C, resulting in the formation of two interconnected phases, one which is up to 96% silica, and the other one being an alkali-rich borate phase which is soluble in hot mineral acids and alcohols, or even water. Leaching out the soluble phase gives porous glass. Apart from the initial glass composition, the texture and structure of the final porous glass is governed by parameters such as the heating time and temperature, as well as the leaching conditions.[65,66] Generally, long heat treatment times at high temperatures and slower cooling rates result in more phase separation, increased silica solubility in the borate phase, and hence larger pores upon acid leaching.

One of the most suitable porous glasses for the new concept of light transmitting membrane substrates is Vycor 7930, which is a porous glass that was first developed by the Corning Glass Company. Vycor 7930 is mostly silica, which makes up about 96% of its composition, followed by 3% boron trioxide and 0.4% sodium oxide. Its pore diameters range from 1 nm[67] to 20.3 μm.[68] It also has an internal surface area of approximately 250 m^2/g and an average porosity of 28%.[67,69] Its UV and visible light transmittance is very high, averaging about 90%. Vycor's chemical composition is similar to dense glasses that have been used as support for titania films, such as borosilicate glass. Because of these similarities, it would be straightforward to immobilise photocatalysts on the Vycor surface using the same techniques used to make photocatalyst films on dense glass slides.

Yazawa *et al.*[70] describe a procedure to fabricate porous glasses with pore sizes as low as 0.5 nm from initial glasses containing Na_2O–B_2O_3–SiO_2. Asymmetric membrane support substrates have also been prepared by dip-coating α-Al_2O_3 tubes in suspensions of borosilicate glass particles consisting of 61.2% SiO_2, 29.7% B_2O_3 and 9.1% Na_2O, followed by sintering to induce phase separation and hydrochloric acid leaching to obtain a composite with pore sizes of 1–4 nm and nitrogen permeance up to 100 times more than Vycor tubing.[71] Such glasses could easily be adopted for water treatment membrane applications.

13.3.3.2.2. Potential advantages and disadvantages

The excellent light transmission properties of porous glasses such as Vycor 7930 in the wavelengths of interest for activation of titania makes it an attractive substrate for photocatalytic membranes. Its structure, which includes surface hydroxyl groups, also makes it suitable for strong adhesion of titania through the formation of oxygen bridges. The manufacturing process allows for control of pore sizes, therefore these can be tailored to suit the type of membrane needed and target water contaminants. Vycor 7930's mechanical strength and chemical inertness, as well as its resistance to dusting and flaking also make it a good candidate for membrane water treatment applications.[72] Porous glasses can be prepared in several geometries which include plates and tubes, which make them attractive for targeted applications[66] particularly membrane filtration.

The heat intensive manufacturing process of the phase separated glass, as well as the need for leaching by alcohols or acids, could be a drawback in the wider adaptation of Vycor 7930, or indeed other porous glasses manufactured using the same process, since this makes the final product costly.

13.3.3.2.3. Application in membrane water treatment

Vycor 7930 has been previously considered as membrane material for desalination (after modification with SiO_2 to reduce the pore sizes) and gas separations.[67] Zhakarov studied the permeability of Vycor capillary porous borosilicate glass membranes with a pore size distribution of 1.8–2.8 nm, 5.0–5.4 nm and over 10.0 nm by forcing through solutions of different salts at pressures of 15 MPa and above.[73] They found that the glass structure does not change with time, and the permeate flux remains constant after 4–5 hours. The permeate flux of 0.059 M NaCl, for example, is reported as $0.29\,Lm^{-2}h^{-1}$. Coating porous glasses with titania could further reduce the pore size and allow exclusion of even the smallest organic compounds and other water contaminants, while the photocatalyst provides oxidative degradation, disinfection as well as anti-fouling capability to the membrane.

13.3.3.3. *Shirasu porous glass*

13.3.3.3.1. Preparation and properties

Shirasu porous glass (SPG) is another example of a porous glass prepared from phase separated glass. Shirasu, a volcanic ash from Kyushu Island, Japan, is added to the glass formulation as a source of silica (SiO_2) and alumina. The primary glass used to make SPG contains CaO, Al_2O_3, B_2O_3 and SiO_2. This is heat treated to 650–750°C to induce separation of the Al_2O_3–SiO_2 and CaO–B_2O_3 phases, followed by hydrochloric acid leaching of the borate phase to

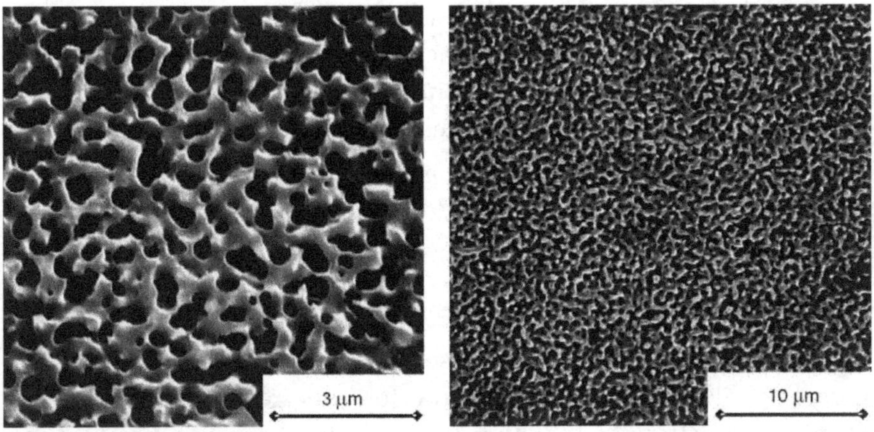

Fig. 13.3. Scanning electron micrographs of a SPG membrane with volume weighted mean pore diameter of 0.262 μm.[75]

obtain SPG.[74,75] Much like the preparation of other porous glasses, control of the pore size is achieved by adjusting the heating time and temperature, as well as the acid leaching conditions. The pore size can range from 40 nm to 20 μm.[75,76] Typically, SPG membrane composition is about 70 wt% SiO_2, 10–15 wt% Al_2O_3 and other minor components such as Na_2O, K_2O, and Fe_2O_3 from Shirasu since it is a natural material.[76] SPG can form a highly uniform porous structure, as shown in Fig. 13.3.

13.3.3.3.2. Practical features for water treatment

SPG membranes are often prepared in tubular form, making them suitable for water filtration applications, but there is also potential to shape them into hollow fibres due the excellent shaping ability that the primary glass possesses. This is quite important because hollow fibre membranes possess higher interfacial area per unit volume.[77] SPG membranes also have potential to operate at high fluxes due to their relatively high porosity of 0.5–0.6, rigid pore structure and high mechanical strength, which allows them to withstand high transmembrane pressures. SPG also has a high affinity for water due to the presence of surface groups such as silanols which are inherently hydrophilic. If needed, hydrophobic modification of the surface is also possible with the addition of functional groups such as *n*-octadecyldimethylchlorosilane.[74]

The potential disadvantages in adoption of SPG lie in the cost of the manufacturing process. Recent interest in SPG is very low, probably because of this concern, whereas other lower cost materials, which include both ceramics and polymers, are available to fabricate membranes.

13.3.3.3.3. Application in membrane water treatment

SPG membranes have been used to generate microbubbles to aerate a pilot-scale biofilm reactor for the treatment of synthetic municipal wastewater.[76] The principal behind the beneficial impact of microbubbles in wastewater treatment is described elsewhere.[78] Even though the performance of the membrane was reduced by external fouling, offline cleaning methods which included ultrasonic, thermal and acid treatment managed to restore performance. The membrane scanning electron microscope (SEM) images before and after the cleaning procedures are shown in Fig. 13.4. The fouling and subsequent cleaning did not cause noticeable membrane pore size distribution changes, although there were small increases in the membrane porosity and pore surface area. Modifying the membrane surface with a titania coating could be undertaken, but no work has been found in literature reporting this.

13.3.3.4. *Other innovative light transmitting substrates*

13.3.3.4.1. Glowing optical fibres

One of the most innovative light transmitting support media found in recent literature is side-glowing optical fibres (SOFs) made from a bare quartz glass fibre core coated with silicone rubber.[79] Although the support's porosity and suitability as a membrane was not evaluated, it provides much needed light transmission to the photocatalyst. SOFs have more uniform light irradiance distribution compared to conventional optical fibres. The SOFs allow light transmission to the inner fibre cores as well as the surface, allowing an economical way of delivering light energy efficiently and uniformly

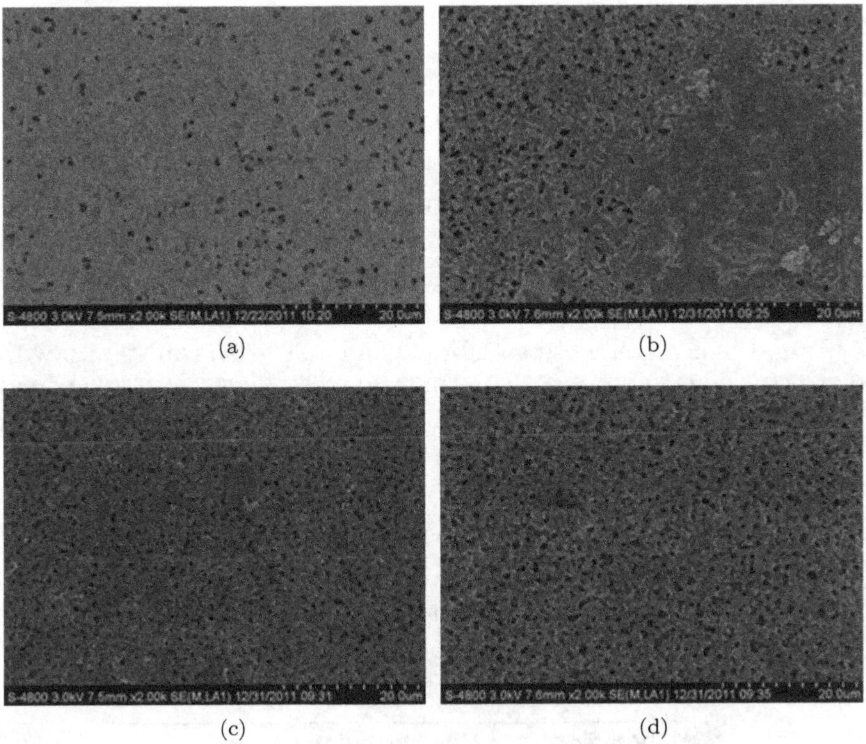

Fig. 13.4. SEM images of a SPG membrane surface used for synthetic municipal wastewater treatment: (a) after fouling and fouling layer removal; (b) after acid treatment; (c) after thermal and acid treatment and (d) the new membrane.

to the photocatalyst. Combining this with membrane functionality could address the challenge of transmitting light in photocatalytic membrane systems. A subsequent study by the same authors found that a titania-reduced graphene oxide (rGO) immobilised on the same optical fibre enhanced pharmaceuticals degradation due to better charge separation, while titania-iron coated on the fibre had good activity under visible light irradiation due to a reduced energy band gap of 2.40 eV.[80]

13.3.3.4.2. Cellulose

Regenerated cellulose (RC) has also been proposed as a light transmitting membrane substrate.[81] Cellulose's superfine network structure provides mechanical support, disperses the photocatalyst and provides a template for nucleate precipitation during nanoparticle growth. Regenerated cellulose can be sourced from sustainable sources such as recycled newspapers, microcrystalline cellulose, cotton lint and soft wood pulp.

Cellulose recovered from recycled newspapers has been combined with titania nanorods to produce a photocatalytic membrane through the phase inversion and casting method.[81] The optical transparency of the membranes was evaluated, as shown in Fig. 13.5. At 550 nm, the neat cellulose membrane had a 90% transmittance, with 0.3% (RC/TiO_2-0.3), 0.5% (RC/TiO_2-0.5) and 0.7% (RC/TiO_2-0.3) titania loading decreasing the transmittance to 75%, 72% and 60% respectively. A decrease in transparency with increased titania loading is consistent with the fact that the titania nanorods used in this study act as scattering centres. The transmittance values are still excellent for the distribution of radiation to photocatalyst particles throughout the membrane, especially if titania is engineered to absorb in the visible light region.

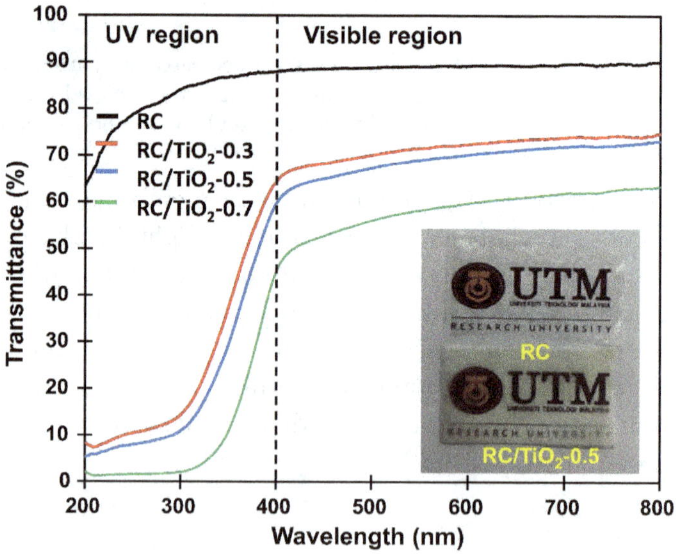

Fig. 13.5. The optical transmittance spectra of a recycled cellulose titania membrane with different loadings of titania nanorods. The inset shows a picture of the RC and RC/TiO_2-0.5 membrane.[81]

A highly transparent cellulose film was also fabricated through free radical polymerization of allyl cellulose (AC).[53] The film had over 90% light transparency in the wavelength range of 520 nm to 800 nm, which makes it suitable for immobilisation of visible light active titania. The cellulose film also had good mechanical properties, highlighted by a tensile strain of 34% and tensile strength of 74 MPa. The degradation half-life of the cellulose film in soil at a temperature of 30°C was determined to be 20 days. The biodegradability of the films, as well as the porosity, would need to be evaluated for applications in water treatment.

Cellulose films with excellent stretchability and transparency were also fabricated from alkali cellulose. Alkali cellulose is usually manufactured by heating wood pulp or cotton waste in sodium hydroxide solutions. Addition of additives such as glucose and urea promoted stretching of the film to up to 233% elongation. Good transparency of 79% at 550 nm was also obtained. The transparency and flexibility of the films can be visualised in Fig. 13.6. These mechanical properties make the film suitable for configuring the film in several membrane module arrangements to suit targeted applications.

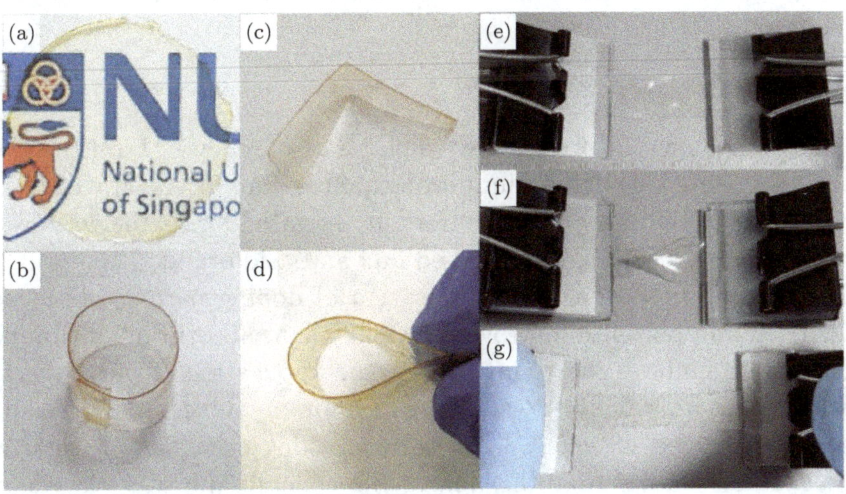

Fig. 13.6. Demonstration of the properties of the cellulose film (a) transparency (b) ability to be rolled (c) ability to be folded and (d) ability to be bent. From its (e) original position, the film could be (f) twisted and (g) stretched.[55]

With the aim of obtaining membranes of high flexibility, mechanical strength and transparency, a nanocomposite of aramid nanofibers (ANFs) and bacterial cellulose (BC) was fabricated via a vacuum-assisted flocculation route.[56] The nanocomposite's mechanical properties could be tuned by varying the amount of ANFs, with the optimum being 4.0 wt% ANFs. ANFs are the nanosized form of the macro-fiber Kevlar™, obtained via deprotonation in potassium hydroxide and dimethyl sulfoxide (DMSO). Adding nanofillers to a material is known to improve its properties, and in this case the mechanical strength of ANFs is imparted to BC. At low concentrations of ANFs, the BC films were transparent due to the good dispersion of nanofibers in BC and the small diameter of the nanofibers, which is lower than visible light wavelengths. At higher ANF concentrations, the appearance of the films begins to turn yellow due to the intrinsic gold colour of ANF as well as nanofiber agglomeration. Agglomeration results in the formation of larger particles with diameters approaching the wavelengths of visible light. The optimum ANF concentration of 4.0 wt% resulted in an increase of tensile strength to 100.36 MPa, compared to 77.13 MPa for the unmodified BC film. The films could also be folded into different shapes, which makes them suitable for shaping into different membrane forms.

13.3.3.4.3. Quartz

Quartz is known for its excellent light transparency in both the UV and visible regions. Thin film porous quartz fibre filters on which sol gel titania was dip-coated have been investigated for textile wastewater treatment, leading to improved flux and acid orange 7 removal.[82] The study focused on the optimization of doping parameters using nitrogen and boron to improve the visible light absorption of titania. It would be interesting to investigate the application of these quartz fibre filters as a light transmitting membrane substrate, especially considering the excellent light transmission properties of quartz. Figure 13.7 shows that the quartz fibre membrane was placed on top of an alumina support for filtration experiments, and light to the photocatalyst was supplied via the conventional way. A study with

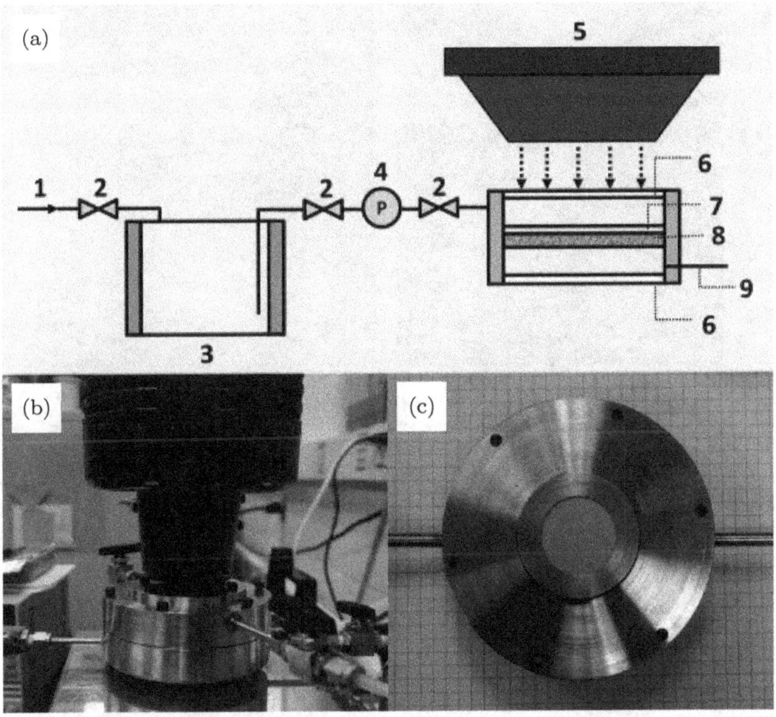

Fig. 13.7. (a) Schematic of the PMR setup containing (1) compressed air line, (2) valve, (3) feed tank, (4) pressure gauge, (5) UV light source, (6) quartz window, (7) quartz fibre membrane, (8) porous alumina support, (9) permeate outlet; (b) collimating funnel for directing light UV light to the quartz fibre membrane; and (c) quartz fibre membrane supported on alumina housed in stainless steel module.

free standing quartz membrane could be used to evaluate the new concept of directing light through the substrate.

13.3.3.4.4. Polylactic acid

A potential candidate for light transmitting membrane substrates is polylactic acid (PLA) composited with chitosan (CS), a polysaccharide obtained from the alkaline treatment of crustacean shells. A porous, hydrophilic, transparent, bactericidal composite of PLA and CS has been prepared via a non-solvent induced phase separation (NIPS) method.[57] The membrane's porosity could be controlled

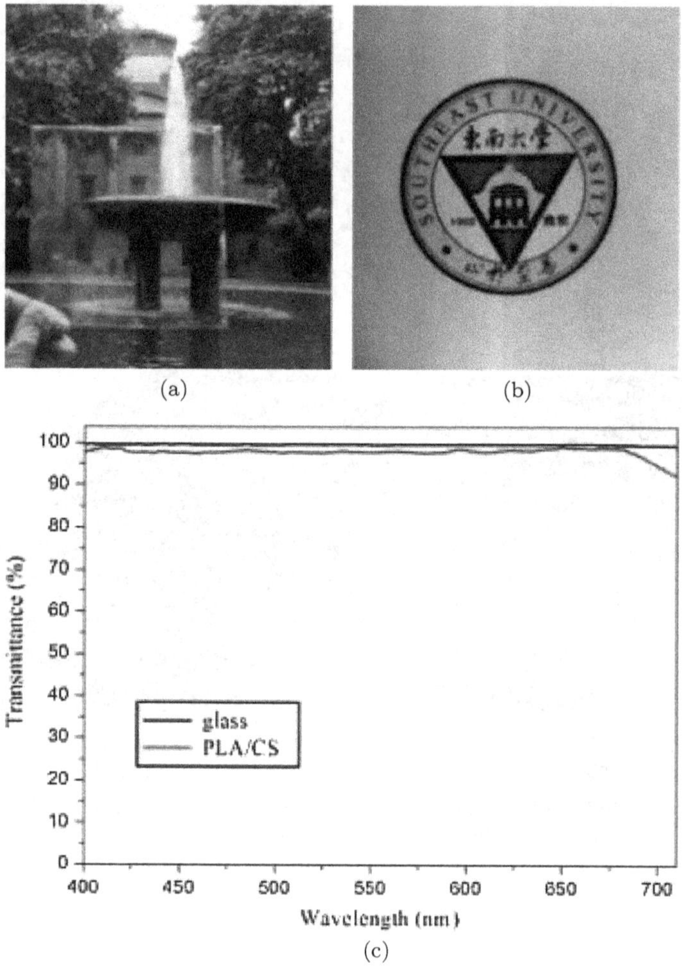

Fig. 13.8. (a) and (b) Photographs of objects viewed through the transparent PLA/CS film and (c) the transmittance of the PLA/CS film, compared to glass.

by varying the PLA/CS ratio. Regardless of the amount of CS, the films had more than 90% transmittance in a wavelength range of 400–700 nm, as shown in Fig. 13.8. A 99.77% bactericidal rate towards *E. coli* was also recorded. Such a high bactericidal rate, even before titania coating makes the material interesting for applications in water disinfection.

13.3.3.4.5. Nylon and single-walled carbon nanotubes

Nylon is one of the most used thermoplastic polymers due to its versatility. On the other hand, single-walled carbon nanotube (SWCNT) have been shown to possess superior optical, electrical, and mechanical properties which are desirable qualities for fabrication of flexible transparent conductive films (TCFs), targeted for application in optoelectronic devices.[59,60] Combining nylon with SWCNTs can result in a hybrid material that utilises the qualities of each of the components. These qualities could also be harnessed for targeted applications in light transmitting membrane substrates for water treatment.

Transparent films fabricated from SWCNTs) and electrospun nylon 6 membranes, followed by dipping of the membranes in a commercial transparent epoxy resin exhibited a light transmittance of 77.4% at 550 nm.[58] The nylon 6 nanofibers acted as a guide to create a two-dimensional, well dispersed network of the nanotubes, while the epoxy resin prevented optical scattering. The improvement in optical transparency is shown in Fig. 13.9.

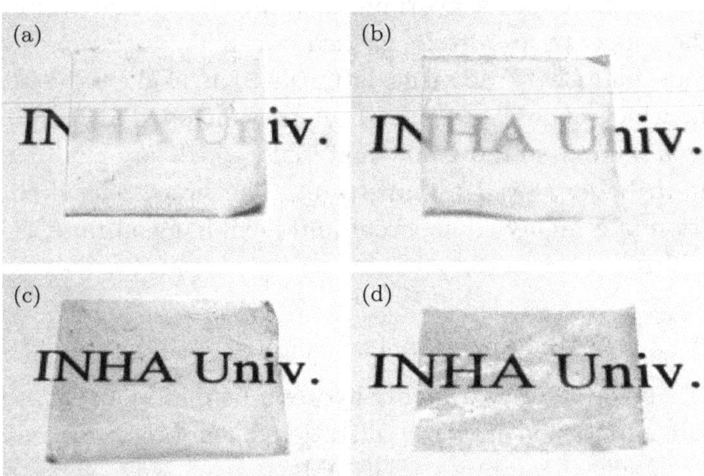

Fig. 13.9. Images of the (a) nylon, (b) SWCNT-adsorbed nylon, (c) transparent epoxy incorporated nylon, and (d) transparent epoxy incorporated SWCNT-adsorbed nylon.[58]

13.3.3.4.6. Silk fibroin

Silk fibroin is a type of slow degrading protein that is obtained from the silkworm from where it can be isolated and generated into membranes. Natural silk fibroin membrane (SFM) has been proposed as a substitute for cation exchange membranes in microbial fuel cells (MFCs).[61] The membranes exhibited high transmittances of 90% in the visible light region of 400–700 nm. Membrane deterioration, however, was observed after one month of operation, so this could be an area of concern.

13.3.3.4.7. Poly(hydroxyethyl acrylamide) (PHEAAm) grafted onto polydimethylsiloxane (PDMS)

Polydimethylsiloxane (PDMS) exhibits high optical transparency and high gas permeability. Transparent and flexible membranes were prepared by depositing Pt/WO_3-thin films on the surface of PDMS membranes for application in hydrogen gas sensing.[62] The PDMS membranes had been previously modified with hydroxyethyl acrylamide (HEAAm) through a UV-induced graft polymerization method to improve hydrophilicity. Modification of the PDMS surface from a hydrophobic to a hydrophilic nature had to be performed to allow the sol gel process to be carried out. The coating protocol did not involve high heat treatment, but utilised an acid catalysed sol–gel method, which allows only mild heat treatments. This is important because it allows soft organic materials which are not high heat tolerant to be employed as substrates. The process resulted in the fabrication of a highly transparent and flexible membrane, as shown in Fig. 13.10.

13.3.3.4.8. Free standing graphene oxide

Graphene oxide (GO) is a highly hydrophilic material that is finding increased application in water filtration membranes. Free standing graphene oxide membranes (FSGOMs) have been investigated for potential applications in several areas that include gas sensing, selective gas transport and water desalination. FSGOMs have been prepared by evaporating GO-methanol or GO-water-methanol on

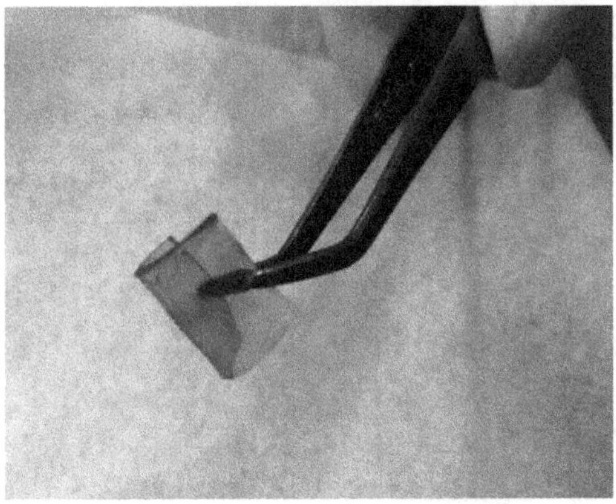

Fig. 13.10. Demonstration of the flexibility of the Pt/WO_3-thin layer-coated PDMS film.[62]

silicon substrates, followed by delamination using 2-propanol, or by mechanical exfoliation.[63] The membranes were evaluated for methylene blue (MB) removal efficiency from MB-propanol and showed ~63% removal up to 4 cycles. This is promising for textile mill wastewater treatment, even before a titania coating is introduced on the membrane surface. More importantly, The FSGOMs displayed 9–35% light transmittance, with the maximum transmittance occurring at ~1000 nm. This transmission ability could be harnessed for light transmitting substrates for photocatalytic membranes.

13.4. Conclusion and Future Perspectives

The concept of combining photocatalysis and membrane filtration into one operational unit to take advantage of each method while addressing the shortcomings of slurry photocatalytic reactors and membrane filtration in water and wastewater treatment is now well established. However, wider adaptation of photocatalytic of membranes beyond laboratory studies is still low due to several limitations which still need to be addressed. One such limitation is the route of supplying light to the photocatalyst, with the traditional approach of directing light to the water being treated leading to obvious inefficiencies and operational challenges. Replacing current opaque membrane substrates with light transmitting substrates offers a new route for supplying light to the photocatalytic layer of the membrane through the substrate itself. This would eliminate the inefficiencies brought by light absorption by the water being treated, as well as simplifying configurational arrangements by simply directing light from the end of a filtration module element.

However, such substrates are either not readily available, require elaborate laboratory equipment to fabricate, or are still too expensive to make. Other suitable substrates with high transparency and required porosity have been applied in other areas, such as conductive films for electrical gadgets, but have not been considered for photocatalytic membranes. As such, future research efforts could be aimed towards fabricating light transmitting porous substrates and studying their application in photocatalytic membranes for water treatment in more detail. Possible substrates such as porous glasses have already been applied in membrane filtration and shown to have good separation functionality as well desirable optical properties, therefore their use as photocatalyst support substrates look promising.

References

1. H. Vatankhah, C. C. Murray, J. W. Brannum, J. Vanneste, C. Bellona, Effect of pre-ozonation on nanofiltration membrane fouling during water reuse applications, *Sep. Purif. Technol.*, **205** (2018) 203–211, 2018/10/31/.
2. L. Jiang, K.-H. Choo, Photocatalytic mineralization of secondary effluent organic matter with mitigating fouling propensity in a submerged membrane photoreactor, *Chem. Eng. J.*, **288** (2016) 798–805, 2016/03/15/.
3. Z. Geng *et al.*, Self-cleaning anti-fouling hybrid ultrafiltration membranes via side chain grafting of poly(aryl ether sulfone) and titanium dioxide, *J. Membr. Sci.*, **529** (2017) 1–10, 5/1/.
4. G. Rao, Q. Zhang, H. Zhao, J. Chen, Y. Li, Novel titanium dioxide/iron (III) oxide/graphene oxide photocatalytic membrane for enhanced humic acid removal from water, *Chem. Eng. J.*, **302** (2016) 633–640, 2016/10/15/.
5. F. Cavani, D. Duprez, S. World, *Handbook Of Advanced Methods And Processes In Oxidation Catalysis: From Laboratory To Industry.* London: Imperial College Press, 2014.

6. S. Malato, M. I. Maldonado, P. Fernández-Ibáñez, I. Oller, I. Polo, R. Sánchez-Moreno, Decontamination and disinfection of water by solar photocatalysis: The pilot plants of the Plataforma Solar de Almeria, *Mater. Sci. Semicond. Process* **42** (2016) Part 1, 15–23, 2//.
7. R. Molinari, C. Lavorato, P. Argurio, Recent progress of photocatalytic membrane reactors in water treatment and in synthesis of organic compounds. A review, *Catal. Today*, **281** (2017) 144–164, 2017/03/01/.
8. S. Mozia, Photocatalytic membrane reactors (PMRs) in water and wastewater treatment. A review, *Sep. Purif. Technol.*, **73**(2) (2010) 71–91, 2010/06/18/.
9. L. Nyamutswa, B. Zhu, D. Navaratna, S. Collins, M. Duke, Proof of concept for light conducting membrane substrate for uv-activated photocatalysis as an alternative to chemical cleaning, *Membranes*, **8**(4) (2018) 122.
10. A. Fujishima and K. Honda, Electrochemical photolysis of water at a semiconductor electrode, *Nature* 10.1038/238037a0 **238**(5358) (1972) 37–38, 07/07/print.
11. F. Quan, Y. Hu, X. Zhang, C. Wei, Simple preparation of Mn-N-codoped TiO_2 photocatalyst and the enhanced photocatalytic activity under visible light irradiation, *Applied Surface Science*, Article **320**, (2014) 120–127.
12. D. A. Keane *et al.*, Solar photocatalysis for water disinfection: materials and reactor design, *Catalysis Science & Technology*, 10.1039/C4CY00006D, **4**(5) (2014) 1211–1226.
13. S. Livraghi, M. Pelaez, J. Biedrzycki, I. Corazzari, E. Giamello, D. D. Dionysiou, Influence of the chemical synthesis on the physicochemical properties of N-TiO_2 nanoparticles, *Catal. Today*, **209** (2013) 54–59, 6/15/.
14. W. Ho, J. C. Yu, S. Lee, Low-temperature hydrothermal synthesis of S-doped TiO_2 with visible light photocatalytic activity, *J. Solid State Chem.*, **179**(4) (2006) 1171–1176, 4//.
15. F. Dong, H. Wang, G. Sen, Z. Wu, S. C. Lee, Enhanced visible light photocatalytic activity of novel Pt/C-doped TiO_2/PtCl4 three-component nanojunction system for degradation of toluene in air, *J. Hazard. Mater.*, **187**(1–3) (2011) 509–516, 3/15/.
16. A. Zaleska, E. Grabowska, J. W. Sobczak, M. Gazda, J. Hupka, Photocatalytic activity of boron-modified TiO_2 under visible light: The effect of boron content, calcination temperature and TiO_2 matrix, *Appl. Catal. B: Environmental*, **89**(3–4) (2009) 469–475, 7/15/.
17. M. Iwase, K. Yamada, T. Kurisaki, O. O. Prieto-Mahaney, B. Ohtani, H. Wakita, Visible-light photocatalysis with phosphorus-doped

titanium (IV) oxide particles prepared using a phosphide compound, *Appl. Catal. B: Environmental*, **132–133** (2013) 39–44, 3/27/.
18. Y. Wu, M. Xing, B. Tian, J. Zhang, F. Chen, Preparation of nitrogen and fluorine co-doped mesoporous TiO_2 microsphere and photodegradation of acid orange 7 under visible light, *Chem. Eng. J.*, **162**(2) (2010) 710–717, 8/15/.
19. E. Bae, W. Choi, J. Park, H. S. Shin, S. B. Kim, J. S. Lee, Effects of surface anchoring groups (carboxylate vs phosphonate) in ruthenium-complex-sensitized TiO_2 on visible light reactivity in aqueous suspensions, *J. Phys. Chem. B*, **108**(37) (2004) 14093–14101, 2004/09/01.
20. H. S. Park, I. M. Jung, G. H. Choi, S. Hahn, Y. S. Yoo, T. Lee, Modification of a rodent hindlimb model of secondary lymphedema: surgical radicality versus radiotherapeutic ablation, *BioMed Res. Int.*, **2013** (2013) 10, Art. no. 208912.
21. W. Choi, Pure and modified TiO_2 photocatalysts and their environmental applications, *Catalysis Surveys from Asia*, journal article **10**(1) (2006) 16–28.
22. R. Kamaludin, A. S. Mohamad Puad, M. H. D. Othman, S. H. S. A. Kadir, Z. Harun, Incorporation of N-doped TiO_2 into dual layer hollow fiber (DLHF) membrane for visible light-driven photocatalytic removal of reactive black 5, *Polymer Testing*, **78**, 105939, 2019/09/01/.
23. Y. Shi, D. Yang, Y. Li, J. Qu, Z.-Z. Yu, Fabrication of PAN@TiO_2/Ag nanofibrous membrane with high visible light response and satisfactory recyclability for dye photocatalytic degradation, *Applied Surface Science*, **426** (2017) 622–629, 2017/12/31/.
24. M. R. Dhananjeyan *et al.*, Photodynamics and Surface Characterization of TiO_2 and Fe2O3 Photocatalysts Immobilized on Modified Polyethylene Films, *J. Phys. Chem. B*, **105**(48) (2001) 12046–12055, 2001/12/01.
25. K. Hofstadler, R. Bauer, S. Novalic, G. Heisler, New reactor design for photocatalytic wastewater treatment with TiO_2 immobilized on fused-silica glass fibers: Photomineralization of 4-chlorophenol, *Environ. Sci. Technol.*, **28**(4) (1994) 670–674, 1994/04/01.
26. A. Haarstrick, O. M. Kut, E. Heinzle, TiO_2-Assisted Degradation of Environmentally Relevant Organic Compounds in Wastewater Using a Novel Fluidized Bed Photoreactor, *Environ. Sci. Technol.*, **30**(3) (1996) 817–824, 1996/02/01.
27. C. H. Ao and S. C. Lee, Enhancement effect of TiO_2 immobilized on activated carbon filter for the photodegradation of pollutants at typical indoor air level, *Appl. Catal. B: Environmental*, **44**(3) (2003) 191–205, 8/30/.

28. X. Meng, Z. Qian, H. Wang, X. Gao, S. Zhang, M. Yang, Sol–gel immobilization of SiO2/TiO_2 on hydrophobic clay and its removal of methyl orange from water, *Journal of Sol-Gel Science and Technology*, journal article **46**(2) (2008) 195–200.
29. L. Song, B. Zhu, S. Gray, M. Duke, S. Muthukumaran, Hybrid processes combining photocatalysis and ceramic membrane filtration for degradation of humic acids in saline water, *Membranes (Basel)*, **6**(1) (2016).
30. S. Leong, A. Razmjou, K. Wang, K. Hapgood, X. Zhang, H. Wang, TiO_2 based photocatalytic membranes: A review, *J. Membr. Sci.*, **472** (2014) 167–184, 12/15/.
31. M. A. Anderson, M. J. Gieselmann, Q. Xu, Titania and alumina ceramic membranes, *J. Membr. Sci.*, **39**(3) (1988) 243–258, 1988/12/01.
32. J. Sabate, M. A. Anderson, H. Kikkawa, M. Edwards, C. G. Hill, A kinetic study of the photocatalytic degradation of 3-chlorosalicylic acid over TiO_2 membranes supported on glass, *J. Catal.*, **127**(1) (1991) 167–177, 1991/01/01.
33. H. Choi, A. C. Sofranko, D. D. Dionysiou, Nanocrystalline TiO_2 Photocatalytic Membranes with a Hierarchical Mesoporous Multilayer Structure: Synthesis, Characterization, Multifunction, *Advanced Functional Materials*, **16**(8) (2006) 1067–1074.
34. H. Zhang, X. Quan, S. Chen, H. Zhao, Fabrication and Characterization of Silica/Titania Nanotubes Composite Membrane with Photocatalytic Capability, *Environ. Sci. Technol.*, **40**(19) (2006) 6104–6109, 2006/10/01.
35. S. Mozia, D. Darowna, R. Wróbel, A. W. Morawski, A study on the stability of polyethersulfone ultrafiltration membranes in a photocatalytic membrane reactor, *J. Membr. Sci.*, **495** (2015) 176–186, 12/1/.
36. A. A. Habibpanah, S. Pourhashem, H. Sarpoolaky, Preparation and characterization of photocatalytic titania–alumina composite membranes by sol–gel methods, *J. Eur. Ceram. Soc.*, **31**(15) (2011) 2867–2875.
37. J. L. Weishaar, G. R. Aiken, B. A. Bergamaschi, M. S. Fram, R. Fujii, K. Mopper, Evaluation of specific ultraviolet absorbance as an indicator of the chemical composition and reactivity of dissolved organic carbon, *Environ. Sci. Technol.*, **37**(20) (2003) 4702–4708.
38. J. Geltmeyer *et al.*, TiO_2 functionalized nanofibrous membranes for removal of organic (micro)pollutants from water, *Sep. Purif. Technol.*, **179** (2017) 533–541, 2017/05/31/.

39. N. Li *et al.*, Precisely-controlled modification of PVDF membranes with 3D TiO_2/ZnO nanolayer: Enhanced anti-fouling performance by changing hydrophilicity and photocatalysis under visible light irradiation, *J. Membr. Sci.*, **528** (2017) 359–368, 2017/04/15/.
40. C. Zhu, G. Liu, K. Han, H. Ye, S. Wei, Y. Zhou, One-step facile synthesis of graphene oxide/TiO_2 composite as efficient photocatalytic membrane for water treatment: Crossflow filtration operation and membrane fouling analysis, *Chem. Eng. Process.: Process Intensification*, **120** (2017) 20–26, 2017/10/01/.
41. Y. Chang *et al.*, Effect of post-heat treatment on the photocatalytic activity of titanium dioxide nanowire membranes deposited on a Ti substrate, *RSC Adv.*, 10.1039/C7RA02092A, **7**(35) (2017) 21422–21429.
42. S. Sanches, C. Nunes, P. C. Passarinho, F. C. Ferreira, V. J. Pereira, J. G. Crespo, Development of photocatalytic titanium dioxide membranes for degradation of recalcitrant compounds, *J. Chem. Technol. Biotechnol.*, **92**(7) (2017) 1727–1737.
43. Y. Zhao, C. Tao, G. Xiao, H. Su, Controlled synthesis and wastewater treatment of Ag2O/TiO_2 modified chitosan-based photocatalytic film, *RSC Advances*, 10.1039/C6RA27295A, **7**(18) (2017) 11211–11221.
44. S. Ramasundaram *et al.*, Photocatalytic applications of paper-like poly(vinylidene fluoride)–titanium dioxide hybrids fabricated using a combination of electrospinning and electrospraying, *J. Hazard. Mater.*, **285** (2015) 267–276, 2015/03/21/.
45. R. M. Huertas, M. C. Fraga, J. G. Crespo, V. J. Pereira, Sol-gel membrane modification for enhanced photocatalytic activity, *Sep. Purif. Technol.*, **180** (2017) 69–81, 2017/06/08/.
46. M. A. Kanjwal, N. A. M. Barakat, I. S. Chronakis, Photocatalytic degradation of dairy effluent using $AgTiO_2$ nanostructures/polyurethane nanofiber membrane, *Ceram. Int.*, **41**(8) (2015) 9615–9621, 2015/09/01/.
47. M. J. Arlos *et al.*, Photocatalytic decomposition of selected estrogens and their estrogenic activity by UV-LED irradiated TiO_2 immobilized on porous titanium sheets via thermal-chemical oxidation, *J. Hazard. Mater.*, **318** (2016) 541–550, 2016/11/15/.
48. B. J. Starr, V. V. Tarabara, M. Herrera-Robledo, M. Zhou, S. Roualdès, A. Ayral, Coating porous membranes with a photocatalyst: Comparison of LbL self-assembly and plasma-enhanced CVD techniques, *J. Membr. Sci.*, **514** (2016) 340–349, 2016/09/15/.
49. X. Zhang, D. K. Wang, J. C. Diniz da Costa, Recent progresses on fabrication of photocatalytic membranes for water treatment, *Catal. Today*, **230** (2014) 47–54, 7//.

50. S. Ramasundaram, H. N. Yoo, K. G. Song, J. Lee, K. J. Choi, S. W. Hong, Titanium dioxide nanofibers integrated stainless steel filter for photocatalytic degradation of pharmaceutical compounds, *J. Hazard. Mater.*, **258–259** (2013) 124–132, 2013/08/15/.
51. M. Zhang *et al.*, Hydrothermal synthesized UV-resistance and transparent coating composited superoloephilic electrospun membrane for high efficiency oily wastewater treatment, *J. Hazard. Mater.*, **383** (2020) 121152, 2020/02/05/.
52. S. H. Kim, S. Y. Kwak, B. H. Sohn, T. H. Park, Design of TiO_2 nanoparticle self-assembled aromatic polyamide thin-film-composite (TFC) membrane as an approach to solve biofouling problem, *J. Membr. Sci.*, Article **211**(1) (2003) 157–165.
53. R. Tong, G. Chen, J. Tian, M. He, Highly transparent, weakly hydrophilic and biodegradable cellulose film for flexible electroluminescent devices, *Carbohydr. Polym.*, **227** (2020) 115366, 2020/01/01/.
54. Z. Li *et al.*, Using cellulose fibers to fabricate transparent paper by microfibrillation, *Carbohydr. Polym.*, **214** (2019) 26–33, 2019/06/15/.
55. D. B. K. Lim and H. Gong, Highly stretchable and transparent films based on cellulose, *Carbohydr. Polym.*, **201** (2018) 446–453, 2018/12/01/.
56. Y. Wu, F. Wang, Y. Huang, Facile and simple fabrication of strong, transparent and flexible aramid nanofibers/bacterial cellulose nanocomposite membranes, *Compos. Sci. Technol.*, **159** (2018) 70–76, 2018/05/03/.
57. W. Han, J. Ren, H. Xuan, L. Ge, Controllable degradation rates, antibacterial, free-standing and highly transparent films based on polylactic acid and chitosan, *Colloids Surf. A Physicochem. Eng. Asp.*, **541** (2018) 128–136, 2018/03/20/.
58. H. Bak, S. Y. Cho, Y. S. Yun, H.-J. Jin, Electrically conductive transparent films based on nylon 6 membranes and single-walled carbon nanotubes, *Curr. Appl. Phys.*, **10**(3), Supplement (2010) S468–S472, 2010/05/01/.
59. O. Urper, İ. Çakmak, N. Karatepe, Fabrication of carbon nanotube transparent conductive films by vacuum filtration method, *Mater. Lett.*, **223** (2018) 210–214, 2018/07/15/.
60. S. Jiang, P.-X. Hou, C. Liu, H.-M. Cheng, High-performance single-wall carbon nanotube transparent conductive films, *J. Mater. Sci. Technol.*, **35**(11) (2019) 2447–2462, 2019/11/01/.
61. G. Pasternak, Y. Yang, B. B. Santos, F. Brunello, M. M. Hanczyc, A. Motta, Regenerated silk fibroin membranes as separators for transparent microbial fuel cells, *Bioelectrochemistry*, **126** (2019) 146–155, 2019/04/01/.

62. R. Ishihara, Y. Yamaguchi, K. Tanabe, Y. Makino, K. Nishio, Preparation of Pt/WO3-coated polydimethylsiloxane membrane for transparent/flexible hydrogen gas sensors, *Mater. Chem. Phys.*, **226** (2019) 226–229, 2019/03/15/.
63. P. K. Narayanam and K. Sundararajan, Organic solvent supported fabrication of transparent free standing graphene oxide membranes, *Ceram. Int.* (2019), 2019/11/02/.
64. L. Wright, Development of a Regional Waste Recovery/Processing Sector Wellington City Council, p. 86, 2003.
65. D. Enke, F. Janowski, W. Schwieger, Porous glasses in the 21st century — a short review, *Microporous Mesoporous Mater.*, **60**(1) (2003) 19–30, 2003/06/19/.
66. A. Marković, D. Stoltenberg, D. Enke, E. U. Schlünder, A. Seidel-Morgenstern, Gas permeation through porous glass membranes: Part I. Mesoporous glasses — Effect of pore diameter and surface properties, *J. Membr. Sci.*, **336**(1) (2009) 17–31, 2009/07/01/.
67. K. Beltsios, G. Charalambopoulou, G. Romanos, N. Kanellopoulos, A Vycor®Membrane with Reduced Size Surface Pores I. Preparation and Characterization, *J. Porous Mater.*, journal article **6**(1) (1999) 25–31.
68. S. Albert, N. Frolet, P. Yot, A. Pradel, M. Ribes, Characterisation of porous Vycor®7930–AgI composites synthesised by electrocrystallisation, *Microporous Mesoporous Mater.*, **99**(1) (2007) 56–61, 2007/02/01/.
69. E. U. Schlünder, J. Yang, A. Seidel-Morgenstern, Competitive diffusion and adsorption in Vycor glass membranes — A lumped parameter approach, *Catal. Today*, **118**(1) (2006) 113–120, 2006/10/30/.
70. T. Yazawa, K. Kuraoka, W. F. Du, Effect of cooling rate on pore distribution in quenched sodium borosilicate glasses, *J. Phys. Chem. B*, Article **103**(45) (1999) 9841–9845.
71. H. Wang and G. R. Gavalas, Mesoporous glass films supported on α-Al2O3, *J. Membr. Sci.*, **176**(1) (2000) 75–85, 2000/08/15/.
72. A. Mitropoulos, Small-angle X-ray scattering studies of adsorption in Vycor glass, (in Eng), *J. Colloid Interface Sci.*, **336**(2) (2009) 679–90, Aug 15.
73. S. L. Zakharov, Borosilicate microporous glasses for reverse osmosis, *Glass Ceram.*, journal article **61**(5) (2004) 178–179.
74. M. Kukizaki and M. Goto, Demulsification of water-in-oil emulsions by permeation through Shirasu-porous-glass (SPG) membranes, *J. Membr. Sci.*, **322**(1) (2008) 196–203, 9/1/.
75. G. T. Vladisavljević, M. Shimizu, T. Nakashima, Permeability of hydrophilic and hydrophobic Shirasu-porous-glass (SPG) membranes

to pure liquids and its microstructure, *J. Membr. Sci.*, **250**(1–2) (2005) 69–77, 3/15/.
76. C. Liu *et al.*, Fouling and structural changes of Shirasu porous glass (SPG) membrane used in aerobic wastewater treatment process for microbubble aeration, *J. Membr. Sci.*, **421–422** (2012) 225–231, 2012/12/01/.
77. M. Kukuzaki, K. Fujimoto, S. Kai, K. Ohe, T. Oshima, Y. Baba, Ozone mass transfer in an ozone–water contacting process with Shirasu porous glass (SPG) membranes — A comparative study of hydrophilic and hydrophobic membranes, *Sep. Purif. Technol.*, **72**(3) (2010) 347–356, 2010/05/11/.
78. A. Agarwal, W. J. Ng, Y. Liu, Principle and applications of microbubble and nanobubble technology for water treatment, *Chemosphere*, Review **84**(9) (2011) 1175–1180.
79. L. Lin, H. Wang, P. Xu, Immobilized TiO_2-reduced graphene oxide nanocomposites on optical fibers as high performance photocatalysts for degradation of pharmaceuticals, *Chem. Eng. J.*, **310** (2017) 389–398, 2017/02/15/.
80. L. Lin, H. Wang, W. Jiang, A. R. Mkaouar, P. Xu, Comparison study on photocatalytic oxidation of pharmaceuticals by TiO2-Fe and TiO_2-reduced graphene oxide nanocomposites immobilized on optical fibers, *J. Hazard. Mater.*, **333** (2017) 162–168, 2017/07/05/.
81. M. A. Mohamed *et al.*, Physicochemical characteristic of regenerated cellulose/N-doped TiO2 nanocomposite membrane fabricated from recycled newspaper with photocatalytic activity under UV and visible light irradiation, *Chem. Eng. J.*, **284** (2016) 202–215, 2016/01/15/.
82. M. Hatat-Fraile *et al.*, Concurrent photocatalytic and filtration processes using doped TiO2 coated quartz fiber membranes in a photocatalytic membrane reactor, *Chem. Eng. J.*, **330** (2017) 531–540, 2017/12/15/.

CHAPTER 14

Transport in Carbon Nanotube Pores: Implications for Next Generation Water Purification Technologies

Aleksandr Noy*, Yun-Chiao Yao, and Xi Chen

Materials Science Division, Physical and Life Sciences Directorate, Lawrence Livermore National Laboratory Livermore, CA 94550, USA
School of Natural Sciences, University of California Merced, Merced, CA 95434, USA

14.1. Introduction: Biomimetic Nanopore Membranes

Membrane-based water treatment processes separate water from salts and contaminants by pushing it through tiny selective pores in thin membrane layers. The performance of such membranes depends on their ability to achieve two key goals that are almost diametrically opposite: they need to maximize flow through the pores, and yet maintain high pore selectivity to ensure the quality of the final product stream. Reverse osmosis (RO) processes, which represent a large part of world's desalination capacity,[1] provide some of the best illustrations of these opposing tensions. A requirement for having pores that are small enough to stop the passage of all ions, including small monovalent species, and yet still enable enough water to pass through to satisfy cost and productivity targets, places severe constraints on the materials that comprise

*Corresponding author: noy1@llnl.gov

these membranes. Current polyamide-based RO membranes utilize the solution-diffusion transport mechanism that defines fundamental permeability-selectivity trade-offs that constrain their performance.

Breaking these performance limitations and addressing other challenges of modern RO desalination, such as inadequate rejection of small solutes and micropollutants, requires new generation of engineered pores that can potentially support transport mechanisms that go beyond solution-diffusion. This is not an easy challenge, as both water molecules and monovalent salt ions, which represent the majority of species that need to be removed from seawater in order to desalinate it, are quite small, the hydrated diameters of those ions on the order of less than 1 nm. Thus, the pores that separate them need to be very small and very precise.

One successful implementation of such a pore is provided by the biological membrane water channels, aquaporins, which are ubiquitous in biological system that require fast and efficient water management. Aquaporins achieve water transport rates that exceeds 1 molecule per nanosecond, which is at least two orders of magnitude higher than current RO membranes. Even more remarkably, aquaporins have extremely high rejection of all other ionic species, providing nearly perfect selectivity.

What are the structural features that are responsible for such a remarkable separation performance of aquaporins? First, the protein pore is quite narrow, ca. 0.3 nm, which is generally smaller than the hydrated diameters of ions. Ion passage through such a small pore would require substantial dehydration of the ion, which carries prohibitively large energy penalty. Second, the channel walls are lined up with hydrophobic residues, which minimize water molecule interactions with the channel walls, reducing channel friction and promoting fast transport. Third, water molecules, which can still squeeze through this opening are forced to form a one-dimensional water chain, which also promotes fast transport through the protein channel. Forth, aquaporin channel length is quite short and roughly equivalent to the membrane thickness (4–5 nm), which also helps to minimize losses. Fifth, and less obvious advantage is that unlike the polymeric membranes, where separation, mechanical, and surface properties are dictated by the same material, biological membranes use a different paradigm based on specialized channels embedded in a phospholipid bilayer that acts as a matrix material. These key features, if reproduced in a robust, manufacturable, and cost effective platform, which comprises defined fast and selective pores embedded in an impermeable matrix, could potentially enable *biomimetic* membranes that could break the permeability-selectivity trade-offs of the current polymer membranes and deliver enhanced desalination performance.

Over the last decade, there has been a number of biomimetic membrane systems reported in the literature based on either natural water channels in a synthetic matrix, such as aquaporin proteins embedded in polymer bilayers, or synthetic channels, such as G-quadruplex channels,[2] and pillar[5]arenes[3] embedded in lipid matrix. Recently, a membrane system based on peptide-appended hybrid[4]arenes reported extremely high ion selectivity and extremely high water permeability.[4] Nevertheless, industrial scale manufacturing and use will likely require a technology platform based on a more robust and scalable nanoporous membrane pore. At this point, most signs point to a carbon nanotube pore as a foundation of such platform.

14.2. Carbon Nanotubes

Carbon nanotubes (CNTs) were first observed by transmission electron microscopy (TEM) in 1991.[5] Ever since the discovery of single wall carbon nanotubes (SWCNTs) in 1993[6,7] these cylinders of carbon atoms with thickness of one atomic layer have found extensive applications in a number of research areas[8] and become a truly iconic molecule of nanotechnology. Subsequent synthetic breakthroughs allowed production of large quantities of high-purity single wall CNTs with a small diameter distribution.[9,10] High resolution TEM, scanning tunneling microscopy (STM), and Raman scattering measurements have shown that SWCNTs exist as rolled-up seamless graphene cylinders with sp^2 bonded carbon atoms organized into a honeycomb structure (Figure 14.1). This roll-up vector (n, m) determines the key parameters of a CNT — diameter and chirality.[11] Depending

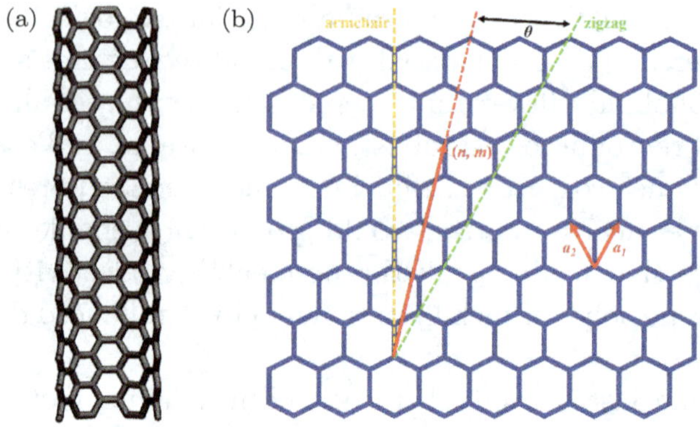

Fig. 14.1. (a) A molecular model of a single-wall carbon nanotube. Half of the nanotube closest to the viewer was removed for clarity. (b). A schematic illustration of honeycomb structure of a graphene sheet that can be folded into single-walled carbon nanotubes along the lattice vectors.

on the chirality (i.e the chiral angle between hexagons and the tube axis), SWCNTs can behave as metallic or semiconducting molecules (with band gaps that range from ca. 10 meV to 0.5 eV for a typical diameter of 1.5 nm) even if they have nearly identical diameters.[11] Intensive effort has been made to study the electronic properties of nanotubes and their atomic structures.[12–16] Metallic CNTs are ideal for studying quantum phenomena in quasi 1D solids, including single electron charging, Luttinger Liquid, weak localization, ballistic transport and quantum interference[12–14,17,18] while semiconducting CNTs are utilized as a perfect building blocks for nanoelectronics, e.g. transistors, logic gates, memory devices and sensors.[12–16]

14.2.1. *Structure*

Structure of single wall CNTs is determined by the diameter d, and chiral angle θ (Fig. 14.1). The chiral angle is commonly defined between 0 and 30 degrees (0° to denote zigzag CNTs, 30° to denote armchair CNTs). The circumference of CNT can be expressed by the chiral vector ($\overrightarrow{C} = n\overrightarrow{a_1} + m\overrightarrow{a_2}$, $\overrightarrow{a_1}$ and $\overrightarrow{a_2}$ are the unit vectors of the hexagonal graphene honeycomb lattice) connecting two crystallographically equivalent sites on the 2D graphene sheet.[19] The superimposition of two edges perpendicular to the vector $\overrightarrow{C}$ yields the chiral CNT structure, which has no distortion of bond angles other than the distortions caused by the cylindrical curvature of the CNT itself.

Alternatively, we can use the (n,m) notation where the chiral vectors (n,0) and (0,m) denote zigzag CNTs; (n,n) denotes armchair CNTs, while the others denote chiral CNTs.[20] The diameter d of a CNT is given by

$$d = \frac{\sqrt{3}a}{\pi}(m^2 + mn + n^2)^{1/2}$$

where the nearest neighboring carbon atoms have a distance of $a =$ 1.44 Å and the chiral angle is given by

$$\theta = tan^{-1}\frac{\sqrt{3}n}{2m + n}.$$

14.2.2. *Synthesis*

CNTs are synthesized by three major methods: arc discharge, laser ablation and chemical vapor deposition (CVD).[16] The earliest CNTs were produced by the Iijima group using arc discharge methods where CNTs were formed as bundles on the negative electrode, and the positive electrode was consumed in a helium atmosphere.[21] Typical lengths of the grown CNTs were in μm ranges, yielding an aspect ratio of 10^2 to 10^6. The very first single wall CNTs were also synthesised in an arc discharge chamber with Fe, Co and other transition metals as catalysts.[6,7] Smalley group introduced the first synthetic breakthrough when they developed a more efficient synthesis route using laser ablation.[9] This method uses a Co-Ni/graphite composite target operating in a furnace at 1200°C yielding 70–90% conversion of graphite to SWCNTs. Argon gases sweep the CNTs from high temperature zone to a water cooled Cu collector downstream outside of the furnace. An even more significant advance came when researchers developed CVD synthesis of SWCNTs using hydrocarbon decomposition over transition metal catalyst nanoparticles.[22,23]

Arc discharge and laser ablation use solid state carbon precursors to provide carbon source for the CNT growth, which involves carbon vaporization at high temperature. They can consistently yield high quality and near-perfect CNT structures, even though large amount of amorphous carbon and other byproducts are inevitable. CVD, on the other hand, utilizes hydrocarbon gases as sources and catalyst particles as seeds for CNT growth. This process happens at comparatively lower temperatures of 500–1000°C.[16] Even though none of these three methods can yet produce large amount of CNTs with identical diameters and chirality, catalytic chemcial vapor depostion is preferred over the other two methods[9,10] due to its ability to produce both individually isolated nanotubes and densely packed vertically aligned arrays.[16,24–29] Previous studies show that the most probable diameter and the width of the diameter distribution are determined by the size and composition of the catalyst particle, growth temperature and other growth conditions,[30–32] even though this control is still far from perfect.

14.2.3. *Mechanical properties*

CNTs have excellent mechanical properties, exhibiting no cross-sectional and twisting distortions. They can be bent and elongated/compressed without fracture. This excellent flexibility is in good agreement with their high tensile strength and bulk modulus.[33] Even CVD-grown CNTs with relatively larger diameters (~10 nm) can bend, twist and kink without fracturing.[34,35] CNTs also show excellent mechanical properties when compressed and can form kink-like structures that can relax elastically then the stress is released.[36,37] These properties bode well for the perspective of incorporating CNTs into membrane structures.

CNTS are also a very hard and tough material. The Young's modulus (Y) of small diameter single wall CNT is in the range of several TPa.[38] The tensile strain to failure can be as high as 40%.[37] Raman spectroscopy has been used to indirectly calculate the Young's modulus in both single-wall[39] and multi-wall[40] CNTs by measuring the shift with strain of the frequency of second-order Raman band in the region of $2600 - 2700\,\text{cm}^{-1}$.

14.2.4. *Vibrational spectra of CNTs*

Phonon dispersion relations in CNTs can be studied via those of graphene using a zone folding approach.[41–43] When the lengths are much larger than the diameters, CNTs can be described in the 1D limit: the length is infinite; k points are continuous; and the contributions from the cap Carbon atoms can be neglected. The same as almost all the other properties, the phonon dispersion relations are diameter and chirality dependent.

For a CNT with N hexagons per unit cell, there are 2 carbon atoms per hexagon and 3 vibrational degrees of freedom per atom, yielding $6N$ total degrees of freedom per the unit cell. For zigzag and armchair CNTs, $N = 2n$, while for chiral CNTs

$$N = \frac{2(m^2 + n^2 + nm)}{d_R},$$

where d_R is the greatest common divisor of $(2n + m, 2m + n)$ and is given by

$$d_R = \begin{cases} d & \text{if } n - m \text{ is not a multiple of } 3d \\ 3d & \text{if } n - m \text{ is a multiple of } 3d. \end{cases}$$

For all symmetry types, CNTs have 4 modes for which the frequency vanishes ($\omega \to 0$) as the wave vector approaches the zone center ($k \to 0$). These modes include rigid rotation around the cylindrical axis and has A_{2g} (A) symmetry; rigid translation mode along the axis and has A_{2u}(A) symmetry; rigid translations along the directions perpendicular to the axis and have E_{1u} (E_1) symmetry (two-fold degenerate).

To be more specific, for (n, n) armchair CNTs with an even n, the vibrational modes are decomposed according to the point group D_{nh}. Armchair CNTs with an odd n in the circumferential direction have D_{nd} symmetry. Zigzag $(n, 0)$ CNTs with an odd n also fall into the D_{nd} point group while when n is even, they fall into D_{nh} point group. For both armchair and zigzag CNTs, the translations are decoupled from the rotations, so that they can be described by symmorphic groups. Chiral CNTs (n, m) are more complicated since the basic symmetry operations involve both rotations and translations. They

Table 14.1. Symmetries of Raman-active and IR-active modes for CNTs[44]

CNTs	Point group	Raman-active modes	IR-active modes
armchair (n, n) with even n	D_{nh}	$4A_{1g} + 4E_{1g} + 8E_{2g}$	$A_{2u} + 7E_{1u}$
armchair (n, n) with odd n	D_{nd}	$3A_{1g} + 6E_{1g} + 6E_{2g}$	$2A_{2u} + 5E_{1u}$
zigzag $(n, 0)$ with even n	D_{nh}	$3A_{1g} + 6E_{1g} + 6E_{2g}$	$2A_{2u} + 5E_{1u}$
zigzag $(n, 0)$ with odd n	D_{nd}	$3A_{1g} + 6E_{1g} + 6E_{2g}$	$2A_{2u} + 5E_{1u}$
chiral (n, m) with $n \neq m \neq 0$	C_N	$4A + 5E_1 + 6E_2$	$4A + 5E_1$

can be described by a nonsymmorphic space group C_j, where $j = (N/2 - 1)$. The irreducible representations A_{2g}, A_{2u} and E_{1u} are relevant to the groups D_{nh} and D_{nd}, while A and E_1 are relevant to group C_j.

The number of vibrational modes increases as the diameter of CNT increases. However the number of Raman-active and infrared-active modes remain constant for each symmetry type. The Raman-active modes transform from A_{1g}, E_{1g}, or E_{2g} for D_{nh} and D_{ng} groups, and from A, E_1 and E_2 for C_j group. Similarly, IR-active modes transform from A_{2u} or E_{1u} for D_{nh} and D_{nd} groups, and from A and E_1 for C_j group. More details can be found in Table 14.1.

Some of the vibrational modes frequencies and Raman cross sections are sensitive to the CNT diameter. Significantly, the radial breathing mode (RBM), A_{1g} is a convenient tool for CNT diameter estimation. The RBM frequency does not depend on the CNT chirality since all the atoms move in phase for the radial breathing mode. Raman spectroscopy of the RBMs has been widely used to provide information on CNT diameter distribution, which is a critical parameter for mass transport studies in CNTs.

14.3. Molecular Dynamic Simulations and Early Experimental Observation of Water in CNTs

After the discovery of CNTs, research in that area grew explosively to thousands of articles in 5 years. However, most of those articles dealt with synthesis, growth control, or mechanical and electronic

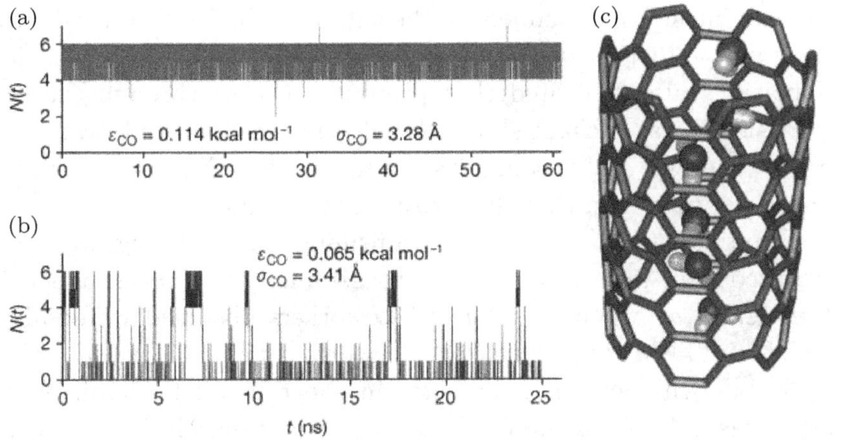

Fig. 14.2. MD simulations of water occupancy in 0.81 nm diameter CNTs. Number N of water molecules inside the tube as a function of time for (a) sp^2 carbon and (b) with reduced carbon-water interactions. (c) Structure of hydrogen-bonded water chain inside the CNT. (Reproduced with permission from Ref. 45.)

properties of CNT. Due to the hydrophobic nature of the CNT interior, none of the scientists initially suspected that CNTs can be filled with water or ions. This situation lasted for a decade, but changed abruptly in 2001 when Hummer and coworkers observed in molecular dynamics (MD) simulations[45] that water can fill the 0.81 nm inner cavity of a (6,6) CNT (Figure 14.2). A particularly striking feature of that simulation was that water inside the CNT cavity changed its arrangement to form a single-file chain, reminiscent of the water arrangement in aquaporin channels.[46] Further analysis also revealed that the enthalpic penalty of losing two hydrogen bonds upon water entry was offset by the narrowed energy state distribution as well as increased rotational entropy of the molecules in the single-file chain.[45]

Applying osmotic pressure to a water-filled nanotube in the simulation revealed an even greater surprise of a nearly friction-less water flow with the transport no longer governed by the Hagen-Poiseuille equation.[45] Moreover, subsequent simulations indicated that the transport barriers arose mostly from entering and exiting the CNT, and were mostly independent of the tube length.[47] Interestingly, fast transport predictions were not limited to just water, subsequent MD simulations also indicated that proton transport through the one-dimensional water chain should be also faster than bulk water due to the single-file water configuration inside the CNT promoting fast proton hopping following the Grotthuss mechanism.[48]

Experimental observations confirming water filling in CNTs appeared not long after that first seminal simulation paper by Hummer *et al.* when Gogotsi and coworkers observed water filling in CNT in TEM micrographs.[49] This study started from the closed nanotubes with an inner diameter in the range 2 to 5 nm, where water was able to fill the CNTs through the sidewall defects during autoclave treatment and remained trapped even under TEM vacuum condition (Figure 14.3). Gogotsi and coworkers further demonstrated the crystalline ice peak in the electron energy loss spectroscopy when they cooled the sample to 80°C. Beside TEM images, neutron scattering also revealed water entry inside CNTs. Zanotti and coworkers showed that 1.4 nm diameter CNTs have an ice-shell and water-chain structure inside the CNT after exposed to saturated water vapor.[50] Later on, Mamontov and coworkers also investigated the water structure inside CNTs using neutron scattering. They studied the temperature dependence of the confined water and

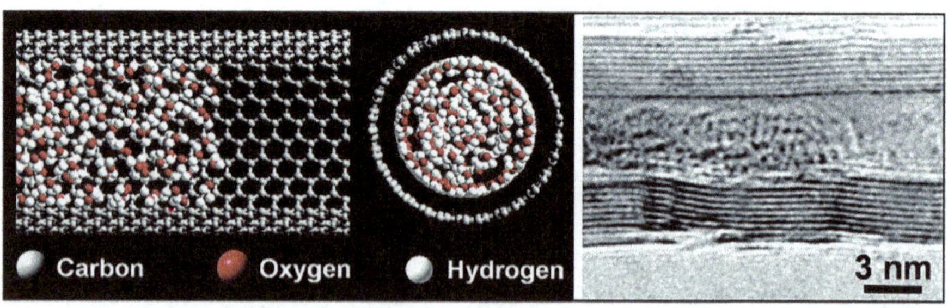

Fig. 14.3. TEM image showing water filling inside a 2.9 nm in diameter multi-walled CNT after autoclave treatment (right). Simulation work illustrating the water arrangement inside the tube with a side view and a cross-sectional snapshot (left). (Reproduced with permission from Ref. 49. © *2004 American Chemical Society.*)

found that the phase transition temperature shifts depend on CNT diameters.[51]

14.4. Experimental Platforms for Observing Transport in CNTs

Computer simulations and experimental results, which showed water filling in CNT pores, prompted researchers to investigate water transport in these channels. The race to demonstrate mass transport in CNTs was won by the group led by Bruce Hinds who in 2004 reported that they had fabricated aligned 4.3 ± 2.3 nm diameter CNT membranes from CVD-grown vertically-aligned CNT forests filled with a polymer matrix and had observed water and ion transport in those membranes.[52] In 2006 a group led by Bakajin and Noy fabricated aligned 1.6 nm pore diameter CNT membranes using SiN_x ceramic fill and demonstrated extremely fast gas and water transport in those channels.[53] A group led by Marand also came up with an innovative approach where CNTs were filtered onto a porous support and sealed with a polymer to form a permeable membrane.[54] We discuss the details of these structures alter in the *Carbon Nanotube Membranes* subsection (Figure 14.4).

Aligned CNT membranes with their μm-long nanotube channels are potentially scalable platforms that can be used for a variety of experiments such as gas flow, pressure-driven water flow, and various electrical measurements. However, a number of significant drawbacks prevented them from being ideal model systems to study fluidic transport in CNTs. A very big concern was the purity of CNTs array, which inevitably contained a mixture of different chiralities and diameters. An even bigger concern was the presence and impact of point defects on the CNT sidewalls. The other concern was the leakage from the gaps between CNTs and defects formed during the filling of the aligned CNT "forest" with the membrane matrix material. Single CNT measurement approaches, developed between 2010 to 2014, partially as a reaction to the problems with aligned CNT membrane platforms, aimed to overcome these drawbacks.[55–58]

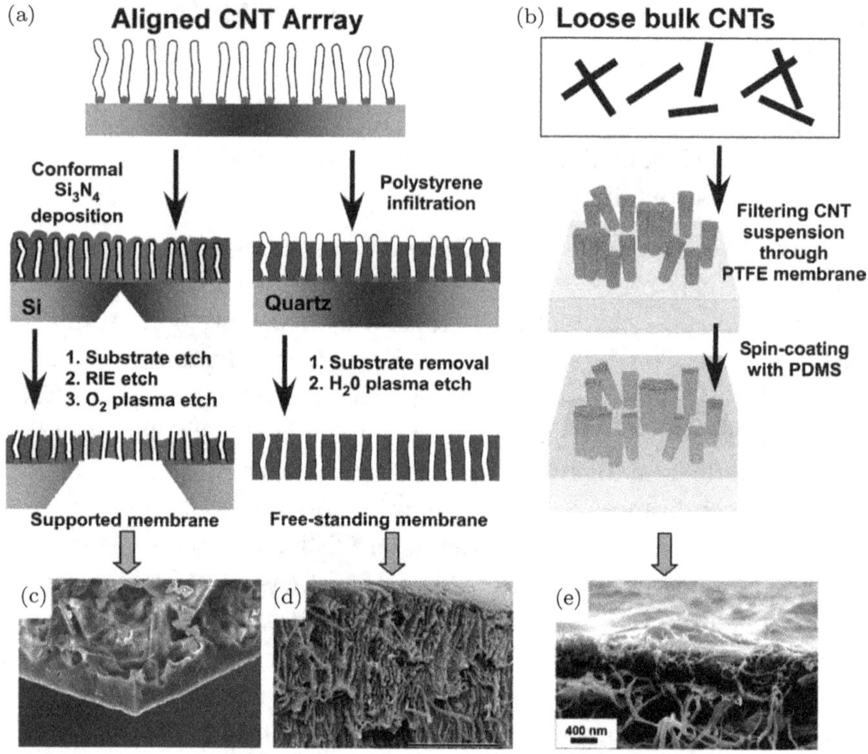

Fig. 14.4. Fabrication of CNT membranes. (a) Aligned CNT array with silicon nitride or polystyrene and (b) semi-aligned CNT with PDMS polymer. (Reproduced with permission from Refs. 62, 58, 52, 54. © *2004, 2006 American Association for the Advancement of Science and 2007 American Chemical Society and 2007 Elsevier.*)

We will introduce these platforms in the subsections 14.4.2 and 14.4.3, respectively.

14.4.1. *Aligned and semi-aligned carbon nanotube membranes*

Hinds group embedded multi-walled CNT (MWNT) arrays in polystyrene (PS) filling.[52] Dense, vertically aligned MWNT arrays were grown from the catalyst on quartz substrates using the CVD method and then filled with PS solution using a spin-coat method. High wetability of CNT with PS leads to complete infiltration of the CNT array. After removal of the top and bottom substrates and etching/opening of the resulting freestanding CNT membranes with H_2O plasma researchers obtained CNT membranes with 4.3 ± 2.3 nm inner pores and 5 to 10 micron in thickness. Hinds group further demonstrated that these membranes can be modified at the pore mouth to introduce different chemical functionality.[52]

Sub-2-nm CNT membranes were fabricated by Holt *et al.* starting from a vertically aligned double-walled carbon nanotube (DWNT) array grown on a surface of a Si chip. Encapsulation with CVD-deposited silicon nitride (Si_3N_4) resulted in a composite that was then opened by plasma etching to produce a membrane.[53,59] This fabrication procedure produced rather thin and fragile CNT membrane patches covering micron-sized apertures that were nonetheless able to withstand over 1 atm pressure gradient during measurements. A recent study from Park's group followed a similar design, but used single-walled carbon nanotube (SWNT) arrays and titanium dioxide (TiO_2) coating.[60,61] A key fabrication step involved the use a well-controlled atomic layer deposition (ALD) to encapsulate the titanium dioxide around SWNT arrays and CNT ends in the membranes were etched away and opened up with argon ion-beam-milling. Crucially, this etching step served also as a quality control step: the etching was stopped once gas flow through the membrane reached a constant value and stopped increasing after further etching, confirming all CNT pores were open and further etching didn't result in creation of additional defect-based leaks through the membrane.[61]

Marand group introduced a hybrid CNT platform in 2007 that featured semi-aligned CNT in polymer membranes.[54] End-group modified CNTs were first treated with acid and filtered through porous membrane, which created semi-aligned CNT on the support membrane. The CNT membranes were later coated with polymers like PDMS. This produced thin-film CNT/polymer nanocomposite with CNT protruding from the top of the membranes, which were used for gas separation and water filtration purposes.[54,63] Hinds group also developed another CNT membrane platform that was based on microtomed slices of an epoxy resin block filled with randomly oriented long carbon nanotubes.[64] If individual CNTs were long enough, they were able to span the cross-section of the membrane and form well-defined pores. A key advantage of these techniques is that they had freed the researchers from the limitation of the CNT pore size imposed by the vertically-aligned CNT "forest" synthesis and allowed them to use smaller diameter pores. On the other hand, these fabrication strategies produced membranes with inherently very low pore densities that placed severe limitations on their functionality.

14.4.2. *Single CNT channel platforms*

One of the first single-CNT measurement platforms was introduced by Lindsay and co-workers in 2010.[55] In that work the measurement device was fabricated by connecting individial 1–2 nm diameter CNTs, grown by ethanol vapor CVD on a Si wafer, to reservoir channels defined by e-beam lithography. The CNT ends were then removed by O_2 plasma etch before the device was sealed. Interestingly, 20% of the devices fabricated with this approach showed relatively large ion current that indicated that the CNT channel was open and conducted ions. Researchers were also able to observe electrophoretic transport of single-strand DNA in this system.

Strano group also reported a platform that enabled measuring ion transport through individual or a small number of single CNTs in 2010 (Figure 14.5).[56] To fabricate these CNT devices, they seeded catalyst nanoparticles at the substrate edge and grew very long aligned CNTs using CVD process.[65] Then, the team protected sections of the nanotubes with lithographically-defined masks and the rest was etched and opened by oxygen plasma before connecting to the two microfabricated fluid reservoirs. Measurements using this platform first focused on electrical potential-driven ion transport.[56,66,67] Subsequently, the same devices platform was used to monitor water transport using Raman spectrum peak shifts, where a shift of the RBM region peak indicated the existence of a continuous water phase inside a nanotube.[68]

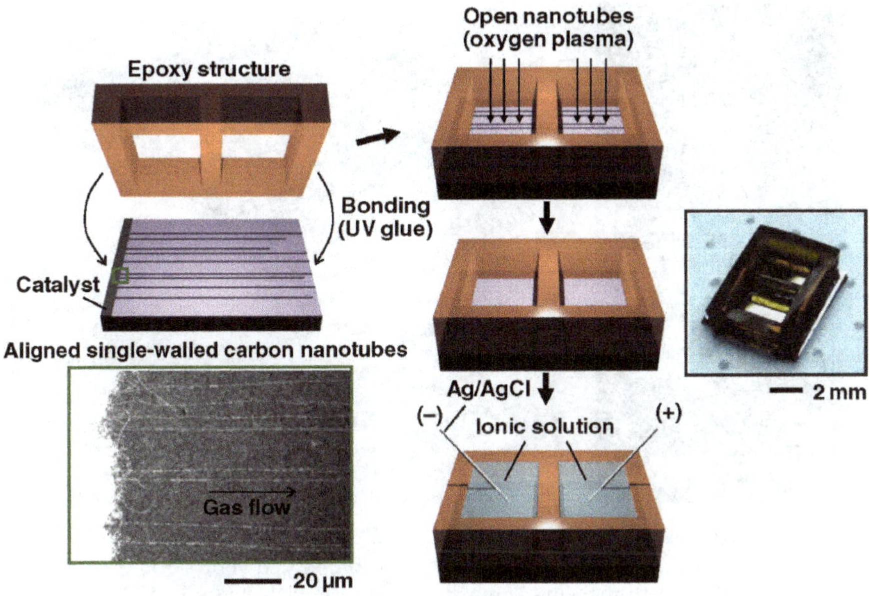

Fig. 14.5. Fabrication procedure used to obtain a single SWCNT channel device. (Reproduced with permission from Ref. 56. © *2010 American Association for the Advancement of Science.*)

Another iconic single nanotube platform was demonstrated by Bocquet group, who utilized a unique glass nanocapillary/ nanotube combination to study nanofluidics in CNTs and boron nitride nanotubes (Figure 14.6).[57,69] To fabricate this nanojet setup, a single, long CNT protruding from a CNT forest was first picked up with a tungsten tip. After that tube was glued to the tip, it was carefully maneuvered into the orifice of a glass nanocapillary. In the next step, local electron beam-induced naphthalene deposition sealed the gap between the CNT and the capillary wall. All steps to make one successful device were done in SEM with well-controlled pressure and position alignment. CNTs used for this procedure were usually in the diameter range from 7 nm to 100 nm and ca. hundred nanometers in length. In the final device a CNT, glued at the capillary tip, bridged two reservoirs: the inside of the capillary and a fluid chamber into which the capillary was placed. This platform

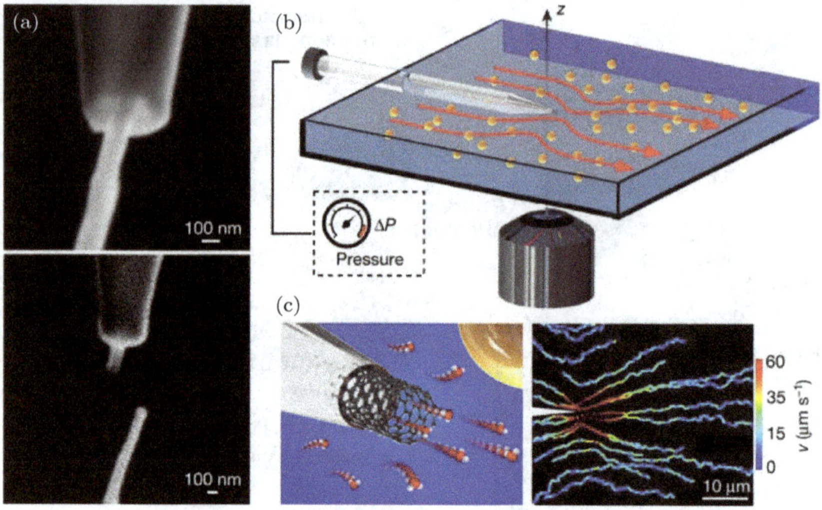

Fig. 14.6. Nanocapillary/CNT experimental set-up. (a) SEM image of a CNT inserted in a nanocapillary (top) and later cut and sealed (bottom). (b) Sketch of the fluidic cell used to measure the water flow rate with "nanojet" CNT tip and tracer particles. (c) Water movement diagram under pressure in this set-up, and the trace of tracer particles, which the flow rate was derived from. (Reproduced with permission from Ref. 57. © *2016 Nature Publishing Group.*)

turned out to be surprisingly versatile: it enabled electrophoretic ion transport experiments[57,70] as well as pressure-driven water transport measurements, which utilized 500 nm polystyrene tracer particles and tracked the motion of these particles to map the flow velocity.[57]

Nuckolls group also created isolated single-walled CNT device with a CNT channel bridging a polymer divider between two reservoirs and used it to measure DNA translocations and ion transport through the channel.[71] Jourdain group fabricated similar devices as well, but with several single CNTs with diameters in the range of 1 to 2 nm, and used them to study electrophoretic ion transport in these channels.[72]

14.4.3. *Carbon nanotube porins: biomimetic CNT membrane pores*

Noy group comes up with a different strategy to measure water and ion transport in CNT channels. Their platform, carbon nanotube

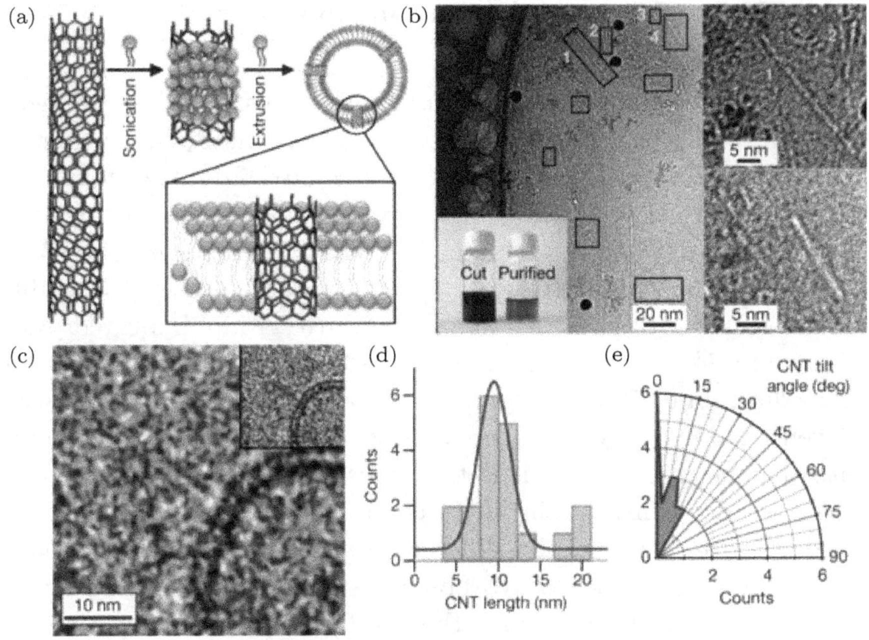

Fig. 14.7. Synthesis and characterization of carbon nanotube porins (CNTPs). (a) Schematic of CNTP preparation and incorporation into lipid bilayer membranes. (b) TEM image of CNTPs. (c) Cryo-EM images of a CNTP inserted in a lipid vesicle. (d) Histogram of CNT length and angle distribution when inserted into the lipid membranes. (Reproduced with permission from Ref. 58. © *2014 Nature Publishing Group.*)

porins (CNTPs), emulated membrane channels, with ultrashort CNTs 10-20 nm in length embedded in lipid or polymer membranes (Figure 14.7).[58,73] To fabricate CNTPs, commercially available, long CNTs were cut by tip-sonication into much shorter pieces.[74] Significantly, this procedure allowed to use CNT stocks of different diameter, producing CNTPs of different size. Specifically, the team made CNTPs with 0.8 and 1.5 nm diameter, allowing direct comparison of transport efficiency between channels with the same geometry but different pore sizes. Another key advantage of this biomimetic platform was that it allowed direct adoption of a number of protocols that were developed for measuring water and ion transport in biological membrane channels, including lipid vesicle and planar lipid bilayer platforms.[75,76]

14.5. Water Transport in Carbon Nanotube Channels

The macroscopic laws of hydrodynamics, given by the Hagen-Poiseuille equation, govern the flow in conventional fluidic channels. From the very first simulations that predicted fast flow in carbon nanotubes it was clear that the conventional fluid flow description would not be applicable anymore.[45,47] What are the physical reasons behind this discrepancy? Further simulations provided some valuable clues. A pioneering work from the Aluru group indicated that continuum flow theory was valid for CNT with 10 nm diameter, but broke down in 0.95 nm nanotubes.[77] Clearly, one of the reasons why conventional hydrodynamics breaks down in the smallest CNT pores with diameters below 1 nm is that those pore sizes squeeze water molecules down into a single-file configuration with the hydrogen bonding pattern drastically different from conventional bulk water arrangement.[45,76] Smooth hydrophobic surface of carbon nanotubes enable inherently low friction at the interface, which is conventionally described as very large slip length[57,78] (the slip length is defined as the depth inside a solid surface where the extrapolated fluidic velocity profile goes to zero). A conventional version of the Hagen-Poiseuille equation assumes a no-slip condition at the interface (i.e. a zero value of the slip length). In the extreme case of a very large slip length compared to the channel diameter, which corresponds to a nearly frictionless surface, the boundary can be considered virtually shear-free, making the Hagen-Poiseuille equation invalid. A simulation by the Aluru group (Figure 14.8) demonstrated some of the effects the nature of the channel surface enacts on the flow inside a nanotube. Of the four nanotubes with similar diameter, a CNT had the highest water velocity and the highest flux, compared to boron nitride tube (BNNT), hydrophilic silicon nanotube (NT Si), and a hypothetical CNT with computationally-engineered rough surface.[79] Consequently, the flux enhancement for the other nanotubes considered in that study was significantly reduced. The study concluded that the origin of the remarkable flux enhancement in CNTs was a combination of the smoothness of the

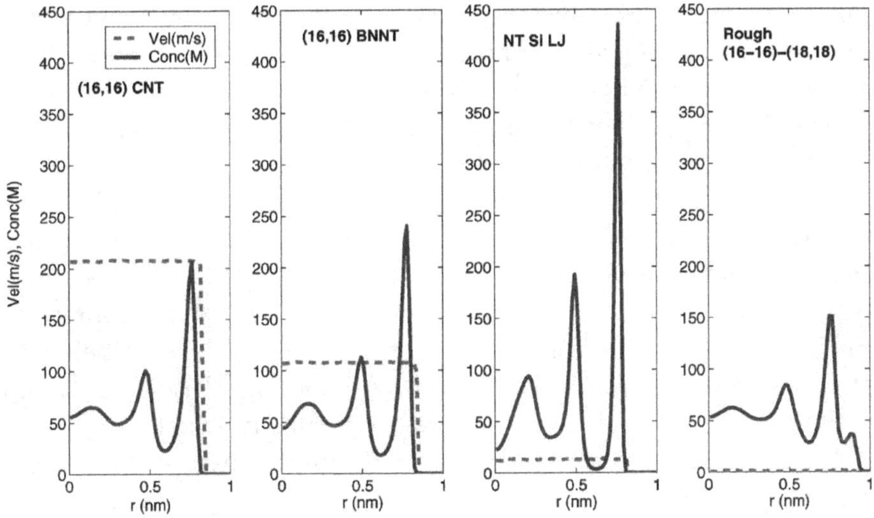

Fig. 14.8. MD simulation of water velocity (red dash line) and water density (blue line) profiles as a function of the radius from the center of a (16, 16) CNT, (16, 16) BNNT, (16, 16) nanotube with silicon, and a rough surface (16, 16)/(18, 18) CNT combination. (Reproduced with permission from Ref. 79. © *2008 American Chemical Society.*)

CNT surface and its hydrophobicity that reduced water molecule interactions with the tube surface.

The first experimental demonstrations of high water flux enhancement in CNT pores were reported shortly after the publication of the first computational predictions. Hinds group first demonstrated high rates of water transport through aligned MWNT membranes with 7 nm in diameter and reported remarkably long slip lengths[78] that indicated more than four orders of magnitude increase in water permeability compared with the Hagen-Poiseuille equation predictions. When they tested the flow of a variety of solvents, including hexane and ethanol, through MWNT membranes, the permeability decreased as the solvents became more hydrophobic. These findings emphasized again that it is not only the smoothness of the CNT surface but also its hydrophobic nature that is responsible for high water velocity. Bakajin, Noy and coworkers also reported water fluxes in aligned sub-2-nm DWNT membranes under pressure-driven flow conditions that exceeded values calculated from continuum hydrodynamics models by more than three orders of magnitude.[53] Later on, Park group demonstrated osmotic water and salt transport across their CNT membranes in forward osmosis (FO) setup again reporting with high water fluxes (Figure 14.9).[61]

Measurements of water filling and flow in individual CNT pores revealed an even more complicated physical picture. Strano group measurements showed evidence of phase transitions between ice, liquid water, and gas phase water in CNTs at near room temperatures,[68] providing an experimental evidence of water behavior that was previously predicted in MD simulations.[80] Bocquet group performed a scientific tour-de-force experiment where they were able to use their "nanojet" platform to observe exceptionally high water transport rates in single CNTs directly.[57] Remarkably, these measurements (Figure 14.6) showed that the values of the slip length became divergent as the size of the carbon nanotubes became smaller and started to approach the "single-digit" nanopore regime (i.e. $d < 10\,\text{nm}$). In contrast, boron nitride nanotubes (BNNTs) of similar diameters showed very small values of slip length (Figure 14.10). Compared to CNTs, BNNTs have a similarly smooth, but much higher charged, surface, which interacts with water molecules quite strongly and slows down the water flow through the tube significantly.

Another interesting comparison could be made with the water transport in aquaporins, which are biological membrane channels responsible for water transport. Very narrow (ca. 3Å) AQP channels, which are lined mostly with hydrophobic residues, enable extremely fast water transport and are considered the gold standard of membrane water channels. The Noy group compared water transport in two types of CNTPs with diameters of 0.8 and 1.5 nm with water transport in aquaporins.[76] While the 1.5 nm diameter CNTP showed bulk water-like transport with the fluxes significantly lower than that of the AQP1 protein channels, the 0.8 nm diameter CNTP showed enhanced water flow rates that exceeded even those of AQP-1 (Figure 14.11). This remarkably strong performance was attributed again to a combination of several factors. Narrow diameter CNTPs

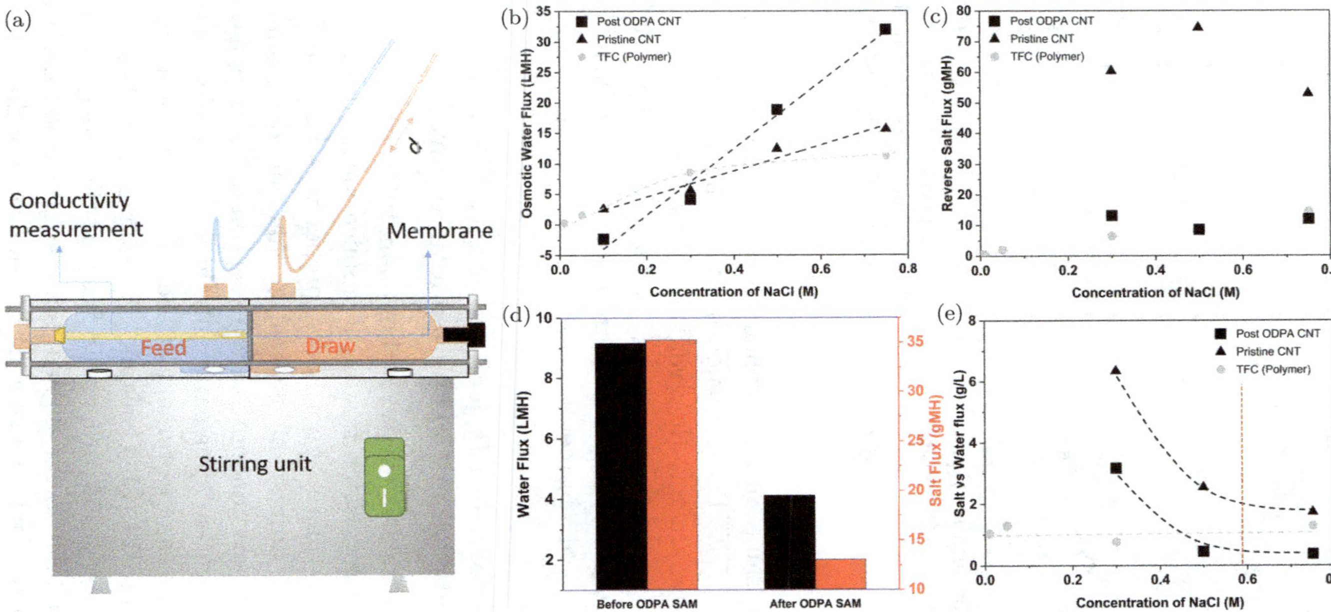

Fig. 14.9. Osmotic water and salt transport across CNT/titanium dioxide membranes in forward osmosis (FO) setup. (a) Sketch of the FO diffusion cell to monitor osmotic water transfer using meniscus capillaries and salt transport using conductivity electrodes. (b, c, and d) Water flux and reverse salt flux through the membrane compared with commercial available polymer membranes and further modified CNT membranes. (e) Membrane performance comparison at different salt conditions. (Reproduced with permission from Ref.[61] ©*2018 American Chemical Society.*)

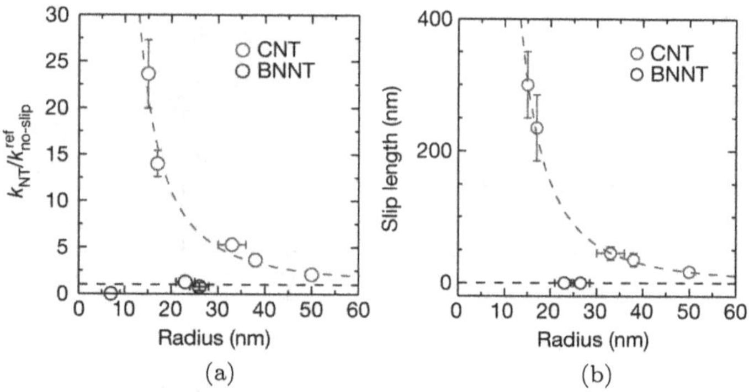

Fig. 14.10. Water permeability and slip length of individual CNTs and BNNTs derived from the nanojet measurements. (a) Normalized permeability and (b) slip length of tubes as a function of nanotube radius. The horizontal dash lines indicate the no-slip prediction. (Reproduced with permission from Ref. 57. © *2016 Nature Publishing Group.*)

benefit from the same physical phenomena that enhance water transport in wider CNTPs: the smoothness of the tube walls and their hydrophobic character. However, in addition to these factors, 0.8 nm CNTPs force water into a single-file configuration where a chain-like hydrogen bonding pattern leads to a further water transport efficiency enhancement.

14.6. Ion Transport, Ion Rejection, and Water/Ion Permselectivity in Carbon Nanotube Pores

High water permeability makes carbon nanotubes an attractive materials platform for water treatment applications; however, the other, and arguably more important,[81] property is ion selectivity and ion rejection. A number of physical mechanisms determine ion selectivity in membrane nanopores: size exclusion at the pore mouth, partial solute dehydration upon entry into the pore, and electrostatic exclusion due to the charges present at the pore mouth or along the pore surface (Figure 14.12). Constant diameter of the CNT pores makes their size exclusion properties fairly straightforward. Holt *et al.* demonstrated sub-2-nm aligned double-wall carbon nanotube membranes excluded 2 nm gold colloids.[53] However, the diameters of

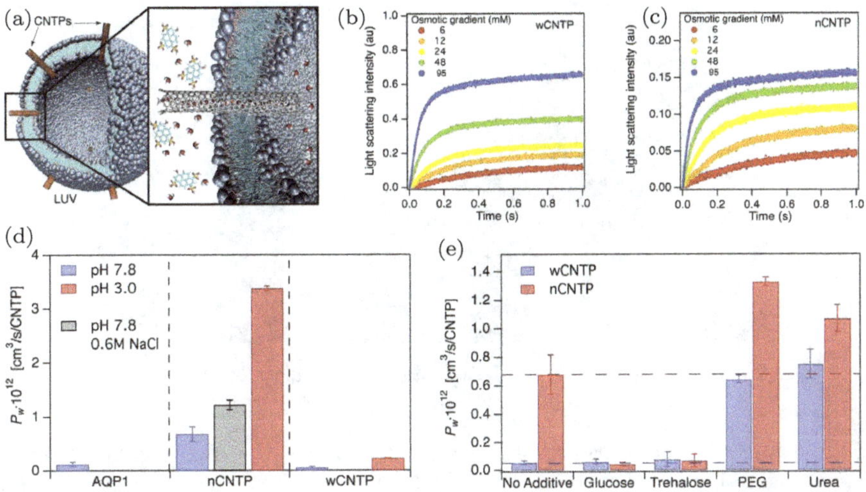

Fig. 14.11. Water transport through 0.8 nm and 1.5 nm diameter CNTPs. (a) Sketch of CNTP in the lipid vesicle platform for water flux measurements under osmotic pressure. (b, c) Light scattering traces for lipid vesicle with CNTPs under various osmotic gradients. (d) Water permeability for 0.8 nm and 1.5 nm diameter CNTPs at pH 7.5 and pH 3.0, compared with AQP-1. (e) Water permeability changes with additives, which change water structure. (Reproduced with permission from Ref. 76. © *2017 American Association for the Advancement of Science.*)

carbon nanotubes are *a-priori* still too large to make the pore exclude all ions based simply on their ionic radii. The situation becomes very different when we consider the hydrated radii of the ions, making the energy costs associated with the partial ion dehydration upon entry into the carbon nanotube pore another powerful mechanism for ion selectivity.[82] Electrostatic repulsion is potentially another powerful mechanism for conveying selectivity to the CNT pores. Generally, the range and the impact of the electrostatic effects is highly dependent on the electrostatic screening due to the presence of mobile charges in solution, which is described by the Debye length. Specifically, when the Debye length of the electrolyte is larger than the radius of the nanopore with charged ends or walls, the electric fields overlap and create a barrier for the ion passage. As the ion concentration increases and the Debye length drops, the electric fields contract and create an opening through which ions can pass. As the Debye

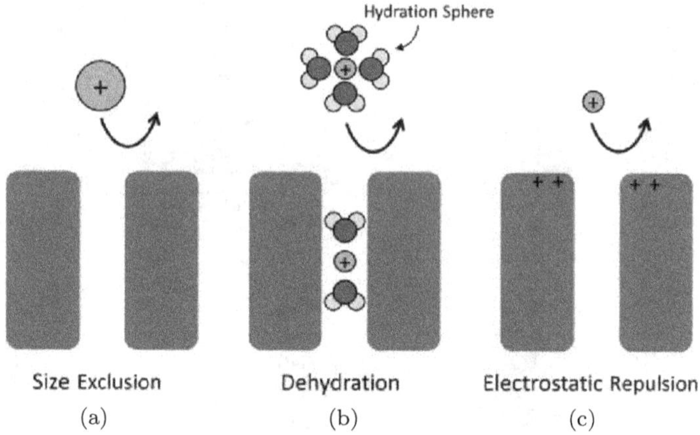

Fig. 14.12. Mechanisms of ion selectivity and rejection for a nano channel, including (a) size exclusion, (b) dehydration barriers, and (c) electrostatic interactions. (Reproduced with permission from Ref. 84. © *2017 Materials Research Society*.)

length at seawater conditions is rather small, ca. 3 Å, we do not expect that pure electrostatic repulsion could represent the dominant rejection mechanism for these types of water feeds. Indeed, the effect of decreasing ion selectivity at higher ion concentration was observed experimentally for both aligned carbon nanotube membranes and carbon nanotube porin experimental platforms.[76,83]

The first observation of ion transport in CNT pores was reported by Hinds and co-workers in their seminal paper on aligned nanotube membranes,[52] where they observed diffusion of $Ru(NH_3)_6^{3+}$ ions across the membrane. The same work demonstrated the possibility of using size exclusion/steric hindrance mechanism for modulating $Ru(NH_3)_6^{3+}$ ion flux by functionalizing the CNT pore entrances with biotin and subsequently attaching a bulky streptavidin molecule to the the biotin.[52] Bakajin, Noy and co-workers studied transport of a variety of ionic compounds in smaller (1.6 nm) diameter aligned DWCNT membranes using pressure-driven reverse osmosis experiments and observed that salt rejection in these pore ranged from nearly 100% for $K_3Fe(CN)_6$ to very low for $CaCl_2$.[83] The degree of rejection was also a strong function of the overall ionic strength of the solution, with KCl rejection ranging from 50% at

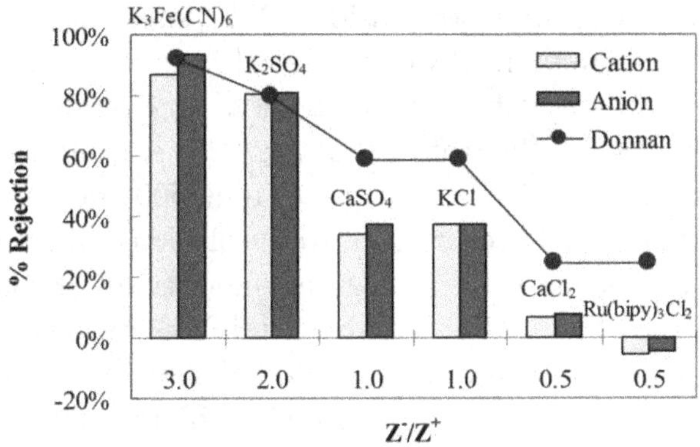

Fig. 14.13. Rejection coefficients (bars) measured for six salt solutions that have the same equivalent concentration but different ion valence. Points (filled circles) indicate rejections calculated from the Donnan model. The calculation assumed charge density corresponding to approximately seven charged groups per nanotube. (Reproduced with permission from Ref. 83. © *2008 The National Academy of Sciences of the USA*.)

low ion concentration to near zero at high concentration. These results provided a very strong evidence that the Donnan electrostatic exclusion was the dominant mechanism for the ion selectivity in CNT pores of this size (Fig. 14.13). Subsequent work by the same group traced the origin of this selectivity to the presence of charged COO^- groups at the rim of the carbon nanotube pores.[85] Interestingly, Marand group reported that membranes with CNTs functionalized with chain-like zwitterion molecules rejected essentially all ions, giving 98.6% of rejection rate on NaCl.[63] The researchers speculated that these bulky functional groups promoted size exclusion and provided multiple charges to enhance electrostatic repulsion.

Observations of ion transport in individual CNT channels took comparatively longer, with the first reports appearing only by 2010.[55,56] Both of these two studies that pioneered this sub-field used ultra-long CNTs incorporated into a microfluidic device platform. First, Lindsay and coworkers reported large electrophoretic current through single SWCNTs, where the measured current exceeded the value calculated from the CNT diameter and KCl bulk solution conductivity.[55] This enhancement was interpreted as another evidence for the friction-less transport inside CNT pores. Interestingly, this work noted that metallic CNTs exhibited larger ion current. Strano and coworkers investigated similar devices and reached a radically-different conclusion; they observed that the current through their devices that had ca. 1.5 nm diameter and 500 μm long CNT was largely protonic.[56] Moreover, this group observed that alkali metal ions could act as as current blockers and create regular current oscillation.[56] A further systematic study of various CNTs with diameters that ranged from 0.94 nm to 2.01 nm showed varied dwell times and blocking currents, in which CNT diameter around 1.58 nm displayed the largest blocking current.[67]

More recently a number of measurements reported ion and proton transport in individual CNT channels with length ranging from several micrometers[70] to ultra-short (10 nm) CNTPs.[75,76] A study from the Noy group reported ultra-fast proton transport in 0.8 nm diameter CNTPs due to the single-file water arrangement inside that channel promoting proton transport via the Grotthuss hopping mechanism,[75] confirming earlier MD simulations predictions.[48] Results from the Bocquet group showed intriguing CNT conductance dependence on salt concentration[70] when they examined KCl transport in CNT with 7 nm, 20 nm, 28 nm, and 70 nm diameters. All CNTs showed power law conductance scaling that transitioned from around linear at high salt concentration to 1/3 power law at low salt (Figure 14.14). The explanation advanced by Bocquet *et al.* relied on an elegant model that assumed that the CNT inner surface carried some negative charge due to adsorption of the OH^- groups. This adsorption process, in turn, was subject to charge regulation at varying ionic strengths, with a combination of this charge regulation and preferential accumulation of the counterions inside the channel producing the observed 1/3 power law scaling of the ion conductance. Some of the main problems with this approach were that there was no direct evidence for the OH^- ions adsorption on the CNT surfaces and that such adsorption could potentially interfere with the fast transport and high water flux in the CNT pores.

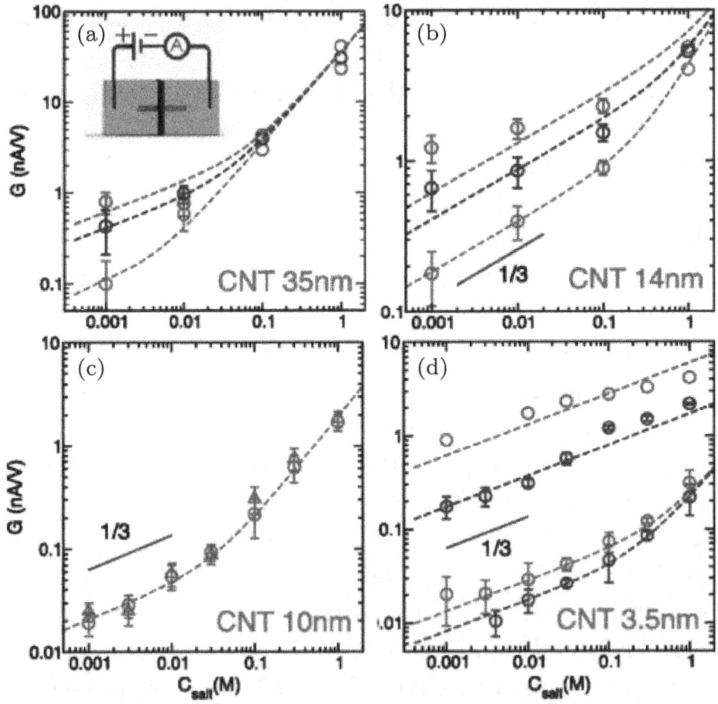

Fig. 14.14. Conductance (G) scaling with a range of electrolyte concentrations measured through single CNTs with various diameters from 70 nm to 7 nm and pH from 10 to 4. (Reproduced with permission from Ref. 70. © *2016 American Physical Society.*)

The 1/3 power law behavior turns out not to be a universal scaling law in CNT pores. Nuckolls group reported non-linear concentration scaling for ion transport in 1.5 nm diameter CNT pores which followed 1/2 power law.[71] That team attributed the observed effects to the charge regulation of the COO^- groups at the pore entrances. Noy group reported an even richer set of behaviors in CNT porins of different diameters. For the ultra-narrow, 0.8 nm diameter CNTPs, the ion conductance initially followed the 1/2 exponent power law and then saturated at higher ionic strengths,[76] mimicking the behavior of small diameter biological ion channels.[86] Ion conductance in larger diameter, 1.5 nm, CNTPs displayed yet another pattern of behavior, showing 2/3 power law and no saturation at high ion concentrations.[87]

Extensive modeling efforts by several groups have clarified some of the origins of these reach set of scaling behaviors. Biesheuvel and Bazant presented an extensive analysis that showed how a solution to the Poisson-Nernst-Planck equations in combination with charge regulation can describe some of the scaling laws observed in the experiments,[88] pointing out that in particular the 1/2 power law could be the evidence of the conductance following the "good co-ion exclusion" limit where the the nanotube pore becomes exclusively selective to one kind of ions. Manghi and coworkers added another layer of complexity to the model by accounting for the possibility of strong fluid slip along the nanotube pore walls.[89] This interpretation in combination with the extensive modeling helped Noy, Aluru and coworkers to rationalize their observation of the 2/3 power law scaling in larger diameter CNT pores and attribute it to extremely strong electroosmotic coupling in these channels.

Strano group, Nuckolls group and Noy group all reported that CNT pores with negatively charged COO^- groups at the ends showed preferential cation selectivity. Nuckolls group compared the transport of KCl and $K_3Fe(CN)_6$ in 1.5 nm diameter CNTs and reported negligible effect on ion transport; however, when they replaced potassium with a bigger cation, tris(bipyridine)ruthenium(II), $[Ru(bpy)_3]^{2+}$, they observed significant reduction of conductance, which they attributed to the increased ion rejection of the larger cation at the nanotube entrance.[71] Noy group performed extensive characterization of ion selectivity in CNTPs using reversal potential measurements in single CNTP conductance experiments.[76] In these experiments CNTPs with 0.8 nm diameter showed near-perfect cation selectivity which started to roll off only at very high ionic strength above the seawater levels (Figure 14.15). This observation is remarkable as it indicates that at small enough diameters CNTPs exhibit monovalent salt rejection that is sufficient enough for seawater desalination. Not surprisingly, larger 1.5 nm diameter CNTPs do not have the same degree of ion selectivity and instead have much weaker cation selectivity with the permselectivity

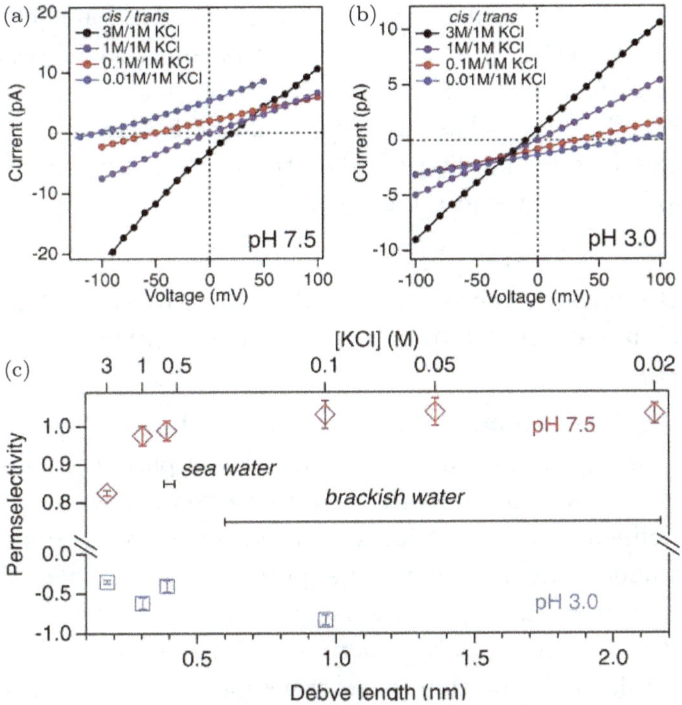

Fig. 14.15. Ion selectivity between cation and anion in 0.8 nm diameter CNTPs. (a and b) Reversal potential measurements showing asymmetric I-V curves at pH 7.5 and pH 3.0. (c) Permselectivity values plotted as a function of the Debye length for higher concentration KCl electrolytes. (Reproduced with permission from Ref. 76. © *2017 American Association for the Advancement of Science.*)

of around 0.5.[87] When the COO^- groups at the ends were neutralized by changing the pH to 3, CNTPs became even less selective, with their permselectivity dropping down to zero.

Overall, despite the significant progress reported in elucidating ion selectivity in CNTs our understanding of the ion rejection mechanisms in these nanochannels remains incomplete. For example, experimental data on CNTP conductance show linear I-V curves even for the smallest 0.8 nm pores, indicating that the barrier for entry into these tubes for at least the majority carrier species is quite low.[76] This observation directly contradicts the results of classical MD simulations that show that ion entry into a small diameter CNT pore requires costly partial dehydration of the ion solvation shell.[90] Even more complicated questions include whether such partial desolvation and strong degree of confinement induce any unusual differential selectivity effects, correlated transport effects. Finally, we still do not have a good understanding of what the concentration polarization effects would look like in these pores.

14.7. Outlook: Carbon Nanotube Pores in the Next Generation Advanced Water Purification Solutions

Overall, carbon nanotube pores represent a drastic departure from the conventional water purification membrane paradigm. The inherent advantage of the smooth, nearly frictionless water transport in well-defined rigid CNT pores over the transport through rough and tortuous polymer membrane pores is undeniable. However, before any practical applications of these pores become reality, researchers will need to solve a number of important problems. First, the permselectivity of the water transport in CNTs needs to be improved to the levels that match or approach those of aquaporins. These efforts will require a deeper understanding of the rejection mechanism in CNT pores, as well as development of targeted high-yield strategies of the pore modification. A particular challenge would be to achieve rejection of uncharged species, organic molecules and micropollutants.

Second, researchers would need to understand the role and impact of defects on the transport and selectivity of longer CNT channels, as those types of channels would likely be present in a membrane that is able to withstand rigors of commercial use. This task is non-trivial, as defects in a closely confined sub-nanometer diameter channel are expected to have a rather out-sized impact and potentially could diminish or even destroy most of the advantages of the carbon nanotube pores. Unfortunately, given the typical defect density in CNTs the likely-hood of having defects in a CNT channel that is longer than 100–200 nm, which is longer than a CNT porin, but shorter than a typical pore in an aligned or semi-aligned CNT membrane, is quite high. Mitigation of these effects would require additional research on controlling or eliminating local defects in the CNT structure.

Third and perhaps the hardest challenge would be to develop approaches for cost-effective scale-up of carbon nanotube assemblies that enable water transport on the scale that is commensurate with water industry applications. Even though there is nothing inherently costly about carbon nanotubes, synthesis of aligned CNT arrays currently requires very high temperatures, low pressure setups, and costly thin-film vacuum deposition approaches for the catalyst. Bulk carbon nanotube synthesis is somewhat more cost-effective, but it also requires facile strategies of producing open-ended CNT channels and assembling them into structures with membrane functionality. Even though the scale and the impact of those challenges should not be minimized, we want to end by once again reiterating the inherent advantages of the CNT pore platform. These pores support some of the fastest water transport rates, they are inherently resistant to fouling, and they potentially offer exquisite control over the pore size and the resulting selectivity that rivals and could potentially exceed those of biological ion channels. Thus, the potential of carbon nanotube pores to revolutionize membrane separations should not be overlooked.

Acknowledgments

This work was supported by the U.S. Department of Energy, Office of Basic Energy Sciences, Division of Materials Sciences and Engineering under Award SCW0972. Work at the Lawrence Livermore National Laboratory was performed under the auspices of the U.S. Department of Energy under Contract DE-AC52-07NA27344.

References

1. M. Elimelech, W. A. Phillip, The future of seawater desalination: Energy, technology, and the environment, *Science* **333** (2011) 712–717.
2. S. L. Forman, J. C. Fettinger, S. Pieraccini, G. Gottarelli, J. T. Davis, Toward artificial ion channels: A lipophilic g-quadruplex, *J. Am. Chem. Soc.* **122** (2000) 4060–4067.

3. W. Si, L. Chen, X.-B. Hu, G. Tang, Z. Chen, J.-L. Hou, Z.-T. Li, Selective artificial transmembrane channels for protons by formation of water wires, *Angew. Chem. Int.* **50** (2011) 12564–12568.
4. W. Song, H. Joshi, R. Chowdhury, J. S. Najem, Y.-X. Shen, C. Lang, C. B. Henderson, Y.-M. Tu, M. Farell, M. E. Pitz *et al.*, Artificial water channels enable fast and selective water permeation through water-wire networks, *Nature Nanotech.* **15** (2020) 73–79.
5. S. Iijima, Helical microtubules of graphitic carbon, *Nature* **354** (1991) 56.
6. S. Iijima, T. Ichihashi, Single-shell carbon nanotubes of 1-nm diameter, *Nature* **363** (1993) 603.
7. D. Bethune, C. H. Kiang, M. De Vries, G. Gorman, R. Savoy, J. Vazquez, R. Beyers, Cobalt-catalysed growth of carbon nanotubes with single-atomic-layer walls, *Nature* **363** (1993) 605.
8. M. Terrones, Science and technology of the twenty-first century: synthesis, properties, and applications of carbon nanotubes, *Ann. Rev. Mater. Res.* **33** (2003) 419–501.
9. A. Thess, R. Lee, P. Nikolaev, H. Dai, P. Petit, J. Robert, C. Xu, Y. H. Lee, S. G. Kim, A. G. Rinzler *et al.*, Crystalline ropes of metallic carbon nanotubes, *Science* **273** (1996) 483–487.
10. C. Journet, W. Maser, P. Bernier, A. Loiseau, M. L. de La Chapelle, d. S. Lefrant, P. Deniard, R. Lee, J. Fischer, Large-scale production of single-walled carbon nanotubes by the electric-arc technique, *Nature* **388** (1997) 756.
11. M. S. Dresselhaus, G. Dresselhaus, P. C. Eklund, *Science of Fullerenes and Carbon Nanotubes: Their Properties and Applications* (Elsevier, 1996).
12. M. S. Dresselhaus, G. Dresselhaus, P. Avouris, Carbon nanotubes: Synthesis, structure, properties, and applications, *Top. Appl. Phys.* **80** (2001) 113–146.
13. C. Dekker *et al.*, Carbon nanotubes as molecular quantum wires, *Phys. World* **52** (1999) 22–30.
14. P. L. McEuen, Single-wall carbon nanotubes, *Phys. World* **13** (2000) 31.
15. H. Dai, Controlling nanotube growth, *Phys. World* **13** (2000) 43.
16. H. Dai, Nanotube growth and characterization, in *Carbon Nanotubes* (Springer, 2001), pp. 29–53.
17. W. Liang, M. Bockrath, D. Bozovic, J. H. Hafner, M. Tinkham, H. Park, Fabry-perot interference in a nanotube electron waveguide, *Nature* **411** (2001) 665.
18. J. Kong, E. Yenilmez, T. W. Tombler, W. Kim, H. Dai, R. B. Laughlin, L. Liu, C. Jayanthi, S. Wu, Quantum interference and

ballistic transmission in nanotube electron waveguides, *Phys. Rev. Lett.* **87** (2001) 106801.

19. M. Dresselhaus, G. Dresselhaus, R. Saito, Carbon fibers based on c 60 and their symmetry, *Phys. Rev.* **45** (1992) 6234.
20. R. Saito, M. Fujita, G. Dresselhaus, U. M. Dresselhaus, Electronic structure of chiral graphene tubules, *Appl. Phys. Lett.* **60** (1992) 2204–2206.
21. T. W. Ebbesen, Carbon nanotubes, *Annu. Rev. Mater. Sci.* **24** (1994) 235–264.
22. M. JosYacamn, M. MikiYoshida, L. Rendn J. G. Santiesteban, Catalytic growth of carbon microtubules with fullerene structure, *Appl. Phys. Lett.* **62** (1993) 657–659.
23. H. Cheng, F. Li, X. Sun, S. Brown, M. Pimenta, A. Marucci, G. Dresselhaus, M. Dresselhaus, Bulk morphology and diameter distribution of single-walled carbon nanotubes synthesized by catalytic decomposition of hydrocarbons, *Chem. Phys. Lett.* **289** (1998) 602–610.
24. J. Kong, H. T. Soh, A. M. Cassell, C. F. Quate, H. Dai, Synthesis of individual single-walled carbon nanotubes on patterned silicon wafers, *Nature* **395** (1998) 878.
25. J. H. Hafner, M. J. Bronikowski, B. R. Azamian, P. Nikolaev, A. G. Rinzler, D. T. Colbert, K. A. Smith, R. E. Smalley, Catalytic growth of single-wall carbon nanotubes from metal particles, *Chem. Phys. Lett.* **296** (1998) 195–202.
26. B. Satishkumar, A. Govindaraj, R. Sen, C. Rao, Single-walled nanotubes by the pyrolysis of acetylene-organometallic mixtures, *Chem. Phys. Lett.* **293** (1998) 47–52.
27. P. Nikolaev, M. J. Bronikowski, R. K. Bradley, F. Rohmund, D. T. Colbert, K. Smith, R. E. Smalley, Gas-phase catalytic growth of single-walled carbon nanotubes from carbon monoxide, *Chem. Phys. Lett.* **313** (1999) 91–97.
28. M. Su, B. Zheng, J. Liu, A scalable cvd method for the synthesis of single-walled carbon nanotubes with high catalyst productivity, *Chem. Phys. Lett.* **322** (2000) 321–326.
29. J.-F. Colomer, C. Stephan, S. Lefrant, G. Van Tendeloo, I. Willems, Z. Konya, A. Fonseca, C. Laurent, J. B. Nagy, Large-scale synthesis of single-wall carbon nanotubes by catalytic chemical vapor deposition (ccvd) method, *Chem. Phys. Lett.* **317** (2000) 83–89.
30. A. Kasuya, Y. Sasaki, Y. Saito, K. Tohji, Y. Nishina, Evidence for size-dependent discrete dispersions in single-wall nanotubes, *Phys. Rev. Lett.* **78** (1997) 4434.
31. H. Dai, Carbon nanotubes: synthesis, integration, and properties, *Accounts of chemical research* **35** (2002) 1035–1044.

32. T. Yamada, T. Namai, K. Hata, D. N. Futaba, K. Mizuno, J. Fan, M. Yudasaka, M. Yumura, S. Iijima, Size-selective growth of double-walled carbon nanotube forests from engineered iron catalysts, *Nature Nanotech.* **1** (2006) 131.
33. M. S. Dresselhaus, G. Dresselhaus, P. Eklund, A. Rao, Carbon nanotubes, in *The Physics of Fullerene-Based and Fullerene-Related Materials* (Springer, 2000), pp. 331–379.
34. M. Endo, K. Takeuchi, S. Igarashi, K. Kobori, M. Shiraishi, H. W. Kroto, The production and structure of pyrolytic carbon nanotubes (pcnts), *J. Phys. Chem. Solids* **54** (1993) 1841–1848.
35. J. Tersoff, R. Ruoff, Structural properties of a carbon-nanotube crystal, *Phys. Rev. Lett.* **73** (1994) 676.
36. J. Mintmire, C. White, Electronic and structural properties of carbon nanotubes, *Carbon* **33** (1995) 893–902.
37. B. I. Yakobson, C. Brabec, J. Bernholc, Nanomechanics of carbon tubes: instabilities beyond linear response, *Phys. Rev. Lett.* **76** (1996) 2511.
38. M. J. Treacy, T. Ebbesen, J. Gibson, Exceptionally high young's modulus observed for individual carbon nanotubes, *Nature* **381** (1996) 678.
39. O. Lourie, H. Wagner, Evaluation of young's modulus of carbon nanotubes by micro-raman spectroscopy, *J. Mater. Res.* **13** (1998) 2418–2422.
40. L. Schadler, S. Giannaris, P. Ajayan, Load transfer in carbon nanotube epoxy composites, *Appl. Phys. Lett.* **73** (1998) 3842–3844.
41. R. Jishi, L. Venkataraman, M. Dresselhaus, G. Dresselhaus, Phonon modes in carbon nanotubules, *Chem. Phys. Lett.* **209** (1993) 77–82.
42. E. Richter, K. R. Subbaswamy, Theory of size-dependent resonance raman scattering from carbon nanotubes, *Phys. Rev. Lett.* **79** (1997) 2738–2741.
43. R. Saito, T. Takeya, T. Kimura, G. Dresselhaus, M. S. Dresselhaus, Raman intensity of single-wall carbon nanotubes, *Phys. Rev. B* **57** (1998) 4145–4153.
44. M. Dresselhaus, G. Dresselhaus, M. Pimenta, P. Eklund, in *Analytical Applications of Raman Spectroscopy* Blackwell Science Ltd., Oxford, UK (1999) pages 367–434.
45. G. Hummer, J. C. Rasaiah, J. P. Noworyta, Water conduction through the hydrophobic channel of a carbon nanotube, *Nature* **414** (2001) 188–190.
46. K. Murata, K. Mitsuoka, T. Hirai, T. Walz, P. Agre, J. B. Heymann, A. Engel, Y. Fujiyoshi, Structural determinants of water permeation through aquaporin-1, *Nature* **407** (2000) 599–605.

47. A. Kalra, S. Garde, G. Hummer, Osmotic water transport through carbon nanotube membranes, *Proc. Natl. Acad. Sci. USA* **100** (2003) 10175–80.
48. C. Dellago, M. M. Naor, G. Hummer, Proton transport through water-filled carbon nanotubes, *Phys. Rev. Lett.* **90** (2003) 105902.
49. N. Naguib, H. Ye, Y. Gogotsi, A. G. Yazicioglu, C. M. Megaridis, M. Yoshimura, Observation of water confined in nanometer channels of closed carbon nanotubes, *Nano Lett.* **4** (2004) 2237–2243.
50. A. I. Kolesnikov, J.-M. Zanotti, C.-K. Loong, P. Thiyagarajan, A. P. Moravsky, R. O. Loutfy, C. J. Burnham, Anomalously soft dynamics of water in a nanotube: A revelation of nanoscale confinement, *Phys. Rev. Lett.* **93** (2004) 035503.
51. E. Mamontov, C. Burnham, S.-H. Chen, A. Moravsky, C.-K. Loong, N. De Souza, A. Kolesnikov, Dynamics of water confined in single-and double-wall carbon nanotubes, *J. Chem. Phys.* **124** (2006) 194703.
52. B. J. Hinds, N. Chopra, T. Rantell, R. Andrews, V. Gavalas, L. G. Bachas, Aligned multiwalled carbon nanotube membranes, *Science* **303** (2004) 62–65.
53. J. K. Holt, H. G. Park, Y. Wang, M. Stadermann, A. B. Artyukhin, C. P. Grigoropoulos, A. Noy, O. Bakajin, Fast mass transport through sub-2-nanometer carbon nanotubes, *Science* **312** (2006) 1034–1037.
54. S. Kim, J. R. Jinschek, H. Chen, D. S. Sholl, E. Marand, Scalable fabrication of carbon nanotube/polymer nanocomposite membranes for high flux gas transport, *Nano Lett.* **7** (2007) 2806–2811, pMID: 17685662.
55. H. Liu, J. He, J. Tang, H. Liu, P. Pang, D. Cao, P. Krstic, S. Joseph, S. Lindsay, C. Nuckolls, Translocation of single-stranded dna through single-walled carbon nanotubes, *Science* **327** (2010) 64–67.
56. C. Lee, W. Choi, J.-H. Han, M. Strano, Coherence Resonance in a Single-Walled Carbon Nanotube Ion Channel, *Science (New York, N.Y.)* **329** (2010) 1320–4.
57. E. Secchi, S. Marbach, A. Nigus, D. Stein, A. Siria, L. Bocquet, Massive radius-dependent flow slippage in carbon nanotubes, *Nature* **537** (2016) 210–213.
58. J. Geng, K. Kim, J. Zhang, A. Escalada, R. Tunuguntla, L. Comolli, F. Allen, A. Shnyrova, K. Cho, D. Munoz, Y. Wang, C. Grigoropoulos, C. Ajo-Franklin, V. Frolov, A. Noy, Stochastic transport through carbon nanotubes in lipid bilayers and live cell membranes, *Nature* **514** (2014) 612–615.

59. J. Holt, A. Noy, T. Huser, D. O. Bakajin, Fabrication of a carbon nanotube-embedded silicon nitride membrane for studies of nanometer-scale mass transport, *Nano Lett.* **4**.
60. C. Guerra, Y. Zhang, M. Li, V. Chawla, R. Erni, J. Michler, H. G. Park, I. Utke, Morphology and crystallinity control of ultrathin tio2 layers deposited on carbon nanotubes by temperature-step atomic layer deposition, *Nanoscale* **7**.
61. M. Lokesh, S. K. Youn, H. G. Park, Osmotic transport across surface functionalized carbon nanotube membrane, *Nano Lett.* **18** (2018) 6679–6685, pMID: 30339023.
62. A. Noy, H. G. Park, F. Fornasiero, J. Holt, C. Grigoropoulos, O. Bakajin, Nanofluidics in carbon nanotubes, *Nano Today* **2** (2007) 22–29.
63. W.-F. Chan, H.-y. Chen, A. Surapathi, M. G. Taylor, X. Shao, E. Marand, J. K. Johnson, Zwitterion functionalized carbon nanotube/polyamide nanocomposite membranes for water desalination, *ACS Nano* **7** (2013) 5308–5319, pMID: 23705642.
64. J. Wu, K. Gerstandt, H. Zhang, J. Liu, B. J. Hinds, Electrophoretically induced aqueous flow through single-walled carbon nanotube membranes, *Nature Nanotech.* **7** (2012) 133–139.
65. S. Huang, X. Cai, J. Liu, Growth of millimeter-long and horizontally aligned single-walled carbon nanotubes on flat substrates, *J. Am. Chem. Soc.* **125** (2003) 5636–5637.
66. W. Choi, C. Y. Lee, M.-H. Ham, S. Shimizu, M. S. Strano, Dynamics of simultaneous, single ion transport through two single-walled carbon nanotubes: Observation of a three-state system, *J. Am. Chem. Soc.* **133** (2011) 203–205, pMID: 21166470.
67. W. Choi, Z. Ulissi, S. Shimizu, D. Bellisario, M. Ellison, M. Strano, Diameter-dependent ion transport through the interior of isolated single-walled carbon nanotubes, *Nat. Commun.* **4** (2013) 2397.
68. K. V. Agrawal, S. Shimizu, L. Drahushuk, D. Kilcoyne, M. Strano, Observation of extreme phase transition temperatures of water confined inside isolated carbon nanotubes, *Nature Nanotech.* **12**.
69. A. Siria, P. Poncharal, A.-L. Biance, R. Fulcrand, X. Blase, S. Purcell, L. Bocquet, Giant osmotic energy conversion measured in a single transmembrane boron nitride nanotube, *Nature* **494** (2013) 455–8.
70. E. Secchi, A. Nigus, L. Jubin, A. Siria, L. Bocquet, Scaling behavior for ionic transport and its fluctuations in individual carbon nanotubes, *Phys. Rev. Lett.* **116**.
71. H. Amiri, K. L. Shepard, C. Nuckolls, R. Hernndez Snchez, Single-walled carbon nanotubes: Mimics of biological ion channels, *Nano Lett.* **17** (2017) 1204–1211, pMID: 28103039.

72. K. Yazda, S. Tahir, T. Michel, B. Loubet, M. Manghi, j. Bentin, F. Picaud, J. Palmeri, F. Henn, V. Jourdain, Voltage-activated transport of ions through single-walled carbon nanotubes, *Nanoscale* **9**.
73. J. Sanborn, X. Chen, Y.-C. Yao, J. Hammons, R. Tunuguntla, Y. Zhang, C. Newcomb, J. Soltis, J. Yoreo, A. Buuren, A. Parikh, A. Noy, Membranes: Carbon nanotube porins in amphiphilic block copolymers as fully synthetic mimics of biological membranes (adv. mater. 51/2018), *Adv. Mater.* **30** (2018) 1870392.
74. R. H. Tunuguntla, A. Escalada, V. A. Frolov, A. Noy, Synthesis, lipid membrane incorporation, and ion permeability testing of carbon nanotube porins, *Nat. Protoc.* **11** (2016) 2029.
75. R. Tunuguntla, F. Allen, K. Kim, A. Belliveau, A. Noy, Ultrafast proton transport in sub-1-nm diameter carbon nanotube porins, *Nature Nanotech.* **11** (2016) 338a.
76. R. H. Tunuguntla, R. Y. Henley, Y.-C. Yao, T. A. Pham, M. Wanunu, A. Noy, Enhanced water permeability and tunable ion selectivity in subnanometer carbon nanotube porins, *Science* **357** (2017) 792–796.
77. R. Qiao, N. R. Aluru, Ion concentrations and velocity profiles in nanochannel electroosmotic flows, *J. Chem. Phys.* **118** (2003) 4692–4701.
78. M. Majumder, N. Chopra, R. Andrews, B. Hinds, Nanoscale hydrodynamics — enhanced flow in carbon nanotubes, *Nature* **438** (2005) 44.
79. S. Joseph, N. R. Aluru, Why are carbon nanotubes fast transporters of water? *Nano Lett.* **8** (2008) 452–458, pMID: 18189436.
80. T. A. Pascal, W. A. Goddard, Y. Jung, Entropy and the driving force for the filling of carbon nanotubes with water, *Proc. Natl. Acad. Sci. USA* **108** (2011) 11794–11798.
81. J. R. Werber, A. Deshmukh, M. Elimelech, The critical need for increased selectivity, not increased water permeability, for desalination membranes, *Environ. Sci. Technol. Lett.* **3** (2016) 112–120.
82. C. Song, B. Corry, Intrinsic ion selectivity of narrow hydrophobic pores, *J. Phys. Chem. B* **113** (2009) 7642–7649, pMID: 19419185.
83. F. Fornasiero, H.-G. Park, J. K. Holt, M. Stadermann, C. Grigoropoulos, A. Noy, O. Bakajin, Ion exclusion by sub 2-nm carbon nanotube pores, *Proc. Natl. Acad. Sci. USA* **105** (2008) 17250–17255.
84. B. Corry, Mechanisms of selective ion transport and salt rejection in carbon nanostructures, *MRS Bulletin* **42** (2017) 306–310.
85. F. Fornasiero, J. B. In, S. Kim, H. G. Park, Y. Wang, C. P. Grigoropoulos, A. Noy, O. Bakajin, ph-tunable ion selectivity in carbon nanotube pores, *Langmuir* **26** (2010) 14848–14853.

86. P. H. Nelson, A permeation theory for single-file ion channels: Corresponding occupancy states produce michaelis–menten behavior, *J. Chem. Phys.* **117** (2002) 11396–11403.
87. Y.-C. Yao, A. Taqieddin, M. A. Alibakhshi, M. Wanunu, N. R. Aluru, A. Noy, Strong electroosmotic coupling dominates ion conductance of 1.5 nm diameter carbon nanotube porins, *ACS Nano* **13** (2019) 12851–12859.
88. P. M. Biesheuvel, M. Z. Bazant, Analysis of ionic conductance of carbon nanotubes, *Phys. Rev. E* **94** (2016) 050601.
89. M. Manghi, J. Palmeri, K. Yazda, F. Henn, V. Jourdain, Role of charge regulation and flow slip in the ionic conductance of nanopores: An analytical approach, *Phys. Rev. E* **98**.
90. B. Corry, Designing carbon nanotube membranes for efficient water desalination, *J. Phys. Chem. B* **112** (2008) 1427–1434, pMID: 18163610.

CHAPTER 15

Carbon Nanomaterials in Desalination Process

Arpita Iddya, Unnati Rao, Jingbo Wang, Yiming Su, and David Jassby*,†

UCLA Samueli School of Engineering, USA
†*jassby@ucla.edu*

15.1. Background

Desalination is a promising approach to augment existing fresh water supplies.[5] Although the process currently accounts for a small fraction of the world's potable water supply, the global desalination market has been increasing over the years, growing from 46.6 million cubic meters per day in 2005 to 67.3 million cubic meters per day in 2009, and projected to increase to approximately 148 million cubic meter per day by 2030.[6,7] A majority of the world's desalination plants are located in the Middle East and North Africa region. Among the Gulf Cooperation Countries, thermally driven desalination technologies dominate over others such as reverse osmosis (RO). However, RO systems contribute to more than half of the global desalination capacity, and continues to be the fastest growing market segment.[8]

More than two thirds of the desalination technologies employed are used to desalinate sea water (59%) and brackish water (23%).[7] Commercial desalination technologies can be primarily divided into

*Corresponding author: davidjassby@gmail.com

two categories: thermal distillation and membrane-based separation.[8] Membrane separation relies on driving forces such as pressure, electric potential, and concentration differences to overcome the natural osmotic pressures and effectively force water through the membrane.[8] Therefore, based on the separation mechanism, desalination processes can be divided into three categories:

(i) thermally driven,
(ii) pressure driven, and
(iii) electrically driven.[9]

Over the past few decades, membrane-based processes for desalination have gained a lot of interest since such processes do not rely on phase change and are thus more energy efficient. Incorporating organic and inorganic particles or fillers, especially nanomaterials, have been studied to improve the mechanical, chemical, and thermal stabilities, and improve separation properties of membranes.[3] Carbon based nanomaterials, such as graphene and carbon nanotubes (CNTs) have received special attention owing to their unique features such as high specific surface area, superior thermal and electrical conductivity, and extraordinary mechanical strength.[4] In this chapter, we provide a review of the advancements in water desalination, with a focus on the incorporation of carbon based nanomaterials in these technologies.

15.2. Desalination Technologies

Water desalination processes separate dissolved salts and other minerals from water. Desalination can be mediated by introducing a driving force one of three ways:

(i) Thermally driven: Separation of water from feed due to phase change
(ii) Pressure driven: Separation of water due to pressure difference (applied, vapor or osmotic)
(iii) Electrically driven: Separation of water due to movement of ions under an applied electric field.

15.2.1. *Temperature*

Thermal based separation processes remove salt from feed water by facilitating phase change of water into vapor or ice.[10] The entire process is similar to the natural water cycle of evaporation and condensation, producing output water with very low salt concentration.[8] Thermal processes offer advantages such as low electricity consumption and easy integration with alternate heat sources, however these processes are also prone to drawbacks such as scaling.[11]

15.2.1.1. *Multi effect distillation*

Multi effect distillation (MED) is the oldest desalination method.[12] The technology is preferred over other desalination methods for the advantages it offers, such as simple pre-treatment and flexible mode of operation. A typical MED consists of four to seven effects with a heat source temperature of about 75–90°C. The heat source for the system can be provided by external streams of other plants such as such as sugar, paper and pulp etc.[13]

MED is highly thermodynamically efficient, involving a set of evaporators and condensers at reduced ambient pressures.[12] The process takes place in consecutive evaporators called effects using the principle of reduced ambient pressure to evaporate water in each of these effects. This allows water to undergo vaporization in multiple effects without providing additional heat after the first effect.[12] Thus, latent heat is recycled in an MED system leading to its high thermal efficiency.[11,14]

Several advances in MED have been proposed in literature, such as,

(i) using alternate energy sources,
(ii) reducing and controlling scaling and fouling,
(iii) alternate materials of construction,
(iv) optimization of process design, and
(v) improvements to component designs.[11,12]

In addition to the aforementioned potential directions, studies have also looked at integrating MEDs with alternate technologies to improve efficiency and reduce overall energy consumption.[15–17]

Scaling and fouling adversely affect the efficiency of MED systems, leading to higher capital and operational costs and increased energy requirement. Their effect can be mitigated by operating at lower brine temperatures, using cheaper materials, and using Thermal Vapor Compression (TVC).[8] Reduction in energy requirement can also be achieved by optimizing operational parameters.[18] Also, optimization of process and component design can help improve the efficiency and performance of MED systems. Recently, Chen *et al.* (2019) developed a model to optimize the operation of MED-TVC systems. The study found that taking fouling into account, the optimization results vary with thermal resistance of fouling, and thus multiple optimizations can improve the efficiency of the MED-TVC.[19] Additionally, utilization of ejectors in MED-TVC systems has been a novel development in component design. MED processes employ ejectors, driven by low-grade energy. The entrainment ratio governs the efficiency of an MED-TVC system, to this effect, Park *et al.* (2010) developed a new design parameter to increase the entrainment ratio to achieve high performance and operational stability.[20] Another study by Han *et al.* (2014) proposed improving TVC performance by reducing the condensation phenomenon inside ejectors by preheating the entrained vapor before entering the ejector.[14] Wang *et al.* (2017) developed a novel adaptive nozzle exit position (ANXP) ejector and optimized the its performance by self-adjusting the position of the primary nozzle.[21]

15.2.1.2. *Multi-stage flash*

Multi-Stage Flash (MSF) is a distillation process that involves evaporation and condensation of water. The process utilizes several "stages" that operate at successively lower pressures to recycle the latent heat of evaporation, thus maximizing recovery and increasing efficiency.[22–25] In an MSF process, seawater moves through a series of vacuumed stages to produce distilled water. The heated brine enters the first stage, wherein the applied vacuum leads to the evaporation of a small amount of vapor and the consequent cooling of the brine. The process is repeated over several stages. In each stage, the product vapor is passed through a demister to remove the entrained droplets.

This vapor then condenses on the outside of the preheater tubes, and is collected over the distillate trays to form the final product water.[26] The repetitive reduction in brine temperature in each stage corresponds with a decrease in stage pressure.

MSF performance and energy consumption are important criteria for its widespread adoption. The performance can be enhanced by using a wide range of working temperatures, increasing the extracting quotient of the flash vapor, and discharging brine at the last stage to increase the gained output ratio.[27,28] High gain ratios however, lead to high energy consumption making thermal desalination less lucrative.[27] To its advantage, MSF can be driven by thermal energy at low temperatures, making the process suitable for integration with cogeneration power plants or other industrial processes. Integration of MSF with existing technologies could help increase efficiency and maximize recovery as MSF can utilize the low quality heat from power cycle or waste heat from other industrial processes.[24,28] This could also reduce the impact of MSF on the environment, which could be further mitigated by optimization of the MSF process using models based on renewable energies.[23]

Recent advances in the MSF process have focused on plant modification, involving design modification, evaporation range expansion, and hybridization and integration with other desalination methods. Studies have also looked at optimizing operational and control parameters. Over the years, thermodynamic design optimization has led to a decrease in the fouling factor.[12] Keyes *et al.* (1986) developed a model for improving product quality, productivity, and operation cost in MSF desalination plants. The model was a multilevel control and optimization technique involving a three pronged approach: 1. Dedicated sequential and regulatory controls, 2. Supervisory and optimizing controls, and 3. Management information control.[23,29] More recently, Alhazmy (2011) modified an MSF plant with brine mixing and cooling, to decrease dependence on seasonal climatic variations and maintain a steady rate of yield throughout.[29]

There has been little research in using carbon-based nanomaterials for process development and improvement in both multi-stage flash and multi effect distillation processes.

15.2.1.3. *Membrane distillation*

15.2.1.3.1. Process description

Membrane distillation (MD) is a membrane-based separation process, in which a hydrophobic polymeric membrane separates the feed (contaminated) side from the permeate (purified) side. A vapor pressure gradient drives the movement of water vapor and volatile compounds from the feed to the permeate side through the pores of the membrane.[30] As a vapor pressure driven process, MD has certain advantages over other membrane-based separation processes that rely on externally applied pressure. For example, MD has no osmotic pressure limitations, and can be used to treat highly contaminated waste streams, resulting in high water recovery.[31] They can also be cost effective, particularly when free waste-heat is available, as they are operated at atmospheric pressures, and can achieve high rejection (>99%).[32–34]

15.2.1.3.2. Configurations

Membrane distillation can typically be operated in a number of different configurations. A few of them are discussed below[30]:

15.2.1.3.2.1. Direct contact membrane distillation (DCMD)

Here, the hot feed side and the cold permeate side are both in contact with the membrane surface on opposite sides. Vapor from the feed side crosses the membrane and condenses, mixing with the cool permeate stream. This is the most common MD configuration.[33]

15.2.1.3.2.2. Air gap membrane distillation (AGMD)

In this mode, the hot feed side is in contact with the membrane surface. On the permeate side, a stagnant film of air separates the membrane from the condensation surface. The vapor crosses the air gap and condenses on the cool surface.[35]

15.2.1.3.2.3. Sweeping gas membrane distillation (SGMD)

In this mode, inert gas is used to sweep the vapor on the permeate side, which is then condensed outside the module.[36]

15.2.1.3.2.4. Vacuum membrane distillation (VMD)

Here, a vacuum pump is used to create vacuum on the permeate side, and the vapor is condensed outside the module. However, VMD systems have a greater susceptibility to membrane wetting.[37]

15.2.1.3.2.5. Challenges

Challenges such as membrane fouling, wetting, high energy requirements and the inability to separate volatile compounds have hindered the commercial application of MD systems.[30,38,39] In order to combat these challenges and improve membrane performance, novel membrane fabrication methods are being developed. Incorporating nanomaterials in membrane synthesis or surface coating enables tuning properties such as hydrophilicity, porosity, charge, bacterial resistance, and thermal and electrical conductivity.[40] Carbon based nanomaterials have exceptional mechanical, electrical, thermal and anti-bacterial properties,[41] and their use in membrane fabrication is a promising technique to improve performance. Carbon nanotubes (CNT), graphene oxide (GO), MXene, carbon black nanoparticles and activated carbon nanoparticles are the most common carbon-based nanomaterials used in membrane synthesis.[40] The following sections will briefly describe their properties, modifications and applications.

15.2.1.4. *Novel materials*

15.2.1.4.1. Carbon nanotubes (CNTs)

CNTs consist of a thin graphene sheet rolled into a hollow cylinder. They are of two types: single wall CNTs and multi wall CNTs, depending on the number of sheets rolled around the hollow core.[42] In addition, CNTs can be functionalized and their surface can be modified to impart attractive properties.[43]

Carbon nanotubes can be assembled into paper like structures known as bucky paper, which have been successfully used as DCMD membranes with a flux of 12 LMH and salt rejection greater than 99%.[44,45] When porous polypropylene membranes were coated with a CNT/PVA layer, DCMD flux was enhanced by nearly 10%, as CNTs offer alternate paths for vapor transport.[46] Further, silicate scaling was effectively dissolved by applying a cathodic electric potential to the conductive CNT membrane.[47] The same CNT/PVA composite, when used in a MD system with applied alternating current successfully reduced silicate and calcium sulfate scaling.[48] Electrospinning is a common technique to incorporate CNTs in the membrane matrix, particularly in MD membranes. Electrospun membranes result in high porosity and open pores. 1–5% by weight of CNTs were incorporated to fabricated robust, hydrophobic, mixed-matric polyvinylidene fluoride co-hexafluoropropylene nanofiber membranes by electrospinning. The presence of CNTs enhanced the mechanical and hydrophobic properties of the mixed matrix membrane. There was a 33–59% increase in flux at 5% CNT loading, depending on the feed concentration.[49] Recently, a novel triple-layer MD membrane was fabricated by the electrospinning technique, using PES/CNT as the primary bulk material, and polyvinylidene fluoride co-hexafluoropropylene/CNT as the outer and inner surface of the membrane. This provided the membrane with high porosity from the bulk material, and surface hydrophobicity. The resulting membrane, when used in DCMD, showed a flux increase of nearly 50%.[50] In another study, CNTs were used in the fabrication of blended PVDF flat sheet membranes. When used in a DCMD system to treat salt water, they resulted in higher rejection and a 15–20% increase in the flux.[51] Dumee *et al.* have published a number of articles describing enhanced flux in DCMD with composite CNT membranes.[52–54] CNTs immobilized on a PTFE backbone demonstrated a 54% increase in flux in DCMD treatment of salt water.[55,56] PVDF blended with MWCNT and MWCNT/SiO_2 nano additives were used to fabricate VMD membranes. These membranes, when used in VMD system, increased the flux due to the synergistic effect of enlarging pores by MWCNTs and the growth of macrovoids by SiO_2 nanoparticles.[57] A CNT/PVA composite, used in an MD system with joule heating, resulted in single-pass recoveries approaching 100%. Joule heating was carried out by applying alternating current to the membrane, eliminating the need for bulk heating. This has the potential to lower energy costs.[58] A composite Fe-CNT/PTFE membrane was fabricated and used in a VMD system. The membrane was directly heated using a high frequency magnetic field, resulting in high salt rejection and improved thermal efficiency.[59] A radio frequency induction heated membrane enhanced flux by nearly four times and showed sustained high fluxes for up to 6 hours during VMD treatment of a $CaSO_4$ solution.[60] A membrane distillation system consisting of a CNT membrane as solar absorption layer achieved thermal efficiency of nearly 85%[61] functionalized A well aligned CNT layer was formed by anchoring CNTs to nanofibers of electrospun membranes. The presence of CNTs reduced the boundary layer effect for viscous flow, assisted surface diffusion and facilitated Knudsen and molecular diffusion. This resulted in superior anti-wetting properties and enhanced flux.[62] Membranes fabricated by immobilizing CNTs by the phase inversion method showed 76% enhancement in flux when used in DCMD of salt water. CNTs and PVDF were dispersed in acetone and methanol, and this mixture was cast on a polypropylene support.[63] When MWCNTs were doped into PSF flat sheet membranes, salt rejection was >99.99%, and the flux was enhanced by nearly 75% over the bare PSF membrane.[64]

15.2.1.4.2. Graphene and graphene oxide

Graphene and graphene oxide (GO) are used as nanofiller materials in MD. Modification with graphene increases surface roughness, resulting in superior anti-wetting properties.[65] A graphene oxide based membrane modified with SiO_2 particles showed excellent salt rejection over a period of 170 hours, along with a stable flux.[66] A solar heated vapor gap MD membrane was fabricated by growing a graphene array on a nickel foam. This system demonstrated energy efficiency of 73.4%.[67] Graphene/PVDF mixed matrix and graphene oxide/PVDF composite-skin membranes, when used in DCMD with salt water as the feed improved flux by 1.8 and 2 times respectively. Rejection for graphene/PVDF was >99.8%, while that with graphene oxide/PVDF was a moderate 80%. Optimal graphene content was 0.87% by weight.[68] A PVDF-graphene composite membrane showed rejection >99.9% in DCMD. This membrane was fabricated by the phase inversion technique.[69] Graphene quantum dots/PVDF nanofibrous membranes showed high flux (10.8 LMH) and rejection (>99%) in AGMD systems.[70] n-Butylamine graphene oxide was incorporated in PVDF hollow fiber membranes, and used for seawater desalination. This modification improves mechanical properties of PVDF, and fluxes as high as 61 LMH at 80°C with rejection >99% was observed.[71] Graphene oxide immobilized on PTFE membranes significantly enhanced flux in DCMD, with 100% salt rejection. Flux observed was 97 LMH at 80°C. This increase in flux was attributed to selective sorption, reduced temperature polarization and the presence of polar functional groups on graphene oxide.[72] Graphene oxide dispersed on PVDF membranes was used in a DCMD system, and showed superior anti-fouling performance. This was hypothesized to be due to the presence of oxygen containing functional groups in graphene oxide.[73]

15.2.1.4.3. Activated carbon nanoparticles

In a DCMD system, a hydrophobic active layer was prepared by electrospinning activated carbon nanoparticles along with PVDF-co-hexafluoropropylene and cast on a thick hydrophilic support layer. When 1.5% by weight of activated carbon nanoparticles were used, a 10% increase in flux over the benchmark PTFE membrane was observed.[74]

15.2.1.4.4. Carbon black nanoparticles

A PVDF membrane was modified using SiO_2/Au nanostructures and unfunctionalized carbon black nanoparticles and used as a photothermal MD membrane. This membrane showed an increase in permeate flux by 33% over the unmodified membrane.[75]

15.2.1.4.5. MXene

MXene is a novel 2D material having several nanoscale layers.[76] When coated on a hydrophobic PVDF membrane and used for DCMD treatment of a solution containing bovine serum albumen and sodium chloride, the composite showed superior anti-fouling properties. By providing localized joule heating, a 12% decrease in heater energy input was observed.[77]

15.3. Pressure

Pressure induced desalination involves separating salt and water by preferential transport of a component across a selective material due to pressure difference across the selective material. These processes can be explained by short-range interactions, which are a combination of steric, dispersion, dipole-dipole and electrostatic interaction between water molecules or ions with another material involving a combination of steric, dispersion, dipole and electrostatic interactions can also be used for desalination.[10]

15.3.1. *Reverse osmosis*

Global climate change, growing population and increasing water and energy demand have exacerbated the complexity of addressing worldwide water scarcity issues.[78,79] Membrane separation technologies stand at the nexus of water and energy: producing clean water requires the input of energy. Among all the desalination processes, reverse osmosis (RO) and nanofiltration (NF) membranes have been demonstrated to be the most energy-efficient method to achieve separation because no phase-change is needed.[80,81] However, RO/NF separation processes suffer from several drawbacks, including high energy consumption when removing high concentration contaminants, membrane fouling, and the generation of concentrated waste streams that require costly disposal.[80,81] Advances in nanotechnology have led to a host of innovation in membrane separations that enabled dramatic performance enhancement in the areas of seawater and groundwater desalination, and wastewater recycling.[82–86] Carbon-based nanomaterials, including carbon nanotubes (CNTs), laser induced graphene (LIG), graphene oxide (GO) and carbon quantum dots (CQD), which have exceptional electrical, mechanical, chemical and thermal properties, are a group of nanomaterials that have been demonstrated to greatly enhance membrane separation processes.

15.3.2. *CNT-based membranes*

15.3.2.1. *CNT-based membrane properties*

The specific physical and chemical features of CNTs, such as their well-defined structure, high aspect ratio, flexible chemical bonding, and stable electro-mechanical properties, all contribute to the unique mechanical, morphological, rheological, electrical, and thermal properties of CNT-based mixed-matrix membranes. Importantly, the ability to form a wide range of covalent bonds between CNT functional groups and host polymers ensures the excellent stability of the material and prevents their release into the environment.[87,88] Based on published studies, two major categories of CNT membranes have been developed: aligned CNT membranes, and randomly oriented CNT membranes.

The majority of CNT desalination membranes reported in the studies are randomly oriented CNT membranes. The incorporation of CNT into polymeric membranes modifies the physical/chemical properties of the composite, and can significantly impact water permeability,[63,89–98] rejection,[91,96,97,99,100] hydrophilicity,[63,89–97] conductivity,[47,88,89,91,93,99–112] mechanical stability[103,105,113,114] and antifouling properties.[47,88,90,93,94,102,104,106–111,115]

Bucky-paper CNT membranes are freestanding mixed-matrix CNT membranes, which are composed of randomly arranged CNTs in the form of a thin mat held together by van der Waals interactions.[116,117] This type of CNT membrane exhibits high adsorption capacity to natural organic matter and strong antimicrobial properties because of a large porous 3D network and high specific surface area.[118–121] Different methods such as vacuum assisted filtration, electrospinning, and layer-by-layer (LBL) deposition have been reported for the fabrication of Bucky-paper CNT membranes.[116,122,123] Bucky-paper CNT membranes were effectively adopted for desalination of seawater via direct contact membrane distillation (DCMD).[44,124] However, one significant drawback of these membranes is their poor mechanical strength, which limits its application especially for the high-pressure processes. Alternatively, the CNTs can be embedded or cross-linked with other polymers, forming a mixed-matrix material to overcome mechanical strength issues.

Because of the inherently hydrophobic and non-reactive nature of CNTs (which often results in incompatibility with the polymer matrix), various chemical or physical modifications have been developed to improve the dispersion of CNTs in the matrix itself.[125] Among these modifications, the most commonly encountered method is the addition of carboxyl and hydroxyl functional groups to the ends and outer layer of CNTs via concentrated acid treatment. These groups help to increase the hydrophilicity and reactivity of the CNTs and enable the CNTs to be covalently bound to polymer functional groups through ester or ether bonds.[114] Other functional groups include acyl chloride, amine, sulfone, and many others.[88,90,95,106,114,115,117,125–131] Once the CNTs are functionalized, they can be cross-linked to surrounding polymers though a wide variety of chemical reactions as demonstrated in many studies.[84,90,92,93,96,99,103,108,119,125,131–135]

Depending on the applications, there are a wide variety of methods to incorporate CNTs into membranes[136] including wet-spinning,[101] dispersion,[47,63,97,98,111,137] phase inversion,[88,92,93,110] interfacial polymerization[94–96,115,138,139] vacuum filtration,[89,90,99,102–104,113,140] pressure deposition,[91,100,107–109] and electrochemical polymerization.[105] The most commonly used desalination membranes, RO/NF, typically have a thin-film composite structure where the thin polyamide active layer is fabricated on a porous support via interfacial polymerization. For these thin-film composite membranes, CNTs have been incorporated throughout the membrane matrix (either in the active layer or the support layer), or as an individual layer, either sitting on top of the active layer or as an intermediate layer between the active layer and the support. The different fabrication processes and applications of CNT-based desalination membranes are summarized in Table 15.1.

Table 15.1. Recent developments of CNT-based desalination membranes.

Type of carbon nanomaterial*	Incorporation method	Application	Added value by carbon nanomaterials	Electrical potential applied	Ref
f-CNT	Pressure deposition on support	RO	Scaling control and removal	Yes	107
f-CNT	Pressure deposition on support	NF	Fouling mitigation	Yes	91
MWCNT	Vacuum filtration on support	NF	Enhanced water flux and salt rejection	No	95
MWCNT-TNT hybrid	Interfacial polymerization into active layer	RO	Enhanced water flux	No	96
MWCNT	Interfacial polymerization into active layer	RO	Enhanced water flux	No	138
f-CNT	Phase inversion into support	RO	Enhanced water flux	No	92
f-CNT	Vacuum filtration on support	RO	Enhanced water flux and biofouling mitigation	No	90
f-CNT	Interfacial polymerization into active layer	RO	Enhanced water flux and biofouling mitigation	No	139
MWCNT	Interfacial polymerization into active layer	RO	Enhanced chlorine resistance	No	141
MWCNT	Interfacial polymerization into active layer	RO	Enhanced water flux, fouling mitigation and chlorine resistance	No	142
f-CNT	Interfacial polymerization into active layer	RO	Enhanced water flux, fouling mitigation and chlorine resistance	No	143
f-CNT	Interfacial polymerization into active layer	RO	Enhanced water flux, fouling mitigation and chlorine resistance	No	144

(Continued)

Table 15.1. (*Continued*)

Type of carbon nanomaterial*	Incorporation method	Application	Added value by carbon nanomaterials	Electrical potential applied	Ref
SWCNT	Pressure deposition on active layer	TFC	Enhanced bacterial cytotoxicity	No	145
f-CNT	Interfacial polymerization into active layer	NF	Enhanced water flux	No	146
f-CNT	Interfacial polymerization into active layer	RO	Enhanced water flux	No	147
Interfacial polymerization	Interfacial polymerization into active layer	RO	Fouling mitigation	No	148
MWCNT-TNT hybrid	Interfacial polymerization in active layer	RO	Enhanced water flux	No	149
f-CNT	Dispersion on active layer	RO	Enhanced boron rejection	Yes	137
f-CNT	Pressure deposition on support	NF	Enhanced water flux and biofouling mitigation	Yes	106

*f-CNT: functionalized CNT; TNT titania nanotube.

Due to the difficulty associated with the preparation of aligned CNT materials, there are only a few studies reported in literature deals with the preparation and use of aligned CNT membrane materials. In the vertically aligned CNT (VACNT) membrane matrix, self-assembled CNTs are vertically aligned on a substrate. A multi-walled VACNT membrane having an internal pore diameter of 6 nm by incorporating CNT fillers into a polystyrene film was first introduced by Hinds *et al.*[150] In general, the fabrication of VACNT membranes involves arranging CNTs perpendicular through the organic or inorganic support layer and the resulting VACNT composite membrane material exhibits excellent sieving capacity for a wide range of contaminants in water.[131,151,152] Another type of VACNT membrane, known as the outer wall-CNT membranes, was recently developed in which the interstices between CNTs, rather than the void channel inside CNT, are utilized to remove contaminants.[153] The sieving performance of VACNTs is limited by the CNT pore-size, therefore, these membranes have not been successfully used to desalinate water as CNT pores are larger than the size of sodium and chloride ions.[154–157]

The physicochemical properties of membranes, including pore size, porosity, roughness, hydrophilicity and electrical conductivity, could be strongly affected by the addition of CNTs to the membrane matrix. Scanning electron microscopy (SEM), field emission scanning electron microscopy (FESEM), transmission electron microscopy (TEM) and atomic force microscopy (AFM) have all been used to evaluate the morphology of CNT membranes. Contact angle analysis is performed to quantify the impact of CNT addition on the material's hydrophilicity.

When CNTs were incorporated in membranes by vacuum filtration, pressure deposition or dispersion, a CNT layer is formed on top of the membranes and the CNT bundles can be observed in SEM images (Fig. 15.1). When CNTs were used as additives in membranes prepared through a phase-inversion process, studies reported the membranes showed larger surface pore sizes, higher porosity, and finger like pores. The mean pore size and porosity increase at the beginning with the addition of an appropriate

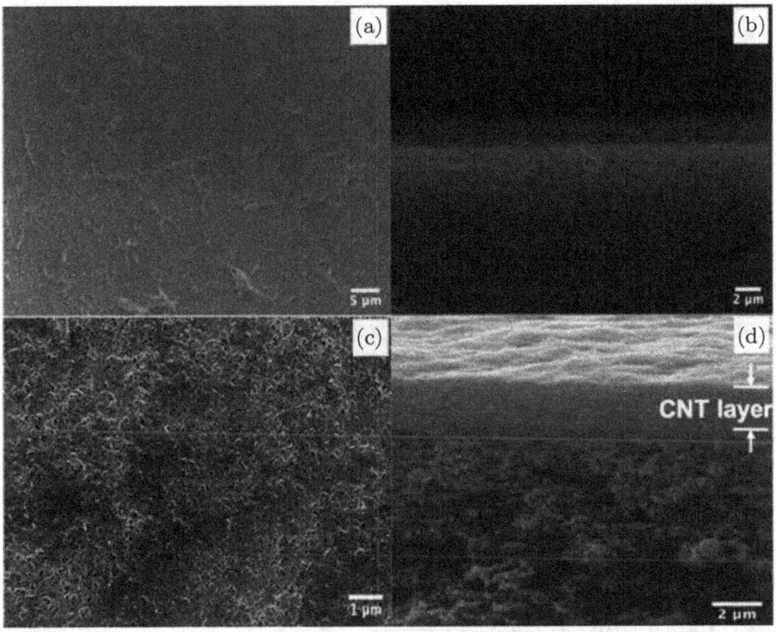

Fig. 15.1. (a) Top and (b) cross-sectional SEM images of pristine RO membrane. (c) Top and (d) cross-sectional SEM images of CNT-PVA-deposited membrane.[137]

amount of CNTs, but slightly decreases when CNT loading is further increased.[84,90,92,101,106,132] Mixed hydrophilic nanoparticles could accelerate the process of phase inversion, resulting in the increase in overall porosity of the modified membranes; however, too many CNTs could increase the viscosity of the casting solution and reduce the number of large pores.[88,125]

The change in surface roughness and membrane hydrophilicity is dependent on the surrounding and substrate polymers, and the mass loading of CNTs also has a profound impact when CNT is incorporated throughout the whole membrane matrix.[90,98,99,102,107,137] De Lannoy *et al.*[158] observed some f-CNT loss from the membrane matrix when they were not cross-linked to polymers, with significantly smaller releases for CNTs with fewer functionalized groups. Considering the fact that high CNT functionalization leads to lower final CNT yields within membranes and greater potential for creating nano-wastes, controlling the CNT type and loading in polymeric membranes is important to achieve the desirable transport, surface, and stability properties for a specific separation process.

15.3.2.2. *Advantages of CNT-based membranes*

15.3.2.2.1. Increased water transport through CNT membranes

CNT membranes have several benefits for advanced desalination due to their intrinsic properties. The slippage effect predicted in VACNT, derived from frictionless transport along the inner pores of CNTs, is expected to result in several orders of magnitude higher water flow through the membrane material, when compared to conventional membranes.[159–161] When incorporated into the active layer, the embedded CNTs not only provide a low-resistance pathway to the water molecules, but also improve the mechanical property of the polymer composites which helped to reduce the compression of the membrane, all of which contribute to higher water flux.[138,139] In addition, Azelee *et al.* ascribe the improvement of water permeability of CNT membrane to the decreasing membrane surface contact angle and increasing membrane surface roughness.[96] When incorporated as the intermediate layer between the active layer and support, Wu *et al.* attributed the higher flux exhibited by the CNT membranes to the robust CNT layer providing a strong mechanical support and preventing the active layer crumpling when exposed to high operation pressure.[95] Chan *et al.* reported that the pore size distribution of the CNT membrane were very similar to that of a pure polymer membrane, and the flux improvement is due to the reduced tortuosity and increased interconnectivity of the water transport channels created by the inclusion of CNTs.[90] When CNTs are incorporated to the support membrane matrix, they serve as pore forming additives that increase the porosity of the material. Therefore, the resulting TFC membranes with CNT-polymer supports showed significantly enhanced water permeance.[92]

15.3.2.2.2. Enhanced chlorine resistance

Most commercial RO/NF membranes are vulnerable to chlorine exposure, but some nanocomposite CNT-TFC membranes exhibit high chlorine resistance due to the CNT incorporation. The presence of CNT within the polyamide active layer matrix in composite membranes acts as anti-chlorination stabilizers: the interactions between polyamide and carbon networks increased the activation energy for chlorination.[141] Moreover, the addition of CNT to a polymer matrix reduced chlorine uptake within the matrix, therefore protecting the polymer from the chlorine oxidative degradation.[142] Zhao *et al.* also proposed that the potential mechanism for the enhanced chlorine resistance of composite membranes may lie in the protection of active site in m-phenylenediamine (one monomer that is widely used to form polyamide active layer) residual by the electron-rich CNTs from being attached by the free chlorine.[143]

15.3.2.2.3. Inherent toxicity of CNTs towards bacteria

Research has shown that CNTs are inherently toxic towards bacteria. The toxicity is attributed to the physical disruption of the bacterial membrane, and/or the CNTs serving as an electron shuttle which oxidizes bacterial cellular machinery.[162–166] In addition, reactive oxygen species generated from bacteria-CNT interaction can induce in vitro apoptosis and cell oxidative stress that encourage biofilm-free conditions during membrane desalination.[109,162,163,167,168]

15.3.2.2.4. Electrochemical reactions at the membrane/water interface

Compare to other nanomaterials, one special property of a percolating network of CNTs is that the CNTs can be used as the conducting element in mixed-matrix membranes, functioning as the electron pathway across the membrane surface and delivering energy directly to the membrane/water interface. The application of an electrical potential to the surface of electro-active CNT-based mixed-matrix membranes has been demonstrated to improve membrane performance in terms of separation, fouling resistance, and permselectivity (Table 15.1)[91,106,107,117,126,132,137,169] Many studies have investigated CNT-based membranes for their electrochemical/electrokinetic anti-fouling and anti-scaling properties. CNT nanocomposites have been demonstrated to mitigate scaling by electrostatic repulsion and proton/hydroxide production from water splitting when an external potential was applied to the electrically conducting membrane surface, which effectively dissolved the accumulated minerals and restored membrane performance.[47,107,113] Electrostatic repulsion and electro-responsive surface morphology changes of CNT membranes also contributes to organic fouling reduction.[91,101,108,110] Electrically conducting CNT membranes also showed reduced fouling during oil/water separation due to field-induced redistribution of surfactants at the oil/water interface;[170] importantly, the anti-fouling effects were independent of salt concentrations, indicating that this method is potentially applicable under a wider range of environmental conditions. In addition, the electrochemical effects of electroactive CNT membranes have been harnessed for biofouling control by driving reactions such as hydrogen evolution (i.e., in-situ bubble formation) and oxygen reduction/H_2O_2 evolution,[88,93,102,104,109,111] or affecting the redox potential across the cell membrane and disrupting redox homeostasis.[171] CNT membranes operated under applied potential enable the coupling of electrochemical degradation reactions with filtration.[100,105] This coupling substantially increases contaminant removal for two main reasons:

(1) the large membrane active surface area resulting from the CNT nanostructure offers increased availability of reactive surface sites; and,
(2) the small pore sizes between CNT strands coupled to the advective transport of the target molecules, reduce the diffusion pathway between the electroactive surface and the molecules, which reduces the mass transport limitations often encountered in batch electrochemical systems.[134,172]

15.3.2.3. *Challenges and strategic research prospects of CNT membranes*

Laboratory studies on the preparation of CNT membranes for water applications have shown promising results. However, more research is still needed before they can be applied in industrial applications and reach commercialization level.

15.3.2.3.1. Scale up from bench to industrial level

Manufacturing scalability is a major hurdle to the commercialization and large-scale applications of CNT membranes. CNT membrane fabrication processes need to overcome several challenges, such as improving the dispersion of CNTs in solvents, improving the compatibility of CNTs with host polymers, and enhancing and stability of CNTs in the host membrane structure. Successful incorporation of CNTs into large-scale membrane materials holds the promise of using these membranes as drop-in replacements in existing treatment trains. While large-scale deposition of randomly aligned CNT mats is relatively straightforward, the formation of large-scale VACNT materials remains challenging.[85,118,121,126,131,132,151–153] The state-of-art fabrication strategies fully take advantage of the tunable physicochemical properties of CNTs, with various studies demonstrating improved performance of CNT membranes after physical/chemical modification of CNTs.[85,90,92,96,99,104,113] This is a promising approach, as maximizing the compatibility of CNTs with existing interfacial polymerization or cross-linking processes ensure the seamless incorporation of CNTs into membranes. Additionally, the large-scale production of CNT-based membrane modules remains to be challenge; while CNT membranes have been fabricated in the forms of hollow-fiber and flat-sheet,[47,55,84,87,97,102,109,111,130,133,173,174] no spiral-wound modules have been tested or fabricated yet.

15.3.2.3.2. Effect on environment and human health

The safety profile of CNTs is a concern, requiring further study with regards to their impact on environmental and human health.[175] Commercialization and public acceptance will be limited without a comprehensive assessment of the potential risks.[176] More investigations are needed on the stability of CNT in membranes and quantification of CNT release during operation. It is recommended to have systematic documents that detailed characterizations of CNTs used in CNT-based membranes and establish constrains to minimize unnecessary exposure to CNT during manufacturing.[177,178] The other issue arises with CNT membrane processes is the generation of potential toxic byproducts mainly associated with electrochemical processes, such as perchlorates or halogenated organic compounds. Special attention should be paid on detailed evaluations of the formation of these hazardous byproducts and development of strategies that can manage, decrease or avoid the production of the toxic byproducts.

15.3.3. *Graphene oxide based membranes*

RO membranes consist of a thin film composite (TFC) configuration, with a fully aromatic polyamide (PA) layer on a microporous support. The PA layer can suffer from fouling, leading to low performance. Moreover, the chlorine-based chemicals used for cleaning the fouled layer can lead to oxidative degradation of the PA layer, therefore, studies have looked into improving the durability and performance of PA based membranes. The unique structure and properties of PA-TFC membranes allow for tailoring through various strategies, such as membrane surface modification or by incorporation of nanomaterials.[179]

CNTs and zeolites used to make nanostructured membranes with nanopores are promising, however, their performance is limited by low permeation. CNTs specifically, can improve flux by providing a frictionless passage, however the surface layer incorporations increase the cost of production. Graphene, a 2-dimensional single layered nanomaterial with sp2 hybridized carbon in a honeycomb lattice has garnered interest owing to their high mechanical, thermal, and electrical properties and lower production costs. Graphene oxide and reduce graphene oxide (rGO) provide an excellent platform for exploring size selective membranes.[179,180]

GO incorporated into PA-TFC membranes improves its antifouling capacity and acts as a protective layer for the PA structure, offering increased chemical resistance to chlorine. Moreover, owing to the several oxygen containing functional groups on its surface, GO imparts superior hydrophilicity to the membrane.[181] Embedding GO can also improve water permeability, surface hydrophilicity and negatively charged surface properties. Song *et al.* demonstrated that by incorporating GO into a PA-TFC membrane, the water permeability increases due to increased diffusion through the nanopores. In addition to increased permeability, the GO flakes also impeded dissolved chlorine from contacting the surface, increasing its chlorine resistance.[182] Incorporating GO into the membrane also prevents adsorption of foulants onto the membrane by increasing the hydrophilicity of the surface. Hydrophilic surfaces form a hydration layer that prevent hydrophobic interactions between the foulants and the membrane surface, thereby inhibiting initial attachment and decreasing its (membrane) fouling propensity.[182,183]

Introducing GO into the membrane can be achieved through various routes such as layer-by-layer deposition, spin coating, or by adding GO to solvent before interfacial polymerization.[179,184,185] The concentration and size of GO has a huge effect on the performance of the produced membrane. Furthermore, GO can successfully form composites with metal oxides to enhance membrane performance. Safarpour *et al.* demonstrated improved performance in terms of water permeability, salt rejection, antifouling property and chlorine resistance for RO TFC membranes by embedding rGO/TiO_2 into the PA layer.[181]

15.4. Electrical Potential

Under an applied electric field, ions migrate towards oppositely charged electrodes. The presence of charge on ions and the overall electroneutrality of bulk water enables use of long-range interactions to selectively exert force on ions. These ions can then be adsorbed on charged surfaces or transported across a selective material to induce separation.[10]

15.4.1. *Electrodialysis*

Electrodialysis (ED) is an electrically mediated desalination process to demineralize, concentrate and/or convert salt containing solutions. The process works on the principle that under an applied electric field, ions in solutions migrate towards the oppositely charged electrode. The setup uses two endplates as electrodes, and a large number of alternating cation- and anion-exchange membranes (CEM and AEM respectively) stacked together, separated by flow spacers.[186–189]

Membrane based processes are preferred for their low energy costs and high effectiveness. ED specifically offers advantages such as, higher recovery rate than reverse osmosis, longer lifetime for membranes, operation at high temperatures, and easy operation.[187,190,191] ED is however plagued by fouling and concentration polarization (CP) at the membrane surface. Fouling adversely affects the flux, and leads to an increase in the electrical resistance.[192–194] The efficiency of ED depends on IEMs and their capacity in removing unwanted charged particles from feed.[195] IEMs have high electrical resistance which can cause high Ohmic losses and as result lead to low energy efficiency.[196] When selecting IEMs for ED applications, the membranes require high permselectivity, low electrical resistance, good chemical, thermal, and mechanical stability. Thus, advancements in ED are driven by the developments in ion exchange membranes.[187]

An attractive method developed to mitigate fouling is surface modification with nanomaterials (e.g., CNT, graphene oxide, nano zeolite, nano SiO_2, nano Fe_2O_3, nano Fe_3O_4, nano TiO_2, and nano Al_2O_3) and/or reactive substance (e.g., chitosan, nafion, sulfonated polyethersulfone, polyvinylidene fluoride, polyvinyl chloride, polyvinyl alcohol).[3,197] Incorporating inorganic particles/fillers, especially nanomaterials, into polymeric matrices has been shown to improve the mechanical strength, thermal, and chemical stabilities, as well as improving separation properties based on synergism between polymer and nanomaterial.[3] While the hydrophobic membrane surfaces derived from the polymers decrease CP, by homogenizing current density distribution for enhanced ion transport, and stimulating electroconvection,[198] nanoparticles provide additional exchangeable ionic groups and contribute to the creation of interconnected ion conducting channels within the membrane matrix.[197] Moreover, it is worth noting that a designed mixed layer of polymer and nanoparticles on ion exchange membrane improves anti-fouling capacity, ion exchange capacity, fixed ionic concentration, ionic permeability and flux.[199]

As mentioned above, various nanomaterials have been used as fillers to improve performance of IEMs in ED. Here, we discuss the carbon based nanomaterials for their extraordinary properties such as high specific surface area, high adsorption characteristics, and potential at improving performance[200]:

15.4.1.1. *Carbon nanotubes (CNT)*

Ion exchange membranes are limited by their low ionic conductivity. As a result, IEMs contribute to the overall electrical resistance which increases the ohmic losses. Low electrical conductivity also slows down the process kinetics by decreasing ion flux through the membrane. Incorporation of fillers or nanomaterials into the polymer matrix is one way to increase conductivity and selectivity, and consequently increase the ion exchange capacity.[196]

Carbon Nanotubes (CNTs) are long carbon cylinders, made from rolled graphite sheets. They can be single walled with a diameter less than 1 nm, or multiwalled with several interlinked nanotubes. Owing to their large aspect ratio and one dimensional structure, CNTs have a greater magnitude of dispersity than microscale fillers or 2 and 3 dimensional nanomaterials.[196,201] Also, their nano-channel tubular structure facilitates the passage of water molecules.[85,125,202]

Incorporating CNTs into polymer matrices has been known to increase the mechanical and thermal stability, permselectivity, hydrophilicity, and electrical conductivity of membranes.[4,85,203–205] One of the crucial steps in fabricating CNT/polymer nanocomposite is the homogenous dispersion of the CNTs in the matrix. A conventional approach for better CNT dispersion is the acid treatment of CNTs to make them more reactive and thus better dispersed.[85,206] Fan *et al.* (2020) successfully fabricated nanocomposite IEMs incorporated with sulfonic acid functionalized CNTs. Functionalization of CNTs allowed for better dispersion of nanotubes without aggregate formation in the polymer matrix. Moreover, the developed membrane demonstrated reduction in the overall membrane resistance while maintaining its permselectivity.[196] Heidary *et al.* (2019) fabricated aminated-CNT embedded membranes into a sulfonated polyphenylene oxide and sulfonated polyvinyl chloride polymer blend. The study noted improved hydrophilicity and performance of the IEMs, however, electrical conductivity decreased with the addition of aminated-CNTs into the polymer matrix.[203] One of the major drawbacks with using CNTs is its high cost, however, significant advances have been made in fabricating high quality CNT at reduced prices.[85]

15.4.1.2. *Graphene and its derivatives*

Graphene is a one atom thick layer of sp2 hybridized carbon atoms linked by σ and π bonds in a hexagonal, honey-comb like crystal lattice structure.[4,10] Compared to CNTs, graphene was found to be more economical and possess superior properties.[4] Graphene and its derivatives, graphene oxide (GO) and reduce graphene oxide (rGO), are effective in improving separation properties of membranes because of their high specific surface area, flexibility, high mechanical strength, and chemical stability.[207] Moreover, graphene has excellent electronic properties with an electrical conductivity of 7200S/m at room temperature.[4,201,202]

Owing to their unique physico-chemical properties, several studies have looked into using graphene to improve the electrical conductivity of membranes for electrodialysis applications.[208–210] Hosseini *et al.* (2020) fabricated a novel cation exchange membrane (CEM) for ED incorporated with graphite nanoparticles. The prepared membrane displayed uniform distribution of graphite nanoparticles which provided conducting areas for membrane, easy flow channels for counter ion transportation, and led to a decline in membrane electrical resistance. The authors stated that the combined effect may strengthen the electric field around the CEM and lead to a possible decrease in concentration polarization.[202]

IEMs for ED applications must especially possess high ionic conductivity and high permselectivity for counter ions. However, nanosized defects in 2D graphene layer can lead to mutual selectivity between co-ions or permselectivity between counter ions in graphene-based membranes. To alleviate this issue, Chaudhury *et al.* (2020) exploited the surface morphology of heterogenous, 3D, macroporous network of Laser Induced Graphene (LIG) to develop LIG embedded cation exchange membranes. The introduction of LIG increased the conductivity of the membrane and effective capacitance in the membrane-electrolyte interface, however the study also noted that the functionality of the membrane, for ion transport, was largely unaffected.[211]

Compared to purely organic polymeric membranes, composite IEMs have superior properties. Moreover, using conducting polymers in the modification of nanomaterials combines the electrochemical properties of the polymer with the properties of the nanomaterials. To prepare compound membranes with improved properties, Hosseini *et al.* (2018) and Zhang *et al.* (2015) developed ion exchange membranes by incorporating polyaniline (PANI)/graphene oxide(GO) functionalized composite nanoplates, and rGO/PANI respectively, into the main polymer matrix. Both studies observed improved dispersion of the PANI/GO and rGO/PANI in the membrane matrix as compared to only GO nanoplates, improving the electrochemical properties of IEM and facilitating ion transport.[209,212] In another study to improve performance of anion exchange membranes, Cseri *et al.* (2018) successfully developed GO–polybenzimidazolium nanocomposite membranes, which demonstrated high mechanical stability and good permselectivity. However, higher GO ratios in the polymer matrix resulted in less durable membranes with high area resistance.[213]

A major challenge with nanocomposite membranes is the decreased stability with increased additive loading, which could be attributed to poor additive dispersion and agglomeration in the polymer matrix. To overcome this problem, Alabi *et al.* (2020) developed novel graphene oxide incorporated cation exchange membrane wherein the ion exchange groups are provided by the nanomaterial additive. The developed membrane demonstrated increased hydrophilicity, and an improved permselectivity and area resistance as compared to commercial CEMs.[214]

15.4.1.3. *Carbon nanofibers (CNFs)*

Carbon Nanofibers (CNFs) are considered great candidates for inorganic fillers for fabrication of mixed matrix membranes, due to their high aspect ratios and high specific surface area. Moreover, CNFs have a highly interconnected fiber structure, with a tunable fiber diameter, and a partial negative charge which help enhance selectivity and ion flux. The high electrical conductivities of CNF also help reduce concentration polarization by enhancing the electrochemical properties of the membrane and the electrical field strength around the membrane.[200]

CNF embedded membranes are commonly prepared by mixing CNF into the casting solution. Jashni *et al.* (2019) prepared CNF incorporated mixed matrix electrodialysis membrane to improve membrane properties. The study found that increasing CNF concentration upto 0.5 wt% decreased the surface roughness, which reduced the boundary layer at the membrane solution interface, and increased surface hydrophilicity. Addition of CNFs also improved membrane potential, transport number, and permselectivity, as well as the mono ionic permeability and flux.[200] Increasing ratio of CNF beyond 0.5 wt% led to a detrimental effect which could be attributed to particle aggregation at higher concentrations.

15.4.1.4. *Activated carbon*

ACNPs are advanced materials with salient features such as high specific surface area, high adsorption capacity and negative ion concentration. Activated carbon is ubiquitous in electrochemical seawater desalination technologies, such as capacitive deionization.[215,216] To prepare membranes with high conductivity Hosseini *et al.* (2018) fabricated super activated carbon nanoparticles (ACNPs) incorporated CEM. The produced membranes had improved ion exchange capacity, membrane potential, fixed ionic concentration and selectivity. Application of these membranes in ED showed high dialytic rate of unwanted ions.[217] Similarly, Ebrahimi *et al.* developed an AC-co-Chitosan composite nanoparticle embedded IEM to produce a membrane with superior electrical properties. The IEM combined the high surface area and high adsorptive property of activated carbon with the hydrophilicity and functionality of chitosan. The developed membranes demonstrated improved properties such as increased IEC, ionic flux, and permeability, owing to the exceptional adsorption capacity of the composite nanoparticles. The study also found that incorporating the AC-co-Chitosan composite nanoparticles also increased the dialytic rate for copper ion removal.[218]

15.4.2. *Capacitive deionization*

Capacitive deionization (CDI) was developed as a membrane-less, low pressure desalination process, mostly used for brackish water (salinity <10,000 ppm). However, since its inception, CDI has been adapted for use in sea water desalination, wastewater remediation, and water softening.[219,220] Globally, CDI contributes to less than one percent of the total capacity of water desalinated, trailing behind conventional desalination processes.[221] Although currently the process is not widely adopted for desalination purposes, it offers advantages such as low energy consumption under low salinity conditions, and low environmental impact making it a promising technology.[222]

CDI operates on the principle of electrical capacitance. When feed water flows between two charged porous electrodes, the applied potential causes migration of the ions towards oppositely charged electrodes. The migrating ions adsorb onto the porous electrode, thus producing desalinated water at the outlet. While the mechanism of desalination in CDI is quite straightforward, the performance of the process is greatly impacted by cell structure (Fig. 15.2). While the previously and currently most widely used cell architectures are flow-through (Fig. 15.2b) and flow between (Fig. 15.2a) respectively, a recent major advancement in CDI cell design is the introduction of membrane capacitive deionization (MCDI), developed by Lee *et al.* in 2006.[223] Introducing ion exchange membranes at both anode and cathode (Fig. 15.2c) largely alleviates the decline of ion adsorption capacity and incomplete regeneration of electrodes. Another two types of CDI cell architectures developed recently are hybrid capacitive deionization (HCDI) (Fig. 15.2e) and the flow capacitive deionization (FCDI) (Figs. 15.2(g–i)). HCDI is an asymmetric system, which employs desalination batteries using faradaic electrode materials as one of the electrodes.[224] FCDI, a

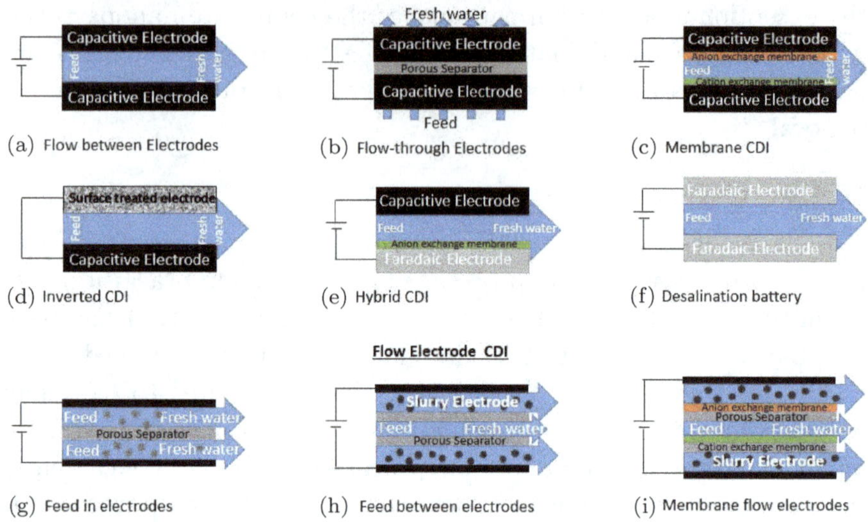

Fig. 15.2. CDI cell flow architectures.

continuous desalination process, uses carbon flow/carbon slurry, pumped through the electrode compartments, as electrodes.

Besides the cell structure, the electrode material used for CDI is also critical to the successful separation of ions from solution. An ideal electrode material requires high specific surface area, high electric conductivity, fast response to electrosorption-desorption changes, chemical and electrochemical stability over a wide range of pH, high tolerance to frequent voltage fluctuations, and low propensity for scaling and fouling.[225] Porous carbon is considered a good candidate for its versatility and properties.[226]

A wide range of carbon electrodes have also been tested for CDI applications, such as activated carbon (AC), carbon aerogels (CA), carbon nanotubes (CNT), carbon nanofibers (CNF), graphene, and carbon composite electrodes.[216,220,227–234] AC electrodes are preferred for their low cost of manufacturing, easy commercial availability, large pore volume, and high available surface area, and tunable surface chemistry.[216,227,235] Notedly, among the critical parameters that dominate the performance of AC electrodes, a highly porous structure, and particularly a good combination of micropores and mesopores, facilitates the transport of ions from aqueous media to the electrode surface and provides large accessible surface area for electrical double layer (EDL) formation. However, poor conductivity and high ion transfer resistance of AC restricts its efficiency.

To address the issues related to AC, researchers turned to one- and/or two-dimensional carbon nanomaterial, i.e., CA,[236,237] CNT, CNF, graphene, ordered mesoporous carbon (OMC), which have high mechanical strength, and chemical stability, in addition to their high surface area, and high electrical and thermal conductivity.[233] However, many of these carbon-based nanomaterials suffer from poor wettability and relatively low electro-sorption capacity (and high manufacturing cost). Recent advance in CDI include the adoption of carbon-based nanocomposites as electrodes. These nanocomposite electrodes mainly include carbon-carbon composites, carbon-metal (oxide) composites, and carbon-polymer composites. With modifications, nanocomposites often demonstrate larger specific surface areas, better pore size, and higher hydrophilicity and conductivity than single component electrodes. Thus, better CDI performance was often reported from systems with nanocomposite electrodes. For instance, Wimalasiri and Zou reported that CNT/graphene composite, with high ion conductivity and large surface area covered by EDL, can obtain a specific capacitance of 220 F/g and an electro-sorption capacity of 26.42 mg NaCl/g, while those of a graphene electrode are 140 F/g and 22.27 mg NaCl/g; OMC allows ions to penetrate deeper into the pores to form the double layer, which has led to development of electrode materials such as hierarchical OMC, hierarchical ACNF, and graphene coated mesoporous C spheres with a balanced mesopore to micropore ratio.[10,228,238] Such materials with increased mesoporosity and well-tuned pore sizes enhance salt adsorption performance.[228] Zaheen *et al.* (2018) fabricated 3D intercalated graphene sheet-sphere nanocomposites, a hierarchically porous composite with interconnected mesopores, to prevent irreversible agglomeration of graphene layers due to strong π-π interactions.[239] Mossad and Zou, and Chen *et al.*, showed that the selectivity of ions follows a hierarchy wherein charge affinity of the ions, i.e., ions having higher affinity are adsorbed preferentially over other ions.[227,240] For the case where charge affinity of the ions is equal, selectivity is determined by the hydration radius.[221,227]

In terms of carbon-metal (oxide) composite electrodes, Yoon *et al.* demonstrated an increased deionization capacity, wherein the capacity of carbon was combined with the charge transfer reaction mediated by the metal (Ag).[234] Among the doping metal and metal oxides, TiO_2, MnO_2, and ZnO are often deposited onto carbon-based nanomaterials for high hydrophilicity to increase accessible EDL surface area, inhibition of physical adsorption of ion species facilitating electro-sorption/electro-desorption, and/or alteration of the surface charge of electrode to increase the ion removal capacity.[232] Similarly, modifying carbon-based nanomaterials with conducting polymer (i.e. chitosan, polyaniline, and ion exchange polymer) also contributes to the improvement of CDI performance via increasing pseudocapacitance and electro-sorption. In addition, there are also a considerable amount of studies that demonstrate the benefit brought with the combination of metal oxide doping and polymer modification onto carbon-based materials.[241–244]

15.5. Summary

Carbon based nanomaterials have gained importance due to their unique properties leading to process enhancements. In this chapter we have summarized the progress made over the last few years in the field of membrane modification using carbon-based nanomaterials, with a specific focus on their structure, fabrication method and influence on desalination performance.

Economic feasibility of these membranes plays a key role in their commercialization. The extra cost of carbon-based membrane fabrication, must be taken into account when considering the wide-scale adoption of these technologies. In a study by De Lannoy *et al.*,[87] the authors found that imparting electrical conductivity to NF membranes would add ~$2.5/m^2 in material costs representing only a 1.5% increase in the total price of traditional tight NF. In the work of Zhang and Vecitis,[110] they reported an additional 15–30%

increase in the capital cost of a MWCNT-PVDF membrane (vs. a commercially available UF membrane material) based on a MWCNT market price of 0.1 US$/g and an estimated UF membrane cost of 15–30 US$/m^2. By accounting for the energy savings from the operation and maintenance costs because of fouling mitigation, they predicted that the time to pay back the additional material cost is between 1.5–5 years. Both Zhu *et al.*[170] and Tang *et al.*[47] determined the additional cost of the CNT UF membrane to be about $9/m^2 based on a CNT cost of $10.00/g. Goh *et al.* estimated prices of CNTs in the range of 0.1–15$/g for MWCNTs and 100–280$/g for SWCNTs.[152] Among all the carbon-based nanomaterials including SWCNTs, graphite, graphene, graphene oxide and reduced graphene oxide, MWCNTs seem to have an economical advantage.

References

1. K. Roy, A. Mukherjee, N. R. Maddela, S. Chakraborty, B. Shen, M. Li, D. Du, Y. Peng, F. Lu, L. C. Garciá Cruzatty, Outlook on the bottleneck of carbon nanotube in desalination and membrane-based water treatment — a review, *J. Environ. Chem. Eng.*, **8**(1) (2020) 103572.
2. G. Amy, N. Ghaffour, Z. Li, L. Francis, R. V. Linares, T. Missimer, S. Lattemann, Membrane-based seawater desalination: Present and future prospects, *Desalination*, **401** (2017) 16–21.
3. S. M. Hosseini, Z. Ahmadi, M. Nemati, F. Parvizian, S. S. Madaeni, Electrodialysis heterogeneous ion exchange membranes modified by SiO_2 nanoparticles: Fabrication and electrochemical characterization, *Water Sci. Technol.*, **73**(9) (2016) 2074–2084.
4. Y. H. Teow, A. W. Mohammad, New generation nanomaterials for water desalination: A review, *Desalination* (2019) November, 2–17.
5. P. G. Youssef, R. K. Al-Dadah, S. M. Mahmoud, Comparative analysis of desalination technologies, *Energy Procedia*, **61** (2014) 2604–2607.
6. D. Cohen-Tanugi, J. C. Grossman, Water desalination across nanoporous graphene, *Nano Lett.*, **12**(7) (2012) 3602–3608.
7. M. W. Shahzad, M. Burhan, L. Ang, K. C. Ng, Energy-water-environment nexus underpinning future desalination sustainability, *Desalination*, **413** (2017) 52–64.

8. T. Mezher, H. Fath, Z. Abbas, A. Khaled, Techno-economic assessment and environmental impacts of desalination technologies, *Desalination*, **266**(1–3) (2011) 263–273.
9. T. Younos, K. E. Tulou, Overview of desalination techniques, *J. Contemp. Water Res. Educ.*, **132** (2005) 3–10.
10. T. Humplik, J. Lee, S. C. O'Hern, B. A. Fellman, M. A. Baig, S. F. Hassan, M. A. Atieh, F. Rahman, T. Laoui, R. Karnik, E. N. Wang, Nanostructured materials for water desalination, *Nanotechnology*, **22**(29) (2011).
11. X. Wang, A. Christ, K. Regenauer-Lieb, K. Hooman, H. T. Chua, Low grade heat driven multi-effect distillation technology, *Int. J. Heat Mass Transf.*, **54**(25–26) (2011) 5497–5503.
12. A. D. Khawaji, I. K. Kutubkhanah, J. M. Wie, Advances in seawater desalination technologies, *Desalination*, **221**(1–3) (2008) 47–69.
13. D. Zhao, J. Xue, S. Li, H. Sun, Zhang, Q. dong. Theoretical analyses of thermal and economical aspects of multi-effect distillation desalination dealing with high-salinity wastewater, *Desalination*, **273**(2–3) (2011) 292–298.
14. B. Han, Z. Liu, H. Wu, Y. Li, Experimental study on a new method for improving the performance of thermal vapor compressors for multi-effect distillation desalination systems, *Desalination*, **344** (2014) 391–395.
15. H. Bamufleh, F. Abdelhady, H. M. Baaqeel, M. M. El-Halwagi, Optimization of multi-effect distillation with brine treatment via membrane distillation and process heat integration, *Desalination*, **408** (2017) 110–118.
16. P. Sharan, T. Neises, C. Turchi, Optimal feed flow sequence for multi-effect distillation system integrated with supercritical carbon dioxide brayton cycle for seawater desalination, *J. Clean. Prod.*, **196** (2018) 889–901.
17. P. Catrini, A. Cipollina, G. Micale, A. Piacentino, A. Tamburini, Exergy analysis and thermoeconomic cost accounting of a combined heat and power steam cycle integrated with a multi effect distillation-thermal vapour compression desalination plant, *Energy Convers. Manag.*, **149** (2017) 950–965.
18. A. M. El-Nashar, Optimizing the operating parameters of a solar desalination plant, *Sol. Energy*, **48**(4) (1992) 207–213.
19. C. bo Chen, L. Sun, D. jun Liu, X. L. Luo, Impact of cumulative fouling characteristics on full-cycle operation optimisation of multi-effect distillation desalination system, *Chem. Eng. Trans.*, **76**(2017) (2019) 649–654.

20. I. S. Park, Numerical investigation of entraining performance and operational robustness of thermal vapor compressor having swirled motive steam inflow, *Desalination*, **257**(1–3) (2010) 206–211.
21. C. Wang, L. Wang, X. Wang, H. Zhao, Design and numerical investigation of an adaptive nozzle exit position ejector in multi-effect distillation desalination system, *Energy*, **140** (2017) 673–681.
22. M. I. T. Opencourseware, Review of Water Resources and Desalination Technologies, *Contract*, (2009).
23. N. Lior, *Advances in Water Desalination*, (2012).
24. G. Raluy, L. Serra, J. Uche, Life cycle assessment of MSF, MED and RO desalination technologies, *Energy*, **31**(13) (2006) 2361–2372.
25. C. Li, Y. Goswami, E. Stefanakos, Solar assisted sea water desalination: A review, *Renew. Sustain. Energy Rev.*, **19** (2013) 136–163.
26. H. T. El-Dessouky, H. M. Ettouney, Y. Al-Roumi, Multi-stage flash desalination: Present and future outlook, *Chem. Eng. J.*, **73**(2) (1999) 173–190.
27. J. Jiang, H. Tian, M. Cui, L. Liu, Proof-of-concept study of an integrated solar desalination system, *Renew. Energy*, **34**(12) (2009) 2798–2802.
28. Y. Junjie, S. Shufeng, W. Jinhua, L. Jiping, Improvement of a multi-stage flash seawater desalination system for cogeneration power plants, *Desalination*, **217**(1–3) (2007) 191–202.
29. M. M. Alhazmy, Multi stage flash desalination plant with brine-feed mixing and cooling, *Energy*, **36**(8) (2011) 5225–5232.
30. A. Alkhudhiri, N. Darwish, N. Hilal, Membrane distillation: A comprehensive review, *Desalination*, **287** (2012) 2–18.
31. F. He, J. Gilron, K. K. Sirkar, High water recovery in direct contact membrane distillation using a series of cascades, *Desalination*, **323** (2013) 48–54.
32. B. L. Pangarkar, M. G. Sane, M. Guddad, Reverse osmosis and membrane distillation for desalination of groundwater: A review. *ISRN Mater. Sci.*, **2011** (2011).
33. K. W. Lawson, D. R. Lloyd, Membrane distillation. II. direct contact MD, *J. Memb. Sci.*, **120**(1) (1996) 123–133.
34. S. Al-Obaidani, E. Curcio, F. Macedonio, G. Di Profio, H. Al-Hinai, E. Drioli, Potential of membrane distillation in seawater desalination: Thermal efficiency, sensitivity study and cost estimation, *J. Memb. Sci.*, **323**(1) (2008) 85–98.
35. G. W. Meindersma, C. M. Guijt, A. B. De Haan, Desalination and water recycling by air gap membrane distillation, *Desalination*, **187**(1–3) (2006) 291–301.

36. M. Khayet, P. Godino, J. I. Mengual, Theory and experiments on sweeping gas membrane distillation, *J. Memb. Sci.*, **165**(2) (2000) 261–272.
37. K. W. Lawson, D. R. Lloyd, Membrane distillation. I. module design and performance evaluation using vacuum membrane distillation, *J. Memb. Sci.*, **120**(1) (1996) 111–121.
38. D. R. L. Kevin W. Lawson, Membrane distillation, *J. Memb. Sci.*, **124**(1) (1997) 1–25.
39. M. Khayet, Membranes and theoretical modeling of membrane distillation: A review, *Adv. Colloid Interface Sci.*, **164**(1–2) (2011) 56–88.
40. Y. Manawi, V. Kochkodan, M. A. Hussein, M. A. Khaleel, M. Khraisheh, N. Hilal, Can carbon-based nanomaterials revolutionize membrane fabrication for water treatment and desalination? *Desalination*, **391** (2016) 69–88.
41. M. S. Mauter, M. Elimelech, Environmental applications of carbon-based nanomaterials, *Environ. Sci. Technol.*, **42**(16) (2008) 5843–5859.
42. K. Balasubramanian, M. Burghard, Chemically functionalized carbon nanotubes, *Small*, **1**(2) (2005) 180–192.
43. S. Qiu, L. Wu, X. Pan, L. Zhang, H. Chen, C. Gao, Preparation and properties of functionalized carbon nanotube/PSF blend ultrafiltration membranes, *J. Memb. Sci.*, **342**(1–2) (2009) 165–172.
44. L. F. Dumée, K. Sears, J. Schütz, N. Finn, C. Huynh, S. Hawkins, M. Duke, S. Gray, Characterization and evaluation of carbon nanotube bucky-paper membranes for direct contact membrane distillation, *J. Memb. Sci.*, **351**(1–2) (2010) 36–43.
45. L. Dumée, J. L. Campbell, K. Sears, J. Schütz, N. Finn, M. Duke, S. Gray, The impact of hydrophobic coating on the performance of carbon nanotube bucky-paper membranes in membrane distillation, *Desalination*, **283** (2011) 64–67.
46. K. Gethard, S. Mitra, Carbon nanotube enhanced membrane distillation for online preconcentration of trace pharmaceuticals in polar solvents, *Analyst*, **136**(12) (2011) 2643–2648.
47. L. Tang, A. Iddya, X. Zhu, A. V. Dudchenko, W. Duan, C. Turchi, J. Vanneste, T. Y. Cath, D. Jassby, Enhanced flux and electrochemical cleaning of silicate scaling on carbon nanotube-coated membrane distillation membranes treating geothermal brines, *ACS Appl. Mater. Interfaces* (2017) acsami.7b12615.
48. U. Rao, A. Iddya, B. Jung, C. M. Khor, Z. Hendren, C. Turchi, T. Cath, E. M. V. Hoek, G. Z. Ramon, D. Jassby, Mineral scale prevention on electrically conducting membrane distillation membranes

using induced electrophoretic mixing, *Environ. Sci. Technol.*, **54**(6) (2020) 3678–3690.
49. L. D. Tijing, Y. C. Woo, W.-G. Shim, T. He, J.-S. Choi, S.-H. Kim, H. K. Shon, Superhydrophobic nanofiber membrane containing carbon nanotubes for high-performance direct contact membrane distillation, *J. Memb. Sci.*, **502** (2016) 158–170.
50. M. R. Elmarghany, A. H. El-Shazly, S. Rajabzadeh, M. S. Salem, M. A. Shouman, M. N. Sabry, H. Matsuyama, N. Nady, Triple-layer nanocomposite membrane prepared by electrospinning based on modified PES with carbon nanotubes for membrane distillation applications, *Membranes (Basel).*, **10**(1) (2020) 15.
51. T. L. S. Silva, S. Morales-Torres, J. L. Figueiredo, A. M. T. Silva, Multi-walled carbon nanotube/PVDF blended membranes with sponge- and finger-like pores for direct contact membrane distillation, *Desalination*, **357** (2015) 233–245.
52. L. Dumée, K. Sears, J. rg Schü tz, N. Finn, M. Duke, S. Gray, Carbon nanotube based composite membranes for water desalination by membrane distillation, *Desalin. Water Treat.*, **17**(1–3) (2010) 72–79.
53. L. F. Dumée, S. Smart, M. C. Duke, S. R. Gray, Next generation membranes for membrane distillation and future prospects. In *Pervaporation, Vapour Permeation and Membrane Distillation*; Elsevier, 2015; pp. 415–447.
54. L. M. Camacho, L. Dumée, J. Zhang, J. de Li, M. Duke, J. Gomez, S. Gray, Advances in membrane distillation for water desalination and purification applications, *Water (Switzerland)*, **5**(1) (2013) 94–196.
55. M. Bhadra, S. Roy, S. Mitra, Flux enhancement in direct contact membrane distillation by implementing carbon nanotube immobilized PTFE membrane, *Sep. Purif. Technol.*, **161** (2016) 136–143.
56. M. Bhadra, S. Roy, S. Mitra, Enhanced desalination using carboxylated carbon nanotube immobilized membranes, *Sep. Purif. Technol.*, **120** (2013) 373–377.
57. R. Zhou, D. Rana, T. Matsuura, C. Q. Lan, Effects of multi-walled carbon nanotubes (MWCNTs) and integrated MWCNTs/SiO_2 Nano-additives on PVDF polymeric membranes for vacuum membrane distillation, *Sep. Purif. Technol.*, **217** (2019) 154–163.
58. A. V. Dudchenko, C. Chen, A. Cardenas, J. Rolf, D. Jassby, Frequency-dependent stability of CNT joule heaters in ionizable media and desalination processes, *Nat. Nanotechnol.*, **12**(6) (2017) 557–563.

59. A. Anvari, K. M. Kekre, A. A. Yancheshme, Y. Yao, A. Ronen, Membrane distillation of high salinity water by induction heated thermally conducting membranes, *J. Memb. Sci.*, **589** (2019) 117253.
60. A. Anvari, K. M. Kekre, A. Ronen, Scaling mitigation in radio-frequency induction heated membrane distillation, *J. Memb. Sci.*, **600** (2020) 117859.
61. E.-D. Miao, M.-Q. Ye, C.-L. Guo, L. Liang, Q. Liu, Z.-H. Rao, Enhanced solar steam generation using carbon nanotube membrane distillation device with heat localization, *Appl. Therm. Eng.*, **149** (2019) 1255–1264.
62. A. K. An, E.-J. Lee, J. Guo, S. Jeong, J.-G. Lee, N. Ghaffour, Enhanced vapor transport in membrane distillation via functionalized carbon nanotubes anchored into electrospun nanofibres, *Sci. Rep.*, **7** (2017) 41562.
63. S. Ragunath, S. Roy, S. Mitra, Carbon nanotube immobilized membrane with controlled nanotube incorporation via phase inversion polymerization for membrane distillation based desalination, *Sep. Purif. Technol.*, **194** (2018) (July 2017) 249–255.
64. M. S. Fahmey, A.-H. M. El-Aassar, M. M. Abo-Elfadel, A. S. Orabi, R. Das, Comparative performance evaluations of nanomaterials mixed polysulfone: A scale-up approach through vacuum enhanced direct contact membrane distillation for water desalination, *Desalination*, **451** (2019) 111–116.
65. S. Sinha Ray, H. Singh Bakshi, R. Dangayach, R. Singh, C. K. Deb, M. Ganesapillai, S.-S. Chen, M. K. Purkait, Recent developments in nanomaterials-modified membranes for improved membrane distillation performance, *Membranes (Basel).*, **10**(7) (2020) 140.
66. Y. Mao, Q. Huang, B. Meng, K. Zhou, G. Liu, A. Gugliuzza, E. Drioli, W. Jin, Roughness-enhanced hydrophobic graphene oxide membrane for water desalination via membrane distillation, *J. Memb. Sci.* (2020) 118364.
67. B. Gong, H. Yang, S. Wu, G. Xiong, J. Yan, K. Cen, Z. Bo, K. Ostrikov, Graphene array-based anti-fouling solar vapour gap membrane distillation with high energy efficiency, *Nano-Micro Lett.*, **11**(1) (2019) 51.
68. C. Athanasekou, A. Sapalidis, I. Katris, E. Savopoulou, K. Beltsios, T. Tsoufis, A. Kaltzoglou, P. Falaras, G. Bounos, M. Antoniou, Mixed matrix PVDF/graphene and composite-skin PVDF/graphene oxide membranes applied in membrane distillation. *Polym. Eng. Sci.*, **59**(S1) (2019) E262–E278.
69. G. Grasso, F. Galiano, M. J. Yoo, R. Mancuso, H. B. Park, B. Gabriele, A. Figoli, E. Drioli, Development of Graphene-PVDF

Composite membranes for membrane distillation, *J. Memb. Sci.* (2020) 118017.
70. A. Jafari, M. R. S. Kebria, A. Rahimpour, G. Bakeri, Graphene quantum dots modified polyvinylidenefluride (PVDF) nanofibrous membranes with enhanced performance for air gap membrane distillation, *Chem. Eng. Process. Intensif.*, **126** (2018) 222–231.
71. K.-J. Lu, J. Zuo, T.-S. Chung, Novel PVDF membranes comprising N-Butylamine functionalized graphene oxide for direct contact membrane distillation, *J. Memb. Sci.*, **539** (2017) 34–42.
72. M. Bhadra, S. Roy, S. Mitra, Desalination across a graphene oxide membrane via direct contact membrane distillation, *Desalination*, **378** (2016) 37–43.
73. H. Qiu, Y. Peng, L. Ge, B. V. Hernandez, Z. Zhu, Pore channel surface modification for enhancing anti-fouling membrane distillation, *Appl. Surf. Sci.*, **443** (2018) 217–226.
74. L. Zhao, C. Wu, X. Lu, D. Ng, Y. B. Truong, Z. Xie, Activated carbon enhanced hydrophobic/hydrophilic dual-layer nanofiber composite membranes for high-performance direct contact membrane distillation, *Desalination*, **446** (2018) 59–69.
75. J. Wu, K. R. Zodrow, P. B. Szemraj, Q. Li, Photothermal nanocomposite membranes for direct solar membrane distillation, *J. Mater. Chem. A*, **5**(45) (2017) 23712–23719.
76. P. J. Bora, T. R. Suresh Kumar, D. Q. Tan, Enhancement of microwave absorption bandwidth of MXene nanocomposites through macroscopic design, *R. Soc. Open Sci.*, **7**(8) (2020) 200456.
77. Y. Z. Tan, H. Wang, L. Han, M. B. Tanis-Kanbur, M. V. Pranav, J. W. Chew, Photothermal-enhanced and fouling-resistant membrane for solar-assisted membrane distillation, *J. Memb. Sci.*, **565** (2018) 254–265.
78. J. Schewe, J. Heinke, D. Gerten, I. Haddeland, N. W. Arnell, D. B. Clark, R. Dankers, S. Eisner, B. M. Fekete, F. J. Colón-González, S. N. Gosling, H. Kim, X. Liu, Y. Masaki, F. T. Portmann, Y. Satoh, T. Stacke, Q. Tang, Y. Wada, D. Wisser, T. Albrecht, K. Frieler, F. Piontek, L. Warszawski, P. Kabat, Multimodel assessment of water scarcity under climate change, *Proc. Natl. Acad. Sci.*, **111**(9) (2014) 3245–3250.
79. K. Hussey, J. Pittock, The energy — water nexus?: Managing the links between energy, *And*, **17**(1) (2018).
80. M. A. Shannon, P. W. Bohn, M. Elimelech, J. G. Georgiadis, B. J. Marïas, A. M. Mayes, Science and technology for water purification in the coming decades, *Nature*, **452**(7185) (2008) 301–310.

81. M. Elimelech, W. A. Phillip, The future of seawater desalination: Energy, technology, and the environment. *Science*, (2011) 712–717.
82. M. M. Pendergast, E. M. V. Hoek, A review of water treatment membrane nanotechnologies, *Energy Environ. Sci.*, **4**(6) (2011) 1946.
83. F. Ahmed, B. S. Lalia, V. Kochkodan, N. Hilal, R. Hashaikeh, Electrically conductive polymeric membranes for fouling prevention and detection: A review, *Desalination*, **391** (2016) 1–15.
84. P. Formoso, E. Pantuso, G. De Filpo, F. P. Nicoletta, Electro-conductive membranes for permeation enhancement and fouling mitigation: A short review, *Membranes (Basel).*, **7**(3) (2017).
85. Z. Z. Chowdhury, S. Sagadevan, R. Bin Johan, S. T. Shah, A. Adebesi, S. I. Md, R. F. Rafique, A review on electrochemically modified carbon nanotubes (CNTs) membrane for desalination and purification of water, *Mater. Res. Express*, **5**(10) (2018).
86. J. R. Werber, A. Deshmukh, M. Elimelech, The critical need for increased selectivity, not increased water permeability, for desalination membranes, *Environmental Science and Technology Letters*, (2016).
87. C. F. de Lannoy, D. Jassby, D. D. Davis, M. R. Wiesner, A highly electrically conductive polymer-multiwalled carbon nanotube nanocomposite membrane, *J. Memb. Sci.*, **415–416** (2012) 718–724.
88. W. Wang, L. Zhu, B. Shan, C. Xie, C. Liu, F. Cui, G. Li, Preparation and characterization of SLS-CNT/PES ultrafiltration membrane with antifouling and antibacterial properties. *J. Memb. Sci.*, **548** (2018) (October 2017) 459–469.
89. Y. Liu, H. Liu, Z. Zhou, T. Wang, C. N. Ong, C. D. Vecitis, Degradation of the common aqueous antibiotic tetracycline using a carbon nanotube electrochemical filter, *Environ. Sci. Technol.*, **49**(13) (2015) 7974–7980.
90. W. F. Chan, E. Marand, S. M. Martin, Novel zwitterion functionalized carbon nanotube nanocomposite membranes for improved ro performance and surface anti-biofouling resistance, *J. Memb. Sci.*, **509** (2016) 125–137.
91. W. Duan, A. Ronen, J. V. de Leon, A. Dudchenko, S. Yao, J. Corbala-Delgado, A. Yan, M. Matsumoto, D. Jassby, Treating anaerobic sequencing batch reactor effluent with electrically conducting ultrafiltration and nanofiltration membranes for fouling control, *J. Memb. Sci.*, **504** (2016) 104–112.
92. T. H. Lee, M. Y. Lee, H. D. Lee, J. S. Roh, H. W. Kim, H. B. Park, Highly porous carbon nanotube/polysulfone nanocomposite supports for high-flux polyamide reverse osmosis membranes. *J. Memb. Sci.*, **539** (2017) (June) 441–450.

93. S. Wang, S. Liang, P. Liang, X. Zhang, J. Sun, S. Wu, X. Huang, In-situ combined dual-layer CNT/PVDF membrane for electrically-enhanced fouling resistance, *J. Memb. Sci.*, **491** (2015) 37–44.
94. H. gyu Choi, M. Son, H. Choi, Integrating seawater desalination and wastewater reclamation forward osmosis process using thin-film composite mixed matrix membrane with functionalized carbon nanotube blended polyethersulfone support layer, *Chemosphere*, **185** (2017) 1181–1188.
95. M. B. Wu, Y. Lv, H. C. Yang, L. F. Liu, X. Zhang, Z. K. Xu, Thin film composite membranes combining carbon nanotube intermediate layer and microfiltration support for high nanofiltration performances, *J. Memb. Sci.*, **515** (2016) 238–244.
96. I. Wan Azelee, P. S. Goh, W. J. Lau, A. F. Ismail, M. Rezaei-DashtArzhandi, K. C. Wong, M. N. Subramaniam, Enhanced desalination of polyamide thin film nanocomposite incorporated with acid treated multiwalled carbon nanotube-titania nanotube hybrid, *Desalination*, **409** (2017) 163–170.
97. O. Gupta, S. Roy, S. Mitra, Enhanced membrane distillation of organic solvents from their aqueous mixtures using a carbon nanotube immobilized membrane, *J. Memb. Sci.*, **568** (2018) (October) 134–140.
98. K. K. Yan, L. Jiao, S. Lin, X. Ji, Y. Lu, L. Zhang, Superhydrophobic electrospun nanofiber membrane coated by carbon nanotubes network for membrane distillation, *Desalination*, **437** (2018) (January) 26–33.
99. F. R. Omi, M. R. Choudhury, N. Anwar, A. R. Bakr, M. S. Rahaman, Highly conductive ultrafiltration membrane via vacuum filtration assisted layer by-layer deposition of functionalized carbon nanotubes, *Ind. Eng. Chem. Res.*, **56**(30) (2017) 8474–8484.
100. W. Duan, G. Chen, C. Chen, R. Sanghvi, A. Iddya, S. Walker, H. Liu, A. Ronen, D. Jassby, Electrochemical removal of hexavalent chromium using electrically conducting carbon nanotube/polymer composite ultrafiltration membranes. *J. Memb. Sci.*, **531** (2017) (March) 160–171.
101. Y. Yang, S. Qiao, R. Jin, J. Zhou, X. Quan, Fouling control mechanisms in filtrating natural organic matters by electro-enhanced carbon nanotubes hollow fiber membranes. *J. Memb. Sci.*, **553** (2018) (January) 54–62.
102. Q. Zhang, P. Arribas, E. M. Remillard, M. C. García-Payo, M. Khayet, C. D. Vecitis, Interlaced CNT electrodes for bacterial fouling reduction of microfiltration membranes, *Environ. Sci. Technol.*, **51**(16) (2017) 9176–9183.

103. G. Gao, M. Pan, C. D. Vecitis, Effect of the oxidation approach on carbon nanotube surface functional groups and electrooxidative filtration performance, *J. Mater. Chem. A*, **3**(14) (2015) 7575–7582.
104. B. S. Lalia, F. E. Ahmed, T. Shah, N. Hilal, R. Hashaikeh, Electrically conductive membranes based on carbon nanostructures for self-cleaning of biofouling, *Desalination*, **360** (2015) 8–12.
105. W. Duan, A. Ronen, S. Walker, D. Jassby, Polyaniline-coated carbon nanotube ultrafiltration membranes: Enhanced anodic stability for in situ cleaning and electro-oxidation processes, *ACS Appl. Mater. Interfaces*, **8**(34) (2016) 22574–22584.
106. C. F. De Lannoy, D. Jassby, K. Gloe, A. D. Gordon, M. R. Wiesner, Aquatic biofouling prevention by electrically charged nanocomposite polymer thin film membranes, *Environ. Sci. Technol.*, **47**(6) (2013) 2760–2768.
107. W. Duan, A. Dudchenko, E. Mende, C. Flyer, X. Zhu, D. Jassby, Electrochemical mineral scale prevention and removal on electrically conducting carbon nanotube-polyamide reverse osmosis membranes, *Environ. Sci. Process. Impacts*, **16**(6) (2014) 1300–1308.
108. A. V. Dudchenko, J. Rolf, K. Russell, W. Duan, D. Jassby, Organic fouling inhibition on electrically conducting carbon nanotube-polyvinyl alcohol composite ultrafiltration membranes, *J. Memb. Sci.*, **468** (2014) 1–10.
109. A. Ronen, W. Duan, I. Wheeldon, S. Walker, D. Jassby, Microbial attachment inhibition through low-voltage electrochemical reactions on electrically conducting membranes, *Environ. Sci. Technol.*, **49**(21) (2015) 12741–12750.
110. Q. Zhang, C. D. Vecitis, Conductive CNT-PVDF membrane for capacitive organic fouling reduction, *J. Memb. Sci.*, **459** (2014) 143–156.
111. C. Cuevas, D. Kim, K. P. Katuri, P. Saikaly, S. P. Nunes, Electrochemically active polymeric hollow fibers based on poly(Ether-b-Amide)/carbon nanotubes, *J. Memb. Sci.*, **545** (2018) (September 2017) 323–328.
112. H. Mao, M. Qiu, X. Chen, H. Verweij, Y. Fan, Fabrication and in-situ fouling mitigation of a supported carbon nanotube/Γ-Alumina ultrafiltration membrane, *J. Memb. Sci.*, **550** (2018) (October 2017) 26–35.
113. F. E. Ahmed, B. S. Lalia, N. Hilal, R. Hashaikeh, Electrically conducting nanofiltration membranes based on networked cellulose and carbon nanostructures, *Desalination*, **406** (2017) 60–66.

114. N. G. Sahoo, S. Rana, J. W. Cho, L. Li, S. H. Chan, Polymer nanocomposites based on functionalized carbon nanotubes, *Prog. Polym. Sci. J.*, **35** (2010) 837–867.
115. X. Song, L. Wang, C. Y. Tang, Z. Wang, C. Gao, Fabrication of carbon nanotubes incorporated double-skinned thin film nanocomposite membranes for enhanced separation performance and antifouling capability in forward osmosis process, *Desalination*, **369** (2015) 1–9.
116. S. Kar, R. C. Bindal, P. K. Tewari, Carbon nanotube membranes for desalination and water purification: Challenges and opportunities, *Nano Today*, **7**(5) (2012) 385–389.
117. Ihsanullah, Carbon nanotube membranes for water purification: Developments, challenges, and prospects for the future. *Sep. Purif. Technol.*, **209** (2019) (July 2018) 307–337.
118. K. Sears, L. Dumée, J. Schütz, M. She, C. Huynh, S. Hawkins, M. Duke, S. Gray, Recent developments in carbon nanotube membranes for water purification and gas separation, *Materials (Basel).*, **3**(1) (2010) 127–149.
119. X. Yang, J. Lee, L. Yuan, S. R. Chae, V. K. Peterson, A. I. Minett, Y. Yin, A. T. Harris, Removal of natural organic matter in water using functionalised carbon nanotube buckypaper, *Carbon N. Y.*, **59** (2013) 160–166.
120. A. A. Alshahrani, H. Al-Zoubi, L. D. Nghiem, in het M. Panhuis, Synthesis and characterisation of MWNT/Chitosan and MWNT/Chitosan-crosslinked buckypaper membranes for desalination. *Desalination*, **418** (2017) (November 2016) 60–70.
121. M. H.-O. Rashid, S. F. Ralph, Carbon nanotube membranes: Synthesis, properties, and future filtration applications, *Nanomaterials*, **7**(5) (2017) 99.
122. A. A. Mamedov, N. A. Kotov, M. Prato, D. M. Guldi, J. P. Wicksted, A. Hirsch, Molecular design of strong single-wall carbon nanotube/polyelectrolyte multilayer composites, *Nat. Mater.*, **1**(3) (2002) 190–194.
123. X. Peng, J. Jin, E. M. Ericsson, I. Ichinose, General method for ultrathin free-standing films of nanofibrous composite materials, *J. Am. Chem. Soc.*, **129**(27) (2007) 8625–8633.
124. L. Dumée, V. Germain, K. Sears, J. Schütz, N. Finn, M. Duke, S. Cerneaux, D. Cornu, S. Gray, Enhanced durability and hydrophobicity of carbon nanotube bucky paper membranes in membrane distillation, *J. Memb. Sci.*, **In Press** (2011) 241–246.

125. M. Sianipar, S. H. Kim, Khoiruddin, F. Iskandar, I. G. Wenten, Functionalized carbon nanotube (CNT) membrane: Progress and challenges, *RSC Adv.*, **7**(81) (2017) 51175–51198.
126. Z. Yang, X. H. Ma, C. Y. Tang, Recent development of novel membranes for desalination, *Desalination*, **434** (2018) (December 2017) 37–59.
127. C. D. Vecitis, M. H. Schnoor, M. S. Rahaman, J. D. Schiffman, M. Elimelech, Electrochemical multiwalled carbon nanotube filter for viral and bacterial removal and inactivation, *Environ. Sci. Technol.*, **45**(8) (2011) 3672–3679.
128. V. Vatanpour, M. Safarpour, A. Khataee, H. Zarrabi, M. E. Yekavalangi, M. Kavian, A thin film nanocomposite reverse osmosis membrane containing amine-functionalized carbon nanotubes, *Sep. Purif. Technol.*, **184** (2017) 135–143.
129. M. Bhadra, S. Roy, S. Mitra, A bilayered structure comprised of functionalized carbon nanotubes for desalination by membrane distillation, *ACS Appl. Mater. Interfaces*, **8**(30) (2016) 19507–19513.
130. M. H. Davood Abadi Farahani, V. Vatanpour, *Polymer/Carbon Nanotubes Mixed Matrix Membranes for Water Purification*; Elsevier Inc., 2019.
131. R. Das, E. Ali, S. Bee, A. Hamid, S. Ramakrishna, Z. Zaman, Carbon nanotube membranes for water purification: A bright future in water desalination, *Desalination*, **336** (2014) 97–109.
132. S. Ali, S. A. U. Rehman, H. Y. Luan, M. U. Farid, H. Huang, Challenges and opportunities in functional carbon nanotubes for membrane-based water treatment and desalination, *Sci. Total Environ.*, **646**(19) (2019) 1126–1139.
133. C. Thamaraiselvan, S. Lerman, K. Weinfeld-Cohen, C. G. Dosoretz, Characterization of a support-free carbon nanotube-microporous membrane for water and wastewater filtration. *Sep. Purif. Technol.*, **202** (2018) (March) 1–8.
134. G. Gao, Q. Zhang, Z. Hao, C. D. Vecitis, Carbon nanotube membrane stack for flow-through sequential regenerative electro-fenton, *Environ. Sci. Technol.*, **49**(4) (2015) 2375–2383.
135. X. Zhao, J. Li, C. Liu, A novel TFC-type FO membrane with inserted sublayer of carbon nanotube networks exhibiting the improved separation performance, *Desalination*, **413** (2017) 176–183.
136. C. Thamaraiselvan, J. Wang, D. K. James, P. Narkhede, S. P. Singh, D. Jassby, J. M. Tour, C. J. Arnusch, Laser-induced graphene and carbon nanotubes as conductive carbon-based materials in environmental technology, *Materials Today.* Elsevier B.V. 2019.

137. B. Jung, C. Y. Kim, S. Jiao, U. Rao, A. V. Dudchenko, J. Tester, D. Jassby, Enhancing boron rejection on electrically conducting reverse osmosis membranes through local electrochemical PH modification, *Desalination* (2020).
138. Y. Baek, H. J. Kim, S. H. Kim, J. C. Lee, J. Yoon, Evaluation of carbon nanotube-polyamide thin-film nanocomposite reverse osmosis membrane: Surface properties, performance characteristics and fouling behavior, *J. Ind. Eng. Chem.*, **56** (2017) 327–334.
139. J. Zheng, M. Li, K. Yu, J. Hu, X. Zhang, L. Wang, Sulfonated multiwall carbon nanotubes assisted thin-film nanocomposite membrane with enhanced water flux and anti-fouling property, *J. Memb. Sci.*, **524** (2017) (August 2016) 344–353.
140. E. M. Remillard, Q. Zhang, S. Sosina, Z. Branson, T. Dasgupta, C. D. Vecitis, Electric-field alignment of aqueous multi-walled carbon nanotubes on microporous substrates, *Carbon N. Y.*, **100** (2016) 578–589.
141. J. Ortiz-Medina, S. Inukai, T. Araki, A. Morelos-Gomez, R. Cruz-Silva, K. Takeuchi, T. Noguchi, T. Kawaguchi, M. Terrones, M. Endo, Robust water desalination membranes against degradation using high loads of carbon nanotubes, *Sci. Rep.*, (2018).
142. S. Inukai, R. Cruz-Silva, J. Ortiz-Medina, A. Morelos-Gomez, K. Takeuchi, T. Hayashi, A. Tanioka, T. Araki, S. Tejima, T. Noguchi, M. Terrones, M. Endo, High-performance multi-functional reverse osmosis membranes obtained by carbon nanotube-polyamide nanocomposite, *Sci. Rep.*, (2015).
143. H. Zhao, S. Qiu, L. Wu, L. Zhang, H. Chen, C. Gao, Improving the performance of polyamide reverse osmosis membrane by incorporation of modified multi-walled carbon nanotubes. *J. Memb. Sci.*, (2014).
144. H. J. Kim, K. Choi, Y. Baek, D. G. Kim, J. Shim, J. Yoon, J. C. Lee, High-performance reverse osmosis CNT/polyamide nanocomposite membrane by controlled interfacial interactions, *ACS Appl. Mater. Interfaces*, (2014).
145. A. Tiraferri, C. D. Vecitis, M. Elimelech, Covalent binding of single-walled carbon nanotubes to polyamide membranes for antimicrobial surface properties, *ACS Appl. Mater. Interfaces* (2011).
146. J. N. Shen, C. C. Yu, H. M. Ruan, C. J. Gao, B. Van der Bruggen, Preparation and characterization of thin-film nanocomposite membranes embedded with poly(methyl methacrylate) hydrophobic modified multiwalled carbon nanotubes by interfacial polymerization, *J. Memb. Sci.*, (2013).

147. L. Zhang, G. Z. Shi, S. Qiu, L. H. Cheng, H. L. Chen, Preparation of high-flux thin film nanocomposite reverse osmosis membranes by incorporating functionalized multi-walled carbon nanotubes, *Desalin. Water Treat.*, (2011).
148. J. Farahbaksh, M. Delnavaz, V. Vatanpour, Investigation of raw and oxidized multiwalled carbon nanotubes in fabrication of reverse osmosis polyamide membranes for improvement in desalination and antifouling properties, *Desalination*, **410** (2017) 1–9.
149. I. Wan Azelee, P. S. Goh, W. J. Lau, A. F. Ismail, Facile acid treatment of multiwalled carbon nanotube-titania nanotube thin film nanocomposite membrane for reverse osmosis desalination, *J. Clean. Prod.*, (2018).
150. B. J. Hinds, N. Chopra, T. Rantell, R. Andrews, V. Gavalas, L. G. Bachas, Aligned multiwalled carbon nanotube membranes, *Science (80-.).*, **303**(5654) (2004) 62–65.
151. J. Lee, S. Jeong, Z. Liu, Progress and challenges of carbon nanotube membrane in water treatment, *Crit. Rev. Environ. Sci. Technol.*, **46** (2016) 99–1046.
152. K. Goh, H. E. Karahan, L. Wei, T. H. Bae, A. G. Fane, R. Wang, Y. Chen, Carbon nanomaterials for advancing separation membranes: A strategic perspective, *Carbon N. Y.*, **109** (2016) 694–710.
153. B. Lee, Y. Baek, M. Lee, D. H. Jeong, H. H. Lee, J. Yoon, Y. H. Kim, A carbon nanotube wall membrane for water treatment, *Nat. Commun.*, **6** (2015) (May 2014) 7109.
154. A. Sabir, M. Wasim, M. Shafiq, T. Jamil, *Carbon Nanotube and Graphene Oxide Based Membranes*; Elsevier Inc., 2019.
155. H. Y. Yang, Z. J. Han, S. F. Yu, K. L. Pey, K. Ostrikov, R. Karnik, Carbon nanotube membranes with ultrahigh specific adsorption capacity for water desalination and purification, *Nat. Commun.*, (2013).
156. X. Zhao, Y. Liu, S. Inoue, T. Suzuki, R. O. Jones, Y. Ando, Smallest carbon nanotube is 3 Å in diameter, *Phys. Rev. Lett.*, **92**(12) (2004) 12–14.
157. M. Carrillo-Tripp, M. L. San-Román, J. Hernańdez-Cobos, H. Saint-Martin, I. Ortega-Blake, Ion hydration in nanopores and the molecular basis of selectivity, *Biophys. Chem.*, **124**(3) (2006) 243–250.
158. C. F. de Lannoy, E. Soyer, M. R. Wiesner, Optimizing carbon nanotube-reinforced polysulfone ultrafiltration membranes through carboxylic acid functionalization, *J. Memb. Sci.*, **447** (2013) 395–402.
159. C. Song, B. Corry, Intrinsic ion selectivity of narrow hydrophobic pores, *J. Phys. Chem. B*, **113** (2009) 7642–7649.

160. B. Corry, Water and ion transport through functionalised carbon nanotubes: Implications for desalination technology, *Energy Environ. Sci.*, **4**(3) (2011) 751–759.
161. J. K. Holt, H. G. Park, Y. Wang, M. Stadermann, A. B. Artyukhin, C. P. Grigoropoulos, A. Noy, O. Bakajin, Fast mass transport through sub — 2-nanometer carbon nanotubes. *Science (80-.)*., **312** (2006) (May) 1034–1038.
162. S. Kang, M. Pinault, L. D. Pfefferle, M. Elimelech, Single-walled carbon nanotubes exhibit strong antimicrobial activity, *Langmuir*, **23**(17) (2007) 8670–8673.
163. S. Kang, M. Herzberg, D. F. Rodrigues, M. Elimelech, Antibacterial effects of carbon nanotubes: Size does matter! *Langmuir*, (2008) November, 6409–6413.
164. K. Busuttil, Single carbon nanotubes prick bacteria to death, *Mater. Today*, **12**(9) (2010) 9.
165. F. Yang, Q. Jiang, W. Xie, Y. Zhang, Effects of multi-walled carbon nanotubes with various diameters on bacterial cellular membranes: Cytotoxicity and adaptive mechanisms, *Chemosphere*, **185** (2017) 162–170.
166. J. Dong, Q. Ma, Advances in mechanisms and signaling pathways of carbon nanotube toxicity, *Nanotoxicology*, **9**(5) (2015) 658–676.
167. D. F. Rodrigues, M. Elimelech, Toxic effects of single-walled carbon nanotubes in the development of E. Coli biofilm, **44**(12) (2010) 4583–4589.
168. C. Yang, J. Mamouni, Y. Tang, L. Yang, Antimicrobial activity of single-walled carbon nanotubes: Length effect, *Langmuir*, **26**(20) (2010) 16013–16019.
169. C. Trellu, B. P. Chaplin, C. Coetsier, R. Esmilaire, S. Cerneaux, C. Causserand, M. Cretin, Electro-oxidation of organic pollutants by reactive electrochemical membranes, *Chemosphere*, **208** (2018) 159–175.
170. X. Zhu, A. V. Dudchenko, C. M. Khor, X. He, G. Ramon, D. Jassby, Field-induced redistribution of surfactants at the oil/water interface reduces membrane fouling on electrically conducting carbon nanotube UF membranes, *Environ. Sci. Technol.* (2018) acs.est.8b02578.
171. S. Pandit, S. Shanbhag, M. Mauter, Y. Oren, M. Herzberg, Influence of electric fields on biofouling of carbonaceous electrodes, *Environ. Sci. Technol.*, **51**(17) (2017) 10022–10030.
172. H. Liu, A. Vajpayee, C. D. Vecitis, Bismuth-doped tin oxide-coated carbon nanotube network: Improved anode stability and efficiency for flow-through organic electrooxidation, *ACS Appl. Mater. Interfaces*, **5**(20) (2013) 10054–10066.

173. L. Ma, X. Dong, M. Chen, L. Zhu, C. Wang, F. Yang, Y. Dong, Fabrication and water treatment application of carbon nanotubes (CNTs)-based composite membranes: A review, *Membranes (Basel).*, **7**(1) (2017).
174. J. Zheng, W. Zhang, X. Zhang, *Carbon Nanotubes for Advancing Separation Membranes*; Elsevier Inc., (2019).
175. S. K. Smart, A. I. Cassady, G. Q. Lu, D. J. Martin, The biocompatibility of carbon nanotubes, *Carbon N. Y.*, **44**(6) (2006) 1034–1047.
176. A. Jafar, Y. Alshatti, A. Ahmad, Carbon nanotube toxicity: The smallest biggest debate in medical care, *Cogent Med.*, **3**(1) (2016) 1–12.
177. P. Jackson, N. R. Jacobsen, A. Baun, R. Birkedal, D. Kühnel, K. A. Jensen, U. Vogel, H. Wallin, Bioaccumulation and ecotoxicity of carbon nanotubes, *Chem. Cent. J.*, **7**(1) (2013) 1–21.
178. A. Mottier, F. Mouchet, C. Laplanche, S. Cadarsi, L. Lagier, J. C. Arnault, H. A. Girard, V. León, E. Vázquez, C. Sarrieu, É. Pinelli, L. Gauthier, E. Flahaut, Surface area of carbon nanoparticles: A dose metric for a more realistic ecotoxicological assessment, *Nano Lett.*, **16**(6) (2016) 3514–3518.
179. W. Choi, J. Choi, J. Bang, J. H. Lee, Layer-by-layer assembly of graphene oxide nanosheets on polyamide membranes for durable reverse-osmosis applications, *ACS Appl. Mater. Interfaces*, **5**(23) (2013) 12510–12519.
180. Q. Liu, G. R. Xu, Graphene oxide (GO) as functional material in tailoring polyamide thin film composite (PA-TFC) reverse osmosis (RO) membranes, *Desalination*, **394** (2016) 162–175.
181. M. Safarpour, A. Khataee, V. Vatanpour, Thin film nanocomposite reverse osmosis membrane modified by reduced graphene oxide/TiO_2 with improved desalination performance, *J. Memb. Sci.*, **489** (2015) 43–54.
182. X. Song, Q. Zhou, T. Zhang, H. Xu, Z. Wang, Pressure-assisted preparation of graphene oxide quantum dot-incorporated reverse osmosis membranes: Antifouling and chlorine resistance potentials, *J. Mater. Chem. A*, **4**(43) (2016) 16896–16905.
183. X. Huang, K. L. Marsh, B. T. McVerry, E. M. V. Hoek, R. B. Kaner, Low-fouling antibacterial reverse osmosis membranes via surface grafting of graphene oxide, *ACS Appl. Mater. Interfaces*, **8**(23) (2016) 14334–14338.
184. H. R. Chae, J. Lee, C. H. Lee, I. C. Kim, P. K. Park, Graphene oxide-embedded thin-film composite reverse osmosis membrane with high flux, anti-biofouling, and chlorine resistance, *J. Memb. Sci.*, **483** (2015) 128–135.

185. F. Shao, L. Dong, H. Dong, Q. Zhang, M. Zhao, L. Yu, B. Pang, Y. Chen, Graphene oxide modified polyamide reverse osmosis membranes with enhanced chlorine resistance, *J. Memb. Sci.*, **525** (2017) (May 2016) 9–17.
186. C. Charcosset, A review of membrane processes and renewable energies for desalination, *Desalination*, **245**(1–3) (2009) 214–231.
187. S. Al-Amshawee, M. Y. B. M. Yunus, A. A. M. Azoddein, D. G. Hassell, I. H. Dakhil, H. A. Hasan, Electrodialysis desalination for water and wastewater: A review, *Chem. Eng. J.*, **380** (2020) (July 2019).
188. T. Xu, C. Huang, Electrodialysis-based separation technologies?: A critical review, *AIChE J.*, **54**(12) (2008) 3147–3159.
189. M. Sadrzadeh, T. Mohammadi, Sea water desalination using electrodialysis, *Desalination*, **221**(1–3) (2008) 440–447.
190. S. K. Adhikary, U. K. Tipnis, W. P. Harkare, K. P. Govindan, Defluoridation during desalination of brackish water by electrodialysis, *Desalination*, **71**(3) (1989) 301–312.
191. P. K. Narayanan, W. P. Harkare, S. K. Adhikary, N. J. Dave, D. K. Chauhan, K. P. Govindan, Performance of an electrodialysis desalination plant in rural area, *Desalinatioon*, **11**(103) (1985) 1984.
192. S. Mikhaylin, V. Nikonenko, N. Pismenskaya, G. Pourcelly, S. Choi, H. J. Kwon, J. Han, L. Bazinet, How physico-chemical and surface properties of cation-exchange membrane affect membrane scaling and electroconvective vortices: Influence on performance of electrodialysis with pulsed electric field, *Desalination*, **393** (2016) 102–114.
193. E. Oztekin, S. Altin, Wastewater treatment by electrodialysis system and fouling problems, *Tojsat*, **6**(1) (2016) 91–99.
194. H. J. Lee, M. K. Hong, S. D. Han, S. H. Cho, S. H. Moon, Fouling of an anion exchange membrane in the electrodialysis desalination process in the presence of organic foulants, *Desalination*, **238**(1–3) (2009) 60–69.
195. C. Jiang, M. M. Hossain, Y. Li, Y. Wang, T. Xu, Ion exchange membranes for electrodialysis: A comprehensive review of recent advances, *J. Membr. Sep. Technol.*, **3** (2014) 185–205.
196. H. Fan, Y. Huang, N. Y. Yip, Advancing the conductivity-permselectivity tradeoff of electrodialysis ion-exchange membranes with sulfonated CNT nanocomposites, *J. Memb. Sci.*, **610** (2020) 118259.
197. A. Alabi, A. AlHajaj, L. Cseri, G. Szekely, P. Budd, L. Zou, Review of nanomaterials-assisted ion exchange membranes for electromembrane desalination, *npj Clean Water*, **1**(1) (2018).

198. M. A. Andreeva, V. V. Gil, N. D. Pismenskaya, V. V. Nikonenko, L. Dammak, C. Larchet, D. Grande, N. A. Kononenko, Effect of homogenization and hydrophobization of a cation-exchange membrane surface on its scaling in the presence of calcium and magnesium chlorides during electrodialysis, *J. Memb. Sci.*, **540** (2017) (June) 183–191.
199. S. M. Hosseini, E. Jashni, S. Amani, B. Van der Bruggen, Tailoring the electrochemical properties of ED ion exchange membranes based on the synergism of TiO_2 nanoparticles-Co-GO nanoplates, *J. Colloid Interface Sci.*, **505** (2017) (September 2019) 763–775.
200. E. Jashni, S. M. Hosseini, J. N. Shen, B. Van der Bruggen, Electrochemical characterization of mixed matrix electrodialysis cation exchange membrane incorporated with carbon nanofibers for desalination, *Ionics (Kiel).*, **25**(11) (2019) 5595–5610.
201. A. E. Burakov, I. V. Burakova, E. V. Galunin, A. E. Kucherova, New carbon nanomaterials for water purification from heavy metals, *Handb. Ecomater.*, **1** (2019) 393–412.
202. S. M. Hosseini, M. Chehreh, E. Jashni, J. N. Shen, Electrochemical characterization of electrodialysis cation exchange membrane incorporated with graphite nanoparticle for deionization, *Ionics (Kiel).*, **26**(3) (2020) 1525–1535.
203. F. Heidary, A. R. Khodabakhshi, D. Ghanbari, Ionic transport properties improvement of a new cation-exchange membrane containing functionalized CNT as a clean technology for refining of saline-liquids, *Environ. Technol. (United Kingdom)*, **0**(0) (2019) 1–16.
204. S. M. Hosseini, S. S. Madaeni, A. R. Khodabakhshi, Preparation and characterization of PC/SBR heterogeneous cation exchange membrane filled with carbon nano-tubes, *J. Memb. Sci.*, **362**(1–2) (2010) 550–559.
205. S. M. Hosseini, F. Jeddi, M. Nemati, S. S. Madaeni, A. R. Moghadassi, Electrodialysis heterogeneous anion exchange membrane modified by PANI/MWCNT composite nanoparticles: Preparation, characterization and ionic transport property in desalination, *Desalination*, **341**(1) (2014) 107–114.
206. J. Yin, G. Zhu, B. Deng, Multi-walled carbon nanotubes (MWNTs)/polysulfone (PSU) mixed matrix hollow fiber membranes for enhanced water treatment, *J. Memb. Sci.*, **437** (2013) 237–248.
207. G. Liu, W. Jin, N. Xu, Graphene-based membranes, *Chem. Soc. Rev.*, **44**(15) (2015) 5016–5030.
208. S. Gahlot, P. P. Sharma, H. Gupta, V. Kulshrestha, P. K. Jha, Preparation of Graphene Oxide Nano-Composite Ion-Exchange Membranes for Desalination Application, *RSC Adv.*, **4**(47) (2014) 24662–24670.

209. Y. Zhang, L. Zou, Y. Wimalasiri, J. Y. Lee, Y. Chun, Reduced graphene oxide/polyaniline conductive anion exchange membranes in capacitive deionisation process, *Electrochim. Acta*, **182** (2015) 383–390.
210. S. M. Hosseini, E. Jashni, M. Habibi, M. Nemati, B. Van der Bruggen, Evaluating the ion transport characteristics of novel graphene oxide nanoplates entrapped mixed matrix cation exchange membranes in water deionization, *J. Memb. Sci.*, **541** (2017) (July) 641–652.
211. S. Chaudhury, A. K. Thakur, R. S. Gojman, C. J. Arnusch, O. Nir, Ion transport in laser-induced graphene cation-exchange membrane hybrids, *J. Phys. Chem. Lett.*, **11**(4) (2020) 1397–1403.
212. S. M. Hosseini, E. Jashni, M. Habibi, B. Van der Bruggen, Fabrication of novel electrodialysis heterogeneous ion exchange membranes by incorporating PANI/GO functionalized composite nanoplates, *Ionics (Kiel).*, **24**(6) (2018) 1789–1801.
213. L. Cseri, J. Baugh, A. Alabi, A. AlHajaj, L. Zou, R. A. W. Dryfe, P. M. Budd, G. Szekely, Graphene oxide-polybenzimidazolium nanocomposite anion exchange membranes for electrodialysis, *J. Mater. Chem. A*, **6**(48) (2018) 24728–24739.
214. A. Alabi, L. Cseri, Al Hajaj, A. G. Szekely, P. Budd, L. Zou, Electrostatically-coupled graphene oxide nanocomposite cation exchange membrane, *J. Memb. Sci.*, **594** (2020) (July 2019) 117457.
215. Y. Li, Z. Ding, J. Li, J. Li, T. Lu, L. Pan, Highly efficient and stable desalination via novel hybrid capacitive deionization with redox-active polyimide cathode. *Desalination*, **469** (2019) (May) 114098.
216. C. Kim, P. Srimuk, J. Lee, S. Fleischmann, M. Aslan, V. Presser, Influence of pore structure and cell voltage of activated carbon cloth as a versatile electrode material for capacitive deionization, *Carbon N. Y.*, **122** (2017) 329–335.
217. S. M. Hosseini, H. Alibakhshi, A. Khodabakhshi, M. Nemati, Enhancing electrochemical performance of heterogeneous cation exchange membranes by using super activated carbon nanoparticles, *J. Pet. Sci. Technol.*, **8**(3) (2018) 14–29.
218. M. Ebrahimi, B. Van der Bruggen, S. M. Hosseini, M. Askari, M. Nemati, Improving electrochemical properties of cation exchange membranes by using activated carbon-co-Chitosan composite nanoparticles in water deionization, *Ionics (Kiel).*, **25**(3) (2019) 1199–1214.
219. L. Wang, S. Lin, Mechanism of selective ion removal in membrane capacitive deionization for water softening, *Environ. Sci. Technol.*, **53**(10) (2019) 5797–5804.

220. G. Luo, Y. Wang, L. Gao, D. Zhang, T. Lin, Graphene bonded carbon nanofiber aerogels with high capacitive deionization capability, *Electrochim. Acta*, **260** (2018) 656–663.
221. F. A. AlMarzooqi, Al Ghaferi, A. A. I. Saadat, N. Hilal, Application of Capacitive Deionisation in Water Desalination: A Review, *Desalination*, **342** (2014) 3–15.
222. J. Lee, P. Srimuk, K. Aristizabal, C. Kim, S. Choudhury, Y. C. Nah, F. Mücklich, V. Presser, Pseudocapacitive desalination of brackish water and seawater with vanadium-pentoxide-decorated multiwalled carbon nanotubes, *ChemSusChem*, **10**(18) (2017) 3611–3623.
223. J. B. Lee, K. K. Park, H. M. Eum, C. W. Lee, Desalination of a thermal power plant wastewater by membrane capacitive deionization, *Desalination*, **196**(1–3) (2006) 125–134.
224. J. Lee, S. Kim, C. Kim, J. Yoon, Hybrid capacitive deionization to enhance the desalination performance of capacitive techniques, *Energy Environ. Sci.*, **7**(11) (2014) 3683–3689.
225. Y. Oren, Capacitive deionization (CDI) for desalination and water treatment — past, present and future (a review) *Desalination*, **228**(1–3) (2008) 10–29.
226. T. J. Welgemoed, C. F. Schutte, Capacitive deionization technology^{TM}: An alternative desalination solution. *Desalination*, **183**(1–3) (2005) 327–340.
227. Z. Chen, H. Zhang, C. Wu, Y. Wang, W. Li, A study of electrosorption selectivity of anions by activated carbon electrodes in capacitive deionization, *Desalination*, **369** (2015) 46–50.
228. C. L. Yeh, H. C. Hsi, K. C. Li, C. H. Hou, Improved performance in capacitive deionization of activated carbon electrodes with a tunable mesopore and micropore ratio, *Desalination*, **367** (2015) 60–68.
229. G. Wang, B. Qian, Q. Dong, J. Yang, Z. Zhao, J. Qiu, Highly mesoporous activated carbon electrode for capacitive deionization, *Sep. Purif. Technol.*, **103** (2013) 216–221.
230. G. Wang, C. Pan, L. Wang, Q. Dong, C. Yu, Z. Zhao, J. Qiu, Activated carbon nanofiber webs made by electrospinning for capacitive deionization, *Electrochim. Acta*, **69** (2012) 65–70.
231. M. W. Ryoo, G. Seo, Improvement in capacitive deionization function of activated carbon cloth by titania modification, *Water Res.*, **37**(7) (2003) 1527–1534.
232. Y. Liu, C. Nie, X. Liu, X. Xu, Z. Sun, L. Pan, Review on carbon-based composite materials for capacitive deionization, *RSC Adv.*, **5**(20) (2015) 15205–15225.

233. P. Liu, T. Yan, L. Shi, H. S. Park, X. Chen, Z. Zhao, D. Zhang, Graphene-based materials for capacitive deionization, *J. Mater. Chem. A*, **5**(27) (2017) 13907–13943.
234. H. Yoon, J. Lee, S. Kim, J. Yoon, Hybrid capacitive deionization with Ag coated carbon composite electrode. *Desalination*, **422** (2017) (May) 42–48.
235. X. Gao, S. Porada, A. Omosebi, K. L. Liu, P. M. Biesheuvel, J. Landon, Complementary surface charge for enhanced capacitive deionization, *Water Res.*, **92** (2016) 275–282.
236. R. W. Pekala, J. C. Farmer, C. T. Alviso, T. D. Tran, S. T. Mayer, J. M. Miller, B. Dunn, Carbon aerogels for electrochemical applications, *J. Non. Cryst. Solids*, **225**(1–3) (1998) 74–80.
237. G. Rasines, P. Lavela, C. MacÍas, M. Haro, C. O. Ania, J. L. Tirado, Electrochemical response of carbon aerogel electrodes in saline water, *J. Electroanal. Chem.*, **671** (2012) 92–98.
238. L. Zou, L. Li, H. Song, G. Morris, Using mesoporous carbon electrodes for brackish water desalination, *Water Res.*, **42**(8–9) (2008) 2340–2348.
239. Z. U. Khan, T. Yan, L. Shi, D. Zhang, Improved capacitive deionization by using 3D intercalated graphene sheet-sphere nanocomposite architectures, *Environ. Sci. Nano*, **5**(4) (2018) 980–991.
240. M. Mossad, L. Zou, A study of the capacitive deionisation performance under various operational conditions. *J. Hazard. Mater.*, **213–214** (2012) 491–497.
241. J. Yang, L. Zou, H. Song, Preparing MnO 2/PSS/CNTs composite electrodes by layer-by-layer deposition of MnO 2 in the membrane capacitive deionisation, *Desalination*, **286** (2012) 108–114.
242. C. Yan, Y. W. Kanaththage, R. Short, C. T. Gibson, L. Zou, Graphene/polyaniline nanocomposite as electrode material for membrane capacitive deionization, *Desalination*, **344** (2014) 274–279.
243. J. J. Wouters, J. J. Lado, M. I. Tejedor-Tejedor, R. Perez-Roa, M. A. Anderson, Carbon fiber sheets coated with thin-films of SiO_2 and γ-Al_2O_3 as electrodes in capacitive deionization: Relationship between properties of the oxide films and electrode performance, *Electrochim. Acta*, **112** (2013) 763–773.
244. Y. Bai, Z. H. Huang, X. L. Yu, F. Kang, Graphene oxide-embedded porous carbon nanofiber webs by electrospinning for capacitive deionization, *Colloids Surfaces A Physicochem. Eng. Asp.*, **444** (2014) 153–158.

CHAPTER 16

Nanomaterials for Pressure Retarded Osmosis

Arvin Shadravan, Pei Sean Goh, and Ahmad Fauzi Ismail*

Universiti Teknologi Malaysia

16.1. Introduction

In the 21st century, human faces a lot of fundamental challenges: energy, water, and food must be made available in increasing amounts for the world's growing population. The quest is open to the broadest exploitation of novel solutions that could turn soon into new, sustainable ways to guarantee the availability of such life-essential items. It should be further observed that the water-energy-food nexus clearly indicates a loop chain where water and energy are both fundamental for food production, water is needed for energy production, and energy is needed for water supply and production (e.g. desalination). Han *et al.* reported that energy consumption globally will rise by about 56% in the year 2040 and the energy usage will increase up to 240 kTWh[1] Renewable energy can be harnessed from different sources: solar, wind, geothermal, biomass, hydro, tidal, wave, and marine currents energy. Beyond these, a lesser-known form of renewable energy is the so-called salinity gradient energy (SGE) or salinity gradient power (SGP).[2,3]

*Corresponding author: afauzi@utm.my

16.2. Energy Generation through Salinity Gradient

Osmotic energy (or salinity-gradient energy) is the energy released when water with different salinities are mixed, like rivers and oceans. By the passage of time, the tremendous potential regarding power generation by the means of salinity gradients has been investigated through many years. Besides, pressure retarded osmosis (PRO) has improved to have a promising role as an approach for energy extraction.[4] Developing osmotic processes that are driven by the membrane may supply sustainable solutions for the needs of either clean water or clean energy. Osmotic processes use the difference of osmotic pressure solutions through a semipermeable membrane for drawing water from a diluted feed solution to a draw solution which is more concentrated.[5] The most efficient advantage of producing energy by the PRO process is that it can be continuous and non-stop and provide constant energy production. Although there are different kinds of methods that are available for energy generation through salinity gradients which contain PRO and RED, the most power generation was harvested by the PRO process.[6]

If two solutions with various sorts of molecular structures such as seawater and river water connect, they will mix instantly and produce energy. Therefore, the concentration of salt is prone to produce energy. Osmotic pressure has been widely used for harvesting energy among renewable energies. In the following, free Gibbs energy of mixing is one of the ways to produce energy.[7]

$$\Delta G_{\text{mix}} = \text{Frictional Losses} + \text{Unutilized energy} + \text{Extractable Work} \quad (1)$$

According to the function of RO and electrodialysis (ED), the PRO process and RED process functions vice versa. Through PRO, molecules of water will spread from a low salinity towards the pressurized and high salinity solution through a membrane. The concentrated solution will go across a hydro-turbine to produce power.[8]

In the process of RED, ion exchange membranes will separate the two kinds of solutions with high salinity and low salinity.[9] The different concentrations of these two solutions will be the reason to spread counter-ions through these membranes. Ion exchange membrane which consists of anion and cation will separate so many channel pairs with high salinity and low salinity during RED.[8] Besides, the major improvements for PRO is about increasing the water permeability and simultaneously decreasing the salt permeability for membranes while in RED the conduction will be increased and there is no need for decreasing perm selectivity (decreasing the thickness of the membrane) and the channel height is enormously decreased. To generate power from salinity gradients, PRO is in better condition according to the power density calculations.[10]

The result of comparison between PRO and RED with each other is that in RED there is more power density production for the low salinity gradients such as facing river water and seawater through a membrane. FO membranes make PRO more trustworthy method and with higher potentiality. The mixing of water and seawater has a power density of about $2.3\,W/m^2$ while in RED the power density is $1.3\,W/m^2$.[11]

Osmotic energy will be produced after the process of mixing feed solution and draw solution with each other.[12] PRO has two significant parts that are PRO membrane module and the turbine by which can generate power.[13] Generally, the structure of PRO and RO membranes are not different from each other but the porosity in PRO membranes is more than RO membranes. Besides, the PRO membrane should have more stability compare to other kinds of membranes in the high concentration solution side because in this side pressure will be increased up to 30 bar and it is related to the osmotic pressure of high concentration solution. Figure 16.1 shows the schematic of this process.[14]

When fresh water left the membrane and passes through the turbine to generate electricity then it will go back to the sea or preserved with a membrane or thermal functionality for reproduction.[15]

Using seawater in the PRO process has some advantages and disadvantages. The first important benefit that it has, is the availability of seawater and the second one is having the free high osmotic pressure. The disadvantages are known as pumping and

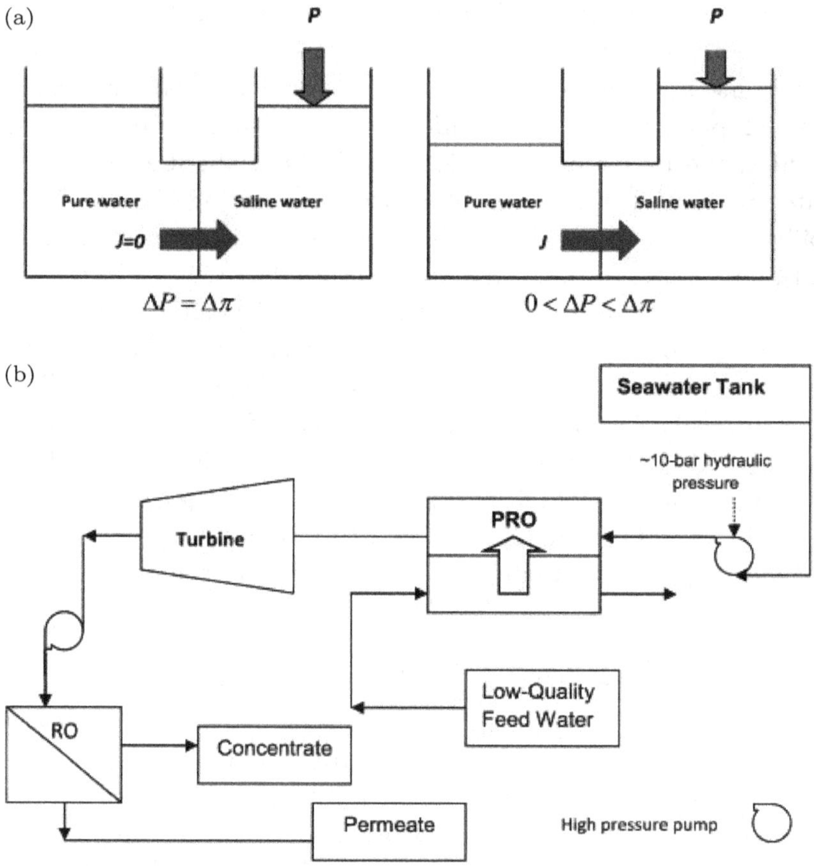

Fig. 16.1. PRO (a) the procedure perception (b) PRO-RO strategy.[14]

the pretreatment.[14] The salt concentration of the seawater is very effective for generating energy while through PRO function the power that was generated after pumping and pre-treatment were 0.2 kWh/m^3.[16]

To calculate the water flux through the PRO process we need to calculate J_w which is the sign of water flux and is shown by the following equation[17]:

$$\text{Jw} = \text{Aw}(\Delta\pi - \Delta P) \quad (2)$$

In this equation, A_w is membrane permeability coefficient ($L/m^2h \cdot bar$) and ΔP is differential feed pressure (bar) and $\Delta\pi$ is the osmotic pressure (bar).[18]

16.3. Pressure Retarded Osmosis

During the period of the osmosis process, permeate water rapidly diluted draw solution, when the feed solution is being concentrated. The hydrostatic pressure is entitled to osmotic pressure such as water transportation passing through the membrane while it is functionalized to draw solution. At any step, while ΔP can be between zero to $\Delta\pi$, water can flow into the saline water because $\Delta\pi$ (Osmotic pressure) is still larger than the pressure which is ΔP. This process is entitled to PRO.[19]

In PRO, two types of solutions are used which the first one is a draw solution that is containing salinity about 0.5M to 2M as seawater and the second one is a feed solution that used RO water to simulate it somehow as river water. In this process, before passing through the membrane, the low concentration solution is pressurized.[20]

PRO was conceptualized for producing salinity gradient energy by[21] in the 1950s and then reinvestigated in the mid1970s due to the world's energy crisis.[22] Loeb and his coworkers conducted theoretical and experimental research on the feasibility of PRO.[23] It was reported that osmotic energy could indeed be harvested under the PRO principles. The power density of PRO membranes is highly dependent on osmotic pressure and water flux. The mechanical stability at high hydraulic pressures is, therefore, one concern for PRO membranes.

16.3.1. *Benefits, drawbacks and challenges in PRO*

It was predicted that all the potential energy sources can harvest almost 2000 TWh every year of power generation that can be more than 10% for the existing demand for energy in the world.

The essential requirements in the membrane for the PRO process are:

(i) high water flux,
(ii) moderate tolerance to pH change and chlorine attack,
(iii) high mechanical strength and stability and
(iv) low internal concentration polarization effect.

Hence, there are several highlights for the significant osmotic process through PRO technology which are as follows:

(a) energy generation through salinity gradient,
(b) basic PRO principles and performance interrelated with the highlights of power density,
(c) novel and promising PRO membranes,
(d) viability and perspectives about the future of the PRO process.

According to osmotic pressure, river water crosses in the same way as same as osmotic pressure because of dilution of draw solution. In addition, the low concentration solution will be depressurized in a turbine for energy production. PRO can be either beneficial or detrimental in industry. Some of the advantages of PRO are as follows: First, the operation of PRO can function 24 hours a day. Secondly, the speed of the wind and the radiation of the sun has no influence on PRO. The third benefit is having a minor footprint and the last is the facilitated scale-up. In addition, if PRO and RO brine matches with each other then it will be good for RO brine because RO brine can be used as both solutions with high concentration and low concentration such as river water and seawater.[24] This characteristic is one of a kind for PRO processes. Moreover, wastewater can be used through PRO as a feed solution. Otherwise, PRO has some limitations that have not been ready yet commercially.[14]

Through PRO, some challenges will be mentioned in the following sentences. The mode of PRO process transport is based on either side of membranes. PRO membranes have a significant role in delivering power density. Due to high salinity in draw region leads to lower water flux which in turn reduce the production of power density. It is mainly due to the adsorption of solutes on the membrane surface. Thus, the PRO membrane should be of thin, hydrophilic and resistant to salt adsorption. Moreover, the salt concentration of the seawater is very effective for generating energy while through PRO function the power that were generated after pumping and pre-treatment was 0.2 kWh/m^3.[14]

Osmotic potential is the main driving force to produce power density under the PRO process. The other challenging part of the PRO process is while increasing water flux (water permeability), the salt permeability will increase too. So finding the optimum point in which water flux and salt permeation has a lot of differences in a range is strongly recommended.[9]

The last challenge is the fouling of PRO membranes. Fouling is one of the major characteristics that results in the decline in the water flux of the membrane. The hydraulic pressure during PRO processes is less than RO processes so membranes that are used for PRO are less likely to foul. Experiments show that cellulose acetate (CTA) membranes are more tolerant of fouling than thin film composite (TFC) membranes because of the CTA which has higher surface hydrophilicity. The structure of the membrane is a very significant characteristic that has an influence on fouling. Especially, the structure of porosity that is coherent to the thickness. Membrane adjustment plays a key role in the trend of fouling.[5]

Membranes with higher selectivity and salt permeability, B, suffered from a lower water permeability and would yield a lower peak power density, while membranes with higher permeability and lower selectivity performed poorly due to severe reverse salt permeation, resulting in a similar projected peak power density. Somehow these days FO membranes conquer this obstacle with decreasing the thickness of the membrane by which is thinner than RO membranes. Furthermore, the problem of FO membranes can be determined as not being so stable under high pressure during PRO processes.[25] The comparison between RO, FO and PRO showed in Fig. 16.2 as followed:

It is worth bearing in mind that PRO has several characteristics which should be compared with other industrialized analyses such as RO and FO. Therefore, in order to conduct research into PRO, all

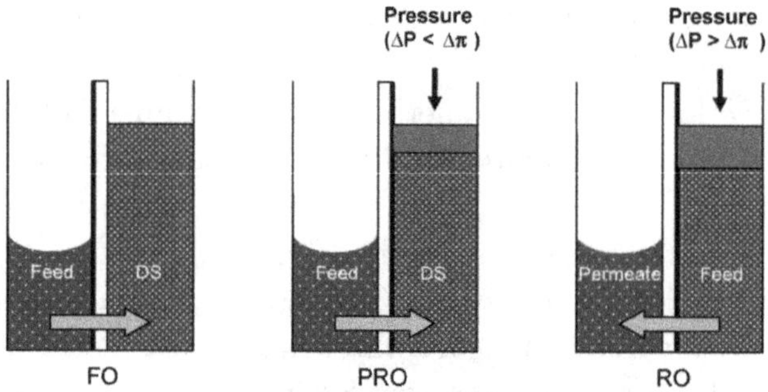

Fig. 16.2. Comparison between RO, FO, and PRO according to $\Delta\pi$ and ΔP.[5]

the merits and demerits of it should be widely discussed which was done in Table 16.1 as follows:

16.3.2. *Internal Concentration Polarization (ICP) impacts on PRO process*

During the usage of the flat sheet RO membrane, the big problem is the shortage of membrane flux which will occur due to the concentration polarization. This will come up with the other problem that is prone to occur which is decreasing osmotic pressure. To be clear, concentration polarization has two kinds: External and internal. Every one of them will happen on one side of the membrane. ICP happens on the feed side but External Concentration Polarization happens on the side which contains draw solution. To decrease the influence of ICP through PRO, draw solution should contact the active layer. Although the connection between the draw solution and active layer supplies more potentiality to withstand in the PRO membranes, delamination will occur if the draw solution is in contact with the support layer.[27]

The membrane development of PRO is one of the most controversial issues that is faced during this process. Through this process gaining high water flux and power density are the most significant part in addition to characteristics of either RO or FO membranes, respectively. Osmotic pressure is equal to the driving force through PRO so the characteristic of the PRO membrane depends on three parameters.[5]

Table 16.1. The comparison between RO, FO, and PRO.[26]

	RO	FO	PRO
Driving Force	• Applying extra pressure (P)	• Osmotic pressure	• Osmotic pressure
Main Application	• Water filtration system • Desalination	• Water filtration system • Desalination	• Harvesting energy
Operating Condition	• P ∼ 10-70 • Brackish water and seawater feed solution • pH 6–7	• P ∼ atmospheric • Brackish, seawater, or some synthetic draw solutions • Impaired water, seawater or other feed solution • pH 6–11	• P ∼ 10–15 bar • River, brackish, seawater, and brine • solution • pH 6–7
Desirable Membrane Property **(1) Physical Morphology**	• Dense upper-layer and porous sub-layer • Thermally and mechanically stable	• Thin films with dense active layer and less torturous sub-layer	• Thin films with dense active layer and less torturous sub-layer
(2) Chemical Property	• Chemically stable to NaCl solution	• So hydrophilic • Chemically stable to chloride solution and synthetic draw solution	• So hydrophilic

(*Continued*)

Table 16.1. (*Continued*)

	RO	FO	PRO
(3) Membrane Requirement	• High water permeability • High solute • Chemically robust for higher pressure	• High water permeability • High solute retention • High stability in synthetic draw solution	• High water permeability • High solute retention to gain osmotic pressure driving force • Strong enough for the external applied pressure
Target Performance	• High flux (almost 4–5 $\mu m\,s^{-1}$)	• High flux and good water recovery	• High power density
Challenges	• Consuming energy • Process price	• Internal Concentration Polarization • Draw solution recovery	• Internal Concentration Polarization • Module design • Membrane cleaning • Feed stream pre-treatment

The general equation to estimate the PRO membrane water flux is Eq (2). Salt permeates across the membrane in the opposite direction of the water flux, from the more concentrated draw solution into the feed solution. The reverse salt flux, Js, is described by[28]:

$$\mathrm{Js} = B(\mathrm{C_{D,m}}, \mathrm{m} - \mathrm{C_{F,m}}) \tag{3}$$

where B is the salt permeability coefficient of the membrane, the active layer, and $C_{D,m}$ and $C_{F,m}$ are the solute concentrations in the solution at the active layer interface on the draw and feed sides.

16.3.3. *PRO process design at estuaries*

One of the most common approaches for considering applications regarding the PRO process is to estimate the net energy which can be extracted at estuaries that rivers meet the sea. The amount of global river discharge is about 37,300 km^3 annually. If dilution of the river water is into the seawater, the potentiality to harvest energy is about 27,200 TW annually (which is equivalent to the continuing power output of 3.1 TW). This is a tremendous amount for energy annually which is feasible to be 20% larger than the global energy harvesting in 2012.[4] The schematic of the PRO process, while the seawater faces the river water, is as follows:

The river source allows the feed solution to be pumped by it, undertakes pretreatment, consequently it permeates through the module of the membrane. The ocean is the source for seawater draw solution to be pumped. The seawater draw solution is pumped in from the ocean, forced to the pretreatment, and then passes across the pressure exchanger (PEX) before passing through the membrane module. The highest energy that was extracted from the system is 0.256 kWh per cubic meter of the entire draw and feed solutions which is equivalent to the Gibbs free energy of mixing. It was estimated that inputs of the leading energy of the system for pumping the water in, pretreating the water, which is influent, pumping the solutions to pass across modules of the membranes are stated along with losses by the operations which are constant-pressure counter-current, concentration polarization (CP), reverse salt flux

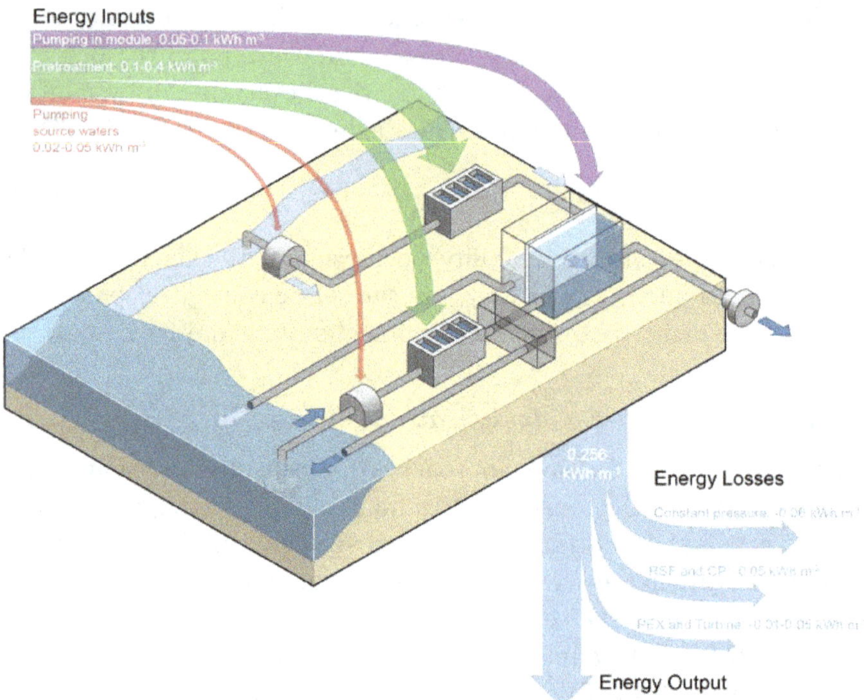

Fig. 16.3. Schematic diagram of the PRO system during the process of mixing seawater and river water.[4]

(RSF) and PEX and inefficiencies of the turbine.[4] Considering the energetic inputs, it will be identified that the inputs will surpass the amount of energy that is feasible to harvest. On the other hand, it is required to achieve revolutionary advancements that contradict the requirement for pretreatment in addition to diminish inefficiencies severely.[4]

16.3.4. *PRO for power generation and desalination*

Sidney Loeb is the first person that used the PRO process with two solutions of high concentration and low concentration in order to produce electricity. Through PRO, river water flows through the way of osmotic pressure or pass the semi-permeable membrane from low concentration to high concentration. To generate electricity from the turbine, the draw solution will be depressurized to modify hydraulic energy. The aim of Sidney Loeb is to check the availability of power production through PRO. To make sure that PRO can produce power, he used freshwater and RO brine as a feed solution and chose Dead-Sea water as the draw solution. Using RO brine for feed solution has two advantages; First, some resources in dry areas are rare like river water and the second one is during the RO brine usage will not deplete the evaporation rate and will refill it.[13]

In the closed-loop PRO process, heat waste is used to produce electricity and PRO has two flows which are flowing in a distillate way as a feed solution and the brine concentrate flow which is used as a draw solution. The benefit of this method is having the zero brine waste release but its availability is due to the feasibility of waste heat which is shown in Fig. 16.4.[14]

For being more cost-effective there is another available method because of using seawater and river water which is an open loop that is shown in Fig. 16.5.

16.3.5. *PRO membranes*

The membrane works as a wall to filter or separate in processes. Membrane technology is a process that has a wide area to search scientifically and covers all the approaches of engineering. Permeable and semi-permeable membranes will allow the substances to pass through them for filtration or separation.[14] Loeb published a disappointing PRO performance when RO membranes were utilized in the PRO module.[29] Concentration polarization was criticized for the underperformance of the PRO process. Loeb and co-workers recognized that the performance of PRO can be developed by decreasing the membrane thickness, for example, by eliminating the support fabric layer of the commercial brackish water reverse osmosisi (BWRO) membrane.[23]

The improvement of the osmotic characterization of the semi-permeable membrane is still one of the significant obstacles through PRO. One of the major issues is related to the low water flux that has consequently resulted in unpromising osmotic power. The desired characteristic of PRO-based on the TFC membrane should be of

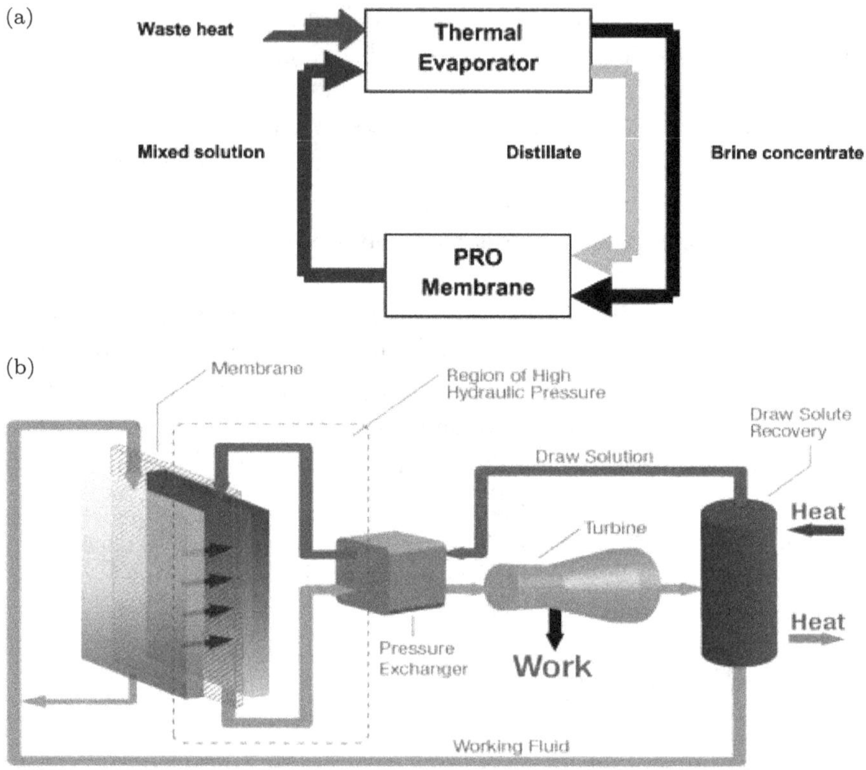

Fig. 16.4. The second kind of PRO process which is closed-loop, (a) Closed-loop method by Sidney Loeb, (b) Closed-loop by McGinnis.[14]

a highly hydrophilic and thin layer. The conventional membranes have high water flux with lower salt rejection properties. It increases the fouling in PRO operation which ultimately deteriorates the harvesting of energy in longer periods.

PRO is an emerging membrane process used for producing power density and desalination applications. It is due to its characteristics that can facilitate, capture and convert the free Gibbs energy or mixing into the useful energy resources. Salinity gradient energy is a zero-emission and sustainable technology that can be practically applied worldwide. PRO is considered a very clean technology which is because of no byproduct formation and no CO_2 emission.[26] For instance, there is potentiality in energy resources in every estuary

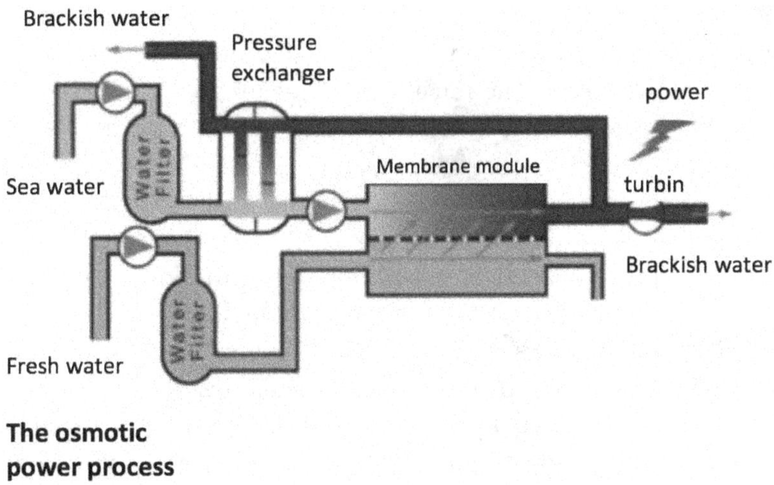

Fig. 16.5. Open-loop system using seawater and freshwater that is more cost-effective.[14]

where the river water faces the seawater or in each desalination plants that release concentrated brines in the sea.[1]

Thin-film nanocomposite (TFN) membranes are widely used in recent years for desalination applications. It is mainly due to the achievement of high flux without compromising the salt rejection. In this regard, nanomaterials such as metal oxides and carbon have gained more attention to improve membrane performance. It can improve the characteristic properties such as hydrophilicity and strength of the material. Hydrophilicity is the main required property in attaining the higher water flux.[30] The conventional TFC RO membranes are dense skin layer, which aids in high salt retention and lower flux. Conversely, the PRO membrane should be of thin skin for the achievement of high flux. PRO-based on the mechanism of the osmotic pressure gradient. Thus, the design of the TFN membrane is the prerequisite for achieving high-power density.

16.3.5.1. *Thin Film Composite (TFC) membranes*

TFC membranes generally possess asymmetric porous support and a top selective skin. The chemistry used in the manufacture of the membrane was basically related to the interfacial polymerization of trimesoylchloride and m-phenylene diamine. The essential difference between reverse and a direct osmosis membrane is the following: a reverse osmosis membrane must withstand pressures of 60 or 80 bar and requires, therefore, a compaction resistant sub-structure. In RO water and salt are flowing in the same direction. The microporous support provides the mechanical strength, while the selective layer performs the separation. In the PRO process described here, the pressures are much lower, and salt and water are flowing in opposite directions. The thickness of the polyamide layer is in the range of hundred nanometers due to the self-terminating nature of the interfacial cross-linking reaction. The advantages of fabricating TFC membranes via interfacial polymerization are that the structure and properties of the substrate and the selective layer can be individually tailored and optimized to achieve desired permeability and salt rejection.[31]

TFC has been widely used for PRO processes to generate power. Thin Film composites are advantageous during PRO processes. As an illustration, these beneficial characteristics can be mentioned have high water flux, mechanically robust and are vigorous to perform well in a wide range of pH. Otherwise, they have some disadvantages too. To illustrate, one of the most challenging obstacles is that they are not enough tolerant to face oxidants and chemicals that contain chlorine.[32] Fabrication of these membranes is possible in the laboratory which is in a similar way to fabricate RO membranes while the substrate and active layer will be prepared by phase inversion and can be done by interfacial polymerization.[33] Hydration Technology Innovations (HTI) is the company that supplies commercial TFC membranes that has high water permeability around 36 L/m^2h using tap water. In addition, the feed and the draw solution are 1M NaCl. Approximately, TFC membranes perform better for having higher water permeability than CTA which is shown in Table 16.2.[14]

However, the salinity gradient resource has an influence on the water flux of the membrane. If the driving force through the membrane increases, then simultaneously the water flux enhances as well. To illustrate, if the water flux goes down from 33 L/m^2h to 15 L/m^2h

Table 16.2. Water permeability through HTI CTA membranes and TFC membranes (* — All the TFC membranes are fabricated by MPD and TMC solution).

Feed solution	Draw solution	Water flux L/m^2h	Membrane orientation	Membrane type	References
DI water	0.5M NaCl	18.6	FO	HTI CTA membrane	34
DI water	1 M NaCl	16.8	FO	HTI CTA membrane	35
50Mm NaCl	4 M NaCl	37.8	PRO	HTI CTA membrane	36
50Mm NaCl	4 M NaCl	27	FO	HTI CTA membrane	36
Tap water	0.5 M NaCl	10.3	FO	HTI CTA membrane	37
DI water	0.5 M NaCl	32.2	PRO	TFC membrane*	36
DI water	1.5 M NaCl	18	FO	TFC membrane*	17
DI water	2 M NaCl	47.5	PRO	TFC membrane*	38
0.5M NaCl	2 M NaCl	11.8	FO	TFC membrane*	38
DI water	2 M NaCl	33	PRO	TFC membrane*	39
3.5% M NaCl	2 M NaCl	1	PRO	TFC membrane*	39

then there is an enhancement for TDS (Total Dissolved Solids) of feed solution that will change from 0 to 3.5% NaCl (Table 16.2) because of much more internal concentration polarization at higher feed solution TDS. From Table 16.2, the results showed a higher water flux can be achieved when the membrane is operating on the PRO mode than on the FO mode. Furthermore, the results in Table 16.2 show that the average water flux in the TFC membranes is higher than in the CTA membranes. Therefore, using the TFC membrane in the PRO process is desirable to increase the membrane flux and hence power density. Table 16.3 shows the TFC membranes which are performing better than CTA and has higher water permeability, power density for generation electricity.[14] However, the controversial issue nowadays about PRO processes is that most of the membranes are tested in laboratories and have not tested in a wide range.[14]

Table 16.3. Result of performing in PRO processes for HTI CTA membranes.

Feed solution	Draw solution	Hydraulic pressure bar	Power density W/m^3	References
0.06% NaCl	Seawater	5	5.7	40
0.2% NaCl	Seawater	8.9	5.6	40
0.5% NaCl	Seawater	8.9	4.1	40
0.06% NaCl	Seawater	12	2.7	41
0.06% NaCl	0.06% NaCl	8.4	11	42
0.23% NaCl	0.06% NaCl	9	10.6	42
0.5% NaCl	0.06% NaCl	9.1	8.4	42
0.06% NaCl	Seawater	9.6	70.4	43
0.06% NaCl	7% NaCl	19.4	30.6	43
0.06% NaCl	4% NaCl	11	9.8	43
0.06% NaCl	24% NaCl	67.2	367	43
DI Water	0.59M NaCl	13	9	44
DI Water	1 M NaCl	15	12	44
10mM NaCl	1 M NaCl	15	10	44
80mM NaCl	1 M NaCl	15	7.5	44
40mM NaCl	1 M NaCl	9	10.6	45
80mM NaCl	1 M NaCl	9.1	8.4	45
40mM NaCl	1 M NaCl	5.1	6.2	45
80mM NaCl	1 M NaCl	5	4.7	45
0.06 NaCl	Seawater	12	5.8	17

16.3.5.2. *Thin Film Nanocomposite (TFN) membranes*

Thin-film nanocomposite (TFN) membrane is a novel type of composite membranes developed via interfacial polymerization (IP) process. Nanoparticles are consolidated within the thin polyamide (PA) dense layer of the thin-film composite (TFC) membrane with the intention of enhancing the characteristics of the interracially polymerized layer. Whether nanocomposite membranes could achieve one or two of the performance enhancements, such as the balance of permeability and selectivity, and nanomaterials incorporated into polymeric membranes for water treatment. Firstly, 0D nanoparticles (TiO_2, Al_2O_3, ZrO_2, SiO_2, ZnO, AgO, etc.), consisting of metal oxides and metals, etc. are usually introduced into polyamide layer to enhance the surface hydrophilicity of membranes. Secondly, in addition to the hydrophilicity of nanoparticles, water transport channels of 1D nanotubes (single/multiwall carbon nanotubes, titanic nanotubes, etc.) also facilitate the permeability and fouling resistance of polymeric membranes. On the other hand, carbon nanotubes could improve the mechanical strength of polymeric membranes in the application of water treatment. Furthermore, 2D silicate sheets (clays) are also adopted to improve the hydrophilicity/hydrophobicity and mechanical strength of water filtration membranes. Finally, the 3D nanomaterials are porous according to their cage-like structure. For example, to improve the hydrophilicity and/or surface charge density, without reducing the separation efficiency of the TFC membrane.

Hydrophilicity is the chief determining parameter of the performance of TFN membranes. As a matter of fact, many types of inorganic nanomaterials have been incorporated with the polyamide layer. These nanoparticle fillers are commonly adopted to varieties of dimensional nanoscale (0D, 1D, 2D, 3D) materials. As an illustration, 3D zeolites are microporous crystalline aluminosilicates with well-defined nanoscale pore structures in their regular frameworks. The hydrophilicity and permeability of 3D zeolites facilitate the permeability/selectivity and fouling resistance of ultrafiltration and reverse osmosis membranes.[46]

Until now, 0D and 1D nanomaterials are prevalently used in the fabrication of TFN membranes. Works of literature on modification of TFN by 3D zeolites are sparsely reported. 3D Nanomaterials have an advantage as higher surface area, water absorption capacity and assembly on the polyamide layer. Thus, the current study is dedicated to investigating the zeolite incorporated in the TFN membrane for PRO application. The foremost approach, unlike other studies, which have dealt with commercial TFC membrane for the power generation. Zeolite is a 3D nanomaterial which is highly hydrophilic as well as the salt rejection properties. It is mainly due to the cage-like structure of the zeolite. Hence, in this study, the zeolite is incorporate in the TFN polyamide layer to improve the membrane surface properties for PRO applications.

Table 16.4 shows the recent nanomaterials with the modification of other materials incorporated with the PRO membranes for

Table 16.4. Recent nanomaterials incorporated with PRO membranes for PRO application process

Membrane	Modification	Water permeability ($L\,m^{-2}h^{-1}\,bar^{-1}$)	Water flux ($L\,m^{-2}h^{-1}$)	Power density (Wm^{-2})	Inference	Reference
TFC Hollow Fiber PES	GO	1.36	—	14.6	GO improved water permeability.	47
TFN membrane	Melamine-based Covalent Organic Framework (COF)	3.87	42.5	12.1	Melamine enhanced hydrophilicity and porosity.	48
TFN membrane	AgNP	9.5	44.8	—	"phase inversion + LbL assembly" process developed water permeability and anti-biofouling property.	7
TFN membrane	CNT	1.42	21.82	1.7	TFN membrane takes advantage of the enhanced porosity and hydrophilicity induced by f-CNT.	49
TFN membrane	GO & HNT	2.4	—	16.7	The fabricated membrane developed power density and greater mechanical strength than commercial PRO membrane.	50

(*Continued*)

Table 16.4. (*Continued*)

Membrane	Modification	Water permeability ($L\,m^{-2}h^{-1}\,bar^{-1}$)	Water flux ($L\,m^{-2}h^{-1}$)	Power density (Wm^{-2})	Inference	Reference
TFC membrane	The tiered structure of polyetherimide (PEI) reinforced by f-CNT	3.04	23.2	17.3	This membrane improved the mechanical properties along with the porosity of the support.	51
TFC membrane	Aminosilane	2.23	36	12.8	Grafting aminosilane culminated in having higher hydrophilicity, membrane strength of PRO membranes.	52
TFC Hollow Fiber membrane	Polyelectrolytes	2	38	16.2	The modified membrane attributed to the intrinsic properties, reducing S parameter due to hydrophilicity.	53
Hollow Fiber membrane	Polyamide-imide (PAI)/ glutaraldehyde (GA)	0.66	27	4.3	The hollow fiber membrane incorporated with PAI/GA has a superior antifouling property while the power density will remain constant.	54

(*Continued*)

Table 16.4. *(Continued)*

Membrane	Modification	Water permeability ($L\,m^{-2}h^{-1}\,bar^{-1}$)	Water flux ($L\,m^{-2}h^{-1}$)	Power density (Wm^{-2})	Inference	Reference
TFC Hollow Fiber PES	$CaCl_2$	3.8	132	38	TFC hollow fiber membrane incorporated with $CaCl_2$, has small pore size along with narrow distribution resulted in having strong support and highest water permeability.	55
TFN membrane	Zeolite (Molecular Sieves)	1.3	—	1.73	TFN membrane incorporated with zeolite improved hydrophilicity of the PRO membranes.	56

increasing the main characteristics for boosting up the performance through PRO analyses which are as follows:

The advancement of TFC membranes is a significant study application in technologies such as NF/RO for water treatment and has gained attention widely since the 1960s.[4] TFN is referred to the incorporation of nanomaterials in the polyamide layer. Recently, TFN membranes are widely used due to their potential to pass the obstacles such as the curve's trade-off among water permeability and salt rejection in comparison with the normal TFC membrane. Moreover, introducing fillers that are inorganic to the polyamide layer has shown impressive characteristics in developing TFN membrane resilience connected with resistance of the chlorine and fouling. Although there are impressive superior accomplishments, some challenges came across during the fabrication of the TFN membranes.[57]

More research in this area is still needed to develop the TFN membrane with greater performance efficiency, reliability, and stability for industrial implementation. The method to improve TFN membrane fabrication can be achieved modification of hydrophilic nanofillers. These can help to develop the defect-free which is organic and the fabrication of nanofillers layer containing PA/inorganic. Subsequently, the usage and the selection of nanofillers in the fabrication of TFN membranes have to depend on the characteristics of the feed.[57]

During fabricating the TFN membrane there are several major obstacles that show the density of nanoparticles in the polyamide layer. The TEM images in Figure 16.6 exhibit particle density in the thin polyamide layer by which decreasing the active surface area that is coherent to nanoparticles.[57]

Many types of research have conducted on nanomaterials regarding their chemical structure. 3D nanomaterials can be the most advantageous for PRO application due to its cage-like structure and it's suitable for salt retention and water transport in the membrane.

In the 1940s after the initial synthesis, zeolites utilized for many purposes such as adsorption or means of ion-exchange and catalysis. Although zeolites have some characteristics which make

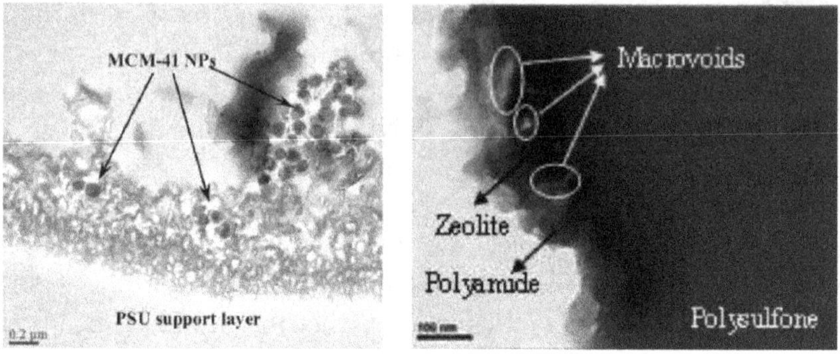

(a) (b)

Fig. 16.6. TEM images for sections which are crossed (a) Thin film nanocomposite with silica nanoparticles, (b) Thin film nanocomposite for NaA zeolite.[57]

it superior due to its uniform pore size, chemically and thermally resistant, it recommended that it can be used as an excellent material which is continuous polycrystalline zeolite layers for the separation purposes.[58] Future studies are prone to have a better understanding which can lead to improve zeolite membranes and eventually facilitate the large-scale application of zeolite membranes to industrial separations.[59] Zeolite consists of many types according to their characteristics and sizes. It can be microporous, mesoporous and microporous. Nano zeolites have been used for mixed matrix membrane in the substrate or incorporated with the polyamide layer.[60] In this study, molecular sieves used as a nanomaterial in the polyamide layer because of its hydrophilic behavior to increase water permeability.[30]

In the range of 0.02–0.1 wt./v% zeolite loading, the incorporation of zeolite-polyamide exhibited enhanced water permeability of membrane suitable due to the porous nature of zeolite. However, a further increase in zeolite loading led to a reduction in water permeability, possibly because of the formation of a thicker polyamide layer. The most permeable TFN membrane (TFN0.1, with wt./v% zeolite loading) had a water permeability approximately 80% higher compared to the baseline TFC membrane.[30] Zeolite-polyamide TFN membranes were prepared similarly to TFC membranes, except that zeolite nanoparticles were added in the 0.05 wt./v% TMC in n-hexane solution beforehand. Varied amount of zeolite nanoparticles (0.02, 0.05, 0.1, 0.2 and 0.4 wt./v%) were dispersed in TMC-n-hexane solution by ultrasonication (Fisher Scientific, Singapore) for 30 min at 20°C. The resultant solution was immediately used for interfacial polymerization with MPD-soaked PSF supports to form the TFN membranes. These membranes are denoted as TFN0.02, TFN0.05, TFN0.1, TFN0.2, and TFN0.4, respectively, where the number corresponds to their zeolite loading.[30]

Based on the three-dimensional channels, varieties of zeolites have been dispersed into the thin polyamide film of nanofiltration and reverse/forward osmosis membranes since the new concept was introduced in 2007.[61] The study of these nanocomposite membranes concerned with and effects of zeolites in the processes of interfacial polymerization on support layers. However, the behaviors and effects of zeolites in the processes of classical phase inversion are also of importance in the fabrication of asymmetric polymeric membranes, focused on the zeolite particles on the structures and performances of poly (phthalazinone ether sulfone ketone) composite ultrafiltration membranes.[62] It is found that hydrophilic zeolites could aid these composite membranes to maintain excellent water permeability, hydrophilicity, and antifouling properties with enhanced solute rejections as a low content (3wt%) of zeolites, while these additives were apt to agglomerate to each other at a high content and in turn affect resultant performances. The strong hydrophilicity of these zeolites could facilitate the in-flow velocity of nonsolvent (water) in the phase inversion process that results in the formation of finger-like pores in the porous support layers. On the other hand, hydrophilic zeolites could absorb water vapor in the evaporation process, which caused the formation of dense skin layers on the membrane surface and in turn improved the selectivity of these nanocomposite membranes. To alleviate the agglomeration of zeolite nanoparticles, external dispersants or pore-forming agents should be involved in the process of membrane formation.[45]

Tian *et al.* carried out a research on PRO applications utilizing thin-film nanocomposite PRO membranes along with the tiered PEI nanofiber support which was reinforced by the functionalization of carbon nanotubes (f-CNT) and consisted of a thin fiber layer.[50] The study showed that through the incorporation of f-CNTs in both layers of nanofiber, it was feasible to have improvement for mechanical properties along with porous support. Moreover, the membranes culminated in having 10 times higher water permeability at the same salt retention in comparison with HTI-CTA membranes. The membrane achieved a peak of its power density which was reported as $17.3\,W/m^2$ at 16.9 bar and the water permeability achieved $3.04\,L\,m^{-2}\,h^{-1}\,bar^{-1}$.[50]

Liu *et al.* reported that silver nanocomposite membranes (AgNP) have novel characteristics that can be modified and increased anti-biofouling property by the "phase inversion + LbL assembly" application.[7] The anti-biofouling performance of the membrane was examined systematically, and the results culminated in having an improvement in mechanical properties along with the porosity of the support layer. Furthermore, its enhanced water permeability was about 24.4% through PRO test mode at the AgNP loading concentration by 0.02 wt.%. Moreover, the water flux was enhanced from 66.9% to 88.2% in addition to live cells in biofilms which had a sharp fall from $95.9 \pm 3.2\%$ to $69.2 \pm 5.3\%$.[7]

In the other study, Zhang *et al.* stated that aminosilane has significant roles to enhance TFC membranes.[52] The exceptional role of aminosilane is about its capability as a grafting agent for improving anti-fouling TFC membranes regarding PRO application has shown successfully by the means of this study. Furthermore, the feasibility of offering hydrophilicity to either the interior part of the support layer or selective layer for PRO membranes is one of the significant characteristics of it. Additionally, it improved the mechanical strength of PRO membranes through the grafting. The presence of aminosilane improves the water flux and power density remarkably. The maximum water permeability achieved was $2.23\,L\,m^{-2}\,h^{-1}\,bar^{-1}$ and the peak of the power density was $12.8\,W/m^2$.[52]

Son *et al.* synthesized a CNT-embedded polyethersulfone (PES) as a support layer and utilized PA as an active layer in a TFN membrane that could culminate in having a novel membrane along with excellent characteristics.[49] The water flux and maximum power density of the developed TFN membrane were found to be augmented by 87% and 110% compared to TFC membranes, respectively. It is due to the formation of the voids at the surrounding of CNT has increased the porosity, volume and diameter meanwhile the carboxylic and carbonyl groups which were found on the CNT surface has improved the hydrophilicity of the membrane. The maximum water flux is $21.82\,\mathrm{L\,m^{-2}\,h^{-1}}$ and the peak of the power density was $1.7\,\mathrm{W/m^2}$.[49]

Li *et al.* carried out a research on the modification of thin-film composite hollow fiber membrane incorporated by polyelectrolytes.[53] This effective method was stated as an advantageous way of being environmentally friendly, facile fabrication and capability for maintaining the performance of the membrane. Furthermore, it attributed to a well-maintained property intrinsically and decreased the structural parameter due to the hydrophilicity sharp rise. The maximum water permeability achieved was $2\,\mathrm{Lm^{-2}\,h^{-1}\,bar^{-1}}$ and the peak of the power density was $16.2\,\mathrm{W/m^2}$.[53]

Li *et al.* reported that integral hollow fiber membrane with chemical cross-linking for PRO processes are applicable by the unconventional orientation.[54] Two different hollow fiber substrates with different porosity were fabricated to examine the appropriate structure of the integral membrane that could be operated in the selective layer facing feed orientation. The PAH/GA cross-linked membranes show high salt retention. The highest achieved water permeability was $0.66\,\mathrm{L\,m^{-2}h^{-1}bar^{-1}}$ and the peak of the power density was $4.3\,\mathrm{W/m^2}$.[54]

Wan *et al.* reported that TFC-PES hollow fiber membranes were engineered with having high power densities for meeting the requirements by adding $CaCl_2$ into the dopes of the polymer.[55] $CaCl_2$ was capable to improve the strength of tensile and bursting pressures for the membranes significantly. This substrate has the smallest amount of pore size by a narrow distribution, which not only findings in the thinnest polyamide selective layer but also offers the strongest support to the selective layer against pressure. Eventually, the TFC-PES hollow fiber membrane culminated in having the maximum water permeability along with the highest power density which their values are $3.8\,L\,m^{-2}\,h^{-1}\,bar^{-1}$ and $38\,W/m^2$.[55]

Lim *et al.* carried out research on dual-layered nanocomposite membrane incorporating Graphene Oxide (GO) and Halloysite Nanotube (HNT).[50] Generally, the new well-developed dual-layered GO/HNT nanocomposite membranes showed higher power density, low specific reverse solute flux (SRSF), and better mechanical strength compared to the membranes including pure PSF substrate. The highest achieved water permeability was $2.4\,L\,m^{-2}\,h^{-1}\,bar^{-1}$ and the peak of the power density was $16.7\,W/m^2$.[50]

Park *et al.* conducted research on the fabrication of a TFC hollow fiber membrane which was an incorporation of hydrophilic GO nanosheets in the PES support layer for PRO applications.[47] The study indicated that adding GO with ($\leq$0.2 wt%) into the PES hollow fiber membrane culminated in having improved structural morphologies along with surface chemistry within the support layer. These outcomes resulted in having an increase in the porosity of membrane with large pore size on the membrane's surface and improved hydrophilicity in addition to higher water permeability. Furthermore, the membrane with stated features achieved the highest water permeability which is $1.36\,L\,m^{-2}h^{-1}\,bar^{-1}$. The highest power density achieved and is $14.6Wm^{-2}$ at 16.5 bar utilizing 1M NaCl as draw solution and deionized water water as feed solution.[47]

By contrast, Gonzales *et al.* utilized the melamine-based covalent organic framework which was incorporated with the thin film nanocomposite (TFN) PRO membrane.[48] The SNW^{-1} was also examined utilizing in two various precursors which was culminated in: 1) As hydrophilic and porous nanomaterial was incorporated in the membrane it showed an improvement in the hydrophilicity of the surface along with having high water flux of TFN membranes regarding the characteristic of amine functional groups for SNW^{-1}, 2) The incorporation of SNW^{-1} is not feasible into TMC because of the reaction among the two, that is extremely alike to the formation of polyamide layer, 3) The highest water permeability achieved that was 3.87 $Lm^{-2}\,h^{-1}bar^{-1}$ for the membrane with 0.1 wt% SNW^{-1} incorporated through the MPD solution but 0.02 wt%

was the optimum loading of the nanomaterial in this study, 4) The maximum power density and water flux for this membrane which was incorporated along with 0.02 wt% SNW^{-1} in MPD were 12.1 Wm^{-2} (at 20 bar) and 42.5 $Lm^{-2}\,h^{-1}$, respectively.[48]

16.4. Conclusion and Perspective

PRO is one of the most promising applications for power generation and membranes are the heart of this method. In order to improve this process, there are a variety of nanoparticles that are somehow promising to be incorporated with composite membranes. In addition, the recent results have shown that they are more promising to use in the PRO process in comparison with the control TFC membrane. Improving the performance of TFN membranes through PRO process is mostly rely on the nanomaterials using in the substrate or active layer (polyamide layer). It has been observed that using nanomaterials in TFN membranes has led into having higher water flux and consequently to have a higher power density compared to the control TFC membrane. Working under different testing conditions will result in having various performance ranges which are between the 1.7–38.0 W/m^2. In addition, achieving better performance through the PRO process relies on the parameters such as different concentrations in the draw and feed solution and operation pressure.

The application of these nanomaterials incorporated with TFN membranes for power generation through PRO is still a new field to be explored. Despite some promising findings obtained from this work, there is always room for improvement. Therefore, some suggestions are recommended as the future directions of this research to enhance TFN and hollow fiber membranes for this application. In order to further improve the membrane flux and power density of the PRO system, the desired membrane should be of a very thin selective layer and the substrate layer. Triethylamine and diamine-based monomers act as a catalyst in PA layer formation under interfacial polymerization. Thereby, high power density can be achieved in PRO techniques. The thickness and the structure of the substrate also play a critical role in determining the flux and power density. Hence it is recommended to fabricate a thin substrate to facilitate the transport of water.

References

1. G. Han, S. Zhang, X. Li, T. S. Chung, Progress in pressure retarded osmosis (PRO) membranes for osmotic power generation, *Prog. Polym. Sci.*, **51** (2014) 1–27. https://doi.org/10.1016/j.progpolymsci.2015.04.005
2. B. E. Logan, M. Elimelech, Membrane-based processes for sustainable power generation using water, *Nature*, **488**(7411) (2012) 313–319. https://doi.org/10.1038/nature11477
3. G. O'Toole, L. Jones, C. Coutinho, C. Hayes, M. Napoles, A. Achilli, River-to-sea pressure retarded osmosis: Resource utilization in a full-scale facility, *Desalination*, **389** (2016) 39–51. https://doi.org/10.1016/j.desal.2016.01.012
4. A. P. Straub, A. Deshmukh, M. Elimelech, Pressure-retarded osmosis for power generation from salinity gradients: is it viable? *Energy Environ. Sci.*, **9**(1) (2016) 31–48. https://doi.org/10.1039/C5EE02985F
5. A. Achilli, T. Y. Cath, A. E. Childress, Power generation with pressure retarded osmosis: An experimental and theoretical investigation, *J. Membr. Sci.*, **343**(1–2) (2009) 42–52. https://doi.org/10.1016/j.memsci.2009.07.006
6. F. Helfer, C. Lemckert, The power of salinity gradients: An Australian example, *Renew. Sust. Energ. Rev.*, **50** (2015) 1–16. https://doi.org/10.1016/j.rser.2015.04.188
7. X. Liu, L. X. Foo, Y. Li, J. Y. Lee, B. Cao, C. Y. Tang, Fabrication and characterization of nanocomposite pressure retarded osmosis (PRO) membranes with excellent anti-biofouling property and enhanced water permeability. *Desalination*, **389** (2016) 137–148. https://doi.org/10.1016/j.desal.2016.01.037
8. Feinberg. (2014). Electronic Thesis and Dissertations Los Angeles Antiadhesive and Antibacterial Coatings for Biofouling Control A dissertation submitted in partial satisfaction of the requirements for the degree Doctor of Philosophy in Civil Engineering by Catalina S. Ma. University of California.
9. N. Y. Yip, D. A. Vermaas, K. Nijmeijer, M. Elimelech, Thermodynamic, energy efficiency, and power density analysis of reverse electrodialysis power generation with natural salinity gradients, *Environ. Sci. Technol.*, **48**(9) (2014) 4925–4936. https://doi.org/10.1021/es5005413

10. Y. Emami, S. Mehrangiz, A. Etemadi, A. Mostafazadeh, S. Darvishi, A brief review about salinity gradient energy, *Int. J. Smart Grid Clean Energy*, **2**(2) (2013) 295–300. https://doi.org/10.12720/sgce.2.2.295-300
11. J. W. Post, J. Veerman, H. V. M. Hamelers, G. J. W. Euverink, S. J. Metz, K. Nymeijer, C. J. N. Buisman, Salinity-gradient power: Evaluation of pressure-retarded osmosis and reverse electrodialysis. *J. Membr. Sci.*, **288**(1–2) (2007) 218–230. https://doi.org/10.1016/j.memsci.2006.11.018
12. R. Rica, R. Ziano, D. Salerno, F. Mantegazza, R. van Roij, D. Brogioli, Capacitive mixing for harvesting the free energy of solutions at different concentrations. *Entropy*, **15**(4) (2013) 1388–1407. https://doi.org/10.3390/e15041388
13. S. Loeb, Energy production at the dead sea by pressure-retarded osmosis: Challenge or chimera? *Desalination*, **120**(2) (1998) 247–262. https://doi.org/10.1016/S0011-9164(98)00222-7
14. A. Altaee, *et al.* Pressure retarded osmosis: Advancement in the process applications for power generation and desalination. *Desalination*, **356** (2015) 31–46. https://doi.org/10.1016/j.desal.2014.09.028
15. L. D. Banchik, M. H. Sharqawy, J. H. Lienhard, Limits of power production due to finite membrane area in pressure retarded osmosis. *J. Membr. Sci.*, **468** (2014) 81–89. https://doi.org/10.1016/j.memsci.2014.05.021
16. A. Altaee, G. Zaragoza, A. Sharif, Pressure retarded osmosis for power generation and seawater desalination: Performance analysis. *Desalination*, **344** (2014) 108–115. https://doi.org/10.1016/j.desal.2014.03.022
17. Yip, Ngai Yin, A. Tiraferri, W. A. Phillip, J. D. Schiffman, L. A. Hoover, Y. C. Kim, M. Elimelech, Thin-film composite pressure retarded osmosis membranes for sustainable power generation from salinity gradients, *Environ. Sci. Technol.*, **45**(10) (2011) 4360–4369. https://doi.org/10.1021/es104325z
18. Q. She, X. Jin, C. Y. Tang, Osmotic power production from salinity gradient resource by pressure retarded osmosis: Effects of operating conditions and reverse solute diffusion, *J. Membr. Sci.*, **401–402** (2012) 262–273. https://doi.org/10.1016/j.memsci.2012.02.014
19. H. Kim, J. S. Choi, S. Lee, Pressure retarded osmosis for energy production: Membrane materials and operating conditions. *Water Sci. Technol.*, **65**(10) (2012) 1789–1794. https://doi.org/10.2166/wst.2012.025
20. D. I. Kim, J. Kim, S. Hong, (2016). Changing membrane orientation in pressure retarded osmosis for sustainable power generation with low

fouling. *Desalination*, 389, 197–206. https://doi.org/10.1016/j.desal.2016.01.008

21. R. E. Pattle, Production of electric power by mixing fresh and salt water in the hydroelectric pile. *Nature*, **174** (1954) 660–660. https://doi.org/10.1038/174660a0
22. S. Loeb, Production of energy from concentrated brines by pressure-retarded osmosis. *J. Membr. Sci.*, **1** (1976) 49–63. https://doi.org/10.1016/S0376-7388(00)82257-7
23. G. D. Mehta, S. Loeb, Internal polarization in the porous substructure of a semipermeable membrane under pressure-retarded osmosis. *J. Membr. Sci.*, **4**(C) (1978) 261–265. https://doi.org/10.1016/S0376-7388(00)83301-3
24. L. F. Greenlee, D. F. Lawler, B. D. Freeman, B. Marrot, P. Moulin, Reverse osmosis desalination: Water sources, technology, and today's challenges. *Water Res.*, 43(9) (2009) 2317–2348. https://doi.org/10.1016/j.watres.2009.03.010
25. L. G. Palacin, F. Tadeo, C. de Prada, K. Touati, Evaluation of the recovery of osmotic energy in desalination plants by using pressure retarded osmosis. *Desalination Water Treat.*, 51(1–3) (2013) 360–365. https://doi.org/10.1080/19443994.2012.715130
26. C. Klaysom, T. Y. Cath, T. Depuydt, I. F. J. Vankelecom, Forward and pressure retarded osmosis: potential solutions for global challenges in energy and water supply. *Chem. Soc. Rev.*, **42**(16) (2013) 6959. https://doi.org/10.1039/c3cs60051c
27. J. Wei, C. Qiu, C. Y. Tang, R. Wang, A. G. Fane, Synthesis and characterization of flat-sheet thin film composite forward osmosis membranes. *J. Membr. Sci.*, **372**(1–2) (2011) 292–302. https://doi.org/10.1016/j.memsci.2011.02.013
28. Y. Cui, X. Y. Liu, T. S. Chung, Enhanced osmotic energy generation from salinity gradients by modifying thin film composite membranes. *Chem. Eng. J.*, **242** (2014) 195–203. https://doi.org/10.1016/j.cej.2013.12.078
29. F. Helfer, C. Lemckert, Y. G. Anissimov, Osmotic power with pressure retarded osmosis: Theory, performance and trends — A review. *J. Membr. Sci.*, **453** (2014) 337–358. https://doi.org/10.1016/j.memsci.2013.10.053
30. N. Ma, J. Wei, R. Liao, C. Y. Tang, Zeolite-polyamide thin film nanocomposite membranes?: Towards enhanced performance for forward osmosis. *J. Membr. Sci.*, **405–406** (2012) 149–157. https://doi.org/10.1016/j.memsci.2012.03.002
31. S. Veríssimo, K. V. Peinemann, J. Bordado, Thin-film composite hollow fiber membranes: An optimized manufacturing method. *J. Membr.*

Sci., **264**(1–2) (2005) 48–55. https://doi.org/10.1016/j.memsci.2005.04.020

32. W. J. Lau, A. F. Ismail, N. Misdan, M. A. Kassim, A recent progress in thin film composite membrane: A review. *Desalination*, **287** (2012) 190–199. https://doi.org/10.1016/j.desal.2011.04.004
33. K. L. Hickenbottom, J. Vanneste, M. Elimelech, T. Y. Cath, Assessing the current state of commercially available membranes and spacers for energy production with pressure retarded osmosis. *Desalination*, **389** (2016) 108–118. https://doi.org/10.1016/j.desal.2015.09.029
34. J. Qin, M. H. Oo, G. Tao, E. R. Cornelissen, C. J. Ruiken, K. F. Korte, K. A. De, Kekre, Optimization of operating conditions in forward osmosis for osmotic membrane bioreactor. *Open Chem. Eng. J.*, **3** (2009) 27–32. https://doi.org/10.2174/1874123100903010027
35. A. Achilli, T. Y. Cath, E. A. Marchand, A. E. Childress, The forward osmosis membrane bioreactor?: A low fouling alternative to MBR processes. *DES*, **239**(1–3) (2009) 10–21. https://doi.org/10.1016/j.desal.2008.02.022
36. Wang, Rong, L. Shi, C. Y. Tang, S. Chou, C. Qiu, A. G. Fane, Characterization of novel forward osmosis hollow fiber membranes. *J. Membr. Sci.*, **355**(1–2) (2010) 158–167. https://doi.org/10.1016/j.memsci.2010.03.017
37. K. Lutchmiah, D. J. H. Harmsen, B. Wols, A. R. D. Verliefde, J. W. Post, E. R. Cornelissen, (2011). Pressure Assisted Osmosis (PAO) to enhance Forward Osmosis (FO) Performance. 9000.
38. K. Y. Wang, T. Chung, G. Amy, Developing thin-film-composite forward osmosis membranes on the PES/SPSf substrate through interfacial polymerization. *AIChE* **58**(2) (2012a) 5–8. https://doi.org/10.1002/aic
39. R. L. McGinnis, M. Elimelech, Global challenges in energy and water supply?: The promise of engineered osmosis. *Environ. Sci. Technol.*, **42**, 23 (2008) 8625–8629.
40. S. Chou, R. Wang, L. Shi, Q. She, C. Tang, A. Gordon, Thin-film composite hollow fiber membranes for pressure retarded osmosis (PRO) process with high power density. *J. Membr. Sci.*, **389** (2012) 25–33. https://doi.org/10.1016/j.memsci.2011.10.002
41. T. Thorsen, T. Holt, (2009). The potential for power production from salinity gradients by pressure retarded osmosis. *J. Membr. Sci.*, **335**(1–2), 103–110. https://doi.org/10.1016/j.memsci.2009.03.003
42. Wang, R, C. Tang, A. G. Fane (2012b). Development of pressure retarded osmosis (PRO) membranes with high power density for osmotic power harvesting. Proceedings of the 3rd Osmosis Membrane Summit, Statkraft, Barcelona, Spain, 26–27.

43. A. Efraty, (2014, January 9). Power generation by pressure retarded osmosis in closed circuit without need of energy recovery. Google Patents.
44. G. Han, S. Zhang, X. Li, T. Chung, High performance thin film composite pressure retarded osmosis (PRO) membranes for renewable salinity-gradient energy generation. *J. Membr. Sci.*, **440** (2013) 108–121. https://doi.org/10.1016/j.memsci.2013.04.001
45. Q. She, Y. Kong, W. Wong, S. Zhao, C. Y. Tang, Organic fouling in pressure retarded osmosis?: Experiments, mechanisms and implications. *J. Membr. Sci.*, **428** (2013) 181–189. https://doi.org/10.1016/j.memsci.2012.10.045
46. P. Wang, J. Ma, F. Shi, Y. Ma, Z. Wang, X. Zhao, Behaviors and effects of differing dimensional nanomaterials in water filtration membranes through the classical phase inversion process: A review. *Ind. Eng. Chem. Res.*, **52**(31) (2013) 10355–10363. https://doi.org/10.1021/ie303289k
47. M. J. Park, S. Lim, R. R. Gonzales, S. Phuntsho, D. S. Han, A. Abdel-Wahab, H. K. Shon, Thin-film composite hollow fiber membranes incorporated with graphene oxide in polyethersulfone support layers for enhanced osmotic power density. *Desalination*, **464**(April) (2019) 63–75. https://doi.org/10.1016/j.desal.2019.04.026
48. R. R. Gonzales, M. J. Park, T. H. Bae, Y. Yang, A. Abdel-Wahab, S. Phuntsho, H. K. Shon, Melamine-based covalent organic framework-incorporated thin film nanocomposite membrane for enhanced osmotic power generation. *Desalination*, **459**(March) (2019) 10–19. https://doi.org/10.1016/j.desal.2019.02.013
49. M. Son, H. Park, L. Liu, H. Choi, J. H. Kim, H. Choi, Thin-film nanocomposite membrane with CNT positioning in support layer for energy harvesting from saline water. *Chem. Eng. J.*, **284** (2016) 68–77. https://doi.org/10.1016/j.cej.2015.08.134
50. S. Lim, M. J. Park, S. Phuntsho, A. Mai-Prochnow, A. B. Murphy, D. Seo, H. Shon, Dual-layered nanocomposite membrane incorporating graphene oxide and halloysite nanotube for high osmotic power density and fouling resistance. *J. Membr. Sci.*, **564**(June) (2018) 382–393. https://doi.org/10.1016/j.memsci.2018.06.055
51. M. Tian, R. Wang, K. Goh, Y. Liao, A. G. Fane, Synthesis and characterization of high-performance novel thin film nanocomposite PRO membranes with tiered nanofiber support reinforced by functionalized carbon nanotubes. *J. Membr. Sci.*, **486** (2015) 151–160. https://doi.org/10.1016/j.memsci.2015.03.054

52. L. Zhang, Q. She, R. Wang, S. Wongchitphimon, Y. Chen, A. G. Fane, Unique roles of aminosilane in developing anti-fouling thin film composite (TFC) membranes for pressure retarded osmosis (PRO). *Desalination*, **389** (2016) 119–128. https://doi.org/10.1016/j.desal.2015.12.024
53. Y. Li, S. Qi, Y. Wang, L. Setiawan, R. Wang, Modification of thin film composite hollow fiber membranes for osmotic energy generation with low organic fouling tendency. *Desalination*, **424**(August) (2017) 131–139. https://doi.org/10.1016/j.desal.2017.10.005
54. Y. Li, S. Zhao, L. Setiawan, L. Zhang, R. Wang, Integral hollow fiber membrane with chemical cross-linking for pressure retarded osmosis operated in the orientation of active layer facing feed solution. *J. Membr. Sci.*, **550**(March 2017) (2018) 163–172. https://doi.org/10.1016/j.memsci.2017.12.074
55. C. F. Wan, T. Yang, W. Gai, Y. Lee, De, T. S. Chung, Thin-film composite hollow fiber membrane with inorganic salt additives for high mechanical strength and high power density for pressure-retarded osmosis. *J. Membr. Sci.*, **555**(January) (2018) 388–397. https://doi.org/10.1016/j.memsci.2018.03.050
56. A. Shadravan (2019). Thin Film Nanocomposite Membranes Incorporated with Zeolite for Power Generation through Pressure Retarded Osmosis. A dissertation submitted in partial fulfilment of the requirements for the award of the degree of Master of Engineering — Chemical Engineering, School of Chemical and Energy Engineering, Universiti Teknologi Malaysia (UTM).
57. W. J. Lau, S. Gray, T. Matsuura, D. Emadzadeh, J. Paul Chen, A. F. Ismail, A review on polyamide thin film nanocomposite (TFN) membranes: History, applications, challenges and approaches. *Water Res.*, **80** (2015) 306–324. https://doi.org/10.1016/j.watres.2015.04.037
58. M. E. Davis, Ordered porous materials for emerging applications. *Nature*, **417**(6891) (2002) 813–821. https://doi.org/10.1038/nature00785
59. M. Yu, R. D. Noble, J. L. Falconer, Zeolite membranes: Microstructure characterization and permeation mechanisms, *Acc. Chem. Res.*, **44**(11) (2011) 1196–1206. https://doi.org/10.1021/ar200083e
60. P. S. Goh, A. F. Ismail, A review on inorganic membranes for desalination and wastewater treatmwent. *Desalination*, **434**(August 2017) (2018) 60–80. https://doi.org/10.1016/j.desal.2017.07.023

61. B. H. Jeong, E. M. V. Hoek, Y. Yan, A. Subramani, X. Huang, G. Hurwitz, A. Jawor, Interfacial polymerization of thin film nanocomposites: A new concept for reverse osmosis membranes. *J. Membr. Sci.*, **294**(1–2) (2007) 1–7. https://doi.org/10.1016/j.memsci.2007.02.025
62. R. Han, S. Zhang, C. Liu, Y. Wang, X. Jian, Effect of NaA zeolite particle addition on poly(phthalazinone ether sulfone ketone) composite ultrafiltration (UF) membrane performance. *J. Membr. Sci.*, **345**(1–2) (2009) 5–12. https://doi.org/10.1016/j.memsci.2009.07.052

Index

Materials and Energy

(Continuation of series card page)

Vol. 16 *Handbook of Porous Materials: Synthesis, Properties, Modeling and Key Applications (In 4 Volumes)*
edited by Vitaly Gitis (Ben Gurion University of the Negev, Israel), Gadi Rothenberg (University of Amsterdam, The Netherlands), Anton A. Kiss (The University of Manchester, UK) and David Eisenberg (Technion — Israel Institute of Technology, Israel)

Vol. 15 *The World Scientific Reference of Amorphous Materials Structure, Properties, Modeling and Main Applications (In 3 Volumes)*
edited by Alexander V. Kolobov (Herzen State Pedagogical University of Russia, Russia), Koichi Shimakawa (Gifu University, Japan), Ivar E. Reimanis (Colorado School of Mines, USA), Nikolas J. Podraza (University of Toledo, USA) and Robert W. Collins (University of Toledo, USA)

Vol. 13 *Theory and Methods of Photovoltaic Material Characterization: Optical and Electrical Measurement Techniques*
by Richard K. Ahrenkiel (Colorado School of Mines, USA and National Renewable Energy Laboratory, USA) and S. Phil Ahrenkiel (South Dakota School of Mines & Technology, USA)

Vol. 12 *World Scientific Handbook of Organic Optoelectronic Devices (Volumes 1 & 2)*
edited by Franky So (North Carolina State University, USA)

Vol. 11 *Industrial Applications of Ultrafast Lasers*
by Richard Haight (IBM TJ Watson Research Center, USA) and Adra V. Carr (IBM TJ Watson Research Center, USA)

Vol. 10 *World Scientific Reference on Spin in Organics (In 4 Volumes)*
edited by Zeev Valy Vardeny (University of Utah, USA) and Markus Wohlgenannt (University of Iowa, USA)

Vol. 9 *Conjugated Polymers and Oligomers: Structural and Soft Matter Aspects*
edited by Matti Knaapila (Technical University of Denmark, Denmark)

Vol. 8 *Thin Films on Silicon: Electronic and Photonic Applications*
edited by Vijay Narayanan (IBM Thomas J Watson Research Center, USA), Martin M. Frank (IBM Thomas J Watson Research Center, USA) and Alexander A. Demkov (The University of Texas at Austin, USA)

Vol. 7 *The WSPC Reference on Organic Electronics: Organic Semiconductors (In 2 Volumes)*
edited by Jean-Luc Brédas (King Abdullah University of Science & Technology, Saudi Arabia & Georgia Institute of Technology, USA) and Seth R. Marder (Georgia Institute of Technology, USA)

CPSIA information can be obtained
at www.ICGtesting.com
Printed in the USA
JSHW031440070621
15637JS00001B/1

9 789811 226977